Climate Change and the Coast

Building Resilient Communities

Climate Change and the Coast

Building Resilient Communities

Bruce C. Glavovic • Mick Kelly
Robert Kay • Ailbhe Travers

CRC Press
Taylor & Francis Group
Boca Raton London New York

CRC Press is an imprint of the
Taylor & Francis Group, an **informa** business

Reprint 2025

CRC Press
Taylor & Francis Group
6000 Broken Sound Parkway NW, Suite 300
Boca Raton, FL 33487-2742

ISBN: 9780367240257 (hbk)

Visit the Taylor & Francis Web site at
http://www.taylorandfrancis.com

and the CRC Press Web site at
http://www.crcpress.com

Printed and bound in India

For sale in India, Pakistan, Nepal, Bhutan, Bangladesh and Sri Lanka only.

Bruce dedicates this book to Peter and Faith Glavovic
Mick thanks Sarah for her patience and her sage advice
Robert dedicates this book to Lizzy Kay
Ailbhe dedicates this book to Lily

Contents

Foreword

Ever since the beginning of the negotiations on the Framework Convention on Climate Change, I have had the privilege of being deeply involved in this fundamental effort of tackling one of the most serious global problems facing us. Climate change is one of the main components of the human impact on the immensely large natural systems that are essential for the long-term survival of our species. We realize that we have entered a new geological epoch, the Anthropocene.

Emissions of greenhouse gases have to be dramatically reduced over the coming decades. However, this is a time-consuming and politically laborious process. Therefore, the negotiations have increasingly had to facilitate adaptation to the negative effects of global warming. Among these consequences, the expected long-term sea-level rise and the increasing risks for climate disasters affecting coastal communities are among the most serious. The editors of this book rightly emphasize that up to 50% of the planet's population live in these zones and that a large number of the megacities are located at or close to the coast. In fact, what we call human civilization and human culture to a large extent developed in these areas of encounter and dynamic transitions.

My native country of Sweden has a long coastline and is at present seriously working on methods to reduce our vulnerability. Conditions are obviously very different in other parts of the world, but, faced with a global problem with so serious consequences for governments at all levels and for individuals in all countries, it is essential that we profit from adaptation practices and experiences in different parts of the world.

The case studies included in this book will be of great interest for all those who will be involved in the needed action to improve resilience to the effects of climate change on the coastal zones. In particular, I am struck by the ambition to provide practical guidance at the community level. As a former negotiator, I believe strongly in the essential role of multilateral agreements to mobilize global action on problems that are eminently global.

But the women and men who face storms, flooding, and destruction at the coastal front line need to be given the practical tools provided by experiences from other parts of the world.

Bo Kjellen
Stockholm Environment Institute

Preface

This book is about climate change and the coast. It focuses on our global coastal communities that face the prospect of rising sea levels, changes in the intensity of extreme events such as coastal storms, ocean acidification, and changes in ocean circulation patterns at a time when they are already stressed by existing environmental pressures and adverse social and economic trends. A key theme of the book is the urgent need to ready ourselves to deal with these coming waves of adversity and learn to plan for an uncertain future.

We embarked on this book because there is an urgent need to integrate insights from research and practice in an accessible way so that coastal communities can plan proactively for this future. Our collective academic and professional experience gives us a distinctive advantage for contributing to this goal and hence our collaboration. Bruce draws on his experience as the project manager for the Republic of South Africa's coastal policy formulation process that culminated in the passage of the Integrated Coastal Management Act that has transformed coastal management in that country. For over a decade, based at Massey University in New Zealand, Bruce has conducted research on how to build community resilience and sustainability by exploring the intersection of coastal governance, climate change adaptation, disaster risk reduction, and natural hazards planning. Mick draws on his career as a climate scientist at the University of East Anglia where, among other things, he researched mechanisms of climate change and climate vulnerability. More recently, he has worked with local communities in diverse settings to tackle a range of environmental issues. Educated in the United Kingdom, Rob held an academic position in New Zealand and has since worked for nearly two decades as a consultant to governments, communities, and companies around the world to translate best practices in coastal management and climate change adaptation into local reality. Ailbhe's contribution is grounded in her expertise as a coastal geomorphologist and environmental scientist and over a decade working as a consultant to coastal stakeholders, governments, and communities on a range of coastal management and climate change adaptation matters.

Our aim is to provide insight on learning to live with climate change through practical guidance for coastal communities. We draw from current theory and first-hand experience to show how to reduce vulnerability to climate change impacts and build adaptive capacity and resilience. This practical advice is based on consideration of a series of stories from real-world communities in diverse coastal settings from around the globe. These stories reveal the nature of climate change and explore the complex set of challenges and opportunities it poses for different coastal communities. We are privileged to have highly regarded specialists from different fields share their perspectives with us based on their first-hand experience. This ensemble of stories enables us to provide up-to-date, grounded insights about how to overcome the barriers and unlock the opportunities facing coastal communities in this era of climate change.

Our book is intended for a diverse audience. It will be of particular interest to professionals working in the public and private sectors of coastal communities, including coastal planners and managers, consultants, and scientists, as well as those who work for a range of international bodies. It provides practical guidance for coastal decision-makers, including elected and appointed government officials, business leaders, and leaders of non-governmental and community-based organizations. The book will be relevant to students and scholars in fields ranging from environmental science to geography, planning, development studies, politics, public administration, policy analysis, emergency management, and emerging interdisciplinary fields such as sustainability studies and adaptive management. It will be of special interest to people in coastal communities who are searching for innovative and practical ways to achieve more sustainable outcomes, especially those elected or appointed to make community decisions and those that advise them.

Bruce C. Glavovic
Massey University

P. Mick Kelly
Tanelorn Associates

Robert Kay
Coastal Zone Management Ltd

Ailbhe Travers
Coastal Geomorphologist

Acknowledgments

Many people and organizations contributed to and supported us in developing and finalizing this book. We especially appreciate the invaluable contributions of all those who wrote case studies. Contributing authors generously shared their knowledge and experience with us and helped to develop a fresh perspective on how coastal communities might tackle the challenge of adapting to a changing climate.

We would like to especially thank Katharine Moody for working closely with authors to finalize the chapters and for her valuable contribution in the technical editing of the manuscript.

We gratefully acknowledge a financial contribution by the Land Oceans Interactions in the Coastal Zone (LOICZ) program that enabled us to meet in Perth to distil the essence of the case studies and develop the reflexive adaptation approach outlined in the concluding chapter.

Bruce would like to thank Massey University and his Head of School in particular for supporting the research that underpins his contributions to this book, including financial support to employ Katharine to assist with technical editing and undertake replacement teaching. He would also like to thank the Earthquake Commission for support and funding research that he conducted and is presented in this book.

Editors

Bruce C. Glavovic holds the EQC Chair in Natural Hazards Planning at Massey University, New Zealand. His research explores the role of governance and land-use planning in building resilient and sustainable communities. He has over 25 years of experience in academia, private consulting, and government, mainly in the Republic of South Africa, the United States, and New Zealand. He is currently vice-chair of the Land–Ocean Interactions in the Coastal Zone (LOICZ) Scientific Steering Committee. He is coeditor of *Adapting to Climate Change: Lessons from Natural Hazards Planning* (2014, Springer), and a contributing author to Working Group II's chapter on Australasia in the IPCC's 5th Assessment Report.

Mick Kelly is a consultant with Tanelorn Associates based in New Zealand's winterless north. Having retired from the Climatic Research Unit at the University of East Anglia in the United Kingdom, where he specialized in research on mechanisms of climate change and climate vulnerability, he now manages 69 acres of regenerating bush and is committed to community-based science projects.

Robert Kay is principal consultant of Coastal Zone Management Pty (Ltd) and of Adaptive Futures, Claremont, Australia, two niche consulting companies advising governments, communities, and companies worldwide on the challenges posed by climate change impacts. Dr. Kay has 25 years of experience in climate change impact assessment, coastal zone management and planning through work in government, consulting, and academic sectors. He holds a position of Visiting Adjunct Professor at the Sustainability Research Center at the University of the Sunshine Coast, Australia. He has an Honours degree in Geology (Wales) and a PhD in Environmental Science (East Anglia, UK).

Ailbhe Travers is a coastal geomorphologist with over 10 years of experience in environmental studies, focusing specifically on the coastal realm. She holds an Honours in Environmental Science (University of Ulster) and a PhD in Geography and Environmental Systems Engineering at the University of Western Australia.

Contributors

Derek Armitage is an Associate Professor with the Department of Environment and Resource Studies, University of Waterloo, Ontario, Canada. His research interests center on the human dimensions of environmental change and the formation of adaptive, multilevel governance systems, with a primary focus on coastal, marine, and freshwater aquatic contexts.

Scott Baum is a Professor at Griffith University, Brisbane, Queensland, Australia. He is trained in economics and sociology and researches the conditions impacting on communities and their residents including climate change. He has published across many areas including climate change adaptation, community disadvantage, and labor market analysis.

Marcus Bussey is a researcher and futurist with the Sustainability Research Centre, University of the Sunshine Coast, Queensland, Australia. He is currently a Visiting Fellow at Nanyang Technology University, Singapore, with the Centre of Excellence in National Security.

Gillian Cambers is a coastal scientist who has worked for island governments and international organizations in the Caribbean, Indian Ocean, and Pacific regions. Recently, she led the Pacific Climate Change Science Program and is also Director of the Sandwatch Program.

R.W. (Bill) Carter is an Associate Director of the Sustainability Research Centre, University of the Sunshine Coast, Queensland, Australia. His research focuses on evaluation of resilience of communities and institutions to change, especially in the context of tourism and natural resource management.

Ralph Chapman is Environmental Studies Director, Victoria University of Wellington, Wellington, New Zealand. His research focuses on climate change policy. He has worked with the NZ Environment Ministry, NZ Treasury, UK Treasury, and OECD. He has degrees in engineering, public policy, and economics.

Darryl Low Choy is Professor of Environmental and Landscape Planning at Griffith University, Brisbane, Queensland, Australia. His current research is focused on values-led planning and indigenous landscape values, peri-urbanization and landscape resilience, and climate change adaptation and postdisaster planning for human settlements.

Florence Crick is a Research Fellow in the Urban Research Program at Griffith University, Brisbane, Queensland, Australia. She has previously worked as a project coordinator at the National Climate Change Adaptation Research Facility, Queensland, Australia, and as a climate change adaptation consultant at the OECD, France.

Luke Dalton is an environmental engineer in the Water Corporation of Western Australia, Stirling, Australia. Since graduating in 2009, he has worked on both environmental management and civil construction projects in Australia and abroad. He holds an Honours degree in Environmental Engineering from RMIT University, Victoria, Australia.

Carmen Elrick-Barr is an environmental scientist with over eight years of experience in environmental management and research. Over this time, she has delivered climate change adaptation programs, projects, and policies in Australasia, the Pacific, and Africa.

Timoteo Caetano Ferreira is a retired Professor from the University of Trás-os-Montes and Alto Douro (UTAD), Vila Real, Portugal and since 2006 he has been collaborating as International Consultant and Chief Technical Adviser with a number of United Nations Organizations including WMO, UNIDO, UNDP, and UNEP. Tim acted as the International Project Development Specialist in the Mozambique LDCF project development process.

Annie George is Chief Executive Officer of BEDROC (Building and Enabling Disaster Resilience of Coastal Communities), an NGO based out of Nagapattinam, Tamil Nadu, India. Previously, she was CEO of the NGO Coordination Resource Centre, Nagapattinam, Tamil Nadu, India, that coordinated recovery activities after the 2004 tsunami.

Mónica Gómez-Erache is a researcher at the School of Sciences, Universidad de la República (UdelaR), Montevideo, Uruguay, and senior advisor for the Climate Change Office of the Uruguayan Environment Ministry, Uruguay. Previously, she was the coordinator of the EcoPlata Program on integrated coastal zone management.

Vivien Gornitz is a geologist and special research scientist with the Columbia University Center for Climate Systems Research and NASA GISS, New York. Her research examines recent and Holocene global sea-level change and extreme climate events in the NY Metro Region.

Ben Harman is an environmental planner with CSIRO Ecosystem Sciences based in Brisbane, Queensland, Australia. His research interests include urban and regional planning, natural resource management, periurbanization, and climate change adaptation. He has a keen interest in understanding these issues and developing strategies to provide sustainable solutions.

Georgina Hart is a researcher at Landcare Research, Auckland, New Zealand and recently completed her Master's thesis that explored the challenges of managing coastal adaptation to sea-level rise in developed areas along the New Zealand coastline.

Radley Horton is an associate research scientist at the Center for Climate Systems Research, Columbia University, New York. He is a convening lead author for the 2013 National Climate Assessment and served as climate science lead for the NYC Panel on Climate Change.

Saleemul Huq is a Senior Fellow with the Climate Change Group at the International Institute for Environment and Development in London. He specializes in the links between climate change and sustainable development, particularly the perspective of developing countries.

Andreas Kannen is a research scientist at Helmholtz-Zentrum Geesthacht, Germany. His work since 1996 has focused on integrated coastal management, coastal and marine governance, spatial planning, and sea-use constellations. He is involved in several national and international research projects and networks.

Ahana Lakshmi is a scientist–consultant at the National Centre for Sustainable Coastal Management and guest faculty at the Institute for Ocean Management, Anna University, Chennai, Tamil Nadu, India, and has worked on collating information on climate change issues. She is on the advisory board for BEDROC (Building and Enabling Disaster Resilience of Coastal Communities), an NGO based out of Nagapattinam, Tamil Nadu, India.

Judy Lawrence is an Adjunct Research Associate at the Climate Change Research Institute, Victoria University of Wellington, Wellington, New Zealand. Previously, she was Director of the NZ Climate Change Office, Ministry for the Environment and Convenor of the National Science Strategy Committee on Climate Change.

Joshua A. Lewis is a doctoral student at the Department of Systems Ecology at Stockholm University, Stockholm, Sweden, and a researcher at the Stockholm Resilience Centre, Stockholm, Sweden. Before his work in Stockholm, Joshua was a research analyst at Tulane University in New Orleans, Louisiana.

Tiziana Luisetti is a Senior Research Associate at the Centre for Social and Economic Research on the Global Environment (CSERGE), University of East Anglia, Norwich, UK.

David C. Major is Senior Research Scientist at the Columbia University Earth Institute's Center for Climate Systems Research, New York. His principal scientific research focus at Columbia is the adaptation of water, transportation, and other critical infrastructure to climate change.

Andrew A. Mather has a doctorate in sea-level rise planning and adaptation and is responsible for strategic coastal planning and management at eThekwini Municipality, Durban, Republic of South Africa. He is a reviewer for Chapter 5 on Coastal Systems and Low Lying Areas of the IPCC's "Working Group II" AR5.

Julie Matthews is Associate Professor in social science at the University of the Sunshine Coast, Queensland, Australia. She is an interdisciplinary researcher with a background in education, sociology, and cultural studies. She is currently an Associate Director of the Sustainability Research Centre and was previously Director of Research in the Faculty of Arts and Social Sciences.

Douglas J. Meffert is Vice President and Louisiana Executive Director of the National Audubon Society, New York. Previously, he was Director of project development for Tulane University's Payson Center for International Development, New Orleans, Louisiana, and executive director of Tulane's RiverSphere project.

Dieter Muehe is a former Full Professor (retired) for coastal and marine geography at Universidade Federal do Rio de Janeiro, Rio de Janeiro, Brazil, now working as Senior Visiting Professor at the graduate program in geography at Universidade Federal do Espírito Santo (UFES) in Vitória, Brazil.

Stephen Myers is a Research Fellow at the University of Ballarat, Victoria, Australia. Stephen has worked with the Adaptation Research Network Marine Biodiversity and Resources, University of Tasmania, Tasmania, Australia, and the CSIRO Coastal Collaboration Cluster on coastal management and adaptive capacity.

Gustavo J. Nagy is Associate Professor of global change at the School of Sciences, Universidad de la República (UdelaR), Montevideo, Uruguay, and a senior advisor in adaptation for the Climate Change Office of the Uruguayan Environment Ministry, Uruguay. He was a lead author of IPCC AR-4. He has degrees in oceanography.

Lesley Patrick is the Program Manager of the City University of New York (CUNY) Institute for Sustainable Cities and a doctoral candidate at the CUNY, New York. She served on the Science Planning Team of the NYC Panel on Climate Change.

Marcus Polette is a Professor in coastal zone management at the Center for Technological Earth and Sea Sciences, Universidade do Vale do Itajaí (UNIVALI), Brazil. He specializes in urban and coastal regional planning and coastal zone management.

R. Purvaja is a Scientist at the National Centre for Sustainable Coastal Management, Chennai, Tamil Nadu, India, specializing in climate change and coastal management. She has published several research papers in international journals and her research interests include studies on trace gas fluxes and biogeochemical cycles in coastal ecosystems.

M. Golam Rabbani is a Senior Researcher at the Bangladesh Centre for Advanced Studies, Dhaka, Bangladesh, and has an academic background in environmental science and risk assessment. He has been working on environment and climate change issues at the national and regional level for close to 10 years.

R. Ramesh is Director at the National Centre for Sustainable Coastal Management, Chennai, Tamil Nadu, India. His expertise includes coastal ecosystem biogeochemistry and coastal zone management. He is the current chair of Land–Ocean Interactions in the Coastal Zone (LOICZ). Ramesh's expertise has evolved from a pure science-based research focus to a science-policy interface for effective coastal zone management in India.

Beate M.W. Ratter is Professor of geography at the University of Hamburg, Hamburg, Germany, and holds a joint position as Head of the Department of Human Dimensions in Coastal Areas, Helmholtz-Zentrum Geesthacht, Germany. Her research focuses on the analysis of environmental management strategies in coastal areas and islands.

Andy Reisinger is Deputy Director of the NZ Agricultural Greenhouse Gas Research Centre, New Zealand. Previously, he was Senior Research Fellow at Victoria University of Wellington, Wellington, New Zealand, led the production of the IPCC Synthesis Report, and was Senior Advisor for the NZ Environment Ministry.

Russell Richards is a Research Fellow at the Sustainability Research Centre, University of the Sunshine Coast, Queensland, Australia, and the Griffith Centre for Coastal Management, Griffith University, Brisbane, Queensland, Australia. His research focuses on developing and applying modeling techniques that allow insight into the vulnerability and adaptive capacity of social and biophysical systems.

Debra C. Roberts has a doctorate in urban ecology and biogeography and is responsible for the biodiversity and climate protection planning at eThekwini Municipality, Durban, Republic of South Africa. She is a lead author for Chapter 8 on Urban Areas contribution to the IPCC's "Working Group II" AR5.

Sharon Roberts-Hodge is a land use planner who works with the Department of Physical Planning, within the Government of Anguilla, Anguilla. She holds the post of Deputy Director within the department and manages the Development Control Section.

Anne Roiko is a Senior Lecturer in environmental health and a member of the Sustainability Research Centre at the University of the Sunshine Coast, Queensland, Australia. Her research focuses on predicting and managing health risks associated with climate change and water-related health risks.

Cynthia Rosenzweig is Senior Research Scientist at NASA GISS, New York, and the Columbia Earth Institute, New York. She was co-chair of the NYC Panel on Climate Change, convening lead author on IPCC AR4, and is Co-director of the Urban Climate Change Research Network.

Marcello Sano is a Research Fellow at the Griffith University Centre for Coastal Management, Queensland, Australia. He is currently involved in different projects related with coastal management and climate adaptation. He holds a PhD from the Universidad de Cantabria, Cantabria, Spain.

Silvia Serrao-Neumann is a Research Fellow in the Urban Research Program, Griffith University, Brisbane, Queensland, Australia. Her research focuses on climate change adaptation related to urban planning and management, adaptation of coastal communities to natural hazards, and interjurisdictional challenges to adaptation.

Vigya Sharma is a postdoctoral fellow at the Centre for Social Responsibility in Mining at the University of Queensland, Brisbane, Queensland, Australia. Her research investigates impacts of climatic variability and extreme weather events in Central Queensland on resource development and the wider socioecological landscape.

Tim F. Smith is Professor of Sustainability and Director of the Sustainability Research Centre at the University of the Sunshine Coast, Queensland, Australia. His research interests focus on the transformation of society toward sustainability.

Mario L.G. Soares is a Professor in the Faculty of Oceanography at the Universidade do Estado do Rio de Janeiro (UERJ), Rio de Janeiro, Brazil, where he coordinates the Laboratory on Mangrove Studies and the Graduate Program in Oceanography. He specializes in mangrove ecology and coastal management, with research interests in the response and vulnerability of coastal systems to climate change and variability and the role of mangrove forests in the global carbon cycle.

William Solecki is the Director of the CUNY Institute for Sustainable Cities, New York, and a Professor of geography at Hunter College, New York. He was co-chair of the NYC Panel on Climate Change and is a lead author on IPCC AR5.

Arame Tall is a climate risk management, adaptation, and sustainable development specialist based in Washington, DC. She holds a PhD in African studies from the Paul H. Nitze School of Advanced International Studies (SAIS) division of The Johns Hopkins University and a master's degree in Climate and Society from Columbia University, New York.

Dana C. Thomsen is a Senior Lecturer in sustainability advocacy and Coordinator of Sustainability Programs at the University of the Sunshine Coast, Queensland, Australia. She is also a member of the Sustainability Research Centre. Her expertise is focused on community-based research, social learning, and societal transformation.

Jessica Troni is a climate change, environment, development, and public policy specialist. At UNDP Jessica has developed over 20 adaptation projects in 13 different countries in the East and Southern African region. She has advised a number of governments in the region on the development of national adaptation strategies including the mainstreaming of climate change into government processes and policies.

R. Kerry Turner is a Professorial Research Fellow in the Centre for Social and Economic Research on the Global Environment (CSERGE), School of Environmental Sciences, University of East Anglia, Norwich. Previously, he was Director of CSERGE and advisor to numerous national and international environmental bodies.

Chen Wen is the Director and Professor of Urban and Regional Development Research Center, Nanjing Institute of Geography & Limnology, Chinese Academy of Sciences, Nanjing, People's Republic of China. Her research focuses on urban and regional development and planning, greening of industry, and sustainable development.

Ann M. Yoachim is Program Manager at the Tulane Institute on Water Resources Law and Policy, New Orleans, Louisiana. Her work focuses on providing technical assistance to Louisiana communities and policy makers on climate change adaptation strategies.

Rae Zimmerman is Professor of Planning and Public Administration at New York University's Graduate School of Public Service and the Director of the Institute for Civil Infrastructure Systems. She studies the resiliency of urban infrastructure services to extreme events.

Part I

Coastal communities and the climate change imperative

Chapter 1

Introduction

Bruce C. Glavovic, P. Mick Kelly, Robert Kay, and Ailbhe Travers

AIM OF THE BOOK

This book is about climate change adaptation and the coastal zone. The coastal zone is the place where land, sea, and air meet. The encounter can be abrupt, waves crashing against towering cliffs, or it can be gentle, the subtle meanderings of tides over coastal mudflats. Coastal ecosystems provide abundant resources, goods, and services that sustain coastal communities. We cherish the coast because of its incredible diversity, richness, and beauty (Figure 1.1). Ironically, because they are such attractive locations, the world's coastal zones are increasingly jeopardized by ill-considered development, by unsustainable resource use, and by, arguably the most significant threat posed by human activity, climate change. As the majority of the world's population live at the land–sea interface or depend on the coastal zone, the coast is the frontline in humanity's endeavor to learn to live with climate change (Box 1.1).

The aim of this book is to assist the process of adaptation by providing practical guidance to coastal communities in preparing for the challenges and unlocking the opportunities presented by climate change. To this end, a set of case studies has been assembled that demonstrate *adaptive journeys* from coastal localities around the globe. Set for the most part in the first decade of the twenty-first century, these case studies provide examples of real-world experience in dealing with the impacts of climate change on the coastal zone at a range of scales and from diverse perspectives. The case studies were mostly compiled in 2011 and, where possible, updated in 2013. Consequently, they present a snapshot of experiences at that point in time. Many of the lessons learned are, however, timeless. These case studies provide tales of success and failure, innovative ideas and obstacles to be overcome, that will assist coastal managers, community members, and decision-makers as they embark on their own adaptive journeys in building resilient coastal communities.

In this chapter, we discuss the key attributes of the coastal zone and outline the important role that it plays in maintaining social, environmental,

Figure 1.1 Iconic coastal locations: (left to right) Sydney, New South Wales, Australia; Lily Beach, Maldives; Juhu Beach, Mumbai, India. (Courtesy of C. Elrick-Barr, C. Elrick-Barr and R.C. Kay, respectively.)

BOX 1.1 DEFINING CLIMATE CHANGE

Climate is commonly defined as the average weather (i.e., locality-specific atmospheric conditions such as air temperature, humidity, wind speed, and precipitation) over a length of time (typically 30 years). Climate varies naturally on all timescales, from interannual to geological timescales. Climate change refers to shifts in the mean state of the climate or in its variability that persist for a long time (at least several decades). These shifts may be due to natural variability or a variety of forms of human action, such as land-use change or activities that pollute the atmosphere. The United Nations Framework Convention on Climate Change uses the term *climate change* selectively to refer to climate trends generated directly or indirectly by human activity, to be distinguished from natural climate variability over similar timescales. In this book, the term *climate change* is used in this latter sense, as shorthand for anthropogenic climate change, the effect on the global climate system of increasing amounts of greenhouse gases in the atmosphere resulting from human activity.

and economic well-being. We provide a summary of the range of challenges that coastal communities are exposed to, placing the potential impact of climate change in the context of the many existing tensions and conflicts in an already pressurized and often unsustainably developed environment. Finally, the structure of the book is summarized with brief chapter outlines.

A DYNAMIC ENVIRONMENT UNDER SIEGE

The coastal zone has been defined formally as "the band of dry land and adjacent ocean space (water and submerged land) in which terrestrial processes and land uses directly affect oceanic processes and uses, and vice versa"

(Ketchum 1972: 4). The zone of direct land–sea influence is relatively narrow, but, as far as ecological processes are concerned, the area of indirect influence extends far out to sea and well up into the catchments that drain the coastal hinterland. Socioeconomic linkages extend much further. The coastal zone is an extremely dynamic environment, characterized by complex interactions between the maritime, terrestrial, and atmospheric processes that converge at the land–sea interface. Shorelines display long-term trends of erosion and accretion and the short-term impact of extreme events such as the landfall of hurricanes presents an unpredictable challenge to those living at the coast (Figure 1.2).

The coast is one of the most biologically productive systems on earth. Coastal ecosystems, such as mangroves, dune systems, estuaries, coral reefs, and sea grass beds, provide an array of goods and services that sustain coastal communities: acting as a natural defense against coastal storms; mitigating floods and controlling erosion; storing water; providing food; abating pollution; cycling and retaining vital nutrients; and providing a range of recreational and cultural assets. Costanza et al. (1997) estimated that the total economic (market plus nonmarket) value of global ecosystem services (including regulatory functions) amount to some USD 33 trillion per annum. Coastal ecosystem services account for 77% of this value, well over twice the value of terrestrial economic services (Martínez et al. 2007). The coast, as a base for maritime transport, also provides an important commercial focus. It should, therefore, not be a surprise that around half

Figure 1.2 A coast under siege—erosion of existing infrastructure causing restricted coastal access at Zavora along the high-energy coast of Mozambique. (Courtesy of A. Travers.)

the human population lives within 100 km of the sea; 21 of the world's 33 megacities are in the coastal zone; and 41%–45% of global economic activity takes place in the coastal zone (Hinrichsen 1999; Martínez et al. 2007; Patterson 2008). The coastal zone is the primary habitat of humanity, a vibrant meeting place where diverse influences and interests converge.

Tension and conflict are inevitable in this narrow and increasingly resource-constrained zone. Transformation of coastal ecosystems to open up opportunities for urban development, agriculture, coastal infrastructure, and a myriad of competing activities is now commonplace with consequent impacts on human well-being. There has been an increasing awareness and recognition on a global scale that over population, poverty, unsustainable resource use, environmental degradation, weak governance systems, and inadequate public infrastructure are undermining local livelihoods and making coastal communities increasingly vulnerable (see Box 1.2). The extent, intensity, and rate of development at the land–sea interface have resulted in levels of coastal degradation that exceed that of any terrestrial system (Brown et al. 2007; MEA 2005; Nellemann and Corcoran 2006). Although innovations in technology, business, and governance arrangements have enabled people to flourish in coastal localities around the world, past and current practices are undermining coastal ecosystems and their ability to sustain human development (Glavovic 2013; Moser et al. 2012).

A wide range of management approaches and governance arrangements has been instituted to reconcile user conflicts and divergent public and private and short- and longer-term interests in the coastal zone (Cicin-Sain and Knecht 1998; Kay and Alder 2005). Many nations govern the coastal zone as a public heritage. Providing public access to the coastal zone is thus one of the important goals of coastal management. Traditionally, though, coastal management efforts have been organized according to sectoral interests and to address locality-specific issues. For example, those responsible for road infrastructure have typically planned and carried out their activities independently of those responsible for fisheries management or nature conservation. A cost-effective coastal road alignment that might be appropriate from the perspective of traffic engineering criteria may be undesirable from a fisheries management or nature conservation perspective if critical habitat or ecosystem functions are adversely affected by the road alignment. Clearly, to be effective, any management strategy must consider potential cross-scale interactions, as well as a wide range of compounding and synergistic interactions (Figure 1.3).

The coastal zone is a distinctive, interconnected social-ecological system that requires a dedicated and integrated management approach to secure and sustain the manifold benefits of coastal ecosystems and communities.

BOX 1.2 WHAT IS VULNERABILITY?

The Intergovernmental Panel on Climate Change (IPCC) (Parry et al. 2007: 883) defines vulnerability as "the degree to which a system is susceptible to, and unable to cope with, adverse effects of climate change, including climate variability and extremes. Vulnerability is a function of the character, magnitude and rate of climate variation to which a system is exposed, its sensitivity and its adaptive capacity." Exposure is the extent to which a coastal community, locality, or region experiences the effects of climate change. Sensitivity is the degree to which a system is affected by or responsive to climate change. Biophysical systems have physiological tolerances to change, for example, temperature and pH. The sensitivity of social systems is dependent on cultural, political, economic, and institutional characteristics. Those dependent on resources vulnerable to climate change are more likely to be sensitive to change. For example, coastal communities dependent on coral reef resources are likely to be more vulnerable than communities whose livelihoods are dependent on less climate-sensitive resources. Adaptive capacity is the ability of a system to adjust to climate change, moderate potential damage, and take advantage of beneficial opportunities or cope with the consequences. Adaptive capacity is thus the ability to deal with climate risk and manage impacts by learning from experience, testing new ideas, and devising new practices. Adaptive capacity in biophysical systems is rooted in biodiversity and landscape heterogeneity. In social systems, adaptive capacity may be conscious or inadvertent and is shaped by the extent to which social institutions, networks, and actors learn and retain knowledge and experience, are flexible and creative in problem-solving, and sustain the capacity to cope with and adapt to change and surprise.

Some coastal settings are significantly more susceptible to climate change impacts than others. Coastal vulnerability hot spots include the Arctic (Forbes 2011); low-lying islands, deltas, and coastal lowlands (e.g., the Maldives and Kiribati, the Mississippi delta, and the coastlands of Bangladesh); coastal communities living in the path of tropical cyclones or hurricanes (e.g., Caribbean islands); coastal megacities (Pelling and Blackburn 2013); and coastal communities characterized by rapid population growth, high dependence on climate-sensitive coastal ecosystems (such as coral reefs), already degraded and stressed coastal ecosystems, weak governance systems, widespread poverty and inequality, and inadequate public infrastructure (World Bank 2013).

Figure 1.3 The interface between land and sea: (clockwise) Taiwan; New Zealand; Norfolk Island; Albania. (Courtesy of R.C. Kay.)

One cannot simply transfer management approaches from either the terrestrial or marine environments and assume that they will work at the land–sea interface. Integrated coastal management is necessary to sustain the productivity, health, and integrity of the coupled social-ecological systems that make up the coastal zone.

Integrated approaches to coastal management facilitate dialogue, cooperation, and coordination of sectoral, locality- and issue-specific activities, recognizing interdependencies, and reconciling contending interests. Pernetta and Elder (1993) defined Integrated Coastal Management* as a single management framework for combining all biophysical and human dimensions of the coastal zone. The pursuit of integrated management raises major issues of governance. Integrated modalities of coastal management and governance are crucial in this era of climate change, an insidious threat that does not respect sectoral or geographical boundaries and whose influence extends beyond conventional planning timescales.

* Often referred to as Integrated Coastal Zone Management.

ON THE FRONTLINE OF A CHANGING CLIMATE

Over the past 50 years, the issue of climate change has evolved from a rather neglected branch of the science of climatology to a leading item on the international political agenda. This evolution has been surrounded by controversy. On the one hand, climate skeptics argue that the scientific basis of climate concern is not well-founded and that the economic cost of climate action is unwarranted. On the other hand, climate idealists claim that climate change threatens the very foundations of society and that radical solutions are urgently required. Scientific uncertainties are inevitable when attempting to understand and predict the long-term behavior of a system as complex as the global climate system. There are genuine issues to be resolved regarding the most effective means of responding to the climate change problem and its ranking on the scale of societal priorities. The climate problem is extremely complex, characterized by uncertainty, ambiguity, and inevitable surprise. It has been described as *super wicked* by Levin et al. (2012). This presents a genuine challenge to society's decision-making systems. Nonetheless, the risks associated with climate change are profound and the problem cannot be ignored.

There is broad scientific agreement that the accumulation of greenhouse gases such as carbon dioxide and methane in the atmosphere is driving a change in global climate that is unprecedented in modern human history (Pachauri et al. 2007). The impact of climate change could lower global gross domestic product by as much as 20% in the future and cause major environmental, social, and economic disruption that would rival the world wars and economic depression of the first half of last century (Stern 2007). It could alter the physical and human geography of the planet (Stern 2009) and may be the most serious foreseeable threat to human development, potentially undermining development gains made to date (UNDP 2008; World Bank 2010) (Box 1.3).

The gravity of the threat posed by climate change has been recognized, and the United Nations Framework Convention on Climate Change, the climate treaty, is guiding the international response to the climate problem. Since the treaty came into force in 1994, there has been a gradual evolution in the focus of attention. Initially, the emphasis was on mitigation, reducing the scale of the problem by limiting greenhouse gas emissions or through other measures to curb the changing composition of the atmosphere. It was apparent from the outset that, even with a concerted international effort to limit emissions, unacceptable impacts could still occur as climate change would not be halted overnight. It is, however, only in recent years, since the landmark Copenhagen Climate Change Summit in 2009, that the imperative of adaptation has received focused political attention.

Not only is there a growing concern about the rate and scale of the climate impacts likely to take place over coming decades and beyond,

BOX 1.3 CLIMATE CHANGE IMPACTS

Notwithstanding the difficulties in distinguishing between the effects of anthropogenic climate change and natural effects, or the impacts of other forms of human activity, it is widely recognized that anthropogenic climate change is already having significant impacts on coastal communities through, for example, intensified coastal erosion, altered flood–drought regimes, salt-water intrusion of groundwater, and impacts on coastal ecosystems. For much of the twentieth century, sea level rose at a rate of 1.7–1.8 mm per annum (Pachauri et al. 2007). The IPCC's Fourth Assessment (AR4) could not place a firm upper bound on the potential rise in sea level by the end of the twenty-first century because of unresolved issues related to ice loss in the Greenland and Antarctic ice sheets (see Chapter 2). Setting the possibility of accelerated loss of the major ice sheets to one side, AR4 concluded that a rise of up to 0.6 m by 2100, primarily due to thermal expansion of the oceans was possible—even this represents a substantial acceleration of the historical rate. In a worst-case scenario, with sea level rising by a meter or more, hundreds of millions of people would be displaced as low-lying coastal areas are inundated; coastal aquifers become saline; climate-sensitive coastal ecosystems such as coral reefs, estuaries, and mangroves are adversely impacted disrupting vital fisheries; and other coastal ecosystem goods and services that sustain coastal communities (Nicholls et al. 2007). Low-lying small island states, such as the Maldives, might disappear under the rising waters. Ocean acidification due to dissolution of carbon dioxide will profoundly impact marine ecosystems and threatens the survival of coral reef ecosystems and their dependent communities. By 2100, ocean pH levels might reach their lowest point in 20 million years. It is estimated that 30% of the world's coral reefs have already been destroyed by nonclimate stressors such as overexploitation, pollution, and that climate change impacts could accelerate this dire situation resulting in more than 80% being lost within decades. Some coastal regions are already experiencing more frequent and intense extreme weather events. Increased sea surface temperatures and rising sea levels will lead to even more severe storms, which will compound storm surges and exacerbate coastal erosion. With more people living at the coast, the risk of coastal disasters increases: 120 million people are already exposed to tropical cyclones every year, and future prospects are cause for deep concern. It is estimated that about 250 million people live within 5 m of the high tide mark and more than 150 million people live within 1 m of high tide. Those most at risk are typically located in the densely

(Continued)

populated low-lying mega-deltas of Asia, including the Ganges–Brahmaputra (Bangladesh), Mekong (Cambodia/Vietnam), and Yangtze (People's Republic of China). Countries with dense populations in low-lying coastal localities in the path of tropical cyclones are especially vulnerable to the intensification of extreme events. Changes in precipitation will lead to some regions experiencing more flooding, whereas other areas will be subjected to more frequent and severe droughts. Changing rainfall patterns may be compounded by shifts in El Niño and La Niña patterns. Altered rainfall changes runoff, and coastal erosion and sedimentation processes, which significantly affect ecosystems such as mangroves, estuaries, and coral reefs. Coastal hypoxia may become more commonplace with nutrient-rich runoff under higher sea surface temperatures. These changing conditions affect the distribution and health of coastal ecosystems and their constituent species, including bacteria, viruses, and disease vectors that affect human health as well as those species that we rely on for food and other ecosystem services. Changing climatic conditions thus profoundly affect ecosystem health and productivity and consequently the health, well-being, and livelihoods of coastal communities.

the scale of vulnerable coastal populations and associated assets is unprecedented. In coming decades, coastal communities will experience intensifying climate change impacts, such as rising sea levels and changes in the intensity of extreme events such as coastal storms, at a time when they are already stressed by environmental degradation and social and economic challenges (Figure 1.4). Practical steps need to be taken by coastal communities to anticipate and adapt to climate change in the face of conflicting interests and incomplete knowledge and understanding.

The Intergovernmental Panel on Climate Change (IPCC), the scientific body charged with advising the international community on the climate issue defines adaptation as the adjustment in natural or human systems in response to actual or expected climatic stimuli or their effects (Smit et al. 2001). Different types of adaptation have been defined (see, e.g., Adger et al. 2007). Anticipatory adaptation occurs prior to observed climate change effects—it is proactive. Reactive adaptation occurs after the impacts of climate change have been observed. Individuals, households, or private companies undertake private adaptation, whereas public adaptation is initiated and implemented by government, potentially with other key actors, to achieve desired collective outcomes. Autonomous adaptation is not a deliberate or conscious response to climate change but is triggered spontaneously by natural or social systems. Planned adaptation

Figure 1.4 Local fishers gather via boat to buy and exchange produce in a small and degraded fishing harbor in Pebane, a remote coastal community in northern Mozambique. (Courtesy of A. Travers.)

is a deliberate choice, based on awareness and anticipation of change and charting a course of action to return to, to maintain, or to achieve a desired state. Finally, maladaptation refers to a response in a social-ecological system that increases vulnerability to climate change or an adaptation measure that paradoxically increases vulnerability instead of reducing it.

In this book, we locate the challenge climate change poses for the inhabitants of the coastal zone in the broader context of the quest for sustainable modes of development. We view climate change not only as a threat but also as an opportunity. Climate trends may have consequences that are potentially beneficial, and any effective adaptation response will promote these opportunities. Climate-proofing existing development plans can provide a stringent test of sustainability. Many climate measures have collateral benefits, in conserving resources and improving the overall health of human and ecological systems. Moreover, the particular characteristics of the climate issue raise fundamental questions about the way society is organized and the manner in which decisions are made. Addressing the long-term coevolution of climate and society provides an opportunity for us all to reflect on our hopes and aspirations for the future. Our approach is to learn from the experiences of coastal communities around the world and distil practical guidance that can be used to build resilience and sustainability (see Box 1.4).

BOX 1.4 WHAT IS RESILIENCE?

Some have defined resilience as the flip side of vulnerability (e.g., Gallopin 2006): the ability to cope with change as opposed to inability to cope with change. Resilience to climate change reflects the ability of social-ecological systems to cope with and adapt to change and deal with uncertainty. Reducing exposure and sensitivity and increasing adaptive capacity builds resilience. Resilience can be thought of as the amount of change that a system can undergo without changing state (Pachauri et al. 2007). Resilience constitutes, first, the amount of change that a social-ecological system can absorb and still retain key structures and functions; second, the extent to which the system is able to reorganize; and finally, the extent to which the system can increase capacity for learning and adaptation (Anderies et al. 2013; Holling 1973; Holling and Gunderson 2002; Walker and Salt 2006; Walker et al. 2004).

LEARNING FROM THE EXPERIENCE OF COASTAL COMMUNITIES

Studies of the implications of climate change for the coast have been undertaken for several decades now. They have been based on a range of assumptions, taken different approaches, and tackled the issue from a range of perspectives. Some have addressed the issue from the point of view of biophysical analysis. Others see the world through a different lens, primarily a lens of human endeavor. Social, political, and economic analysts frame the climate change issue as one caused by humans, and they stress the need to locate impact assessment and adaptation studies in the human dimension. Clearly, there are a number of such lenses or perspectives through which the climate change issue can be viewed.

The editorial strategy for this book has been shaped by our respective backgrounds in coastal governance and natural hazards planning (Bruce C. Glavovic), climate science (Mick Kelly), coastal planning (Robert Kay), and coastal geomorphology (Ailbhe Travers). The editorial process has been informed by a shared commitment to interdisciplinary approaches and our experience, as scholars and practitioners, in coastal areas on all inhabited continents. It is our view that the sheer magnitude and complexity of the climate change *problématique* requires more open and collaborative, integrative, and adaptive approaches. In this book, we have selected and juxtaposed a range of case studies on climate change and the coast to help make sense of this complex issue. Leading specialists share the stories of communities in coastal settings as diverse as the Canadian Arctic, New York City, Bangladesh, Australia, New Zealand, Republic of South Africa, the Caribbean, Kiribati islands, and many more, from highly industrialized societies where economic assets are concentrated in the coastal zone to communities for whom subsistence coastal resource use is the key concern (see Figure 1.5; Table 1.1).

The case studies were selected to ensure broad geographical coverage, with all inhabited continents represented as well as a diverse range of approaches and focal issues. This is in keeping with our view that the most effective ways forward in adapting to climate change will be found by sharing the rich experience of researchers, analysts, and diverse coastal communities (Figure 1.6, 1.7, and 1.8).

The case studies range in theme, from the development of government policy through the effectiveness of community consultation to practical measures to protect livelihoods (see Table 1.1). Some approaches are top-down in nature and others from the grassroots. The authors are drawn from academia, government service, consultancies, and nongovernmental organizations, and all authors have first-hand knowledge and experience of the case studies they write about. The aim has been to present a variety of perspectives from a wide range of circumstances and through the eyes of specialists from diverse disciplines and professional backgrounds. What these perspectives do have in common is that they are decision-oriented or outcome-focused, drawing on current scholarship and real-world experience to provide practical advice for assessing and reducing climate vulnerability, and strengthening adaptive capacity and community resilience. The case studies are framed by the following questions:

- *What makes this place special?* To describe the distinctive social-ecological-governance features of each coastal locality, including measures taken to cope with past and present environmental and societal drivers of change.
- *Looking forward, what makes this place at particular risk from future climate change and other potential threats?* To outline the vulnerability of the locality, including the potential consequences of climate change and other environmental threats and socioeconomic pressures.
- *What can be learnt from past experience?* To highlight key successes and failures evident in existing measures to build community resilience, adaptive capacity, and sustainability.
- *What will help and what will hinder?* To identify critical barriers and opportunities for mainstreaming climate change adaptation measures into prevailing planning and decision-making and identifying the most significant aspects.
- *How can we move forward?* To draw out at least three practical recommendations for building community resilience, adaptive capacity, and sustainability.

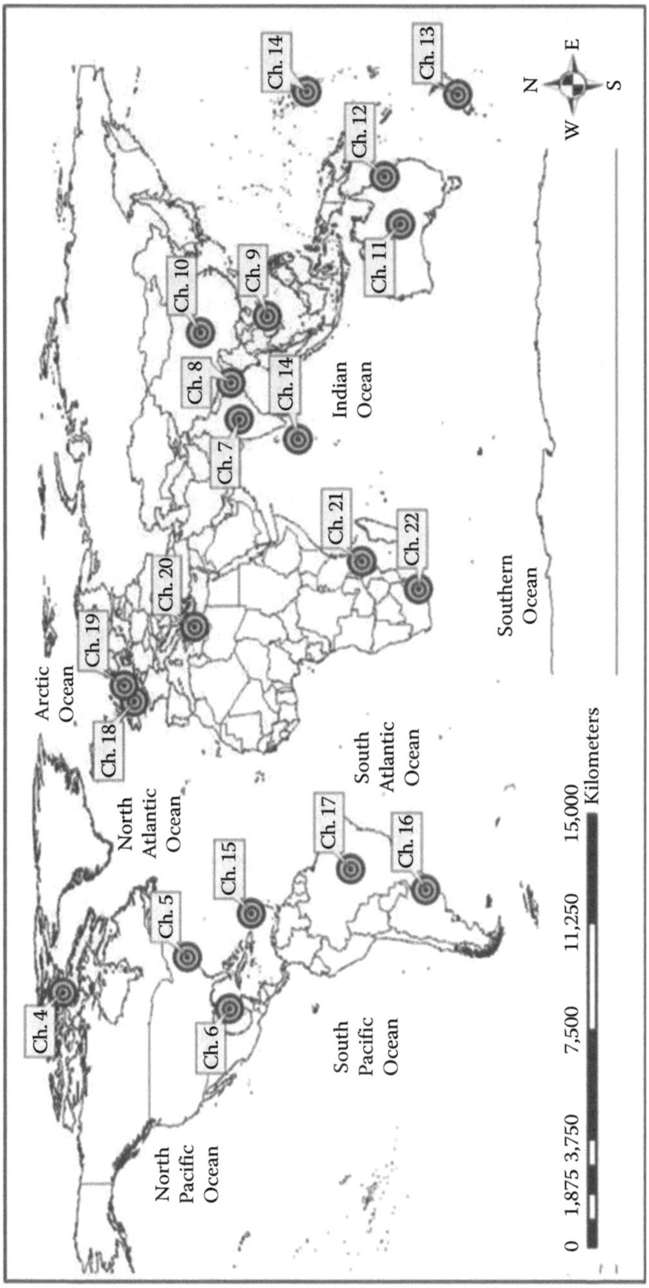

Figure I.5 Case study locations, indicated by chapter.

Table 1.1 The case studies: Stories from coastal communities on the frontline

Region	*Chapter*	*Location*	*Focus*
Part II: North America	4	Canadian Arctic	Coastal communities of the Canadian Arctic demonstrate that collaborative networks and social learning can reduce vulnerability and foster adaptive capacity.
	5	New York City	Documents a process where extensive stakeholder involvement and state-of-the-art scientific projections and mapping were used to produce a climate risk assessment and adaptation strategy for New York City.
	6	Louisiana Coast	A case for novel approaches to coastal planning, development, and ecosystem restoration in a complex coastal area where transformation has previously been driven by crisis.
Part III: South and Southeast Asia	7	India	Discussion of a participatory, grassroots approach to irrigation management in an Indian deltaic environment as a means to facilitate climate change adaptation for a community largely dependent on agriculture.
	8	Bangladesh	Description of Bangladesh's National Adaptation Programmes of Action priorities and identification of existing barriers to adaptive decision-making identified.
	9	Vietnam	Explores the strengths and weaknesses of climate policy and coastal zone management in Vietnam underlining the importance of coordination, the need for full participation and the desirability of a flexible, step-by-step approach.
	10	People's Republic of China and Australia	Consideration of strategic growth management planning initiatives in the Yangtze Delta, People's Republic of China, and South East Queensland, Australia with respect to climate change adaptation. Practical steps to build resilience and adaptive capacity in both locations are discussed.
Part IV: Australasia	11	Australia	Overview of vulnerability and adaptation studies and policy initiatives that have been carried out in Australia over the past 20 years.

(*Continued*)

Table 1.1 (Continued) The case studies: Stories from coastal communities on the frontline

Region	*Chapter*	*Location*	*Focus*
	12	South East Queensland	A case study of local government responses in the South East Queensland region highlighting key adaptive capacity issues and the role of biophysical, socioeconomic, and political drivers of change in shaping future adaption options.
	13	New Zealand	Examines two case study sites in New Zealand, revealing the technical and institutional challenges that need to be faced in efforts to build resilience through managed retreat in highly developed regions.
Part V: Small islands	14	Kiribati and Maldives	A comparison of climate change issues and adaptive capacity for two small island states. Discussion on the impacts that divergent adaptive pathways may have on potential long-term resilience.
	15	Anguilla, Caribbean	Examination of the evolution of climate change planning in the Caribbean with specific focus on the development of coastal planning guidelines in Anguilla.
Part VI: South America	16	Uruguay	Focus on the EcoPlata and Adaptation Measures to Climate Change in Coastal Areas of Uruguay initiatives involved in developing activities to assess and reduce vulnerability to climate change impacts and strengthen institutional arrangements, governance, community resilience, and adaptive capacity.
	17	Brazil	Review of the challenges facing sustainable management of the Brazilian coastal zone in light of climate change. Specific attention is given to the issues associated with large-scale development in the coastal strip.
Part VII: Europe	18	England	North Sea coast of England considered as a case study to assess past coastal management principles and practices and the emergence of new coastal policy practice in England, drawing lessons for future, more adaptive management strategies such as managed realignment.

(*Continued*)

Table 1.1 (Continued) The case studies: Stories from coastal communities on the frontline

Region	*Chapter*	*Location*	*Focus*
	19	North Sea	Explores how experience with Maritime Spatial Planning in North Sea countries can offer contextual insights for future *climate adaptation governance* using offshore wind farm development as a case study. Discusses challenges to harmonize adaptation across different styles of planning and policy development without overriding nationally and regionally accepted specifics in process design and planning culture.
	20	Mediterranean	Deals with trans-boundary management of the Mediterranean coastal zone in the face of climate change focusing on recommendations from the Priority Actions Programme/Regional Activity Centre of the United Nations Environment Programme's Mediterranean Action Plan.
Part VIII: Africa	21	Mozambique	Outlines the decision-making process to select potential adaptation options for pilot site adaptation projects in a developing African nation. Work was undertaken as part of UNDP's programming activities for a GEF 5 Project to facilitate climate change adaptation in the coastal zone of Mozambique.
	22	Durban, Republic of South Africa	Deals with the approaches, challenges, and lessons learnt in Durban during the development and implementation of the Municipal Climate Protection Programme and the possible way forward in terms of the further development of the program.

These stories explore the complex mix of challenges and opportunities that climate change poses for coastal communities. Together, they reveal important lessons for preparing for climate change that can be drawn upon to help coastal communities develop innovative and practical ways to build resilience and sustainability. The latest research findings in coastal management, environmental science and governance, climate change science, and resilience and sustainability studies, as well as other areas of scholarship, are woven into these narratives.

Figure 1.6 Lagoon coastline, Kiribati. (Courtesy of A. Travers.)

At an editorial meeting in November 2011, the editors reviewed the case studies, identified key issues and themes, to determine whether or not a unifying conceptual framework could be defined. Drawing on the lessons from the case studies, as well as previous research and practical experience, we argue in Chapter 23 that there is a need to develop a reflexive approach to adaptation that is participatory in nature and inclusive, innovative, exploratory, and learning-oriented, flexible yet robust.

STRUCTURE OF THE BOOK

This book comprises eight main parts made up of this introductory part comprising three chapters that introduce readers to this book, our understanding of climate change and anticipated coastal zone impacts, and a conceptual framework for understanding the nature of the coast and the climate change adaptation imperative. Subsequent parts present case studies organized geographically to explore selected community experiences in North America, South and Southeast Asia, Australasia, small island nations, South America, Europe, and Africa. In the concluding part, a final chapter distils key lessons from these case studies and makes recommendations for future action.

In Chapter 2, Mick Kelly presents an overview of our current understanding of climate change and its likely impacts on the coastal zone in the twenty-first century. This account rests on the Fourth Assessment of the IPCC, the science behind which most of the case studies reported in this book are based, and highlights scientific developments since that review was published in 2007. The rate and magnitude of future climate change cannot be forecast with precision, and this is particularly the case at the local level. The optimal adaptation pathway in any specific set of circumstances may not be clear. A sustainable response to climate change should retain flexibility so that it can adapt to changing knowledge and circumstances. The chapter underscores the need for coastal communities to continue to address the pressing immediate challenges of nonclimate drivers of coastal change while building capacity to deal with longer-term climate change.

In Chapter 3, Bruce C. Glavovic explains why coastal communities are the frontline in the global sustainability crisis and the pivotal arena for learning how to adapt to climate change. He describes the waves of adversity that face coastal communities as well as the many ways in which communities can build layers of resilience for adapting to climate change. He develops a conceptual framework to comprehend the complexity of the climate change *problématique*, its magnitude, timescale, insidious and far-reaching nature, and inherent uncertainty and ambiguity. This framework for deliberative coastal governance is intended to develop the critical capacities and competencies needed by coastal communities to build layers of resilience to navigate coming waves of adversity and is a foil to reflect upon in light of the case studies that follow.

In Part II, the first suite of case studies explores climate change in diverse coastal zones of North America, from the Canadian Arctic to the challenges of adapting to anticipated sea-level rise in New York City and the global vulnerability hot spot of the Mississippi delta. In Chapter 4, Derek Armitage reveals how the distinctive coastal communities of the Canadian Arctic are drawing on longstanding cultural approaches to adaptation but are having to develop new coping strategies in the face of unprecedented rates of change and uncertainty. New institutional arrangements and multilevel comanagement practices have emerged that, by facilitating collaborative networks and social learning, can reduce vulnerability and foster adaptive capacity. The importance of identifying and acting on policy windows of opportunity is stressed. In Chapter 5, William Solecki and his coauthors describe how a comprehensive climate risk assessment and adaptation strategy for New York City has been developed with extensive stakeholder involvement and state-of-the-art scientific projections and mapping. Flexible adaptation strategies that promote the co-benefits of protection against existing and future climate risks are being considered. The importance of regular monitoring and reassessment of key coastal indicators is

Figure 1.7 Coastal recreation: Manhattan Beach, Los Angeles, California. (Courtesy of R.C. Kay.)

emphasized. Finally, in Chapter 6, Joshua Lewis, Ann Yoachim, and Douglas Meffert present a case study of the Louisiana coast, showing how historical choices shape vulnerability to climate change. In the present-day, Hurricane Katrina and the British Petroleum (BP) oil spill in the Gulf of Mexico have triggered political action and spurred novel approaches to coastal planning, development, and ecosystem restoration. As far as the future is concerned, the authors argue that, if current sea-level rise projections are accurate, this century must see a transformation in the region. Will that transformation be driven by foresight or, as too often in the past, by crisis?

The case studies in Part III focus on the coastal zones of South and Southeast Asia, drawing on experiences from India, Bangladesh, and Vietnam as well as a comparative analysis of urbanizing coastal areas in People's Republic of China and Australia. In Chapter 7, Ramesh and his coauthors describe national-level efforts to address climate change in India. They juxtapose these efforts with grassroots action to empower a deltaic community dependent on agriculture vulnerable to climate change and, in particular, sea-level rise and associated hazards. Cross-scale interactions and interdependencies emerge as critical considerations in adaptation endeavors. Lessons learnt from the participatory approach to irrigation management included the need for greater social mobilization than anticipated. The Bangladesh coastal zone is one of the most vulnerable regions in the world. Coastal communities in this low-lying, densely populated region are dependent on climate-sensitive coastal resources, and poverty and limited access to public services is widespread. In Chapter 8, Saleemul Huq and Golam Rabbani describe the National Adaptation Programme of Action for Bangladesh and present examples of adaptation technologies and practices in agriculture and the water supply and sanitation sectors. The need for integration of the climate issue in the development of policies, plans, and programs in climate-sensitive sectors and a lack of adequate tools, knowledge, and methodologies for guidance and advice in decision-making are identified as key concerns. In Chapter 9, Mick Kelly explains how the Vietnam coastal zone has experienced rapid transformation due to environmental change and socioeconomic reforms and argues that climate change will compound the stress already facing coastal communities. This chapter explores the strengths and weaknesses of climate policy and coastal zone management in addressing these challenges and opportunities, underlining the importance of coordination, the need for full participation and the desirability of a flexible, step-by-step approach. Finally, in Chapter 10, Darryl Low Choy, Chen Wen, and Silvia Serrao-Neumann provide a bridge between continents, focusing on growth management planning initiatives being undertaken in the rapidly urbanizing coasts of the Yangtze delta region in the People's Republic of China and the South East Queensland region of Australia, among the most rapidly growing metropolitan regions in their respective countries. They explore how climate change is being addressed in the strategic planning and growth management efforts of both regions and the practical steps that need to be taken to build resilience, adaptive capacity, and sustainability in rapidly urbanizing regions in the face of climate change. Both regions have existing planning processes that could facilitate the inclusion of climate change science and adaptation considerations into their respective regional plans and policies.

Part IV comprises a suite of case studies from Australia and New Zealand. In Chapter 11, Robert Kay, Ailbhe Travers, and Luke Dalton provide an overview of vulnerability and adaptation studies and policy initiatives that have been carried out in Australia over the past 20 years. The overview shows how adaptation endeavors are shaped by prevailing societal mores and political priorities, in particular. Notwithstanding concerted efforts at all levels of government, much remains to be done to develop and implement an integrated and robust adaptation approach. With over 85% of Australia's population living within 50 km of the coastline, particular attention needs to be focused on climate change impacts on urbanized areas. In Chapter 12, Tim Smith and coauthors present a case study of local government responses in the South East Queensland region. The study highlights adaptive capacity issues related to infrastructure provision, emergency response, and rapid socioeconomic change and shows that biophysical, socioeconomic, and political drivers of change shape future adaption options. Although fragmented management is hindering the response to climate change, some coastal local governments have become key advocates for progressive adaptation policies. Finally, in Chapter 13, Andy Reisinger and his coauthors examine two case study sites in New Zealand, revealing the technical and institutional challenges that need to be faced in efforts to build resilience through managed retreat in highly developed regions. They question the effectiveness of present responses to climate change if they are employed in a static way and are not supported by additional policies that recognize the dynamic nature of coastal hazards. The chapter demonstrates the need to build a robust information base and develop appropriate policy tools to address the inherent uncertainty regarding sea-level rise and associated impacts on particular localities and facilitate managed retreat over long time frames.

The challenges facing small island nations are discussed in Part V. Chapter 14 features a comparative analysis of Kiribati and the Maldives by Carmen Elrick-Barr, Bruce C. Glavovic, and Robert Kay. The analysis reveals that, despite apparently similar climate issues and challenges, the biophysical setting; political history; and cultural, social, and economic contexts of these islands differ markedly and profoundly affect their adaptation possibilities in the short-, medium-, and longer-term. Different adaptation pathways have profoundly different implications over these time frames

Figure 1.8 Canal estates, Mandurah, Western Australia. (Courtesy of R.C. Kay.)

and lead to potentially more restrictive or resilient futures. The authors call for the construction of collaborative partnerships between government, international collaborators, and key actors in the private sector and civil society in order that robust social choices can be made. In Chapter 15, Gillian Cambers and Sharon Roberts-Hodge examine how planning for climate change in the Caribbean has evolved over the past 20 years. Attention is focused on the development and implementation of coastal planning guidelines for Anguilla, highlighting the key roles played by key actors in government, the private sector, and local communities in translating rhetoric into reality. Opportunities for mainstreaming climate change and climate variability in the government planning agenda include incorporating wise practices into the planning process; using visual images of local phenomena, such as beach erosion, to illustrate long-term processes such as climate change; involving the private sector, especially the insurance industry, in development planning; maximizing windows of opportunity after extreme events; and highlighting the economic impacts of poor planning practices.

Part VI concerns South American experience, with two case studies providing insight into climate change adaptation challenges in Uruguay and Brazil. With more than 70% of the Uruguayan population living in the coastal zone, nonclimate and climate stressors pose massive short-, medium-, and longer-term challenges. In Chapter 16, Gustavo Nagy, Monica Gómez-Erache, and Robert Kay show that work being undertaken in Uruguay on climate change vulnerability and adaptation sheds light on climate risks and the challenges of improving coastal governance regarding climate change across municipal, regional, and national scales. They conclude that coastal adaptation efforts need to build on, and support, existing frameworks for Integrated Coastal Zone Management. Enhanced coordination in the assessment of extreme event-related impacts has proved the main driver in increasing awareness of the climate issue. In Chapter 17, Marcus Polette and his coauthors underscore the challenges of linking national coastal zone management efforts in Brazil with regional and local efforts and establishing effective linkages between legislative provisions and sectoral management practices to better understand and address climate change. These challenges are, however, dwarfed by the relentless pressure of development projects that are transforming large areas of the coastal zone. Many coastal communities are increasingly exposed to coastal hazards that are likely to be amplified by climate change.

In Part VII, attention turns to European experience with case studies from England, the North Sea, and the Mediterranean. In Chapter 18, the case study from the eastern coastal zone of England by Sir Kerry Turner and Tiziana Luisetti shows how progressive land-use change, urbanization, industrialization, and extreme storm events have transformed the coastal zone. Climate change impacts pose significant additional threats to coastal communities in this region. Past adaptive strategies relied heavily on *hard* engineering solutions that are not likely to be sufficient in the face of anticipated climate change. Resilience to climate change will require governance reforms and new approaches to stakeholder engagement and compensation, as well as social justice and equity considerations. The North Sea case study by Andreas Kannen and Beate Ratter in Chapter 19 focuses on the challenges arising from emerging climate risks in a region with multiple jurisdictions, intensive conflicting uses, and new activities, such as off-shore wind energy, that necessitate novel trans-boundary governance arrangements. Maritime spatial planning is a vital tool for engaging key stakeholders to reconcile contending uses and make provision for a changing but uncertain future. Trans-boundary challenges come to the fore in Chapter 20 on the Mediterranean by Ailbhe Travers and Carmen Elrick-Barr. The Mediterranean has a longstanding regional approach to coastal governance and, more recently, climate change adaptation. Particular attention is focused on the experience of and lessons learnt from the Priority Actions Programme/Regional Activity Centre of the United Nations Environment Programme's Mediterranean Action Plan.

In Part VIII, the final set of case studies focuses on Africa, drawing on experiences in Mozambique and the Republic of South Africa. Mozambique is one of the least developed countries in the world, and many coastal communities are extremely vulnerable to climate change. In Chapter 21, the case study by Ailbhe Travers and her coauthors outlines experiences and lessons learned from an initiative to secure funds from the Least Developed Country Fund to build national institutional capacity and pilot community-scale ecosystem-based adaptation. In Chapter 22, the closing case study, by Andrew Mather and Debra Roberts, concerns the city of Durban, Republic of South Africa. Durban is a global biodiversity hot spot, major port city, and tourist destination and a place subject to a range of socioeconomic, environmental, and governance challenges, including inequitable development with affluent neighborhoods juxtaposed against impoverished communities. The chapter describes the approaches, challenges, and lessons learned from developing and implementing the Municipal Climate Protection Programme, which has particular relevance for cities in developing nations that face significant climate risks but are already struggling with high levels of unemployment, poverty, and HIV/AIDS. Ironically, the adoption of a more sectoral (rather than a cross-sectoral or integrated) approach to the development of municipal adaptation plans has allowed *champions* to emerge within line functions, facilitated the development of more realistic work plans, and encouraged more cross-sectoral dialogue and engagement.

In Part IX, the concluding chapter of the book distils key insights from these case studies and develops the framework of reflexive adaptation, and outlines practical guidance that can be used to help communities navigate the challenges and opportunities presented at the coast by climate change complexity, uncertainty, ambiguity, and surprise.

REFERENCES

Adger, W.N., Agrawala, S., Mirza, M.M.Q., Conde, C., O'Brien, K., Pulhin, J., Pulwarty, R., Smit B., and Takahashi, K. (2007). "Assessment of adaptation practices, options, constraints and capacity," in M.L. Parry, O.F. Canziani, J.P. Palutikof, P.J. van der Linden, and C.E. Hanson (eds.), *Climate Change 2007: Impacts, Adaptation and Vulnerability. Contribution of Working Group II to the Fourth Assessment Report of the Intergovernmental Panel on Climate Change*, Cambridge, Cambridge University Press. http://www.ipcc.ch/ (accessed August 2013).

Anderies, J.M., Folke, C., Walker, B., and Ostrom E. (2013). "Aligning key concepts for global change policy: Robustness, resilience, and sustainability," *Ecology and Society* 18(2): 8.

Brown, C., Corcoran, E., Hekerenrath, P., and Thonell, J. (eds.) (2007). *Marine and Coastal Ecosystems and Human Well-Being: A Synthesis Report Based on the Findings of the Millennium Ecosystem Assessment*. New York: UNEP.

Cicin-Sain, B. and Knecht, R. (1998). *Integrated Ocean and Coastal Management: Concepts and Practices*. Washington, DC: Island Press.

Costanza, R., d'Arge, R., De Groot, R., Farber, S., Grasso, M., Hannon, B., Limburg, K. et al. (1997). "The value of the world's ecosystem services and natural capital," *Nature*, 387: 253–60.

Forbes, D.L. (ed.) (2011). *State of the Arctic Coast 2010—Scientific Review and Outlook*. Geesthacht, Germany: Land-Ocean Interactions in the Coastal Zone, 178p.

Gallopin, G.C. (2006). "Resilience, vulnerability and adaptation: A cross-cutting theme of the International Human Dimensions Programme on Global Environmental Change," *Global Environmental Change*, 16(3): 293–303.

Glavovic, B.C. (2013). "The coastal innovation paradox," *Sustainability*, 5: 912–33.

Hinrichsen, D. (1999). *Coastal Waters of the World: Trends, Threats, and Strategies*. Washington, DC: Island Press.

Holling, C.S. (1973). "Resilience and stability of ecological systems," *Annual Review of Ecology, Evolution, and Systematics*, 4: 1–23.

Holling, C.S. and Gunderson, L.H. (2002). "Resilience and adaptive cycles," in L.H. Gunderson and C.S. Holling (eds.), *Panarchy: Understanding Transformations in Human and Natural Systems*. Washington, DC: Island Press.

Kay, R. and Alder, J. (2005). *Coastal Planning and Management*, 2nd edn. London; New York: Taylor & Francis.

Ketchum, B.H. (ed.) (1972). *The Water's Edge: Critical Problems of the Coastal Zone*. Cambridge, MA: MIT Press.

Levin, K., Cashore, B., Bernstein, S., and Auld, G. (2012). "Overcoming the tragedy of super wicked problems: Constraining our future selves to ameliorate global climate change," *Policy Science*, 45: 123–52.

Martínez, M.L., Intralawan, A., Vázquez, G., Pérez-Maqueo, O., Sutton, P., and Landgrave, R. (2007). "The coasts of our world: Ecological, economic and social importance," *Ecological Economics*, 63: 254–72.

Millennium Ecosystem Assessment (MEA). (2005). *Ecosystems and Human Well-Being: Synthesis*. Washington, DC: Island Press.

Moser, S.C., Williams, S.J., and Boesch, D.F. (2012). "Wicked challenges at land's end: Managing coastal vulnerability under climate change," *Annual Review of Environment and Resources*, 37: 51–78.

Nellemann, C. and Corcoran, E. (eds.) (2006). *Our Precious Coasts: Marine Pollution, Climate Change and the Resilience of Coastal Ecosystems,* Arendal, Norway: UNEP.

Nicholls, R.J., Wong, P.P., Burkett, V.R., Codignotto, J.O., Hay, J.E., McLean, R.F., Ragoonaden, S., and Woodroffe, C.D. (2007). "Coastal systems and low-lying areas," in M.L. Parry, O.F. Canziani, J.P. Palutikof, P.J. van der Linden, and C.E. Hanson (eds.), *Climate Change 2007: Impacts, Adaptation and Vulnerability. Contribution of Working Group II to the Fourth Assessment Report of the Intergovernmental Panel on Climate Change*, Cambridge: Cambridge University Press. http://www.ipcc.ch/ (accessed August 2013).

Pachauri, R.K. and Reisinger, A. (eds.) (2007). *Climate Change 2007: Synthesis Report. Contribution of Working Groups I, II and III to the Fourth Assessment Report of the Intergovernmental Panel on Climate Change.* Geneva, Switzerland: IPCC. http://www.ipcc.ch/ (accessed August 2013).

Parry, M.L., Canziani, O.F., Palutikof, J.P., van der Linden, P.J., and Hanson, C.E. (eds.) (2007). *Climate Change 2007: Impacts, Adaptation and Vulnerability. Contribution of Working Group II to the Fourth Assessment Report of the Intergovernmental Panel on Climate Change,* Cambridge: Cambridge University Press. http://www.ipcc.ch/ (accessed August 2013).

Patterson, M. (2008). "Ecological shadow prices and contributory value: A biophysical approach to valuing marine ecosystems," in M. Patterson and B. Glavovic (eds.), *Ecological Economics of the Oceans and Coasts,* Cheltenham: Edward Elgar.

Pelling, M. and Blackburn, S. (2013). *Megacities and the Coast: Risk, Resilience and Transformation.* London: Earthscan.

Pernetta, J.C. and Elder, D.L.E. (1993). *Cross-Sectoral, Integrated Coastal Area Planning: Guidelines and Principles for Coastal Area Development.* Gland, Switzerland: IUCN.

Smit, B., Pilifosova, O., Burton, I., Challenger, B., Huq, S., Klein, R.J.T., and Yohe, G. (2001). "Adaptation to climate change in the context of sustainable development and equity," in J.J. McCarthy, O. Canziani, N.A. Leary, D.J. Dokken, and K.S. White (eds.), *Climate Change 2001: Impacts, Adaptation and Vulnerability. Contribution of the Working Group II to the Third Assessment Report of the Intergovernmental Panel on Climate Change.* Cambridge: Cambridge University Press.

Stern, N. (2007). *The Economics of Climate Change: The Stern Review.* Cambridge; New York: Cambridge University Press.

Stern, N. (2009). *Managing Climate Change and Overcoming Poverty: Facing the Realities and Building a Global Agreement.* London; Leeds: Centre for Climate Change Economics and Policy.

United Nations Development Programme (UNDP). (2008). *Human Development Report 2007/2008. Fighting Climate Change: Human Solidarity in a Divided World.* New York: UNDP.

Walker, B.H., Holling, C.S., Carpenter, S.C., and Kinzig, A.P. (2004). "Resilience, adaptability and transformability," *Ecology and Society*, 9(2): 5. http://www.ecologyandsociety.org/vol9/iss2/art5/ (accessed August 2013).

Walker, B.H. and Salt, D. (2006). *Resilience Thinking: Sustaining Ecosystems and People in a Changing World*. Washington, DC: Island Press.
World Bank. (2010). *World Development Report 2010: Development and Climate Change*. Washington, DC: World Bank.
World Bank. (2013). *Turn Down the Heat: Climate Extremes, Regional Impacts, and the Case for Resilience. A Report for the World Bank by the Potsdam Institute for Climate Impact Research and Climate Analytics*. Washington, DC: World Bank.

Chapter 2

Climate drivers in the coastal zone

P. Mick Kelly

Abstract: Current understanding of the potential course of climate change, sea-level rise, and related impacts on the coastal zone over the twenty-first century is defined in this chapter. The assessment is based on the conclusions of the Fourth Assessment of the Intergovernmental Panel on Climate Change, published in 2007. Subsequent developments in understanding are also discussed and full consideration is given to uncertainties in the climate projections. As far as adaptation is concerned, the critical issue is how to plan for an uncertain future. At one end of the spectrum of possibilities, the change in the planetary environment over the twenty-first century may be unprecedented in recent human history. At the other end of the range, the change in climate may prove not much greater than humanity has experienced in recent decades. The degree of uncertainty increases over time, so a process of staged, or nested, implementation, allowing for plans to evolve as experience dictates, appears advisable.

INTRODUCTION

Taking full account of the climate dimension when planning for the future must rest on the foundations of scientific understanding of the potential threat of climate change and sea-level rise and the diverse implications for coastal processes. What can be said with confidence regarding the future environment that coastal inhabitants will face? As the late Steve Schneider (2008) observed, forecasting climate is like gazing into a partly cloudy crystal ball. Though it is possible to define the broad picture regarding future trends in global climate, details, particularly at the local level, remain unclear. This situation reflects, in part, limitations in scientific understanding and the complexity of the climate system. But it is also the case that long-term trends in global climate will always be subject to some degree of uncertainty, dependent as they are on the myriad social, economic, and political processes that influence greenhouse gas emissions, including the international political response to the climate problem itself. Planning for an uncertain future is the critical challenge that climate change poses for decision-makers, a challenge rendered more difficult by the polemical debate between cynics and zealots.

The Intergovernmental Panel on Climate Change (IPCC) was established by the World Meteorological Organization and the United Nations Environment Programme on the basis of a resolution of the United Nations General Assembly in 1988. The role of the IPCC is to provide authoritative guidance to politicians and decision-makers on the ever-developing science of climate change: to provide an assessment of published scientific, technical, and socioeconomic information that builds understanding about the nature and risks associated with climate change (IPCC 2007a). To ensure credibility, assessments are prepared by international teams of scientists, and all draft reports are reviewed stringently by experts and by governments as an essential part of the reporting process. The IPCC has been criticized and, as is inevitable in an endeavor of this scale, errors have been made (Hulme and Mahony 2010; Shapiro et al. 2010). Regardless, the IPCC is, beyond doubt, the most credible source of scientific information on the climate issue.

The Fourth Assessment of the IPCC, commonly referred to as AR4, was published in 2007. The deliberations of three working groups, on the physical science basis (Solomon et al. 2007), impacts, adaptation and vulnerability (Parry et al. 2007), and mitigation of climate change (Metz et al. 2007), formed the basis of the review, which focused on recent developments in scientific and technical understanding. AR4 contains a wealth of detail concerning the strengths and weaknesses in the scientific community's understanding of the climate problem. In this chapter, an assessment of trends in the climate of coastal zone and in global and regional sea level over the twenty-first century based on the IPCC Fourth Assessment is presented. It is the science on which AR4 was based, if not the Assessment itself, that has informed the case studies presented in this book. The potential for revision to the AR4 estimates as the IPCC Fifth Assessment is finalized in 2013/2014 is also considered, and implications of the uncertainty in the projections for adaptation policy are briefly discussed.

GLOBAL TRENDS IN TEMPERATURE AND SEA LEVEL

Global mean surface air temperature and sea level are key indicators of the state of the climate system. Table 2.1 presents the AR4 projections for the end of the twenty-first century, the decade 2090–2099, for these two parameters (Meehl et al. 2007). The baseline period is 1980–1999. Projections are given for six greenhouse gas emissions scenarios, plausible futures, published by the IPCC in the Special Report on Emissions

Table 2.1 Projected global mean surface warming and sea level at the end of the twenty-first century for the six marker scenarios

		Temperature change (degree Celsius at 2090–2099 relative to 1981–1999)		*Sea-level rise (meter at 2090–2099 relative to 1981–1999)*
Scenario		*Best estimate*	*Likely range*	*Model-based range, excluding dynamical ice flow changes*
B1	Global environmental sustainability	1.8	1.1–2.9	0.18–0.38
A1T	Globalized world, rapid economic growth with reliance on nonfossil energy resources	2.4	1.4–3.8	0.20–0.45
B2	Local environmental sustainability, intermediate growth	2.4	1.4–3.8	0.20–0.43
A1B	Globalized world, rapid economic growth with balance between fossil and nonfossil energy sources	2.8	1.7–4.4	0.21–0.48
A2	Regionalized world, slow economic growth	3.4	2.0–5.4	0.23–0.51
A1FI	Globalized world, rapid economic growth, fossil-energy intensive	4.0	2.4–6.4	0.26–0.59

Source: IPCC, "Summary for policymakers," in S. Solomon et al. (eds.), *Climate Change 2007: The Physical Science Basis. Contribution of Working Group I to the Fourth Assessment Report of the Intergovernmental Panel on Climate Change*, Cambridge University Press, Cambridge; New York, http://www.ipcc.ch/, 2007b.

Note: The IPCC Special Report on Emissions Scenarios (Nakićenović and Swart 2000) groups possible futures into four main scenario families (A1, A2, B1, and B2) based on divergent development pathways. The four main scenario families differ in the degree to which the world economy is globalized/regionalized (A1 and B1/A2 and B2) and its economic or environmental focus (A1 and A2/B1 and B2). The A1 scenarios are further divided into those with an emphasis on nonfossil fuel use, fossil fuel use, or a balance between the two (A1T/A1FI/A1B). The A1 scenario storyline assumes a world of very rapid economic growth, a global population that peaks in mid-century and rapid introduction of new and more efficient technologies. A1 is divided into three groups that describe alternative directions of technological change, as indicated in the table. The B1 scenario describes a convergent world, with the same global population as A1, but with more rapid changes in economic structures toward a service and information economy, underpinning global environmental security. B2 describes a world with intermediate population and economic growth, emphasizing local solutions to economic, social, and environmental sustainability. A2 describes a very heterogeneous, or regionalized, world with high population growth, slow economic development, and slow technological change.

Scenarios (SRES) in 2000* (Nakićenović and Swart 2000). The scenarios were based on the existing literature, modeling exercises, and the participation of expert individuals and groups. Projections of the social, economic, and technological trends that will determine the levels of polluting activities were translated into estimates of emissions of all the significant greenhouse gases and of sulfur (which can have a cooling effect). The *marker* scenarios in Table 2.1 were selected as a representative sample of families of scenarios associated with different storylines covering socioeconomic and technological trends over coming decades. All of the scenarios exclude policies directed at the climate problem beyond those in existence at the time they were developed (and this excludes actions related to the Kyoto Protocol).

A number of sources of uncertainty arise in projecting global trends in climate and sea level. There is the inevitable uncertainty in future emissions of greenhouse gases, dependent as they are on social, economic, and technological trends, and, hence, in the change in the composition of the atmosphere that drives global warming. Then, understanding of the effect of any change in atmospheric composition on global climate is incomplete (see Box 2.1). Finally, there are aspects of the link between global climate change and sea-level change that are not firmly established. That there are uncertainties does not mean that nothing useful can be said about future trends; rather that any forecast has to be expressed in terms of a range of possibilities and not as a single definitive estimate.

In the case of global temperature, the IPCC projection for each scenario is in the form of a *best estimate* of the rise and a *likely range* (likely, in the IPCC terminology, means a greater than 66% probability of occurrence). In the case of sea level, the physical science working group had limited confidence in the estimates of one factor, enhanced discharge from the Greenland and Antarctic ice sheets as a result of dynamical ice flow changes. For this reason, no best estimate could be defined and, as the higher value in the stated range does not incorporate accelerated discharge, the possibility of a greater rise than indicated in Table 2.1 could not be excluded. Both the temperature and sea-level rise projections were based on coupled atmosphere–ocean general circulation model (AOGCM) simulations undertaken for the World Climate Research Programme Coupled Model Intercomparison Project (Meehl et al. 2005, 2007).

Considering first global surface air temperature, the overall range of possibilities evident in Table 2.1 indicates that the change in global temperature between the final decade of the late twenty-first century and the closing

* AR4 found that baseline emissions scenarios developed since SRES was published in the 2000, though differing in detail, were similar in terms of overall emissions levels (Metz et al. 2007). It should be noted that, for much of the first decade of the present century, real-world emissions were close to the *worst-case* SRES scenario (Abraham 2011).

BOX 2.1 PREDICTING GLOBAL CLIMATE TRENDS

It is a particular characteristic of greenhouse gases that they trap, in the lower atmosphere, longwave radiation (infrared energy) emitted from the earth's surface before it can escape to space. This greenhouse effect arises because of a correspondence between the frequency bands of the outgoing terrestrial radiation and the vibrational frequencies of the greenhouse gas molecules. As the greenhouse gas content of the atmosphere increases, surface temperatures will rise. The greenhouse effect is a well-established physical mechanism. What is not so certain is just how much temperatures will increase given a particular change in the greenhouse gas content of the atmosphere. The key parameter is the so-called climate sensitivity, which links the change in the distribution of energy within the atmosphere, the radiative forcing, to the resultant change in global surface air temperature once a steady-state equilibrium has been reached. The sensitivity of the global climate system cannot be measured directly, and values for this parameter are derived from the results of global climate models and representative observational or proxy climate data (Hegerl et al. 2007).

The precise value of the climate sensitivity has proved difficult to pin down (Maslin and Austin 2012); uncertainty in the climate sensitivity is a major contributor to the range of projections of global mean surface air temperature seen in Table 2.1. The 1986 review of climate science by the Scientific Committee on Problems of the Environment, a standing committee of the International Council of Scientific Unions, concluded that the climate sensitivity likely fell within the range 1.5°C–4.5°C warming for a doubling of atmospheric carbon dioxide, or the equivalent, once equilibrium had been reached (Bolin et al. 1986). No best estimate could be given. Some 20 years later, AR4 concluded that the likely range was 2.0°C–4.5°C, only a minor adjustment, though it did define the most likely value to be about 3°C (Hegerl et al. 2007). The AR4 review was based on a far greater range of evidence, and confidence in the assessment has, therefore, improved. Projections of the development of climate change over time must factor in the relatively slow response of the oceans compared to the atmosphere, which moderates the change in surface air temperatures, and other factors that determine the *transient* response of the climate system to the changing atmospheric composition. Uncertainties in the transient response also contribute to the range of the global temperature projections.

That climate predictions should be subject to uncertainty is inevitable. Simulating the behavior of a system as complex as the climate system, even on

(*Continued*)

a supercomputer, presents a substantial challenge. Alongside uncertainties in scientific understanding, computational considerations limit model accuracy. Moreover, because of nonlinearity and instability in the equations that govern the modeled behavior of the climate system, predictability cannot be extended indefinitely (although it does appear that long-term forecasts tend to cluster around particular states). See Randall et al. (2007) for further discussion of climate model performance.

20 years of the twentieth century could be as low as 1.1°C (B1 scenario) or it may be as high as 6.4°C (A1F1 scenario). To place these trends in context, the global warming that occurred over the period 1906–2005 has been estimated at 0.74°C with a 5%–95% probability range of 0.56°C–0.92°C (Trenberth et al. 2007). At the lower end of the range of possibilities, then, we face a change in climate not much greater than humanity has recently experienced. This, arguably, could be managed without too much difficulty. At the other end of the range, the potential change in climate is almost an order of magnitude greater than experienced during the twentieth century. Indeed, it rivals in scale, and most likely surpasses in rate, the warming that brought the planet out of the most recent ice age (Jansen et al. 2007) and that constituted a substantial change in the planetary environment. At this end of the spectrum of possibilities, humanity faces a substantial challenge (see Box 2.2).

The estimates of global surface air temperature trends to the final decade of the twenty-first century provide a basis for the projections of sea-level rise. Figure 2.1 breaks down the projections for the period 2090–2099 into five components: expansion of the warming oceans; the changing surface mass balance of small glaciers and ice caps (mostly the result of melting); the surface mass balance of the Greenland and of the Antarctic ice sheets (largely melting and snow accumulation); and enhanced discharge from the large ice sheets, termed *scaled-up dynamical imbalance*. Thermal expansion is the major contributor, the central estimate for this factor accounting for 70%–75% of the overall central estimate for each scenario (Meehl et al. 2007). The thermal expansion estimates are derived directly from the AOGCM simulations. Changes in the mass balance of ice on land other than the Greenland and Antarctic ice sheets, referred to as glaciers and ice caps in Figure 2.1, are dependent on local conditions and estimates for individual areas are usually derived by *downscaling* larger-scale climate model projections and using the resulting local-scale climate parameters to perturb the observed mass balance (cf. Schneeberger et al. 2003). As it is not possible to estimate the sensitivity of every individual area of ice to climate variability, a global average sensitivity to global average temperature is

BOX 2.2 THE WORLD BEYOND 4°C

Participants at the 2009 conference "Four Degrees and Beyond," held in Oxford in the United Kingdom, considered what the world might look like if global temperature rises by 4°C or more. "Even with strong political will, the chances of shifting the global energy system fast enough to avoid 2°C [the goal recognized by the Copenhagen Accord] are slim," New et al. (2011: 9) argue, "... eventual temperatures rise of 3°C or 4°C are much more likely." In their contribution to the Oxford conference, Nicholls et al. (2011) conclude that, with warming at this level, global sea level could rise by between a half-meter and 2 m by the end of the twenty-first century (though a rise of more than a meter is considered less likely than a lower increase). Sea level will continue to rise beyond 2100 as the ocean moves to equilibrium with the atmospheric warming and substantial increases are possible if thresholds are crossed resulting in irreversible melting of the Greenland ice cap or some degree of breakup of the West Antarctic ice sheet. Whether or not an aggressive attempt is made to protect the world's coastlines, large numbers of people would be placed at risk by sea-level rises of this scale, with the population of South and Southeast Asia particularly vulnerable (Nicholls et al. 2011). With no protection, Nicholls et al. (2011) estimate that 72 million people could be displaced by a half-meter rise in sea level and 182 million by a rise of 2 m. In the case of a 2 m rise in sea level, the cost of protecting all but the most sparsely populated coastal areas would be USD 270 billion a year to 2100. Stafford-Smith et al. (2011) argue that adaptation to global warming of 4°C or more cannot be seen simply as an extension of the incremental response to lesser degrees of climate change; it will have to be a more fundamental, transformative process.

calculated based on area-weighting representative local sensitivities (Meehl et al. 2007). Precipitation has been shown not to be a significant factor on the global scale. As far as the larger ice sheets are concerned, recent studies confirm that snow accumulation over the Antarctic ice sheet will dominate melting through the twenty-first century, whereas the reverse applies in the case of the Greenland ice sheet (Meehl et al. 2007). Estimates of the changing mass balance require high-resolution climate data to simulate processes on the ice margins where the processes that affect the net balance are most active. Data are derived from high-resolution climate model experiments or by downscaling the output from low-resolution models.

As well as melting and accumulation, the major ice sheets will be affected by changes in ice flow, ice dynamics. It is considered that the

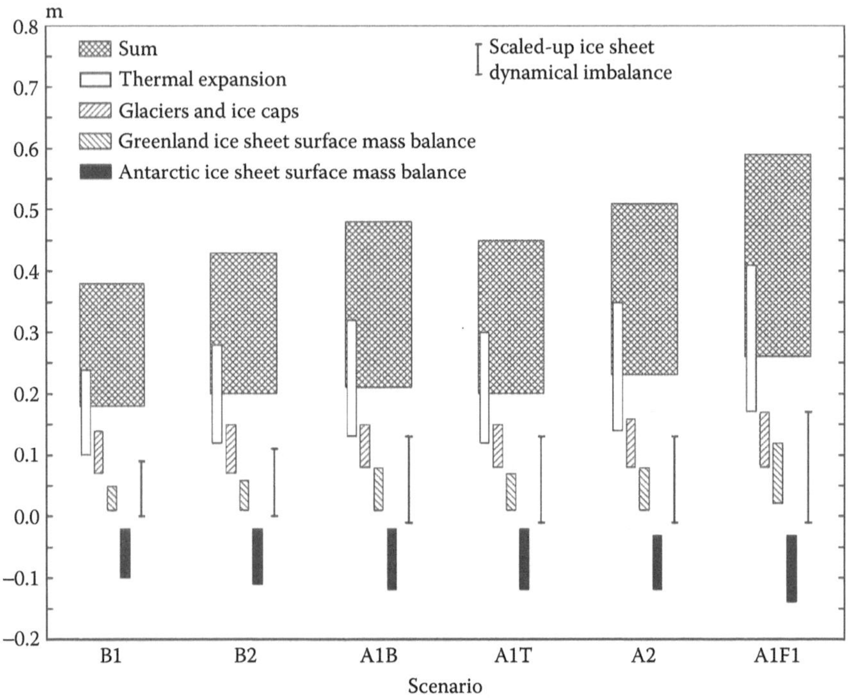

Figure 2.1 Projections and uncertainties (5%–95% probability ranges) for global sea-level rise and its contributing factors for the period 2090–2099 (relative to 1980–1999) for the six marker scenarios. (Data from Meehl, G.A. et al., Global climate projections," in S. Solomon et al., eds., *Climate Change 2007: The Physical Science Basis. Contribution of Working Group I to the Fourth Assessment Report of the Intergovernmental Panel on Climate Change*, Cambridge University Press, Cambridge; New York, http://www.ipcc.ch/, 2007.) (Note: The sum does not include any contribution from rapid acceleration of flow within the Greenland and Antarctic ice sheets [scaled-up ice sheet dynamical imbalance]).

consequences of general changes in topography at the continental scale (steepening of slopes and thinning of outlet glaciers, for example) will be relatively minor compared to that of changes in mass balance (Meehl et al. 2007), and these effects are included in the mass balance estimates presented in Figure 2.1. There is, however, the possibility of localized disruption of the dynamics of the Greenland and West Antarctic ice sheets and this could result in greatly accelerated mass loss as, for example, warming water erodes the floating ice shelves that buttress the ice sheet behind. There is evidence of recent acceleration in the flow rate of outlet glaciers (cf. Lemke et al. 2007), and the estimates for changes in the surface mass balance of the Greenland and Antarctic ice sheets in Figure 2.1 (and Table 2.1) include a constant, scenario-independent contribution, amounting to 0.32 ± 0.35 mm a year, derived from recent data (Meehl et al. 2007). The contribution of ice dynamics might, however, increase or decrease over time depending, for example, on whether the recent acceleration is the first stage of a long-term response to global warming or is short-lived, being the result of natural variability. At the time AR4 was prepared, no model was available to simulate the complex atmospheric, oceanic, and cryospheric processes involved at the ice sheet margin. The estimate of the contribution of scaled-up dynamic imbalance given in Figure 2.1 was presented for illustrative purposes only. In the case of a prolonged acceleration as global warming develops, the estimated response was based on a simple linear scaling of the recent observations with global temperature. The estimate for the case of a short-term, transient response was based on decay-time calculations. In any event, there was not sufficient confidence in these figures to include them in the net projections of sea-level change given in Table 2.1.

The sea-level rise projections over the period 1980–1999 to 2090–2099 presented in Table 2.1 range from a lower estimate of 0.18 m (B1 scenario) to an upper estimate of 0.59 m (A1F1 scenario). Including the illustrative contribution of scaled-up ice sheet dynamics, an additional 0.17 m rise by 2090–2099 for the A1F1 scenario increases the upper estimate to 0.76 m. There is no change in the lower estimate of 0.18 m. The scenario-dependent ranges are a rise of 0.18–0.26 m by 2090–2099 at the lower end of the range of possibilities (0.18–0.25 m if the notional estimates of scaled-up ice discharge are included) and, at the upper end, a rise of 0.38–0.59 m by 2090–2099 (increasing to 0.47–0.76 m if the estimates of scaled-up ice discharge are included). To place these trends in context, global mean sea level rose at a rate of 0.17 ± 0.05 m a year averaged over the twentieth century (Bindoff et al. 2007; see also Cazenave and Remy 2011). There has been considerable variability on the decadal timescale, with sea level rising at 3.1 ± 0.5 mm a year over the period 1993–2003 (Bindoff et al. 2007).

Figure 2.2, from Church et al. (2011), shows the evolution of the projected rise in sea level over time based on the AR4 analysis.* The range for the year 2100 is 0.19–0.63 m (0.18–0.80 m including scaled-up ice sheet discharge) from the 1990 baseline. It is notable that, because of lags in the response of the climate system and cryosphere to the change in atmospheric composition, and given that time is needed for the different development pathways to diverge substantially, it is only during the second half of the

* Note that a different baseline is used for the projections in Table 2.1 than for those in Figure 2.2.

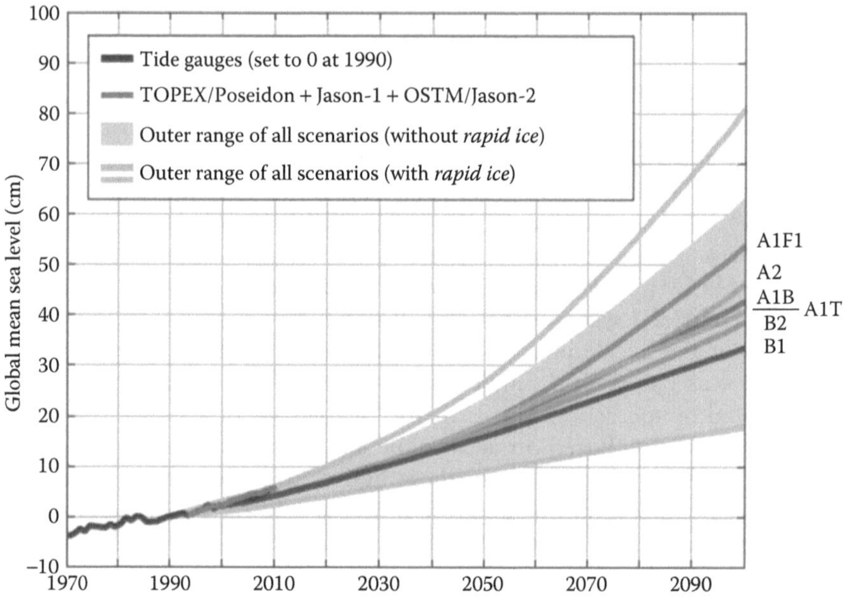

Figure 2.2 Global averaged projections of sea-level rise with respect to 1990 for the six marker scenarios. (Adapted from Church, J.A. et al. *Oceanography*, 24(2), 130–43, 2011.) (Note: The shaded region/outer light lines show the full range of projections, not including/including any scaled-up ice sheet discharge component. The continuous lines from 1990 to 2100 indicate the central value of the projections associated with the marker scenarios, including the scaled-up discharge contribution [*rapid ice*]. The observational estimates of globally averaged sea level based on tide-gauge measurements and satellite altimeter data are also shown. The tide-gauge data are set to zero at the start of the projections in 1990, and the altimeter data are set equal to the tide gauge data at the start of the record in 1993.)

twenty-first century that substantial differences between the projections associated with the various emissions scenarios occur. At 2050, the central values of the projections of sea-level rise associated with the six marker projections (the solid lines in Figure 2.2) fall within the range 0.15–0.20 m above the 1990 level. The overall range of uncertainty in the projections at this time is around 0.09–0.27 m (including the estimate of scaled-up ice sheet discharge). During the second half of the twenty-first century, the projections associated with the different emissions scenarios do deviate. Scientific uncertainty continues to make a substantial contribution to the range. Figure 2.2 highlights another important point. Comparison between the sea-level rise back-projections and estimates of global average sea level derived from tide gauge and satellite altimetry data indicates that the observed rise over the recent period has been toward the upper end of the IPCC projections (Church et al. 2011).

AT THE LOCAL LEVEL

Global temperature and sea level provide useful indicators of the overall state of the climate system, but estimates of the likely spatial distribution of changes in a range of climate and oceanographic parameters are needed to support development and implementation of remedial policy. Local changes in sea level induced by climate change will differ from the global mean, dependent, among other things, on the regional distribution of ocean density changes and circulation (Meehl et al. 2007). Changes in relative sea level associated with the past glacial cycle, isostatic adjustment, must be considered. Redistribution of mass from glaciers and ice sheets as a result of global warming will alter the loading on the underlying ocean bed, causing vertical motion in the crust and changes in the gravitational field (Tamisiea and Mitrovica 2011), resulting, broadly, in a fall in sea level near the major ice sheets and a rise further away.

Slangen et al. (2012) model the various factors determining changes in local sea level, drawing on the AR4 conclusions, to produce maps of twenty-first century trends for three SRES emissions scenarios. The results show that many areas will experience changes that differ substantially from the global average. For the central A1B scenario, the projected rise in global mean sea level is 0.47 m over the twenty-first century, but local values range from –3.91 to 0.79 m. The patterns of change are similar for the three emission scenarios, with large falls in sea level in the Arctic and rises close to the areas glaciated during the latest glacial maximum (including both coasts of North America). The local sea-level rise is also greater than the global average in lower middle latitudes of both hemispheres and in the western Indian Ocean, the major contributing factors being the change in ocean density and the change in mass distribution.

Alongside the threat of more extensive, more frequent, and, in some cases, permanent inundation, sea-level rise brings with it the risk of other adverse consequences for the coastal zone (Nicholls 2011; Nicholls et al. 2007). Changing patterns of erosion and accretion, greater landward intrusion of saltwater intrusion, wetland loss, and effects on water tables and drainage are likely, if not inevitable (Table 2.2). The broader change in ocean and atmospheric climate will have other impacts (Table 2.3). Particularly in the case of a change in the frequency or severity of extreme events, these impacts may be as significant as and, likely, more immediate than the change in sea level. For example, as global climate alters, changes in the atmospheric circulation will influence storm frequencies and tracks, affecting surge and wave characteristics, extreme water levels, and the potential for storm damage and flooding. The warming ocean may also generate more frequent and/or more violent storms in some areas. Any consequent change in wave climate will influence patterns of erosion and accretion and could alter beach contours. Rising sea surface temperature will affect

Table 2.2 The main natural system effects of relative sea-level rise and climatic and nonclimatic factors that could influence these effects

Natural system effect		*Possible interacting factors* — *Climatic*	*Possible interacting factors* — *Nonclimatic*
Inundation/flooding	Surge (flooding from sea)	Wave/storm climate, erosion, sediment supply	Sediment supply, flood management, erosion, land reclamation
	Backwater effect (flooding from rivers)	Runoff	Catchment management and land use
Wetland loss (and change)		Carbon dioxide fertilization, sediment supply, migration space	Sediment supply, migration space, land reclamation (i.e., direct destruction)
Erosion (of *soft* morphology)		Sediment supply, wave/storm climate	Sediment supply
Saltwater intrusion	Surface waters	Runoff	Catchment management (overextraction), land use
	Groundwater	Rainfall	Land use, aquifer use (overpumping)
Impeded drainage/ high water tables		Rainfall, runoff	Land use, aquifer use, catchment management

Source: Nicholls, R.J., *Oceanography*, 24(2), 144–57, 2011.

the ocean circulation, increasing stratification, and, as patterns of oceanic temperature evolve, changing the distribution of coastal currents. At higher latitudes, sea ice cover will be reduced. As the ocean absorbs more carbon dioxide, carbon dioxide fertilization will be enhanced and coastal waters will become more acidic. Finally, there will be changes in the rainfall, temperature, and other climate parameters both within the coastal zone and neighboring areas. Changing rainfall and temperature patterns could affect flood risk and water quality and salinity as well as the supply of sediments and nutrients (see Christensen et al. 2007; Field et al. 2012; Meehl et al. 2007; and Nicholls et al. 2007 for further details).

A full assessment of the threat at the local level must, of course, also take account of the impact of human activities, such as water extraction and various forms of land-use change and coastal development (cf. Table 2.2), that may affect local sea level. "Whether the ocean is rising or the land is subsiding makes little difference to a person who is up to [their] waist in water," as Milliman and Haq (1996: 3) observe. To take just one example, the Chao Phraya Delta in Thailand, subsidence of more than 0.2 m occurred along

Table 2.3 Main climate drivers for coastal systems, their trends due to climate change, and their main physical and ecosystem effects

Climate driver (trend)	*Main physical and ecosystem effects on coastal systems*
Carbon dioxide concentration (↑)	Increased carbon dioxide fertilization and decreased seawater pH (or *ocean acidification*) negatively impacting coral reefs and other pH-sensitive organisms
Sea surface temperature (↑, R)	Increased stratification/changed circulation; reduced incidence of sea ice at higher latitudes; increased coral bleaching and mortality; (poleward) species migration; increased algal blooms
Sea level (↑, R)	Inundation, flood and storm damage; erosion; saltwater intrusion; rising water tables/impeded drainage; wetland loss (and change)
Storm intensity (↑, R)	Increased extreme water levels and wave heights; increased episodic erosion, storm damage, risk of flooding, and defense failure
Storm frequency (?, R) Storm track (?, R)	Altered surges and storm waves and hence risk of storm damage and flooding
Wave climate (?, R)	Altered wave conditions, including swell; altered patterns of erosion and accretion; reorientation of beach plan form
Runoff (R)	Altered flood risk in coastal lowlands; altered water quality/salinity; altered fluvial sediment supply; altered circulation and nutrient supply

Source: Nicholls, R.J. et al., "Coastal systems and low-lying areas," in M.L. Parry et al. (eds.), *Climate Change 2007: Impacts, Adaptation and Vulnerability. Contribution of Working Group II to the Fourth Assessment Report of the Intergovernmental Panel on Climate Change*, Cambridge University Press, Cambridge; New York, http://www.ipcc.ch/, 2007.

Trend key: ↑ = increase; ? = uncertain; R = regional variability

the coast near Bangkok over the period 1992–2000 due to water extraction (Winterwerp et al. 2005). Saito et al. (2007) conclude that a relative sea-level rise of only 0.1 m can induce coastal erosion along this muddy coast. The shoreline has retreated more than one km near the river mouth over the past 50 years. Trends such as these, resulting from other anthropogenic drivers, provide a context for the projected changes in sea level and other aspects of the local environment induced by global warming, providing a means of assessing the significance of the climate trends. They also provide a clear indication of the first steps to be taken in reducing vulnerability to future climate change by strengthening the resilience of the coastal zone.

TOWARD THE FIFTH IPCC ASSESSMENT

It is certain that the IPCC Fifth Assessment will revise the projections of future changes in sea level. But how substantial will that revision be? There are three main considerations: changes in the emissions scenarios; improved understanding of the response of the climate system to the changing atmospheric composition; and further research into the factors determining sea-level change. The Fifth Assessment will base its projections on an expanded set of emissions scenarios, expressed in terms of *representative concentration pathways* (RCPs) (Moss et al. 2010). One notable difference between these RCPs and the AR4 scenarios lies in the inclusion of mitigation considerations toward the lower end of the emissions range. Given the development of new emissions scenarios, it is inevitable that the Fifth Assessment will produce revised projections of global sea level. It seems unlikely, though, that there will be any substantial change in the upper limit of the temperature projections as a result of the adoption of the new set of scenarios (cf. Rogelj et al. 2012). Hence, this development is not likely to have a substantial effect on the upper range of the sea-level estimates. As far as the response of the climate system is concerned, the rate of change in the estimates of, for example, the climate sensitivity over the course of the series of IPCC assessments since 1991 (see Box 2.1) suggests that substantial revision in this area is unlikely, though confidence in the global temperature projections may well improve.

Understanding of the processes that determine changes in global sea level has improved since AR4 was published. Church et al. (2011) review progress since AR4 and the potential for further developments, highlighting the contribution of new and improved observational data sets in improving understanding of past trends and calibrating and testing models. The fact that AR4 did not present a complete assessment of potential sea-level rise, deliberately omitting the uncertain estimate of accelerated ice sheet dynamics from the projections for the end of the twenty-first century, has led to a considerable effort to improve understanding of, and model, relevant processes. The critical issue is the contribution of accelerated discharge, dynamic imbalance, in the large ice sheets. Improved models of ice sheet dynamics are being developed (cf. Joughin et al. 2010), and it is likely that the result of these experiments will replace the *illustrative* estimates presented in AR4. The model results from Joughin et al. (2010) are not inconsistent with the conservative AR4 assumption of a constant contribution from scaled-up ice sheet discharge, but this is a single pioneering study.

There has been concern that the AR4 projections might have underestimated the potential rise in sea level (Rahmsdorf 2010), particularly given the fact that the recent observational record is falling toward the upper end of the range of model simulations (cf. Figure 2.2), and this has led to the adoption of more direct approaches in estimating the potential rise in sea level. Various investigators have used semiempirical methods to derive a direct link between sea level and global temperature or some other parameter, though there are grounds for questioning this approach (Lowe and Gregory 2010). These semiempirical approaches have resulted in a greater range of projected rates of sea-level rise (Church et al. 2011; Nicholls et al. 2011). Rahmsdorf (2007), for example, derived a range of 0.5–1.4 m above the 1990 level by 2100, while Grinsted et al. (2010) projected a range of

0.3–2.2 m by 2100. Pfeffer et al. (2008), on the basis of modeling and analogies with past conditions, concluded that a sea-level rise of 0.8 m over the twenty-first century was plausible, with 2.0 m the maximum that was physically tenable. Reviewing post-AR4 sea-level rise projections, Nicholls et al. (2011) conclude that an extreme upper limit, albeit of low probability, for the rise over the twenty-first century would be of the order of 2 m.

On the basis of research to date, it seems unlikely that there will be a dramatic revision of the AR4 projections in the IPCC's Fifth Assessment. It is reasonable to expect, given research developments since AR4 (see, e.g., Church et al. 2011), that various inconsistencies between the observational record and simulated changes will be resolved and that there will be greater confidence overall in the estimation of the various contributions to global sea-level rise and, perhaps, in the definition of the regional pattern of change. It would be surprising if the next IPCC assessment failed to deliver a more robust estimate of the contribution of ice flow dynamics and this may permit the definition of a plausible upper bound and, perhaps, a *best estimate* for sea-level rise by the end of the twenty-first century. Where might a plausible upper bound lie? Perhaps the crystal ball should remain shrouded until the Fifth Assessment reports, but, on the basis of the evidence available in early 2012 (cf. Cazenave and Remy 2011; Church et al. 2011; Nicholls et al. 2011),* it might be anticipated that the quoted upper bound for the rise in sea level over the twenty-first century will fall closer to a meter than AR4 estimated, perhaps around 0.8 m, with a central estimate of a 0.4–0.5 m rise.

PLANNING FOR AN UNCERTAIN FUTURE

A rise in global sea level as a result of global warming is inevitable. The scale of that rise is, however, subject to considerable uncertainty. At one end of the spectrum of possibilities, global sea level may rise around 20 cm over the present century. In a worst-case scenario, the rise could approach a meter. The uncertainty in the projections poses a vexing problem for planners and decision-makers. In a world of constrained resources and competing demands, how much weight should we attach to the worst-case projection? Should we base current planning on the central estimates, ignoring the range of uncertainty? Or perhaps we should simply tread water until the forecasts become more definite. None of these options is viable on its own; a more sophisticated response is needed. Kelly (2000) argues that a sustainable response to an uncertain future should be characterized by caution and staged implementation, based on flexibility, diversity, continual evaluation of performance, and an informed and engaged community.

* See also Bamber and Aspinall (2013) for a concurrent assessment of expert opinion.

As noted earlier in this chapter, due to inertia in the social and physical systems determining sea level, the projections of global sea level over the coming 30–40 years are less dependent on emissions scenario than over the subsequent 50 years. This is likely to be a lasting characteristic of the climate projections. The degree of certainty in the forecasts will always be greater in the shorter term, with inherent unpredictability increasing over time whatever the state of scientific understanding. This suggests a staged, or nested, approach to adaptation, dependent on planning timescales.

The range in the existing projections of global sea-level rise for the year 2050 is around 0.09–0.27 m with respect to 1990 (Figure 2.2). There are three priority areas for action. Even allowing for local departures from the global average, it is not unreasonable to conclude that, to a planning horizon of the year 2050, the effect of sea-level rise in many coastal areas will be comparable to the immediate impacts of other socioeconomic and environmental drivers. This would suggest that an immediate priority should be to address existing problems so that the coast and its inhabitants are better placed to deal with the emerging threat of climate change. The second priority must be to respond to the potential impact of sea-level rise within this time frame. Finally, it is imperative that the other threats and opportunities accompanying the change in climate might affect the coastal zone over this period, such as the potential for more frequent and/or more violent storms. An integrated approach is essential. It is also imperative that adaptive measures adopted in the near future do not restrict options at a later date, flexibility must be ensured, and that thorough performance evaluation ensures that future generations of planners, and coastal communities at risk, can benefit from an evolutionary process of learning through experience.

Over longer planning timescales, to the year 2100, the critical issue is the degree to which the rise in global sea level accelerates. Optimistically, with robust and prompt international action to curb greenhouse gas emissions and a low sensitivity of the climate system to the changing atmospheric composition, the rate of change during the second half of the twenty-first century might be not too different from that projected for the period to 2050. Without that political response, and assuming a high sensitivity of sea level to global warming in line with the worst-case IPCC projections, the rate of sea-level rise could be more than triple. Planning over this time frame must deal not only with a broader range of possible futures, but also with the possibility of environmental change on a scale unprecedented in recent human experience.

POSTSCRIPT

The physical sciences report of the IPCC Fifth Assessment (AR5; IPCC 2013) was released in September 2013 as this book was in the final stages of preparation. The report contains new estimates of global surface air temperature and sea-level change over the twenty-first century, updating those discussed in this chapter. As far as the increase in global surface air temperature is concerned, the AR5 report projects a likely range of between 0.3°C and 4.8°C from 1986–2005 to 2081–2100. The likely range in the rise in global mean sea level is 0.26–0.82 m over the same period. The derivation and presentation of the AR5 projections differs from that used in AR4. In the case of global temperature, the downward shift in the overall range is partly the result of the use of concentration rather than emissions scenarios, as discussed in this chapter, which eliminates uncertainties in modeling the link between emissions and atmospheric concentrations, and is also attributable to a widening of the likely range in the equilibrium climate sensitivity, a return to the pre-AR4 range of 1.5°C–4.5°C (cf. Box 2.1). The revised sea-level rise projections are also dependent on these factors, but now include an estimate of accelerated discharge, dynamic imbalance, in the large ice sheets. There is greater confidence in the AR5 sea-level rise projections as a result of improved understanding.

ACKNOWLEDGMENT

I thank Mike Hulme and my coeditors for their helpful comments.

REFERENCES

Abraham, J. (2011). *"Very worried" about escalating emissions? You should be*, The Conversation. http://theconversation.edu.au/ (accessed March 2012).

Bamber, J.L. and Aspinall, W.P. (2013). "An expert judgement assessment of future sea level rise from the ice sheets," *Nature Climate Change*. doi:10.1038/nclimate1778.

Bindoff, N.L., Willebrand, J., Artale, V., Cazenave, A., Gregory, J., Gulev, S., Hanawa, K. et al. (2007). "Observations: Oceanic climate change and sea-level," in S. Solomon, D. Qin, M. Manning, Z. Chen, M. Marquis, K.B. Averyt, M. Tignor, and H.L. Miller (eds.), *Climate Change 2007: The Physical Science Basis. Contribution of Working Group I to the Fourth Assessment Report of the Intergovernmental Panel on Climate Change*, Cambridge; New York: Cambridge University Press. http://www.ipcc.ch/ (accessed August 2013).

Bolin, B., Döös, B.R., Jäger, J., and Warrick, R.A. (eds.) (1986). *The Greenhouse Effect, Climatic Change and Ecosystems*. London: Wiley.

Cazenave, A. and Remy, F. (2011). "Sea level and climate: Measurements and causes of changes," *WIREs Climate Change*, 2: 647–62.

Christensen, J.H., Hewitson, B., Busuioc, A., Chen, A., Gao, X., Held, I., Jones, R. et al. (2007). "Regional climate projections," in S. Solomon, D. Qin, M. Manning, Z. Chen, M. Marquis, K.B. Averyt, M. Tignor, and H.L. Miller (eds.), *Climate Change 2007: The Physical Science Basis. Contribution of Working Group I to the Fourth Assessment Report of the Intergovernmental Panel on Climate Change*, Cambridge; New York: Cambridge University Press. http://www.ipcc.ch/ (accessed August 2013).

Church, J.A., Gregory, J.M., White, N.J., Platten, S.M., and Mitrovica, J.X. (2011). "Understanding and projecting sea-level change," *Oceanography*, 24(2):130–43.

Field, C.B., Barros, V., Stocker, T.F., Qin, D., Dokken, D.J., Ebi, K.L., Mastrandrea, M.D. et al. (eds.) (2012). *Managing the Risks of Extreme Events and Disasters to Advance Climate Change Adaptation.* Cambridge; New York: Cambridge University Press.

Grinsted, A., Moore, J.C., and Jevrejeva, S. (2010). "Reconstructing sea level from paleo and projected temperatures," *Climate Dynamics*, 34: 461–72.

Hegerl, G.C., Zwiers, F.W., Braconnot, P., Gillett, N.P., Luo, Y., Orsini, J.A.M., Nicholls, N., Penner, J.E., and Stott, P.A. (2007). "Understanding and attributing climate change," in S. Solomon, D. Qin, M. Manning, Z. Chen, M. Marquis, K.B. Averyt, M. Tignor, and H.L. Miller (eds.), *Climate Change 2007: The Physical Science Basis. Contribution of Working Group I to the Fourth Assessment Report of the Intergovernmental Panel on Climate Change*, Cambridge; New York: Cambridge University Press. http://www.ipcc.ch/ (accessed August 2013).

Hulme, M. and Mahony, M. (2010). "Climate change: What do we know about the IPCC?" *Progress in Physical Geography*, 34(5): 705–18.

IPCC. (2007a). "IPCC history." Geneva: IPCC Facts. http://www.ipccfacts.org/history.html (accessed November 2011).

IPCC. (2007b). "Summary for policymakers," in S. Solomon, D. Qin, M. Manning, Z. Chen, M. Marquis, K.B. Averyt, M. Tignor, and H.L. Miller (eds.), *Climate Change 2007: The Physical Science Basis. Contribution of Working Group I to the Fourth Assessment Report of the Intergovernmental Panel on Climate Change*, Cambridge; New York: Cambridge University Press. http://www.ipcc.ch/ (accessed August 2013).

IPCC. (2013). "Summary for policymakers," in T.F. Stocker, D. Qin, G.-K. Plattner, M. Tignor, S.K. Allen, J. Boschung, A. Nauels, Y. Xia, V. Bex, and P.M. Midgley (eds.). *Climate Change 2013: The Physical Science Basis. Contribution of Working Group I to the Fifth Assessment Report of the Intergovernmental Panel on Climate Change*, Cambridge; New York: Cambridge University Press. http://www.ipcc.ch/ (accessed November 2013).

Jansen, E., Overpeck, J., Briffa, K.R., Duplessy, J.-C., Joos, F., Masson-Delmotte, V., Olago, D. et al. (2007). "Palaeoclimate," in S. Solomon, D. Qin, M. Manning, Z. Chen, M. Marquis, K.B. Averyt, M. Tignor, and H.L. Miller (eds.), *Climate Change 2007: The Physical Science Basis. Contribution of Working Group I to the Fourth Assessment Report of the Intergovernmental Panel on Climate Change*, Cambridge; New York: Cambridge University Press. http://www.ipcc.ch/ (accessed August 2013).

Joughin, I., Smith, B.E., and Holland, D.M. (2010). "Sensitivity of 21st century sea-level to ocean induced thinning of Pine Island Glacier, Antarctica," *Geophysical Research Letters*, 37, L20502. doi:10.1029/2010GL044819.

Kelly, P.M. (2000). "Towards a sustainable response to climate change," in M. Huxham and D. Sumner (eds.), *Science and Environmental Decision Making*, Harlow: Pearson Education.

Lemke, P., Ren, J., Alley, R.B., Allison, I., Carrasco, J., Flato, G., Fujii, Y. et al. (2007). "Observations: Changes in snow, ice and frozen ground," in S. Solomon, D. Qin, M. Manning, Z. Chen, M. Marquis, K.B. Averyt, M. Tignor, and H.L. Miller (eds.), *Climate Change 2007: The Physical Science Basis. Contribution of Working Group I to the Fourth Assessment Report of the Intergovernmental Panel on Climate Change*, Cambridge; New York: Cambridge University Press. http://www.ipcc.ch/ (accessed August 2013).

Lowe, J.A. and Gregory, J.M. (2010). "A sea of uncertainty," *Nature Reports Climate Change*. http://www.nature.com/climate/index.html (accessed June 2012).

Maslin, M. and Austin, P. (2012). "Uncertainty: Climate models at their limit?," *Nature*, 486: 183–84.

Meehl, G.A., Covey, C., McAvaney, B., Latif, M., and Stouffer, R.J. (2005). "Overview of the coupled model intercomparison project," *Bulletin of the American Meteorological Society*, 86(1): 89–93.

Meehl, G.A., Stocker, T.F., Collins, W.D., Friedlingstein, P., Gaye, A.T., Gregory, J.M., Kitoh, A. et al.(2007). "Global climate projections," in S. Solomon, D. Qin, M. Manning, Z. Chen, M. Marquis, K.B. Averyt, M. Tignor, and H.L. Miller (eds.), *Climate Change 2007: The Physical Science Basis. Contribution of Working Group I to the Fourth Assessment Report of the Intergovernmental Panel on Climate Change*, Cambridge; New York: Cambridge University Press. http://www.ipcc.ch/ (accessed August 2013).

Metz, B., Davidson, O.R., Bosch, P.R., Dave, R., and Meyer, L.A. (eds.) (2007). *Climate Change 2007: Mitigation of Climate Change. Contribution of Working Group III to the Fourth Assessment Report of the Intergovernmental Panel on Climate Change*, Cambridge; New York: Cambridge University Press. http://www.ipcc.ch/ (accessed August 2013).

Milliman, J.D. and Haq, B.U. (eds.) (1996). *Sea-Level Rise and Coastal Subsidence: Causes, Consequences, and Strategies*. Dordrecht, the Netherlands: Kluwer.

Moss, R.H., Edmonds, J.A, Hibbard, K.A., Manning, M.R., Rose, S.T., van Vuuren, D.P., Carter, T.R., Emori, S. et al. (2010). "The next generation of scenarios for climate change research and assessment," *Nature*, 463: 747–56.

Nakićenović, N. and Swart, R. (eds.) (2000). *Special Report on Emissions Scenarios. A Special Report of Working Group III of the Intergovernmental Panel on Climate Change*. Cambridge; New York: Cambridge University Press. http://www.ipcc.ch/ (accessed August 2013).

New, M., Liverman, D., Schroder, H., and Anderson, K. (2011). "Four degrees and beyond: The potential for a global temperature increase of four degrees and its implications," *Philosophical Transactions of the Royal Society A*, 369: 6–19.

Nicholls, R.J. (2011). "Planning for the impacts of sea-level rise," *Oceanography*, 24(2): 144–57.

Nicholls, R.J., Marinova, N., Lowe, J.A., Brown, S., Vellinga, P., de Gusmão, D., Hinkel, J., and Tol, R.S.J. (2011). "Sea-level rise and its possible impacts given a 'beyond 4°C world' in the 21st century," *Transactions of the Royal Society A*, 369: 161–81.

Nicholls, R.J., Wong, P.P., Burkett, V.R., Codignotto, J.O., Hay, J.E., McLean, R.F., Ragoonaden, S., and Woodroffe, C.D. (2007). "Coastal systems and low-lying areas," in M.L. Parry, O.F. Canziani, J.P. Palutikof, P.J. van der Linden, and C.E. Hanson (eds.), *Climate Change 2007: Impacts, Adaptation and Vulnerability.*

Contribution of Working Group II to the Fourth Assessment Report of the Intergovernmental Panel on Climate Change, Cambridge; New York: Cambridge University Press. http://www.ipcc.ch/ (accessed August 2013).

Parry, M.L., Canziani, O.F., Palutikof, J.P., van der Linden, P.J., and Hanson, C.E. (eds.) (2007). *Climate Change 2007: Impacts, Adaptation and Vulnerability. Contribution of Working Group II to the Fourth Assessment Report of the Intergovernmental Panel on Climate Change*, Cambridge; New York: Cambridge University Press. http://www.ipcc.ch/ (accessed August 2013).

Pfeffer, W.T., Harper, J.T., and O'Neel, S. (2008). "Kinematic constraints on glacier contributions to 21st-century sea-level rise," *Science*, 321: 1340–3.

Rahmsdorf, S. (2007). "A semi-empirical approach to projecting future sea-level rise," *Science*, 315: 368–70.

Rahmsdorf, S. (2010). "A new view on sea-level rise," *Nature Reports Climate Change*. http://www.nature.com/climate/index.html (accessed July 2012).

Randall, D.A., Wood, R.A., Bony, S., Colman, R., Fichefet, T., Fyfe, J., Kattsov, V. et al. (2007). "Climate models and their evaluation," in S. Solomon, D. Qin, M. Manning, Z. Chen, M. Marquis, K.B. Averyt, M. Tignor, and H.L. Miller (eds.), *Climate Change 2007: The Physical Science Basis. Contribution of Working Group I to the Fourth Assessment Report of the Intergovernmental Panel on Climate Change*, Cambridge; New York: Cambridge University Press. http://www.ipcc.ch/ (accessed August 2013).

Rogelj, J., Meinshausen, M., and Knutti, R. (2012). "Global warming under old and new scenarios using IPCC climate sensitivity range estimates," *Nature Reports Climate Change*. http://www.nature.com/climate/index.html (accessed July 2012).

Saito, Y., Chaimanee, N., Jarupongsakul, Th., and Syvitski, J.P.M. (2007). "Shrinking megadeltas in Asia: Sea-level rise and sediment reduction impacts from case study of the Chao Phraya Delta," *LOICZ Inprint*, 2: 3–9.

Schneeberger, C., Blatter, H., Abe-Ouchi, A., and Wild, M. (2003). "Modelling changes in the mass balance of glaciers of the northern hemisphere for a transient 2×CO_2 scenario," *Journal of Hydrology*, 282: 145–63.

Schneider, S.H. (2008). Global warming: Science and policy, Groks Science Radio Show and Podcast. http://grokscience.wordpress.com/transcripts/stephen-schneider/ (accessed March 2012).

Shapiro, H.T., Diab, R., de Brito Cruz, C.H., Cropper, M., Fang, J., Fresco, L.O, Manabe, S. et al. (2010). *Climate Change Assessments: Review of the Processes and Procedures of the IPCC*. Amsterdam, the Netherlands: InterAcademy Council.

Slangen, A.B.A., Katsman, C.A., van de Wal, R.S.W., Vermeersen, L.L.A., and Riva, R.E.M. (2012). "Towards regional projections of 21st century sea-level change based on IPCC SRES scenarios," *Climate Dynamics*, 38, 1191–209.

Solomon, S., Qin, D., Manning, M., Chen, Z., Marquis, M., Averyt, K.B., Tignor, M., and Miller, H.L. (eds.) (2007). *Climate Change 2007: The Physical Science Basis. Contribution of Working Group I to the Fourth Assessment Report of the Intergovernmental Panel on Climate Change*, Cambridge; New York: Cambridge University Press. http://www.ipcc.ch/ (accessed August 2013).

Stafford-Smith, M., Horrocks, L., Harvey, A., and Hamilton, C. (2011). "Rethinking adaptation for a 4°C world," *Transactions of the Royal Society A*, 369: 196–216.

Tamisiea, M.E. and Mitrovica, J.X. (2011). "The moving boundaries of sea-level change: Understanding the origins of geographic variability," *Oceanography*, 24(2): 24–39.

Trenberth, K.E., Jones, P.D., Ambenje, P., Bojariu, R., Easterling, D., Tank, A.K., Parker, D. et al. (2007). "Observations: Surface and atmospheric climate change," in S. Solomon, D. Qin, M. Manning, Z. Chen, M. Marquis, K.B. Averyt, M. Tignor, and H.L. Miller (eds.), *Climate Change 2007: The Physical Science Basis. Contribution of Working Group I to the Fourth Assessment Report of the Intergovernmental Panel on Climate Change*, Cambridge; New York: Cambridge University Press. http://www.ipcc.ch/ (accessed August 2013).

Winterwerp, J.C., Borst, W.G., and de Vries, M.B. (2005). "Pilot study on the erosion and rehabilitation of a mangrove mud coast," *Journal of Coastal Research*, 21(2): 223–30.

Chapter 3

On the frontline in the Anthropocene

Adapting to climate change through deliberative coastal governance

Bruce C. Glavovic

Abstract: A deliberative approach to coastal governance is needed to navigate the stormy seas of the Anthropocene. Coasts are the frontline of the global struggle for sustainability and the primary arena for learning how to adapt to the *super-wicked* problem of climate change. Coastal communities need to build *layers of resilience* in the face of *waves of adversity* due to unsustainable practices that are compounded by climate change impacts. Well-intentioned but modest adaptation measures to maintain the *status quo* can reduce climate risks and even mitigate some climate impacts in the short term. However, the root causes and drivers of unsustainable coastal development, institutional inertia, and path-dependent maladaptation need to be confronted. Emerging adaptation efforts, however, reveal persistent barriers for translating theory into practice. How might adaptation barriers be overcome? Much can be learned from decades of coastal management experience. This experience demonstrates that *business as usual* is untenable. Sustainable coastal development is widely espoused but elusive in practice. To break the impasse, new modalities of innovative transitional and even transformative coastal governance need to be envisioned and institutionalized, with climate change adaptation an integral part thereof. Deliberative coastal governance provides a foundation for managing climate risk at the coast, charting adaptive pathways, and building resilience in the face of the contestation, complexity, uncertainty, ambiguity, and surprise that characterize life on the frontline in the Anthropocene.

INTRODUCTION

This chapter explains why the coast is the frontline of the sustainability struggle in the Anthropocene and the primary arena in which humanity must learn how to adapt to climate change. It shows that *layers of resilience* need to be built as a buffer against *waves of adversity* stemming from unsustainable coastal development and escalating coastal disaster risk driven in part by climate change. An exploration of the conceptualization and practice of adaptation reveals little progress on the ground and unrelenting barriers for translating rhetoric into reality. Despite the necessity to protect coastal communities against sudden shocks, such as more frequent and intense coastal storms (see Chapter 2), it is at best myopic to adopt strategies that merely seek to maintain the *status quo* in the face of uncertain but inexorable sea-level rise and related slow onset climate risks. Crafting institutional* reforms that build adaptive capacity and resilience is urgently needed but takes place in the context of path-dependent inertia and even overt resistance. How then might adaptation barriers be overcome in practice? Valuable insights can be gleaned from five decades of coastal management experience. The nature of coastal management is briefly explored, and key features of emerging thinking about coastal governance† highlighted. It is argued that adaptation needs to be institutionalized as a core dimension of coastal governance. But the conceptualization and practice of coastal governance needs to be further developed given the distinctive challenges posed by climate change. A deliberative approach to coastal governance is proposed to enable coastal communities to navigate the stormy seas of the Anthropocene.

THE STORMY SEAS OF THE ANTHROPOCENE

The term *Anthropocene* is used to define a new geological era that reflects the dominant influence that humans now have in shaping global biogeochemical processes (Crutzen 2002; Crutzen and Stoermer 2000; Zalasiewicz et al. 2011). Human activities may have transgressed critical planetary boundaries, including nutrient cycling, biodiversity, and climate change (Rockström et al. 2009), and have already generated pervasive and pernicious ecological and societal impacts that have dire consequences for life on earth (Millennium Ecosystem Assessment 2005). Global change is, thus, a critical issue for humanity, and the transition to sustainable development has long been recognized as a compelling policy imperative (United Nations 1992; World Commission on Environment and Development 1987). Climate change is an integral part of global change and the sustainable development agenda. Climate change impacts are already being experienced, compounding and exacerbating the negative consequences of unsustainable development (see Chapter 2), and are especially pronounced along coastal margins around the world (Lloyd et al. 2013; Moser et al. 2012; Nicholls et al. 2007).

---

* *Institutions* are structures and processes (e.g., the law and social norms) that govern behavior and interactions between key actors in society and mould social choices.

† *Government* is the formal organization of the state, whereas *governance* refers to the interactions of the state, civil society, and the private sector to address societal problems through power sharing, social coordination, and collective action.

The frontline of the sustainability struggle

As shown in Chapter 1, close to half the world's population and economic development is located at the narrow land–sea interface. The coastal population is concentrated in densely populated rural areas and small to medium cities (Small and Nicholls 2003), with population growth and development intensification projected to accelerate, causing distinctive challenges in coastal megacities (Li 2003; Pelling and Blackburn 2013). Coastal ecosystems are subject to *coastal squeeze* (Doody 2004) and unsustainable pressures (Crossland et al. 2005) that imperil their capacity to sustain coast-dependent livelihoods (Moser et al. 2012). To compound matters, the land–sea interface is a frontier of risk, with more and more people and infrastructure exposed to coastal hazards, and this risk is intensifying and could generate cascading failure with the cumulative and synergistic impacts of a changing climate. The 136 largest coastal cities could risk combined annual losses of USD 1 trillion due to flooding by 2050 unless protective measures are improved (Hallegatte et al. 2013). Climate change impacts at the coast range from biogeochemical impacts to impacts on infrastructure, livelihoods, institutions, and distinctive cultures and traditions (see Table 3.1).

The narrow land–sea interface is the locus of countervailing trajectories of population growth and development intensification on the one hand and, on the other, increasing climate risk and impacts that compound unsustainable practices. The coast is thus the frontline of the global struggle for sustainability and the principal arena in which humanity must learn to live with the unfolding reality of climate change (Glavovic 2013a). The current generation of coastal communities will consequently play a pivotal role in charting the sustainability course in the Anthropocene. Coastal communities will need to build layers of resilience as a buffer against waves of adversity as climate change escalates disaster risk and magnifies the negative impacts of unsustainable coastal practices (Figure 3.1).

Waves of adversity, layers of resilience

This section explores the waves of adversity facing coastal communities, the *super-wicked* problem of climate change, and the imperative to build layers of resilience, drawing on insights from the real-world experience of coastal communities in the Mississippi delta.

Waves of adversity

Although climate change is a global phenomenon, impacts will be experienced locally and become entangled with a gamut of locality-specific development and environmental challenges and opportunities. Impacts will be spread

Table 3.1 Impacts of climate change in the coastal zone

Climate change at the coast	*Direct and indirect impacts on coastal ecosystems and communities*
• Higher atmospheric and ocean temperatures • Sea-level rise • Changed oceanic conditions (e.g., ocean acidification due to elevated carbon dioxide levels in the atmosphere; changes in ocean currents) • Changes in coastal storm tracks, frequencies, and intensities • Changes in precipitation patterns and coastal runoff	*Bio-geochemical and ecological impacts:* • More frequent and intense extreme events • Increased coastal erosion • Increased flooding • Salinization of surface and groundwater • Displacement, disruption, and/or loss of climate-sensitive coastal ecosystems (such as coral reefs, mangroves, and wetlands) • Changes in the distribution and abundance of coastal and marine species, including the spread of exotic and invasive species *Socioeconomic and infrastructure impacts:* • Loss of property and land • Safe quality housing at risk • Increased flood risk/loss of life • Damage to coastal protection works and other infrastructure • Loss of commercial, recreational, and subsistence resources • Loss of tourism, recreation, and coast-dependent activities • Impacts on coastal agriculture and aquaculture through changes in temperature and precipitation and soil and/or water quality • Impacts on livelihoods and human health due to declining food sources, exposure to extreme events, heat stress, vector- and water-borne diseases, and reduced access to potable water • Diversion of resources to adaptation responses • Increasing cost of coastal defense/protection measures • Increasing insurance premiums and uninsurability • Disruption of financial and insurance markets • New economic opportunities *Impacts on institutions, cultures, and ways of life:* • Increased uncertainty and changes to coastal lifestyles that intensify coastal conflict • More frequent weather- and climate-related coastal disasters • New livelihood challenges and/or opportunities • Climate change-induced displacement and migration • Political and social instability, and social unrest

(*Continued*)

Table 3.1 (Continued) Impacts of climate change in the coastal zone

Climate change at the coast	*Direct and indirect impacts on coastal ecosystems and communities*
	• Threats to some coast-dependent cultures and ways of life • Impacts on national prestige (opportunity or threat depending on leadership, scientific knowledge and understanding, and institutional capacity)

Source: Abuodha, P.A. and Woodroffe, C.D. *International Assessments of the Vulnerability of the Coastal Zone to Climate Change, Including an Australian Perspective.* Final Report submission to Australian Greenhouse Office in response to RFQ 116/2005DEH, 2006; Parry, M.L. et al. *Climate Change 2007: Impacts, Adaptation and Vulnerability. Contribution of Working Group II to the Fourth Assessment Report of the Intergovernmental Panel on Climate Change*, Cambridge University Press, Cambridge, 2007; Nicholls, R.J. and Kebede, A.S., "Indirect impacts of coastal climate change and sea-level rise: the UK example," *Climate Policy*, 12(suppl), S28–S52, 2012; USAID, *Adapting to Coastal Climate Change: A Guidebook for Development Planners*, USAID, Washington, DC, 2009.

Figure 3.1 Malé, capital of the Maldives, an urbanized island protected by a seawall. (Courtesy of Bruce C. Glavovic.)

unevenly between and within nations and communities, creating vexing climate inequities (Adger et al. 2006). Consequently, there are complex linkages between global change, including climate change, and human development, and much remains to be done to understand better this complexity (Boyd and Juhola 2009; Pelling 2011). It is nonetheless clear that many individuals and entire communities are vulnerable to climate variability and shocks because of their marginalization and exclusion from political processes and market access, not simply because of exposure to climate change impacts (see Chapter 1: Box 1.2 for a discussion about vulnerability). In many cases, climate change will cause extra stress, compound existing problems, create new problems, and increase risk, especially for those who are poor, socially excluded, and disadvantaged (World Bank 2013). For those already experiencing the adverse impacts of inequitable and unsustainable coastal development, vulnerability to climate change is a formidable prospect that portends waves of adversity—notwithstanding the possibility that a changing climate may also open up new livelihood opportunities. It raises vexing questions about *inter alia* sustainability; climate justice; coastal governance; and how to mobilize, integrate, and apply scientific and local knowledge. Adapting to climate change should therefore be framed in the context of the sustainable development agenda, with an explicit focus on the sociopolitical and economic drivers of poverty and inequity (Adger et al. 2009a; Cannon and Müller-Mahn 2010).

Adaptation presents coastal communities with an opportunity to reflect critically upon prevailing development trajectories and chart new pathways that lead to outcomes that are more economically, socially, culturally, and ecologically sustainable. Many coastal communities already exposed and vulnerable to climate impacts have limited resources and adaptive capacity and will need appropriate support, including from higher levels of government and even from the international community, in their efforts to adapt to climate change. Building resilience in the face of climate change-driven waves of adversity at the coast is thus a profoundly challenging multidimensional and cross-scalar undertaking, with significant ethical, cultural, social, political, economic, technological, and ecological implications. Better understanding of this complexity is central to developing and implementing effective adaptation strategies.

The super-wicked problem of climate change

Climate change, the risk it poses, and the adaptation imperative facing coastal communities constitute the archetypal *wicked problem*. Wicked problems are characterized by complexity, contradictory, and changing requirements that are hard to recognize and the prospect that promising solutions generate unanticipated but even more pernicious problems, making it difficult to define the problem unambiguously or untangle it from other intractable societal problems (Rittel and Weber 1973). The climate issue at the coast involves a myriad of interacting biogeochemical linkages and ethical, cultural, social, political, technological and economic interactions (Crossland et al. 2005), and teleconnections that extend from the local to global scale (Liu et al. 2013). These interactions are nonlinear, and minor changes can result in disproportionately significant consequences. Coastal ecosystems sustain varied and distinctive coastal livelihoods, cultures, and ways of life that together comprise coupled social-ecological systems that have system thresholds, discontinuities, and tipping points that climate change may transcend, especially if abrupt, extreme changes take place (World Bank 2013). There are no panaceas for the climate *problématique* and choices will need to be made in the context of systemic uncertainty, nonlinearity, evolving dynamics with intricate feedback loops, real social-ecological limits and barriers, and inevitable surprise. Levin et al. (2012) describe climate change as a super-wicked problem that presents a policy-making tragedy because prevailing institutional structures and processes are poorly equipped to generate timely solutions despite the necessity to avert potentially catastrophic impacts.

To further compound this complexity, climate change exacerbates the deleterious impacts of stubbornly unsustainable patterns of coastal development. Incremental improvements may help to alleviate some impacts, but the underlying drivers and root causes of unsustainability, complacency about climate change, institutional inertia, and maladaptive policies and practices need to be comprehended and addressed (Adger et al. 2009b). This super-wicked problem cannot be understood by reliance on traditional disciplinary sciences alone, and scientific understanding cannot be converted simplistically into rational adaptation policies and strategies. There are gaps in our scientific knowledge. Moreover, potentially viable solutions are imbued with ambiguity because there are multiple legitimate views framed by contested values, perceptions, and interests, and the power to effect change is unevenly distributed across society and even within communities. The climate change science–policy–practice interface is consequently a contested arena. Coastal communities therefore need to reflect critically on how to frame the issue of climate change, learn from real-world experience, while recognizing the limits of relying on past experience, and explore alternative pathways to reduce climate risk and adapt to interacting climate and nonclimate stressors and drivers of change. The extent to which coastal communities are able to anticipate and withstand coming waves of adversity will be determined by the extent to which they can transition away from practices that make them vulnerable to climate stresses and navigate toward pathways that build layers of resilience.

Layers of resilience

Resilience has been defined as the amount of change that a social-ecological system can undergo without changing state, which means that the system maintains its characteristic structure and key functions (see Chapter 1: Box 1.4). It reflects the ability of social-ecological systems to respond to change and, in doing so, plan for an uncertain future that includes slow-onset change as well as sudden shocks due to natural hazard events or anthropogenic perturbations. Resilience is thus a desirable property because communities want to avoid transitioning into undesirable states (Walker et al. 2002). The scale of climate change, however, means that changes in state are inevitable in some circumstances and localities. A resilient community may therefore need to manage transitions between states while endeavoring to adopt desirable configurations, including securing the livelihoods of the most vulnerable individuals and groups. Resilience thus encompasses a wide range of perspectives that all deal to some extent with the interplay between persistence and transformation (Carpenter and Brock 2008).

There are varying contestable and even contradictory notions of resilience (e.g., Folke 2006; Gallopín 2006), and parallel streams of resilience scholarship, including engineering, ecology, psychology, and interdisciplinary approaches focusing on social-ecological systems, that are only relatively recently being integrated (e.g., Berkes and Ross 2013). The engineering (or *resistance*) and ecological framings of resilience are insufficient for understanding and addressing the coupled social-ecological character of climate change at the coast. Resilience from a social-ecological perspective constitutes, first, the amount of change that a social-ecological system can absorb and still retain key structures and functions; second, the extent to which the system is able to reorganize; and third, the extent to which the system can increase capacity for learning and adaptation in the face of change (Carpenter and Gunderson 2001; Holling 1973; Walker et al. 2004). Social-ecological resilience scholars argue that adaptive capacity reflects learning capacity and the flexibility to experiment and adopt novel solutions that are responsive to a variety of challenges, including climate change, and it is thus one dimension of resilience (Norris et al. 2008; Walker et al. 2002) (see Chapter 1: Box 1.2). According to Berkes and Ross (2013), the social-ecological systems perspective on resilience has deepened understanding about adaptive relationships and learning across nested levels and focused attention on feedbacks, nonlinearity, unpredictability, scale, renewal cycles, system memory, disturbance events, and windows of opportunity. Those working on resilience from the perspective of the psychology of development and mental health have tended to concentrate more on individual and community resilience, with attention focused on capabilities and strengths and building resilience through agency and self-organization, and have highlighted the importance of people–place connections, values and beliefs, knowledge and learning, social networks, collaborative governance, economic diversification, infrastructure, leadership, and outlook. An integrative perspective can help to advance understanding of social-ecological systems and identify locality-specific capabilities and connections that can be activated by agency and self-organization (Berkes and Ross 2013) and thus has particular relevance to the coast.

The livelihoods of coastal communities are dependent upon and entwined with coastal ecosystem health and integrity and the coast therefore needs to be viewed as a complex, coupled, dynamic, and evolving social-ecological system. An integrative resilience approach recognizes this systems perspective and the need to develop individual and collective capacity to mobilize diverse sources of resilience in the face of uncertainty, change, and sudden shocks (Adger et al. 2005b). This vantage point underscores the importance of stakeholder agency and the processes through which alternative framings of resilience and coastal sustainability are developed (Larsen et al. 2011). An integrative resilience approach recognizes that change is normal and that incomplete knowledge and understanding is inevitable. It seeks to understand vulnerability and anticipated impacts and to identify opportunities for adapting to climate change. It acknowledges that coastal resources need to be managed flexibly in the face of change, informed by dynamic limits and internal feedbacks, and based on monitoring, learning, and adaptation (Berkes and Folke 2008). Building resilience therefore requires an understanding of the exposure and sensitivity of coastal communities to climate change across scales—recognizing that communities are embedded in a globalized world (Wilson 2012), identification of climate risks and potential impacts, and development of adaptive strategies that are appropriate for particular circumstances. Crucially, building resilience must take into account interacting climate and nonclimate drivers of change. Some adaptive actions might yield short-term benefits but turn out to be maladaptive in the long-term and vice versa. Similarly, adaptation measures at one scale might inhibit adaptation at other scales or adversely affect vulnerability and adaptive capacity elsewhere. Different approaches will therefore need to be pursued depending on locality-specific circumstances and changes over time. There are multiple sources of resilience in social-ecological systems and these sources need to be identified (Adger et al. 2011). Resilience invariably means different things to different communities, and consequently, social choices need to be made about the purpose of resilience-building efforts. Resilience is thus socially constructed and imbued with sociopolitical and ethical dimensions that compel reflection, public dialogue, and deliberation (Keessen et al. 2013). Magis defines community resilience as

> ... the existence, development, and engagement of community resources by community members to thrive in an environment characterized by change, uncertainty, unpredictability, and surprise. Members of resilient communities intentionally develop personal and collective capacity that they engage to respond to and influence change, to sustain and renew the community, and to develop new trajectories for the communities' future.
>
> Magis 2010: 402

Although resilience varies between communities, those characteristics that are key to strengthening community resilience, mobilized through agency and self-organizing, include (see, e.g., Amundsen 2012; Berkes and Ross 2013; Buikstra et al. 2010; Magis 2010) the following:

- People–place connections
- Values and beliefs
- Community resources, including resource entitlements, accessibility, and development
- Social and community networks and support
- Collective action through engaged governance facilitated by inclusive and collaborative institutions
- Community infrastructure and services
- Diverse and innovative economy
- Community leadership and active agents
- Knowledge, skills, and learning
- A positive outlook, including the capacity to accept change

Community resilience cannot be taken for granted. Prevailing resilience may be insufficient to cope with future climate change that may be unprecedented in recent human history. Extant resilience may therefore be illusionary and lead to complacency about the need to strengthen adaptive capacity and resilience, making active community engagement and reflexive learning vitally important (Amundsen 2012). Building resilience thus necessitates community-relevant adaptation strategies and pathways that are adjusted over time in anticipation of and in response to predictable and unpredictable waves of adversity.

Although the concept of resilience has many strengths as a key organizing construct for the pursuit of sustainability, *inter alia* fostering an integrated approach to understanding the dynamics and governance dimensions of global change in social-ecological systems (Duit et al. 2010), the limits of the concept are recognized, chiefly because its roots in the physical sciences and ecology constrain its application to social systems, in particular with respect to power relationships and addressing poverty, inequity, and injustice (Adger et al. 2009a, 2009b; Béné et al. 2012; Berkes and Ross 2013; MacKinnon and Derickson 2013)—which are key dimensions of coastal vulnerability in the Anthropocene. Hence the need for more focused attention on the ethics and politics of climate change, and social constructs such as agency, self-efficacy, empowerment, optimism, and self-esteem in shaping how individuals and communities deal with change, shocks, and stresses, as well as the values and behavioral traits that enable community bonding, connection to place, and cross-cultural resilience (Berkes and Ross 2013). Pelling (2011) argues that framing *adaptation as resilience* is a point of departure but that adaptation also needs to be viewed as a sociopolitical transition and even transformative process in this era of global change. This critical framing of resilience recognizes the central role of power in shaping adaptation options and prospects for society and underscores the value-laden nature of endeavors to reduce vulnerability, build adaptive capacity and resilience, and pursue sustainability in the Anthropocene.

What does this mean in practice? The case studies in this book shed light on practical challenges and opportunities for building the layers of resilience necessary to buffer coastal communities from waves of adversity and navigate new pathways toward adaptive and sustainable coastal development. The experience of and prospects for communities living in the Mississippi delta is emblematic of the challenges facing coast-dependent communities in the Anthropocene (Glavovic 2008a, 2013b, 2014). In Chapter 6, Lewis, Yoachim, and Meffert explore the vulnerability to climate change of coastal Louisiana and show that the resilience of many communities in this region is dependent upon, among other things, healthy coastal wetlands to attenuate the impact of coastal storms and sustain livelihoods, functional levees or stopbanks to prevent flooding, effective civic leadership, social capital, and equitable political and economic opportunities. Figure 3.2 illustrates notional layers of resilience that buffer the communities of coastal Louisiana from sudden shocks, such as hurricanes, and the slow-onset impacts of climate

Figure 3.2 Waves of adversity, layers of resilience.

change. The vulnerability of delta communities and many in New Orleans in particular was cruelly exposed by devastating hurricanes in 2005, especially Hurricane Katrina, the 2010 BP-Deepwater Horizon oil spill, the flood risk from the Mississippi River—which came perilously close to overtopping the levees in 2011, and the prospect of sea-level rise along a subsiding deltaic coastline.

Delta communities need to mobilize and secure the material and human resources or assets of the region (that can be framed as different forms of *capital*, e.g., Stokols et al. 2013) to build layers of resilience (see Glavovic 2014). The delta experience shows that deep social vulnerability can exist alongside community strengths and resilience (see Chapter 6). Communities that have *thick* layers of resilience, with minimal vulnerability fissures and fractures, will be better able to cope with sudden shocks and plan for and respond to change in ways that minimize exposure and sensitivity to climate change. The capacity to adapt to a changing climate will be enabled by, *inter alia*, the development and institutionalization of an ethic of social equity and ecological stewardship that is mobilized by political will, public support, and inclusive and flexible governance institutions; deeper levels of public trust and supportive social networks; robust public infrastructure and protective works; stimulating and diversifying the economy and creating sustainable livelihood opportunities; social learning, and the requisite knowledge and skills to navigate the challenges of the Anthropocene.

It behooves every coastal community to build resilience. In practice, however, this is a fraught and vexatious undertaking. The rate and scale of global change, and compounding climate-driven waves of adversity, pose a daunting challenge for communities with fragile resilience in climate hot spots like the Mississippi delta. The uncertainties inherent in climate change make it difficult to define which dimensions of resilience might be more or less important at a particular point in time for a specific community, underscoring the need to facilitate authentic public engagement in governance processes that foster reflexivity and social learning to enhance adaptive capacity—as revealed in a study of adaptation in Norway (Amundsen 2012). This finding was corroborated in a study of coastal tourism-dependent communities in Thailand following the 2004 Indian Ocean tsunami: resilience was framed as a negotiated normative governance process in which the role of stakeholder agency and processes are key to developing legitimate resilience visions and practices (Larsen et al. 2011). A study of community resilience and vulnerability to tsunamis in the Solomon Islands found that communities have to negotiate trade-offs between some adaptive capacities that sustain general system resilience at the expense of increased vulnerability to low-probability events like tsunami, making resilience a dynamic and multiscale challenge that is interwoven with patterns of vulnerability (Lauer et al. 2013). Complex multiscalar resilience interactions are not new. Communities have always had to navigate interacting sociopolitical, trade, economic, technological, and environmental changes. For example, the demise of historic Norse Greenland communities may have been driven by changing regional economic and trading patterns compounded by environmental change that resulted in cascading collapse of interconnected settlements (Dugmore et al. 2007; cf. Diamond 2005). The study of climate change adaptation is, however, relatively recent. What has been learned from this scholarship and real-world experience to inform adaptation efforts at the coast?

CLIMATE CHANGE ADAPTATION: PRACTICE, LIMITS, AND BARRIERS

This section provides a succinct overview of evolving adaptation practices and explores limits and barriers to adaptation. As explained in Chapter 1 (see Box 1.2), adaptive capacity is simply the potential to adapt. It is the ability of a system to respond or adjust to the consequences of changes and stresses or prepare in advance for anticipated changes and stresses (Smit et al. 2001). It reflects the capacity of actors (from individuals to organizations, communities, and beyond, e.g., Brown and Westaway 2011) in social-ecological systems to build resilience (Folke et al. 2010) by mobilizing resources and community strengths to address perceived and prevailing stresses (reflecting agency or the capacity to act independently) (Berkes and Ross 2013; Engle 2011). It varies over time and within and between systems, contexts, and communities (Adger et al. 2007). How can adaptive capacity be mobilized, strengthened, and institutionalized so that coastal communities can anticipate and respond to a changing climate? Increasing attention has been focused on this issue in recent years, notably through a variant of environmental governance-labeled adaptive comanagement that has particular application in situations characterized by complexity, interconnectedness, persistent and potentially dramatic change, and high degrees of uncertainty (Scarlett 2013)—which characterize climate change at the coast. Core components of adaptive comanagement include (see, e.g., May and Plummer 2011; Olsson et al. 2004; Plummer and Armitage 2007; Plummer and Baird 2013; Plummer et al. 2013; Scarlett 2013) the following:

- Pluralism (i.e., a diversity of viewpoints) and variety
- Actors who are autonomous, that is, self-aware and reflexive
- Linkages within and between levels
- Information sharing and communication to foster shared knowledge and understanding
- Social learning and ability to develop, test, reflect upon, and revise actions

- Capacity to modify viewpoints and behavior in light of changing circumstances
- Shared authority and decision-making

How have adaptation practices evolved over time and to what extent has the adaptive capacity of coastal communities been strengthened?

Evolving adaptation practices

Over time, adaptation approaches have evolved from a focus on impacts and making a choice between alternative strategies to recognizing that adaptation is an iterative process that requires stepwise actions with locality-specific application. Increasing attention has been focused on situating adaptation in the wider context of sustainable development and to the key roles of disaster risk reduction and local-level land-use or spatial planning in crafting adaptive policies and pathways. Central to this endeavor is turning barriers into enablers of adaptation.

From adaptation impacts and options to adaptation pathways

Three broad alternative adaptation strategies to sea-level rise were initially identified by the Intergovernmental Panel on Climate Change (IPCC) (Dronkers et al. 1990) and elaborated upon in subsequent guidance on assessing climate change impacts and adaptation at the coast (e.g., Carter et al. 1994; IPCC 1992): (1) *Protect* involves the use of hard-engineered structures, such as seawalls, and soft-engineered approaches, such as beach nourishment or dune rehabilitation, to protect land and physical infrastructure from climate-driven sea level-related impacts such as intensified storms, erosion, and sea-level rise (see Figure 3.3). (2) *Accommodate* involves continued occupation and physical development along the seashore in the face of climate impacts and adoption of mitigation measures such as elevating buildings, emergency shelters, warning systems, growing flood and/or salt-tolerant crops, and so on. (3) *Retreat* is a risk reduction strategy that involves the abandonment of activities along exposed areas of the seashore and a progressive move landward. It accepts that the costs of sea level-related impacts outweigh the benefits of protective or accommodative measures in high-risk localities. Retreat can be achieved through land-use planning regulations, building codes, and/or economic incentives. It could involve phased retreat of existing development or the withdrawal of subsidies that otherwise encourage occupation of the seashore despite climate risks. *Avoid* can be an integral part of a retreat process or it could be regarded as a fourth strategy that precludes new physical development in areas exposed to sea level-related impacts.

Figure 3.3 Cape Town, Republic of South Africa. (Courtesy of Bruce C. Glavovic.)

Each of the above strategies involves potentially significant economic, social, institutional, technological, and environmental impacts that need to be carefully assessed to determine the most appropriate strategy, in particular circumstances. In practice, however, adaptation goes well beyond making a one-off choice between protect, accommodate, and retreat (and/or avoid), and the early focus on assessing impacts and selecting a preferred strategy has shifted toward understanding barriers and opportunities for strengthening adaptive capacity, locating adaptation within the wider context of human development and governance, and exploring adaptation pathways to reduce risk, build resilience, and promote sustainability (e.g., Adger et al. 2007; Füssel 2007; Haasnoot et al. 2013; Klein et al. 1999; Larsen et al. 2011).

Planning adaptation policies and pathways has only recently become the focus of attention (e.g., Haasnoot et al. 2013; Levin et al. 2012), but draws on long-standing traditions in planning and policy studies that go back to work such as that of Dewey (1927) who framed policy-making as a social learning process based on experimentation and adaptation. These approaches recognize that a single *optimal* policy or plan cannot be designed credibly, let alone implemented successfully, for a future that is characterized by deep uncertainty and inevitable but unpredictable change.

Adaptive policy-making and planning typically adopt a strategic vision of the future, identify short-term actions, and then establish a framework to guide reflection and lesson learning to structure future actions in a stepwise fashion as triggers or tipping points are reached that no longer meet the plan's objectives, and enable new courses of action to be pursued. To what extent are these insights being translated into reality?

Adaptation in practice

Adaptive capacity and resilience prospects for coastal communities are shaped by individual, community, and societal adaptations already underway, including adaptations in insurance and reinsurance markets, coastal planning regimes, public health provisions, and an array of livelihood-specific adaptations (Adger et al. 2007). Not all such initiatives are labeled *adaptation* but together they are significant in strengthening adaptive capacity and resilience. Initiatives such as the Adaptation Fund and Pilot Program in Climate Resilience and the UN Framework Convention on Climate Change National Adaptation Programmes of Action, together with a variety of mechanisms put in place by international development agencies, donors, and nongovernmental organizations, have stimulated adaptation planning worldwide, especially in so-called developing countries. There are also many region-wide or transnational (e.g., in the European Union, Caribbean, and South Pacific) and national adaptation planning initiatives now underway, and a plethora of subnational and local-level adaptation initiatives, as this book shows.

Despite growing attention being focused on adaptation through recent scholarship (Berrang-Ford et al. 2011) and a dramatic increase in adaptive actions in diverse settings (Lesnikowski et al. 2013; Preston et al. 2011; Tompkins et al. 2010), compared to climate mitigation efforts, planned adaptation initiatives are relatively recent and evidence of effective implementation is limited. In a study of 4,104 adaptive actions taken by 117 UNFCC parties, Lesnikowski et al. (2013) found scant evidence of effective translation of rhetoric into reality and a stubborn gap in understanding. Ford et al. (2011) found limited evidence of adaptive action in developed nations other than those in climate-sensitive sectors and institutional actions at the municipal level, typically with support from higher levels of government. In a review of 57 adaptation plans from Australia, the United Kingdom, and the United States, Preston et al. (2011) found that plans were largely underdeveloped with critical weaknesses arising because nonclimatic factors were not properly considered and key aspects of adaptive capacity, such as entitlements to key assets, were ignored. Adaptation implementation challenges prevail in many coastal nations and localities as is clearly demonstrated in the case studies in this book.

There is concerted pressure on local government to build hard-engineered protective structures to protect private property and investment along the seashore even in the face of escalating risk. However, as waves of adversity intensify, accommodative, retreat, and avoidance strategies will become increasingly compelling. Judging what constitutes successful adaptation across different scales will need to be assessed in the context of particular coastal communities and circumstances. Adger et al. (2005a) suggest that the normative criteria of effectiveness, efficiency, equity, and legitimacy may be useful for judging success from a resilience and sustainability vantage point. Prevailing distributions of power and inequity, however, make realization of such ideals seemingly remote. The challenge is to bring about behavioral and institutional change in face of systemic patterns of unsustainable development, institutional rigidity, and strong vested interests to maintain the *status quo* in which long-term concerns about public safety, equity, and sustainability are discounted relative to short-term, private interests. Prevailing responses to climate change are thus embedded in institutional procedures, technological and economic pathways, and cultural practices that are typically unsustainable and maladaptive, hence the need to target sources of path dependency, identify barriers to action, and unlock opportunities for social learning and adaptation.

Adaptation limits, barriers, and enablers

Emerging scholarship sheds light on the nature of adaptation limits and barriers and identifies challenges and opportunities for building adaptive capacity and resilience (e.g., Adger et al. 2005a, 2007; Burch 2010; Gupta et al. 2010; Jones and Boyd 2011; Juhola and Westerhoff 2011; Lebel et al. 2011; Moser and Ekstrom 2010; Peñalba et al. 2012; Storbjörk 2010; Termeer et al. 2012). It is constructive to distinguish between limits and barriers to adaptation. Limits can be framed as absolute obstacles or thresholds beyond which existing ecosystems, species, activities, livelihoods, or system states cannot be maintained—even in a modified form. Beyond such thresholds, state changes in social-ecological systems, including potentially irreversible changes, are inevitable. There are natural or physical and ecological limits and barriers to adaptation. The rate and scale of climate change may surpass critical biophysical thresholds beyond which ecological systems and their dependent coastal communities may no longer be viable especially when climate change impacts are superimposed on unsustainable practices. Human, resource, and informational barriers include constraints in knowledge, financial resources, and technology. Knowledge of and understanding about adaptation is constrained by the complexity of climate change. Although there is potential for developing a variety of adaptation technologies, uncertainty and social resistance may inhibit technological development, diffusion, and adoption. Technologies may not be economically feasible, and they may be socially undesirable or location-specific and not widely transferable. Social barriers include cognitive, normative, and institutional barriers and stem from different perceptions, worldviews, interpretations, experiences, entitlements, and responses to climate change. Tolerance to risk varies within and between communities, as will preferences for adaptive options and alternative pathways. Adger et al. point out that

> ... diverse and contested values—underpinned by ethical, cultural, risk and knowledge considerations—underlie adaptation responses and thus define mutable and subjective limits to adaptation. ... [N]otwithstanding physical and ecological limits affecting natural systems, climate change adaptation is not only limited by such exogenous forces, but importantly by societal factors that could possibly be overcome. ... [W]e suggest that an adaptable society is characterized by awareness of diverse values, appreciation and understanding of specific and variable vulnerabilities to impacts, and acceptance of some loss through change. The ability to adapt is determined in part by the availability of technology and the capacity for learning but fundamentally by the ethics of the treatment of vulnerable people and places within societal decision-making structures.
>
> Adger et al. 2009a: 350

A range of studies shows that the ability of communities to act collectively is constrained by prevailing levels of public trust and social capital (Adger 2003) and that the legitimacy of governance institutions, in particular, plays a pivotal role in adaptation prospects (Adger et al. 2009a, 2009b; Bierbaum et al. 2013; Brooks et al. 2005; Brown et al. 2010; Eakin and Lemos 2006; Engle 2011; Engle and Lemos 2010; Gupta et al. 2010; Haddad 2005; Ivey et al. 2004; Jones and Boyd 2011; Yohe and Tol 2002). For example, Termeer et al. (2012) analyze the National Adaptation Strategies of the Netherlands, the United Kingdom, Finland, and Sweden and identify five institutional weaknesses or barriers: lack of openness toward learning and variety; strong one-sided reliance on scientific experts; tension between top-down policy development and bottom-up implementation; distrust in the problem-solving capacity of civil society; and the difficulty of reserving funding for long-term action. Some of these institutional barriers are mirrored in a study by Lebel et al. (2011) of the institutional traps that compound vulnerability to climate change and flooding in Thailand. These traps include (1) fragmentation whereby bureaucratic competition and separatism result in poor coordination, lack of institutional capability, and gaps in service provision; (2) rigidity that reflects a preoccupation with control, stability, and certainty that maintains and reinforces inflexible institutions; (3) single-scale focus manifest in a narrow concentration of resources and capacity at one scale that ignores the benefits accruing from cross-scale interactions; (4) elite capture of agendas by those who use expertise and technical tools to serve their own interests and not those of marginalized and vulnerable groups; and (5) a crisis mentality that stems from a lack of effective long-term strategic planning and can result in a knee-jerk reaction to emergencies driven by political pressures and opportunities. Overcoming these institutional barriers or traps requires authentic public participation in disaster risk reduction, prioritizing risk reduction for socially vulnerable groups, a focus on building adaptive capacity at multiple scales and levels, integrating emergency management and climate change adaptation in development planning, and strengthening the links between knowledge and practice (Lebel et al. 2011).

The meager progress made to date in translating adaptation thinking into practice is hardly surprising given these formidable barriers. Levin et al. (2012) argue that innovative path-dependent policy interventions that trigger progressive incremental changes away from business as usual need to be created and institutionalized to address the super-wicked problem of climate change. The challenge is to envisage and institutionalize processes that chart pathways to overcome barriers and unlock enablers of adaptive capacity and resilience at the coast.

A key conclusion of the 2007 IPCC assessment of adaptation in the coastal zone was that isolated efforts to reduce climate risk will be less effective than those that are an integral part of coastal management endeavors and that integrated coastal management (ICM) has the potential to overcome a number of the adaptation barriers prevailing in the coastal zone (Nicholls et al. 2007). How have coastal management efforts evolved over time and what are the prospects for effectively implementing adaptation as a core part of ICM?

COASTAL GOVERNANCE, CLIMATE RISK, AND ADAPTING TO CLIMATE CHANGE

This section briefly traces the evolution of coastal management efforts; outlines key elements of ICM and coastal governance, the alignment of ICM and adaptation best practice, the nature of climate risk at the coast; and explores how to break the prevailing impasse and adapt to climate change at the coast.

From coastal zone management to coastal governance

Tension and conflict at the coast are inevitable as contending interests converge at this narrow interface and increasingly resource-constrained arena. Nations govern the coast as a key part of their national heritage—in the interest of current and future generations. Traditionally, however, coastal management efforts have been organized according to sector-specific interests like transport, mining, and conservation and to address locality-specific issues. Contemporary coastal management efforts recognize that the coast is a distinctive, interconnected social-ecological system that requires a dedicated and integrated management approach to secure and sustain its manifold benefits (Cicin-Sain and Knecht 1998; Kay and Alder 2005; Krishnamurthy et al. 2008).

A range of coastal management approaches has been institutionalized over time to reconcile user conflicts and divergent public and private, and short- and long-term, interests (see Figure 3.4): from coastal zone management to integrated coastal zone management, integrated coastal area management, and ecosystem-based management. For millennia, coastal communities used traditions, taboos, sanctions, and various governance arrangements to manage coastal and marine resources. Coastal management efforts by modern nation-states are more recent and are traced by some to the late 1960s work of the Stratton Commission that shaped three decades of US federal coastal and marine policy and practice, including the passage of the 1972 US Coastal Zone Management Act (Merrell et al. 2001). Soon thereafter, coastal management efforts spread around the world, with attention initially focused on managing natural resources in the coast zone (Sorensen 1997). Integrated approaches to coastal management began to predominate

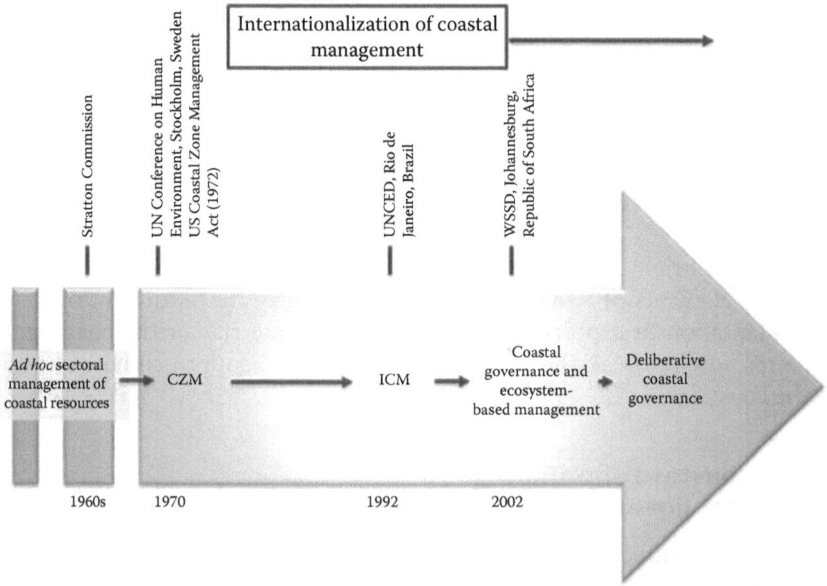

Figure 3.4 Evolution of coastal management.

in the early 1990s, through *inter alia* Agenda 21 which heralded ICM as a key enabler of sustainable development, which was reiterated at the World Summit on Sustainable Development 10 years later. Coastal governance and ecosystem-based approaches took center stage from the late 1990s and early 2000s (see, e.g., Cicin-Sain and Knecht 1998; Lubchenco and Petes 2010; Olsen 2002, 2003) and, more recently, have been built upon by *inter alia* marine spatial planning (Kannen 2012), with attention focused on bridging the science–policy–practice interface (Bremer and Glavovic 2013) through adaptive (Torell 2000; Walters 1997) and deliberative governance modalities (Glavovic 2013c; Jentoft and Chuenpagdee 2009).

While these approaches have different emphases, they typically advocate and seek to promote an integrated approach to collaborative decision-making. A widely cited definition of ICM is

> ... a process that unites government and the community, science and management, sectoral and public interests in preparing and implementing an integrated plan for the protection and development of coastal ecosystems and resources. *The overall goal of ICM is to improve the quality of life of human communities who depend on coastal resources while maintaining the biological diversity and productivity of coastal ecosystems.*
>
> GESAMP 1996: 2, emphasis in original

ICM seeks to overcome compartmentalization of divergent sectoral, administrative, spatial, and temporal interests by promoting inclusive governance arrangements to improve cooperation, coordination, and integration within and between coastal actor networks, sectors, and interests across scales. For example, ICM seeks to establish enabling integrative formal and informal institutional structures and processes. ICM encourages social learning and the integration of coastal science and local and traditional knowledge. ICM also advocates strategic thinking, locality-relevant and locality-informed application supported by intentional learning-from-experience based on monitoring, evaluation, program adjustment, and adaptation.

ICM is increasingly framed as coastal governance wherein divergent goals, interests, and understanding are negotiated in political interactions between coastal stakeholders (Bremer and Glavovic 2013). The governance challenge is to enable key actors from government, civil society, and the private sector to work together in ways that reconcile private and public, and short- and long-term, interests in pursuit of resilience and sustainability (see Figure 3.5).

Governance involves actors and networks from the state, civil society, and the private sector, interacting with the scientific community and media among others, to address societal concerns through social coordination,

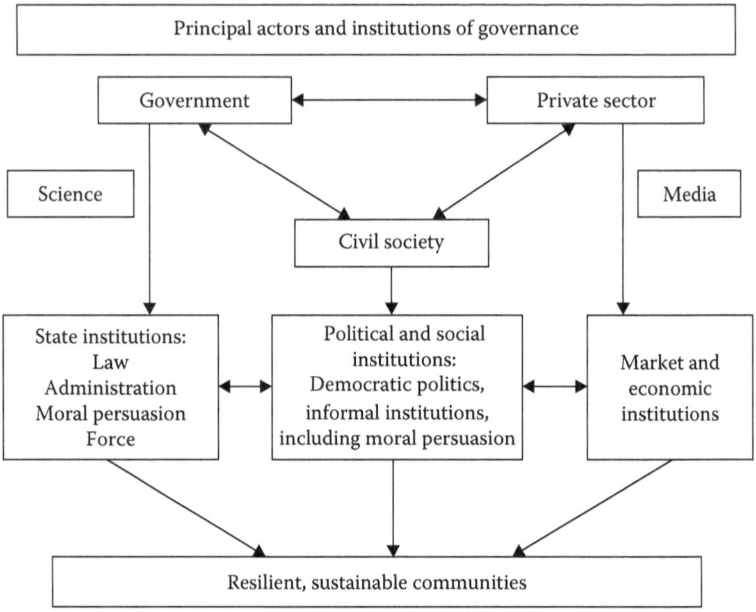

Figure 3.5 Principal governance actors and institutions.

power sharing, and collective action (Kooiman 2003). These principal actors interact through a range of institutions that include both formal (e.g., law and administration) and informal (e.g., norms, mores, and sociocultural *rules*) structures and processes to make social choices that shape adaptation, resilience, and sustainability prospects.

Many ICM initiatives have been started in coastal communities around the world over the past four to five decades, with a wide range locality-specific governance interactions unfolding in different contexts (Burbridge et al. 2012; Sorensen 1997). Notwithstanding diverse approaches, the underlying principles and praxis of ICM as governance is recognized as providing an enabling approach for overcoming adaptation barriers at the coast (Celliers et al. 2013; Falaleeva et al. 2011; Nicholl et al. 2007; Taylor et al. 2013; Tobey et al. 2010; USAID 2009).

ICM and coastal governance as a framework for building adaptive capacity and resilience

Decades of ICM and coastal governance experience have revealed *success factors* or best practices for realizing the goals of ICM and, unsurprisingly, they seek to address the adaptation barriers identified previously and align with and enable the key features of adaptive capacity and resilient communities and social-ecological systems identified earlier (see Table 3.2).

Table 3.2 Institutionalizing ICM *success factors* to overcome adaptation barriers and build adaptive capacity and resilience

Adaptation barriers	*ICM as coastal governance (Glavovic 2008b; Nicholls et al. 2007; Stojanovic et al. 2004)*	*Adaptive capacity*	*Resilient communities*	*Resilient social-ecological systems (see Figure 3.2)*
• *Natural or physical and ecological limits and barriers* • *Human, resource, and informational barriers* • *Cognitive, normative, and institutional barriers* Examples of institutional barriers: • Fragmentation, including single-scale focus; tension between top-down and bottom-up implementation • Rigidity: Lack of openness to learning and variety • Elite capture and strong one-sided reliance on scientific experts • Crisis mentality • Distrust in the problem-solving capacity of civil society • Difficulty of reserving funding for long-term action	• Sustainability-driven, people-centered, and empowering • Participatory • Cooperative • Integrative • Learning-oriented • Long-term perspective • Strategic • Contingent • Incremental • Precautionary • Adaptive	• Pluralism and variety • Autonomous actors • Linkages within and between levels • Information sharing and communication • Social learning • Capacity to change • Shared authority and decision-making	• Values and beliefs • Social and community networks and support • Collective action • Diverse and innovative economy • Knowledge, skills, and learning • People–place connections • Community resources • Community leadership and active agents • Positive outlook and capacity to accept change	• Natural capital • Moral capital • Social capital • Political capital • Physical capital • Economic/financial capital • Human capital

Institutionalizing these ICM success factors will strengthen adaptive capacity and build resilient coastal communities and social-ecological systems. In summary, adaptation prospects are enhanced by institutionalizing coastal governance approaches that make strategic, locality nuanced, and informed decisions based on political will and meaningful public engagement to resolve conflicting interests in pursuit of coastal equity and sustainability; mobilize knowledge across the science–policy–practice domains, integrating expert, local and traditional knowledge, to inform coastal planning and decision-making; establish enabling institutional mechanisms for sustained resourcing for long-term actions, and horizontal and vertical integration to reduce fragmentation, rigidity, duplication, and sectoralism; and promote social learning in the face of complexity and uncertainty through *inter alia* reflexivity, monitoring, evaluation, and stepwise adaptive actions.

Translating ICM and coastal governance best practices into action has proved to be persistently difficult—despite five decades of sustained effort. The ICM governance arena can create an enabling environment for adapting to climate change or, despite laudable intentions, it can hamper and/or delay adaptation or even entrench maladaptation (Celliers et al. 2013). In order to pinpoint and overcome barriers to implementing ICM, Olsen (2002, 2003) and colleagues (Olsen et al. 2009) developed the order of outcomes framework (see Figure 3.6) that recognizes that coastal governance is a long-term undertaking that requires stepwise changes in behavior and institutional reform to overcome prevailing unsustainable path dependencies.

The first order of outcomes has four enabling conditions: (1) agreement among key actors about clear goals for desired coastal outcomes against which progress can be measured; (2) a core group of well-informed and supportive constituencies from government, civil society, and the private sector; (3) adequate institutional capacity to develop and implement a suite of ICM policies and a plan of action; and (4) government commitment to give effect to these policies and provide the necessary authorities and resources for ongoing implementation. The second order of outcomes details actions that signal implementation of the policies and plan of action and demonstrates changes in the behavior of key actors and institutions involved in resource use, regulation, and making investments that affect coastal outcomes. A key factor is sustained financing of ICM implementation. Many donor-funding ICM programs have not been funded for long enough periods of time to transition up to and beyond this stage because it can take many years and possibly a decade or more to make this transition. The third order of outcomes is attained when specified social-ecological goals are being achieved and provides the basis for transitioning to the fourth order of outcomes at which point long-term sustainable coastal development outcomes can be realized.

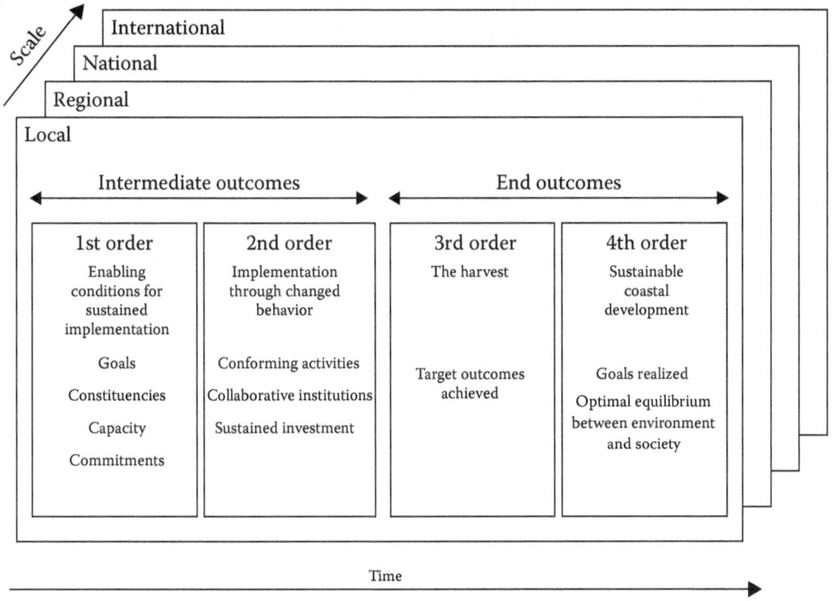

Figure 3.6 ICM order of outcomes. (After Olsen, S.B. et al., *The Analysis of Governance Responses to Ecosystem Change: A Handbook for Assembling a Baseline*, GKSS Research Centre, Geesthacht, Germany, 2009.)

The super-wicked problem of climate change compounds the already difficult challenge of translating ICM best practices into tangible reality and brings to the fore several considerations that deserve more focused attention, notably an even stronger need to prioritize nature-based coastal protection strategies; adopt a very long planning horizon; integrate climate mitigation and adaptation opportunities in the governance process; and address the confounding issue of uncertainty (Tobey et al. 2010). Each of these considerations has been addressed to some degree in past ICM and coastal governance scholarship, but as Tobey et al. (2010) argue, deserve more focused attention in this era of climate change. The case studies in this book reinforce this viewpoint but many bring to the fore the pivotal issues of uncertainty and climate risk, which, compared to the other factors, have received little attention in the ICM and coastal governance literature.

Climate risk at the coast

Prevailing initiatives have yet to resolve the risk problems faced by many coastal communities worldwide (Adger et al. 2003; Gibbs et al. 2013; Hallegatte et al. 2013) or are at best poorly developed (Dale et al. 2013). Risk governance scholars have elucidated the limitations of traditional risk management approaches for dealing with risk-laden wicked problems (Palmer 2012; Renn 2008; Stirling 2010). What can be learned from risk governance scholarship to better understand and address coastal risk in general and climate risk at the coast in particular?

Risk has traditionally been defined as the probability of a hazardous event multiplied by the consequences if the event occurs (after Knight 1921). But climate risk at the coast cannot be reduced to simple *measurable uncertainty* that is amenable to quantification according to probability and consequence metrics. Relying on traditional risk management approaches to deal with such risk can lead to perverse outcomes. In a review of studies on risk and risk governance in the Baltic Sea, Renn et al. observed that

> ... it is a consistent finding that in most of these cases, the risks are treated, assessed and managed as if they were simple. The assessment and management routines in place do not do justice to the nature of such risks. The consequences of this maltreatment ranges from social amplification or irresponsible attenuation of the risk, sustained controversy, deadlocks, legitimacy problems, unintelligible decision-making, trade conflicts, border conflicts, expensive rebound measures, and lock-ins.
>
> Renn et al. 2011: 236

Risk scholars (e.g., Aven and Renn 2009, 2010; Renn 2008; Stirling 2010) distinguish different risk characteristics that stem from lack of knowledge and/or competing knowledge claims about risk problems. Renn and colleagues (e.g., Klinke and Renn 2002, 2010, 2012; Klinke et al. 2006; Renn 2008), for example, distinguish simple risk or measurable uncertainty from complexity (the difficulty of identifying and quantifying causal relationships between potential causal agents and particular observed effects), scientific uncertainty (the difficulty of predicting the occurrence of events and/or their consequences because of inadequate scientific knowledge), and sociopolitical ambiguity (varied legitimate but countervailing views about the same risk phenomena and their circumstances). Most risk problems have elements of these different characteristics. For example, smoking has relatively low complexity and uncertainty, but high ambiguity. Nuclear power has high complexity, low uncertainty, and high ambiguity. Addressing climate risk at the coast involves high levels of complexity, uncertainty, and ambiguity. Understanding the characteristics of different risk problems is necessary to ascertain appropriate risk analysis and intervention approaches.

Renn and colleagues (Klinke and Renn 2012; Renn 2008; Renn et al. 2011; Van Asselt and Renn 2011) have developed an adaptive and integrated risk governance model that situates traditional risk analysis and treatment in contemporary governance settings and is helpful for tackling risk problems that include value and interest conflicts, conflicting evidence claims, and contested views about risk acceptability and tolerability—all of which infuse the coastal climate risk *problématique*. Their model is iterative, inclusive, deliberative, and learning-oriented and takes into account expert, stakeholder and public perspectives, and needs and interests depending on the character of the risk problem(s) under consideration (Klinke and Renn 2012). Given the deep uncertainty that characterizes climate risk, the risk assessment and treatment process needs to be reflexive, integrative, and adaptive over time as understanding of the nature of climate and coupled coastal risks evolves, as risk perceptions and appetites change, and treatment options can be adjusted in light of changing institutional dynamics, informed by structured evaluation and modification, and open to seize *windows of opportunity*. Insights from emerging risk governance scholarship need to be incorporated into climate change adaptation and coastal governance efforts, and this need is beginning to be recognized (e.g., May and Plummer 2011) and is reinforced by recent efforts to bridge hitherto parallel streams of adaptation and disaster risk reduction scholarship (e.g., Field et al. 2012) and natural hazards planning and climate change adaptation (Glavovic and Smith 2014).

Despite convergent scholarship from diverse fields and concerted effort to translate best practice principles into reality at the coast, unsustainable coastal practices are pervasive, climate risk at the coast compounds extant vulnerability, and coming waves of adversity imperil coastal livelihoods and lifestyles. Adaptation plans and actions have proliferated but formidable barriers prevent effective implementation. Prospects appear dismal despite recent advances in risk governance scholarship, emerging insights about conditions for building adaptive capacity and resilience, and over five decades of ICM experience around the world with few initiatives assembling the first order of outcomes let alone transitioning beyond this stage (Burbridge et al. 2012; Glavovic 2008b; Lloyd et al. 2013). How can this impasse be broken?

Breaking the impasse

The pursuit of coastal sustainability has been elusive—long before the prospect and consequences of rapid climate change became readily apparent—even in societies with *good governance* systems founded upon *inter alia* open democracy, efficient markets, a sound legal system, effective administrative systems, and effective compliance with and enforcement of laws that foster the public good. Why this impasse and how might it be broken? Dryzek's (1987) analysis of why prevailing governance institutions (or social choice mechanisms in his terms) typically fail to protect life-sustaining ecosystems provides valuable insights for breaking this impasse. The key finding of Dryzek's analysis is that prevailing social choice mechanisms fail to produce ecologically rational outcomes and environmental problems are displaced from one social choice arena to another without effective resolution (see Figure 3.7).

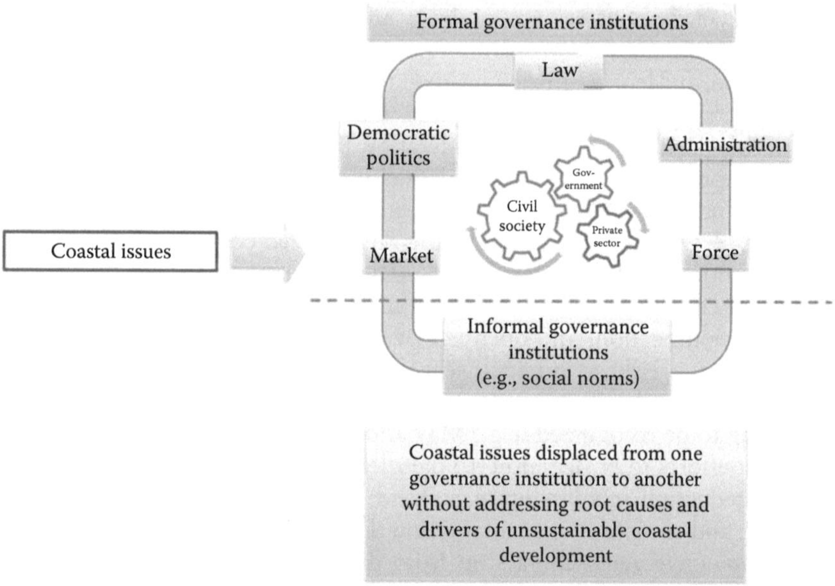

Figure 3.7 Displacement of coastal issues.

Market price signals fail to accord coastal ecosystem services their full societal value. To correct market failure, coastal issues are *shifted* to other institutional arenas. Administrative systems work well when dealing with familiar, routine, and static problems. But they do not deal well with environmental problems characterized by high levels of complexity and deep uncertainty as is typical of coastal issues in this era of climate change. Coastal problems are displaced into the political arena where there is the potential for democratic politics to resolve conflicting interests. However, this potential is difficult to realize in practice because current political structures and processes tend to privilege short-term private interests over long-term public safety, social equity, and ecological sustainability. Legal systems can help to address shortcomings in the market, administrative, and political arenas. But the law responds slowly and tends to be rigid rather than resilient. Furthermore, it is impossible to codify *climate complexity* at the coast, and effective enforcement of *good laws* is difficult to achieve *inter alia* because of the trans-boundary character of environmental problems. Informal social choice mechanisms, such as moral persuasion, offer potential but seem unlikely to yield ecologically rational outcomes without resorting to repressive enforcement. The use of force (e.g., imprisonment of those found guilty of illegal coastal activities), bargaining, and international law can be effective but have failed to stem pervasive unsustainable practices. In short, ecologically rational outcomes are thwarted by prevailing economic, social, legal, and political rationality. Social goals such as equity and poverty eradication are similarly thwarted. Consequently, contemporary governance for coastal sustainability is at an impasse. Coastal issues are locked into a vortex of intractable conflict, with problems displaced from one social choice arena to another without securing outcomes in the long-term public interest. How might the rationality that *locks* in path-dependent unsustainable and maladaptive practices be transmogrified and institutions reformed and if necessary transformed to overcome adaptation barriers and foster adaptive capacity and resilience?

The starting point is to explore the interrelationships between the concepts discussed in this chapter that together frame the climate change adaptation imperative at the coast: trajectories of coastal development, vulnerability, and risk; adaptive capacity, strategies, and pathways; resilience; and adaptive and integrative governance of risk and coastal social-ecological systems. Then, attention can be focused on points of entry and processes to reframe the rationality that underlies the prevailing impasse. These interrelationships are portrayed in Figure 3.8, which draws on insights from *inter alia* development studies, especially the sustainable livelihoods approach initially mooted by the WCED (1987), developed by Chambers and Conway (1991), and applied in diverse settings, including the pursuit of coastal sustainability (e.g., Glavovic and Boonzaaier 2007). Figure 3.8 portrays the vulnerability context of coastal communities; depicts governance interactions between coastal actors as being mediated by formal and informal institutions that enable or hinder access to coastal assets or resources; and shapes the choice of livelihood and adaptation strategies and pathways that together determine coastal livelihood outcomes (e.g., reduced risk, better income), which can be symbolized by layers of resilience.

Coastal communities are situated in a vulnerability context—exposed to shocks and stresses by virtue of their location and their predisposition to suffer harm. Climate risk is one dimension of coastal risk. People living at the coast pursue various livelihood strategies depending in part on available coastal resources or assets and their entitlements to these resources. Chambers and Conway define a *livelihood* as

> … the capabilities, assets (stores, resources, claims and access) and activities required for a means of living: a livelihood is sustainable which can cope with and recover from stress and shocks, maintain or enhance its capabilities and assets, and provide sustainable livelihood opportunities for the next generation; and which contributes net benefits to other livelihoods at the local and global levels in the long and short term.
>
> Chambers and Conway 1991: 6

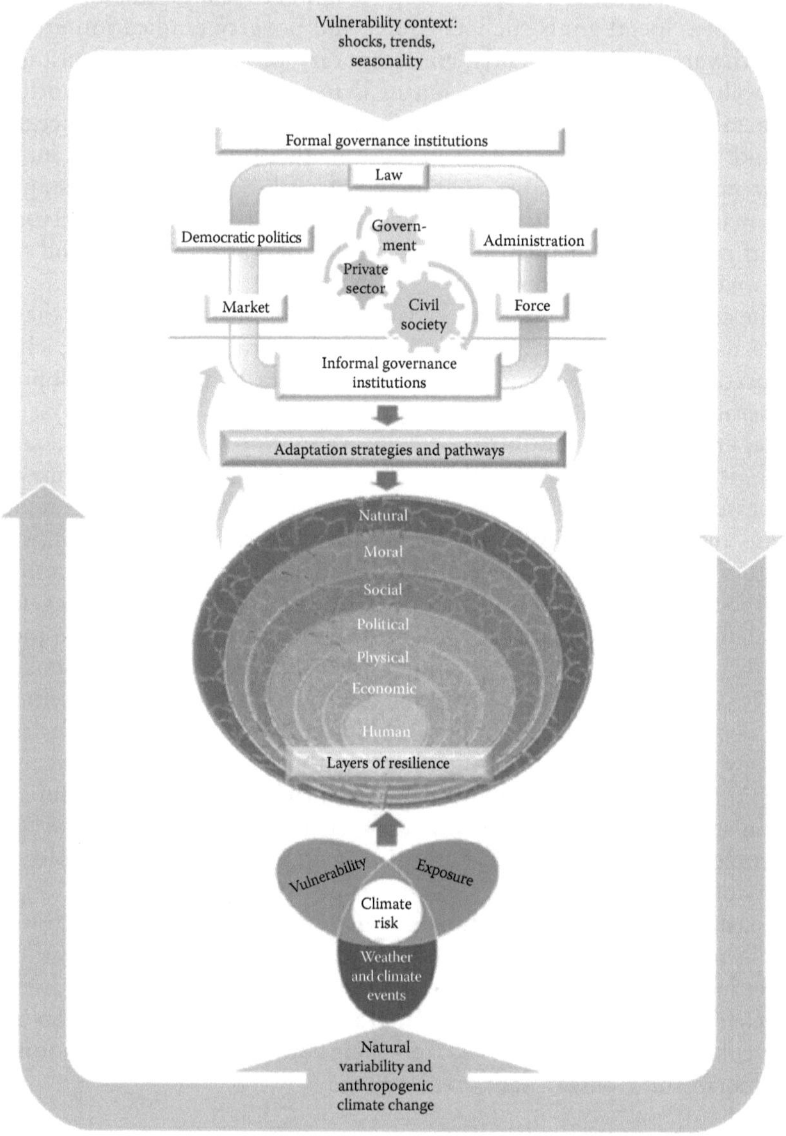

Figure 3.8 Vulnerability, adaptive capacity, resilience, and climate risk.

Coastal assets can be thought of as different forms of capital (see Figure 3.2). Access to these assets is determined by mediating governance structures and processes that include both formal and informal institutions. The resultant transactions and interactions between governance actors and institutions not only structures resource access and use but also determines livelihood outcomes, including coastal livelihood equity and sustainability. The choice of livelihood strategy, including adaptation strategies, is shaped by entitlements and capabilities and results in different strategies being pursued by different individuals, groups, and communities—from fishing to tourism, migration, and so on. Adaptive capacity and livelihood strategies determine adaptation pathways and the extent to which communities mobilize and capitalize coastal assets as layers of resilience. Climate change thus profoundly affects coastal livelihood options through direct and indirect impacts on coastal ecosystems, livelihoods, and lifestyles (see Table 3.1) and climate and nonclimate risks and, over time, the feasibility of different livelihood strategies, and consequently livelihood outcomes.

This portrayal accentuates the pivotal mediating role of governance institutions in enabling or disenabling adaptation and resilience. Prevailing institutional rationality locks in unsustainable and inequitable pathways. There is no panacea for transitioning away from this path dependency but it is nonetheless crucial. How can communities chart a safe passage through the stormy seas of the Anthropocene in the face of the deep uncertainty and super-wicked problem of climate change?

The disparate streams of scholarship and practice explored in this chapter reveal a powerful convergence on the fundamental requirement necessary to make this transition. ICM and coastal governance scholars and practitioners concur: This transition needs to be founded upon the authentic engagement of coastal stakeholders and communities in reflexive (or self-critical learning) deliberation (Glavovic 2013c; Moser et al. 2012; Olsen et al. 2009), informed not only by the predominant technical problem-solving discourse but founded on ethical and deep-rooted emotional, spiritual, and relational perspectives on and connections to the coast and societal choices about its future (Hofmeester et al. 2012). Risk governance scholars corroborate this view: Addressing wicked risk problems necessitates inclusive and deliberative assessment and management processes that can integrate different kinds of knowledge and understanding, including scientific and nonscientific traditions, and the underpinning normative judgments that buttress the moral and ethical justification of different pathways (e.g., Palmer 2012; Renn 2008). Resilience and adaptation scholars agree

> … [D]iverse and contested values—underpinned by ethical, cultural, risk and knowledge considerations—underlie adaptation responses and thus define mutable and subjective limits to adaptation. Given diverse values of diverse actors, there is, we believe, a compelling need to identify and recognize implicit and hidden values and interests in advance of purposeful adaptation interventions. As a consequence, we suggest that there is a requirement for governance mechanisms that

> can meaningfully acknowledge and negotiate the complexity arising from the manifestation of diverse values—for example, deliberative platforms for adaptive action involving wide sets of stakeholders.
>
> Adger et al. 2009b: 350

This convergence of thinking is bolstered by environmental and sustainability governance scholarship (e.g., Duit et al. 2010; Lemos and Agrawal 2006). Innovative modalities of deliberative coastal governance need to be envisaged and institutionalized—with adaptation an integral part thereof—to break the prevailing impasse. The next section outlines a conceptual framework that dovetails insights from the foregoing scholarship but deepens and extends this work by drawing on deliberative democracy scholarship. It locates adaptation in the context of a deliberative governance approach that enables coastal communities to understand and address climate risk, chart adaptation pathways, and build resilience in the face of the contestation, complexity, uncertainty, ambiguity, and surprise that epitomize life on the frontline in the Anthropocene.

CHARTING A SAFE PASSAGE IN STORMY SEAS: DELIBERATIVE COASTAL GOVERNANCE

What is deliberation and why is it foundational for conceptualizing and institutionalizing coastal governance approaches that build adaptive capacity and resilience in the Anthropocene? Deliberation is a noncoercive process of communication that stimulates reflection on societal values, preferences, and interests (Dryzek 2000). It is founded on participant exchange through information sharing, discussion, and debate to stimulate considered and well-informed views about matters of collective concern that evolve with social learning (Chambers 2003). Deliberation can take place in formal and informal public settings but needs to be authentic, inclusive, and consequential if it is to have a legitimate impact on social choices (Dryzek 2009). It is therefore more than public discussion and is foundational for opening up opportunities for transformative change that advances sustainability (Baber and Bartlett 2005; Dryzek 2011; Fischer 2000; Goodin and Dryzek 2006; Gupte and Bartlett 2007). In practice, the challenge is to create multiple interconnected and overlapping arenas of public deliberation wherein the ideas and viewpoints of diverse governance actors are exchanged, debated, and contested in a noncoercive manner. Such deliberative discourse goes beyond *rational arguments* to include other modalities of communication, such as collaborative processes, storytelling, contestation, and dissent. Ultimately, deliberation enables reflexive social learning to build robust knowledge and understanding that is crucial for adapting to a changing climate (Hegger et al. 2012).

Achieving this deliberative ideal is far from simple or straightforward. There is a well-established scholarship that elucidates the barriers to deliberative, communicative, or participatory decision-making in the public realm (e.g., Cooke and Kathari 2001; Hickey and Mohan 2004; Hoppe 2011; King et al. 1998). Such barriers are likely to be especially pronounced when addressing climate risk and the adaptation imperative (Few et al. 2007). Deliberative processes might open up opportunities for integrative and collaborative governance but can be truncated by trenchant institutional resistance and opposition (Darbas 2008). Deliberative governance is thus a contested endeavor. Extensive scholarship (e.g., Baber and Bartlett 2005; Dryzek 2011; Fischer 2000; Goodin and Dryzek 2006; Gupte and Bartlett 2007) and growing empirical evidence (e.g., Menzel and Buchecker 2013; Newig and Fritsch 2009), nevertheless, show that such processes can build human, social, and political capital that enables resolution of novel and complex problems—albeit costly in terms of time and resource investment—in ways that cannot otherwise be achieved. How might a deliberative approach to coastal governance be envisaged and institutionalized?

Figure 3.9 presents a conceptual framework to build adaptive capacity, resilience, and sustainability through a deliberative practice of coastal governance that is founded on four deliberative or *process outcomes*. This framework builds upon and extends a framework by Glavovic (2013c) by making explicit the issue of climate risk and the articulation of adaptation strategies and pathways. The process outcomes are foundational for the *coastal outcomes* described by Olsen and colleagues (see Figure 3.6). The order of outcomes progress sequentially and organically with inherent feedback—not in a linear and inexorable manner. Figure 3.9 is therefore presented as an elliptical cycle rather than as a sequence of *blocks* as previously portrayed. This portrayal emphasizes the iterative and ongoing nature of the deliberative governance process and recognizes that there are no *silver bullet* solutions or fixed *end outcomes*. Rather, interim strategies, temporary solutions, and stepwise adaptive pathways need to be agreed upon and pursued in the face of systemic complexity and what is *known, unknown,* and *unknowable* about climate risk. The process outcomes that underpin the coastal outcomes were defined by Glavovic (2013c) on the basis of a review of the deliberative democracy literature and, in particular, the work of Dryzek (1990, 2000, 2009, 2011), Fischer (2000, 2003, 2006), Fung (2003, 2004, 2006), Fung and Wright (2001, 2003), and the adaptation of a framework by Carcasson (2009). The point of departure is building human, social, and political capital through deliberative issue framing and learning, and enhanced democratic attitudes and skills. These outcomes facilitate a transition to the second-order process outcomes of community-oriented action and an institutional culture that facilitates behavioral change aligned with sustainability goals. Community problem-solving capacity is strengthened in the third-order process outcome. The ultimate

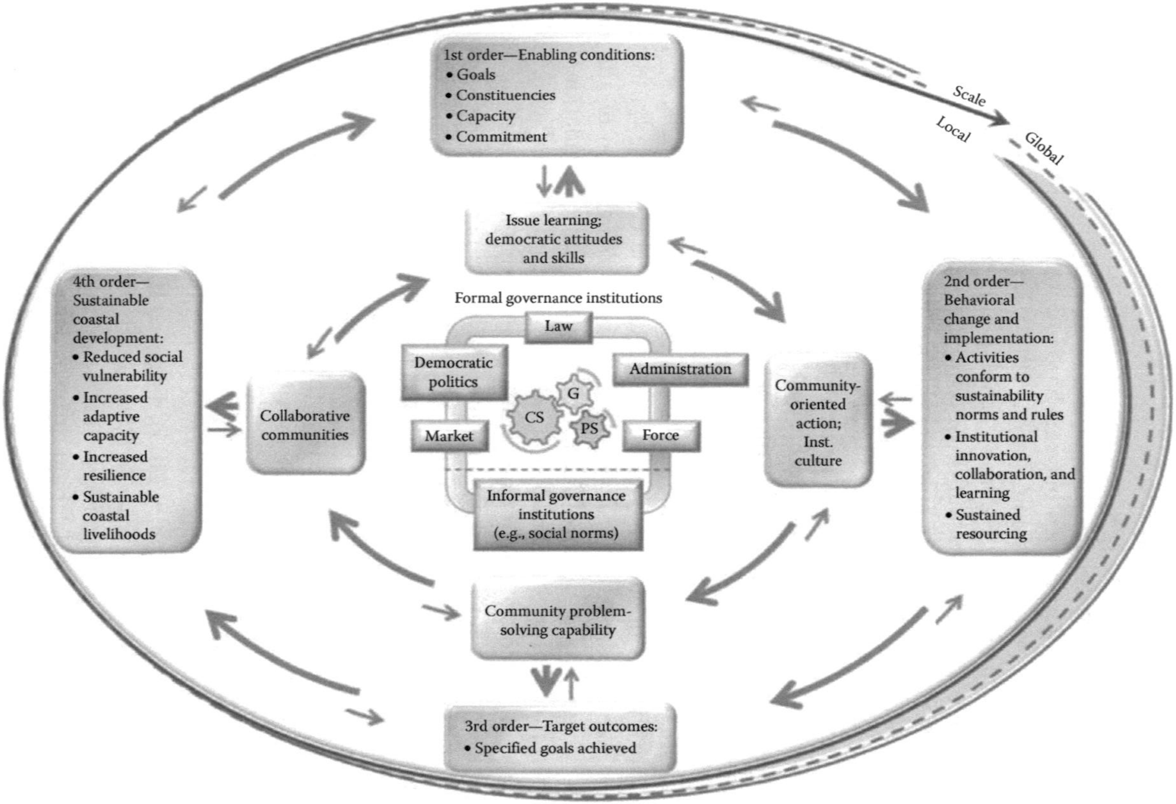

Figure 3.9 Deliberative coastal governance.

process outcome is to build collaborative coastal communities that are enabled to reduce climate risk, chart adaptive pathways, and build layers of resilience in the face of climate and global change. This fourth order of outcomes recognizes that coastal sustainability is not an *optimal equilibrium end state* but a normative position that assumes locality-specific and evolving reality in different social-ecological settings (Glavovic 2013c).

The first-order process outcomes are issue framing and learning, and improved democratic attitudes and skills. Deliberation can help coastal stakeholders and communities transcend the limitations of traditional disciplinary science in understanding complex coastal issues and adaptation challenges. It can help to engage participants in a meaningful process of mutual framing and learning about coastal problems, including climate risk and plausible adaptation strategies and pathways. It can thus help to align science, policy, and practice, including nonscientific understanding. The distinction between complexity, scientific uncertainty, and ambiguity is especially relevant in this context as it highlights the need for expert, stakeholder, and public engagement in risk analysis and management to be contingent on the nature of the issue and associated risk problem under consideration. Deliberation has the potential to unlock *public spiritedness* that can help to deepen and expand awareness, appreciation, and tolerance of divergent perspectives. It can improve skills in communication and judgment that enhance group interactions and decision-making. This first-order stage is the starting point of an ongoing adaptive deliberative process, which recognizes that a range of inherent uncertainties and risk problems need to be considered, short-term targets need to be framed by longer-term community and societal goals, and commitments need to be made to taking short- to medium-term actions while keeping options open for alternative future pathways and to ongoing adjustment when changing course is warranted (e.g., Walker et al. 2013). These first-order process outcomes thus establish a social learning and democratic foundation for realizing the enabling conditions of the coastal order of outcomes: goals, constituencies, capacity, and commitment.

The second-order process outcomes that are fostered by deliberation are community-oriented action and an evolving institutional culture that enables better decision-making in the context of complexity and uncertainty. Transitioning toward this outcome is a perpetual challenge in the face of rampant individualism and alienation from public decision-making processes that make it so difficult to sustain community-oriented action. Deliberation can help stakeholders and community members develop a shared purpose, reconcile private and community interests, and encourage involvement in community activities. People with divergent and contending interests can become part of a community-wide process of communication and shared learning that helps to transcend individualism, resolve conflict, and enhance the legitimacy of community decisions. Such outcomes need to be institutionalized to be effective in reconciling contending interests. These process outcomes are thus necessary for transitioning to and beyond the second-order coastal outcome of implementation through behavioral change and building an institutional culture that fosters innovation, learning, and collaboration that is vital for resolving coastal problems rather than merely displacing them from one institutional arena to another. Albeit challenging, the transition from first to second order of outcomes has been achieved through deliberative processes in diverse coastal contexts—from Republic of South Africa (Glavovic 2008b) to community based post-Katrina recovery efforts (Nelson et al. 2007; Wilson 2009) and in many of the following case studies (Figure 3.10). Progressing beyond this order of outcome remains challenging.

Community problem-solving capacity, the third process outcome, can be bolstered through deliberative processes because coastal issues and risks are reframed and better understood, institutional capacity and decision-making is strengthened, and community members recognize and take on roles and responsibilities that help to improve community relationships. Communicative interactions can foster empathy and tolerance of differences that help to facilitate sustained deliberation, but solving community problems, especially the complex and ambiguous intricacies of adaptation,

Figure 3.10 Louisiana Speaks: Post-Katrina recovery planning workshop, New Orleans, Louisiana, July 2006. (Courtesy of Bruce C. Glavovic.)

is far from straightforward. The institutional rationality that is locked into unsustainable path dependencies can be challenged and redefined over time by deliberative processes that focus collective attention on and enable achievement of targeted community goals. Deliberation thus enhances intracommunity problem-solving capacity and is essential for realizing third-order coastal outcomes.

Collaborative communities are the fourth-order process outcome. They can sustain and are sustained by intra- and intercommunity deliberative processes that are meaningful, inclusive, and enable institutionalization of informal and formal structures and processes that foster cross-scalar and multilevel deliberation. Both intra- and intercommunity deliberation are necessary to build layers of resilience and make headway toward the elusive fourth-order coastal outcome of sustainable coastal development.

This conceptual framework highlights foundational deliberative or process outcomes that underpin the coastal order of outcomes that together facilitate progressive changes in the behavior of coastal actors and networks as they build the cognitive, democratic, and sociopolitical capacity to reflect on and address the fractious social choices that prevail on the frontline in the Anthropocene. Institutionalizing this approach necessitates a sea change in thinking and praxis. Charting adaptation pathways that foster resilience and sustainability cannot be achieved through business as usual or incremental change. Transformative innovations in coastal governance are necessary to build layers of resilience in the face of waves of adversity. All key governance actors need to be actively involved in innovative modalities of deliberative engagement that are appropriate for addressing the complexity, uncertainty, and ambiguity that surrounds climate change and wider coastal sustainability imperatives. Government, with local government supported by higher levels of government, plays a pivotal role in catalyzing transformative change, but important roles are also played by nongovernmental organizations, community-based organizations, and other scientific, civil society, and private sector actors and networks in opening up opportunities to build adaptive capacity and more resilient, sustainable communities.

CONCLUSION

Coastal communities live on the frontline of the global struggle for sustainability, facing escalating coastal disaster risk driven in part by climate change. The experience of many coastal communities demonstrates that the pursuit of sustainability presents obdurate governance challenges even in societies touted to have good governance systems. Coastal issues are locked in a quagmire of intractable conflict, with problems displaced from one social choice arena to another without securing equitable and sustainable outcomes, a situation that is compounded by the super-wicked climate change problem. How might the rationality that locks in unsustainable path dependencies be transmogrified to overcome adaptation barriers and nurture adaptive capacity and resilience in the Anthropocene?

The starting point is to recognize that coastal communities are located in a socially constructed vulnerability context that includes synergistic climate and nonclimate risks. Access to coastal resources is mediated by formal and informal governance institutions that regulate the interactions between and entitlements of coastal actors, and consequently the choice of livelihood and adaptation strategies and pathways that together, yield livelihood outcomes. Symbolized as layers of resilience, these outcomes determine the extent to which coastal communities will be buffered against coming waves of adversity. A review of ICM, risk governance and adaptation and resilience scholarship reveals convergent thinking about the foundational requirement for breaking the current impasse: *Coastal stakeholders and communities need to be engaged in an authentic, inclusive, and consequential process of deliberative governance to navigate the stormy seas of the Anthropocene.* A conceptual framework for adaptive coastal governance, founded on four deliberative outcomes, is outlined. The framework takes into account the multidimensional nature of climate risk at the coast and the need to chart interim strategies, temporary solutions, and stepwise adaptation pathways that are attuned to locality-specific histories and social-ecological characteristics. The point of departure or first-order process outcome is to build human, social, and political capital through deliberative issue framing and learning, and improved democratic attitudes and skills. The second-order process outcomes include community-oriented action and an evolving institutional culture that facilitates behavioral change aligned with coastal sustainability goals. The third-order process outcome is effective community problem-solving capabilities. The fourth-order process outcome is collaborative coastal communities that can reduce climate risk, chart adaptive pathways, and build resilient, sustainable livelihoods in the face of the contestation, complexity, uncertainty, ambiguity, and surprise that epitomize life in the Anthropocene.

ACKNOWLEDGMENTS

I thank my coeditors for their constructive feedback on initial drafts of this chapter. I also thank Paul Schneider for preparing the figures for this chapter. I gratefully acknowledge the financial support of the New Zealand Earthquake Commission that enabled me to undertake the research that underpins this chapter. The views expressed here reflect solely those of the author.

REFERENCES

Abuodha, P.A. and Woodroffe, C.D. (2006). *International Assessments of the Vulnerability of the Coastal Zone to Climate Change, Including an Australian Perspective*. Canberra: Australian Greenhouse Office.

Adger, W.N. (2003) "Social capital, collective action, and adaptation to climate change," *Economic Geography*, 79(4): 387–404.

Adger, W.N., Agrawala, S., Mirza, M.M.Q., Conde, C., O'Brien, K., Pulhin, J., Pulwarty, R., Smith, B., and Takahashi, K. (2007). "Assessment of adaptation practices, options, constraints and capacity," in M.L. Parry, O.F. Canziani, J.P. Palutikof, P.J. van der Linden, and C.E. Hanson (eds.), *Climate Change 2007: Impacts, Adaptation and Vulnerability. Contribution of Working Group II to the Fourth Assessment Report of the Intergovernmental Panel on Climate Change*, Cambridge: Cambridge University Press. http://www.ipcc.ch/ (accessed August 2013).

Adger, W.N., Arnell, N.W., and Tompkins, E.L. (2005a). "Successful adaptation to climate change across scales," *Global Environmental Change*, 15: 77–86.

Adger, W.N., Brown, K., Nelson, D.R., Berkes, F., Eakin, H., Folke, C., Galvin, K. et al. (2011). "Resilience implications of policy responses to climate change," *WIREs Climate Change*, 2: 757–66.

Adger, W.N., Dessai, S., Goulden, M., Hulme, M., Lorenzoni, I., Nelson, D.R., Naess, L.O., Wolf, J., and Wreford, A. (2009a). "Are there social limits to adaptation to climate change?" *Climatic Change*, 93: 335–54.

Adger, W.N., Hughes, T.P., Folke, C., Carpenter, S.R., and Rockström, J. (2005b). "Social-ecological resilience to coastal disasters," *Science*, 309(5737): 1036–9.

Adger, W.N., Huq, S., Brown, K., Conway, D., and Hulme, M. (2003). "Adaptation to climate change in the developing world," *Progress in Development Studies*, 3(3): 179–95.

Adger, W.N., Lorenzoni, I., and O'Brien, K.L. (2009b). *Adapting to Climate Change: Thresholds, Values, Governance*. Cambridge: Cambridge University Press.

Adger, W.N., Paavola, J., Huq, S., and Mace, M.J. (eds.) (2006). *Fairness in Adaptation to Climate Change*, Cambridge, MA: MIT Press.

Amundsen, H. (2012). "Illusions of resilience? An analysis of community responses to change in northern Norway," *Ecology and Society*, 17(4): 46.

Aven, T. and Renn, O. (2009). "The role of quantitative risk assessments for characterising risk and uncertainty and delineating appropriate risk management options, with special emphasis on terrorism," *Risk Analysis*, 29(4): 587–600.

Aven, T. and Renn, O. (2010). *Risk Management and Governance*. Heidelberg, Germany; New York: Springer.

Baber, W.F. and Bartlett, R.V. (2005). *Deliberative Environmental Politics: Democracy and Ecological Rationality*. Cambridge, MA: MIT Press.

Béné, C., Godfrey-Wood, R., Newsham, A., and Davies, M. (2012). Resilience: New utopia or new tyranny? Reflection about the potentials and limits of the concept of resilience in relation to vulnerability-reduction programmes, IDS Working Paper 405, CSP Working Paper Number 006, p. 16.

Berkes, F. and Folke, C. (eds.) (1998). *Linking Social and Ecological Systems: Management Practices and Social Mechanisms for Building Resilience*. Cambridge: Cambridge University Press.

Berkes, F. and Ross, H. (2013). "Community resilience: Toward an integrated approach," *Society & Natural Resources*, 26: 1, 5–20.

Berrang-Ford, L., Ford, J.D., and Paterson, J. (2011). "Are we adapting to climate change?," *Global Environmental Change*, 21: 25–33.

Bierbaum, R., Smith, J.B., Lee, A., Blair, M., Carter, L., Chapin, III, F.S., Fleming, P., Ruffo, S., Stults, M., McNeeley, S., Wasley, E., and Verduzco, L. (2013). "A comprehensive review of climate adaptation in the United States: More than before, but less than needed," *Mitigation and Adaptation Strategies for Global Change*, 18: 361–406.

Boyd, E. and Juhola, S. (2009). "Stepping up to the climate change: Opportunities in re-conceptualising development futures," *Journal of International Development*, 21: 792–804.

Brooks, N., Adger, W.N., and Kelly, P.M. (2005). "The determinants of vulnerability and adaptive capacity at the national level and the implications for adaptation," *Global Environmental Change—Human and Policy Dimensions*, 15(2): 151–63.

Brown, H., Nkem, J., Sonwa, D., and Bele, Y. (2010). "Institutional adaptive capacity and climate change response in the Congo Basin forests of Cameroon," *Mitigation and Adaptation Strategies for Global Change*, 15(3): 263–82.

Brown, K. and Westaway, E. (2011). "Agency, capacity, and resilience to environmental change: Lessons from human development, well-being, and disasters," *Annual Review of Environment and Resources*, 36: 321–42.

Buikstra, E., Ross, J., King, C.A., Baker, P.G., Hegney, D., McLachlan, K., and Rogers-Clark, C. (2010). "The components of resilience—Perceptions of an Australian rural community," *Journal of Community Psychology*, 38(8): 975–91.

Burbridge, P.R., Glavovic, B.C., and Olsen, S.B. (2012). "Practitioner reflections on integrated coastal management experience in Europe, South Africa, and Ecuador," in H. Kremer and J. Pinckney (eds.), *Management of Estuaries and Coasts*, Waltham, MA: Academic Press.

Burch, S. (2010). "Transforming barriers into enablers of action on climate change: Insights from three municipal case studies in British Columbia, Canada," *Global Environmental Change*, 20: 287–97.

Cannon, T. and Müller-Mahn, D. (2010). "Vulnerability, resilience and development discourses in context of climate change," *Natural Hazards*, 55(3): 621–35.

Carcasson, M. (2009). *Beginning with the End in Mind: A Call for Goal-Driven Deliberative Practice*. New York: Centre for Advances in Public Engagement.

Carpenter, S.R. and Brock, W.A. (2008). "Adaptive capacity and traps," *Ecology and Society* 13(2): 40.

Carpenter, S.R. and Gunderson, L.H. (2001). "Coping with collapse: Ecological and social dynamics in ecosystem management," *BioScience*, 51: 451–7.

Carter, T.R., Parry, M.L., Nishioka, S., and Harasawa, H. (eds.) (1994). *Technical Guidelines for Assessing Climate Change Impacts and Adaptations*. Report of Working Group II of the Intergovernmental Panel on Climate Change. London; Tsukuba, Japan: University College London; Centre for Global Environmental Research.

Celliers, L., Rosendo, S., Coetzee, I., and Daniels, G. (2013). "Pathways of integrated coastal management from national policy to local implementation: Enabling climate change adaptation," *Marine Policy*, 39: 72–86.

Chambers, R. and Conway, G. (1991). *Sustainable Rural Livelihoods: Practical Concepts for the 21st Century*. Brighton: IDS.
Chambers, S. (2003) "Deliberative democratic theory," *Annual Review of Political Science*, 6: 307–26.
Cicin-Sain, B. and Knecht, R.W. (1998). *Integrated Coastal and Ocean Management: Concepts and Practices*. Washington, DC: Island Press.
Cooke, B. and Kathari, U. (eds.) (2001). *Participation: The New Tyranny?* New York: Zed Books.
Crossland, C.J., Baird, D., Ducrotoy, J.-P., Lindeboom, H., Buddemeier, R.W., Dennison, W.C., Maxwell, B.A., Smith, S.V., and Swaney, D.P. (2005). "The coastal zone—A domain of global interactions," in C.J. Crossland, H.H. Kremer, H.J. Lindeboom, J.I.M. Crossland, and M.D.A. Le Tissier (eds.), *Coastal Fluxes in the Anthropocene: The Land-Ocean Interactions in the Coastal Zone Project of the International Geosphere-Biosphere Programme*, Berlin, Germany: Springer.
Crutzen, P.J. (2002). "Geology of mankind," *Nature*, 415: 23.
Crutzen, P.J. and Stoermer, E.F. (2000). "The Anthropocene," *Global Change Newsletter*, 41: 17–18.
Dale, A., Vella, K., Pressey, R.L., Brodie, J., Yorkston, H., and Potts, R. (2013). "A method for risk analysis across governance systems: A Great Barrier Reef case study," *Environmental Research Letters*, 8: 1–16.
Darbas, T. (2008). "Reflexive governance of urban catchments: A case of deliberative truncation," *Environment and Planning A*, 40: 1454–69.
Dewey, J. (1927). *The Public and Its Problems*. New York: Holt & Co.
Diamond, J.M. (2005). *Collapse: How Societies Choose to Fail or Succeed*. New York: Penguin.
Doody, J.P. (2004). "'Coastal squeeze'—An historical perspective," *Journal of Coastal Conservation*, 10(1/2): 129–38.
Dronkers, J., Gilbert, J.T.E., Butler, L.W., Carey, J.J., Campbell, J., James, E., McKenzie, C., Misdorp, R., Quin, N., Ries, K.L., Schroder, P.C., Spradley, J.R., Titus, J.G., Vallianos, L., and von Dadelszen, J. (eds.) (1990). *Strategies for Adaption to Sea Level Rise. Report of the IPCC Coastal Zone Management Subgroup: Intergovernmental Panel on Climate Change*. Geneva, Switzerland: Intergovernmental Panel on Climate Change.
Dryzek, J.S. (1987). *Rational Ecology: Environment and Political Economy*. New York: Basil Blackwell.
Dryzek, J.S. (1990). *Discursive Democracy: Politics, Policy, and Political Science*. New York: Cambridge University Press.
Dryzek, J.S. (2000). *Deliberative Democracy and Beyond: Liberals, Critics, Contestations*. Oxford: Oxford University Press.
Dryzek, J.S. (2009). "Democratization as deliberative capacity building," *Comparative Political Studies*. 42(11): 1379–402.
Dryzek, J.S. (2011). *Foundations and Frontiers of Deliberative Governance*. Oxford: Oxford University Press.
Dugmore, A.J., Keller, C., and McGovern, T.H. (2007). "Norse Greenland settlement: Reflections on climate change, trade, and the contrasting fates or human settlements in the North Atlantic islands," *Arctic Anthropology*, 44(1): 12–36.
Duit, A., Galaza, V., Eckerberga, K., and Ebbesson, J. (2010). "Introduction: Governance, complexity, and resilience," *Global Environmental Change*, 20: 363–8.

Eakin, H. and Lemos, M.C. (2006). "Adaptation and the state: Latin America and the challenge of capacity-building under globalization," *Global Environmental Change—Human and Policy Dimensions*, 16(1): 7–18.

Engle, N.L. (2011). "Adaptive capacity and its assessment," *Global Environmental Change*, 21: 647–56.

Engle, N.L. and Lemos, M.C. (2010). "Unpacking governance: Building adaptive capacity to climate change of river basins in Brazil," *Global Environmental Change*, 20: 4–13.

Falaleeva, M., O'Mahony, C., Gray, S., Desmond, M., Gault, J., and Cummins, V. (2011). "Towards climate adaptation and coastal governance in Ireland: Integrated architecture for effective management?," *Marine Policy*, 35: 784–93.

Few, R., Brown, K., and Tompkins, E.L. (2007). "Public participation and climate change adaptation: Avoiding the illusion of inclusion," *Climate Policy*, 7(1): 46–59.

Field, C.B., Barros, V., Stocker, T.F., Qin, D., Dokken, D.J., Ebi, K.L., Mastrandrea, M.D.K., Mach, J., Plattner, G.-K., Allen, S.K., Tignor, M., and Midgley, P.M. (eds.) (2012). *Managing the Risks of Extreme Events and Disasters to Advance Climate Change Adaptation. A Special Report of Working Groups I and II of the Intergovernmental Panel on Climate Change*. Cambridge; New York: Cambridge University Press.

Fischer, F. (2000). *Citizens, Experts, and the Environment: The Politics of Local Knowledge*. Durham, NC: Duke University Press.

Fischer, F. (2003). *Reframing Public Policy: Discursive Politics and Deliberative Practices*. Oxford: Oxford University Press.

Fischer, F. (2006). "Participatory governance as deliberative empowerment. The cultural politics of discursive space," *American Review of Public Administration*, 36(1): 19–40.

Folke, C. (2006). "Resilience: The emergence of a perspective for social–ecological systems analyses," *Global Environmental Change*, 16: 253–67.

Folke, C., Carpenter, S.R., Walker, B., Scheffer, M., Chapin, T., and Rockström, J. (2010). "Resilience thinking: Integrating resilience, adaptability and transformability," *Ecology and Society*, 15(4): 20.

Ford, J.D., Berrang-Ford, L., and Paterson, J. (2011). "A systematic review of observed climate change adaptation in developed nations," *Climatic Change*, 106: 327–36.

Fung, A. (2003). "Recipes for public spheres: Eight institutional design choices and their consequences," *Journal of Political Philosophy*, 11: 338–67.

Fung, A. (2004) *Empowered Participation: Reinventing Urban Democracy*. Princeton, NJ: Princeton University Press.

Fung, A. (2006) "Varieties of participation in democratic governance," *Public Administration Review*, 66(Suppl 1): 66–75.

Fung, A. and Wright, E.O. (2001). "Deepening democracy: Innovations in empowered local governance," *Politics and Society*, 29(1): 5–41.

Fung, A. and Wright, E.O. (eds.) (2003). *Deepening Democracy: Institutional Innovations in Empowered Participatory Governance*. New York: Verso.

Füssel, H.-M. (2007). "Adaptation planning for climate change: Concepts, assessment approaches, and key lessons," *Sustainability Science*, 2: 265–75.

Gallopín, G.C. (2006). "Linkages between vulnerability, resilience, and adaptive capacity," *Global Environmental Change*, 16(3): 293–303.
GESAMP. (1996). *The Contributions of Science to Integrated Coastal Management*. Rome, Italy: Food and Agriculture Organization of the United Nations.
Gibbs, M.T., Thébaud, O., and Lorenz, D. (2013). "A risk model to describe the behaviours of actors in the houses falling into the sea problem," *Ocean and Coastal Management*, 80: 73–9.
Glavovic, B.C. (2008a) "Sustainable coastal communities in the age of coastal storms: reconceptualising coastal planning and ICM as 'new' naval architecture," *Journal of Coastal Conservation*, 12(3): 125–34.
Glavovic, B.C. (2008b). "Sustainable coastal development in South Africa: Bridging the chasm between rhetoric and reality," in R. Krishnamoorthy, B.C. Glavovic, A. Kannen, D.R. Green, A. Ramanathan, Z. Han, S. Tinti, and T.S. Agardy (eds.), *Integrated Coastal Zone Management: The Global Challenge*, Singapore: Research Publishing Services.
Glavovic, B.C. (2013a). "The coastal innovation paradox," *Sustainability*, 5: 912–33.
Glavovic, B.C. (2013b). "Disasters and the continental shelf: Exploring new frontiers of risk," in M.H. Nordquist, J.N. Moore, A. Chircop, and R. Long (eds.), *The Regulation of Continental Shelf Development: Rethinking International Standards*. Leiden, the Netherlands: Martinus Nijhoff Publishers.
Glavovic, B.C. (2013c). "The coastal innovation imperative," *Sustainability*, 5: 934–54.
Glavovic, B.C. (2014). "Waves of adversity, layers of resilience: Floods, hurricanes, oil spills and climate change in the Mississippi Delta," in B.C. Glavovic and G.P. Smith (eds.), *Adapting to Climate Change: Lessons from Natural Hazards Planning*, Berlin, Germany: Springer.
Glavovic, B.C. and Boonzaaier, S. (2007). "Confronting coastal poverty: Building sustainable coastal livelihoods in South Africa," *Ocean & Coastal Management*, 50(1/2): 1–23.
Glavovic, B.C. and Smith, G.P. (eds.) (2014). *Adapting to Climate Change: Lessons from Natural Hazards Planning*, Springer.
Goodin, R.E. and Dryzek, J.S. (2006) "Deliberative impacts. The macro-political uptake of mini-publics," *Politics and Society*, 34(2): 219–44.
Gupta, J., Termeer, C., Klostermann, J., Meijerink, S., van den Brink, M., Jong, P., Nooteboom, S., and Bergsma, E. (2010). "The adaptive capacity wheel: A method to assess the inherent characteristics of institutions to enable the adaptive capacity of society," *Environmental Science and Policy*, 13: 459–71.
Gupte, M. and Bartlett, R.V. (2007). "Necessary preconditions for deliberative environmental democracy? Challenging the modernity bias of current theory," *Global Environmental Politics*, 7(3): 94–106.
Haasnoot, M., Kwakkel, J.H., Walker, W.E., and ter Maat, J. (2013) "Dynamic adaptive policy pathways: A method for crafting robust decisions for a deeply uncertain world," *Global Environmental Change*, 23: 485–98.
Haddad, B.M. (2005). "Ranking the adaptive capacity of nations to climate change when socio-political goals are explicit," *Global Environmental Change—Human and Policy Dimensions*, 15(2): 165–76.

Hallegatte, S., Green, C., Nicholls, R.J., and Corfee-Morlot, J. (2013). "Future flood losses in major coastal cities," *Nature Climate Change*, 3: 802–6.

Hegger, D., Lamers, M., Van Zeil-Rozema, A., and Dieperink, C. (2012). "Conceptualising joint knowledge production in regional climate change adaptation projects: Success conditions and levers for action," *Environmental Science and Policy*, 18: 52–65.

Hickey, S. and Mohan, G. (eds.) (2004). *Participation: From Tyranny To Transformation?* New York: Zed Books.

Hofmeester, C., Bishop, B., Stocker, L., and Syme, G. (2012). "Social cultural influences on current and future coastal governance," *Futures*, 44: 719–29.

Holling, C.S. (1973). "Resilience and stability of ecological systems," *Annual Review of Ecology, Evolution, and Systematics*, 4: 1–23.

Hoppe, R. (2011). "Institutional constraints and practical problems in deliberative and participatory policy making," *Policy & Politics*, 39(2): 163–86.

IPCC. (1992). "A common methodology for assessing vulnerability to sea level rise," 2nd revision, in *IPCC CZMS, Global Climate Change and the Rising Challenge of the Sea*. Report of the Coastal Zone Management Subgroup, Response Strategies Working Group of the Intergovernmental Panel on Climate Change, Ministry of Transport, Public Works and Water Management, The Hague, the Netherlands, Appendix C.

Ivey, J.L., Smithers, J., de Loe, R.C., and Kreutzwizer, R.D. (2004). "Community capacity for adaptation to climate-induced water shortages: Linking institutional complexity and local actors," *Environmental Management*, 33(1): 36–47.

Jentoft, S. and Chuenpagdee, R. (2009). "Fisheries and coastal governance as a wicked problem," *Marine Policy*, 33: 553–60.

Jones, L. and Boyd, E. (2011). "Exploring social barriers to adaptation: Insights from western Nepal," *Global Environmental Change*, 21: 1262–74.

Juhola, S. and Westerhoff, L. (2011). "Challenges of adaptation to climate change across multiple scales: A case study of network governance in two European countries," *Environmental Science and Policy*, 14: 239–47.

Kannen, A. (2012). "Challenges for marine spatial planning in the context of multiple sea uses, policy arenas and actors based on experiences from the German North Sea," *Regional Environmental Change*, 12(3). DOI 10.1007/s10113-012-0349-7.

Kay, R. and Alder, J. (2005). *Coastal Planning and Management*, 2nd edn., London; New York: Taylor & Francis.

Keessen, A.M., Hamer, J.M., Van Rijswick, H.F.M.W., and Wiering, M. (2013). "The concept of resilience from a normative perspective: Examples from Dutch adaptation strategies," *Ecology and Society*, 18(2): 45.

King, C.S., Feltey, K.M., and O'Neill Susel, B. (1998). "The question of participation: Toward authentic public participation in public administration," *Public Administration Review*, 58(4): 317–26.

Klein, R.J.T., Nicholls, R.J., and Mimura, N. (1999). "Coastal adaptation to climate change: Can the IPCC technical guidelines be applied?," *Mitigation and Adaptation Strategies for Global Change*, 4: 239–52.

Klinke, A., Dreyer, M., Renn, O., Stirling, A., and van Zwanenberg, P. (2006). "Precautionary risk regulation in European Governance," *Journal of Risk Research*, 9(4): 373–92.

Klinke, A. and Renn, O. (2002). "A new approach to risk evaluation and management: Risk-based, precaution-based, and discourse-based strategies," *Risk Analysis*, 22(6): 1071–94.

Klinke, A. and Renn, O. (2010). "Risk governance: Contemporary and future challenges," in J. Eriksson, M. Gilek, and C. Ruden (eds.), *Regulating Chemical Risks: European and Global Challenges*, Berlin, Germany: Springer.

Klinke, A. and Renn, O. (2012). "Adaptive and integrative governance on risk and uncertainty," *Journal of Risk Research*, 15(3): 273–92.

Knight, F. (1921). *Uncertainty and Profit*. New York: Houghton Mifflin.

Kooiman, J. (2003). *Governing as Governance*. London: Sage.

Krishnamurthy, R., Glavovic, B.C., Kannen, A., Green, D.R., Ramanathan, A., Han, Z., Tinti, S., and Agardy, T.S. (eds.) (2008). *Integrated Coastal Zone Management: The Global Challenge*, Singapore: Research Publishing Services.

Larsen, R.K., Calgaro, E., and Thomalla, F. (2011). "Governing resilience building in Thailand's tourism-dependent coastal communities: Conceptualising stakeholder agency in social–ecological systems," *Global Environmental Change*, 21: 481–91.

Lauer, M., Albert, S., Aswani, S., Halpern, B.S., Campanella, L., and La Rose, D. (2013). "Globalization, Pacific Islands, and the paradox of resilience," *Global Environmental Change*, 23: 40–50.

Lebel, L., Manuta, J.B., and Garden, P. (2011). "Institutional traps and vulnerability to changes in climate and flood regimes in Thailand," *Regional Environmental Change*, 11: 45–58.

Lemos, M.C. and Agrawal, A. (2006). "Environmental governance," *Annual Review of Environment and Resources*, 31: 297–325.

Lesnikowski, A.C., Ford, J.D., Berrang-Ford, L., Barrera, M., Berry, P., Henderson, J., and Heymann, S.J. (2013). "National-level factors affecting planned, public adaptation to health impacts of climate change," *Global Environmental Change*, 23(5):1153–63.

Levin, K., Cashore, B., Bernstein, S., and Auld, G. (2012). "Overcoming the tragedy of super wicked problems: Constraining our future selves to ameliorate global climate change," *Policy Science*, 45: 123–52.

Li, H. (2003). "Management of coastal mega-cities—A new challenge in the 21st century," *Marine Policy*, 27(4): 333–7.

Liu, J., Hull, V., Batistella, M., DeFries, R., Dietz, T., Fu, F., Hertel, T.W., Izaurralde, R.C., Lambin, E.F., Li, S., Martinelli, L.A., McConnell, W.J., Moran, E.F., Naylor, R., Ouyang, Z., Polenske, K.R., Reenberg, A., de Miranda Rocha, G., Simmons, C.S., Verburg, P.H., Vitousek, P.M., Zhang, F., and Zhu, C. (2013). "Framing sustainability in a telecoupled world," *Ecology and Society*, 18(2): 26.

Lloyd, M.G., Peel, D., and Duck, R.W. (2013). "Towards a social-ecological resilience framework for coastal planning," *Land Use Policy*, 30: 925–33.

Lubchenco, J. and Petes, L.E. (2010). "The interconnected biosphere: Science at the ocean's tipping points," *Oceanography*, 23(2): 115–29.

MacKinnon, D. and Derickson, K.D. (2013). "From resilience to resourcefulness: A critique of resilience policy and activism," *Progress in Human Geography*, 37(2): 253–70.

Magis, K. (2010). "Community resilience: An indicator of social sustainability," *Society & Natural Resources*, 23(5): 401–16.

May, B. and Plummer, R. (2011). "Accommodating the challenges of climate change adaptation and governance in conventional risk management: Adaptive collaborative risk management (ACRM)," *Ecology and Society*, 16(1): 47.

Menzel, S. and Buchecker, M. (2013). "Does participatory planning foster the transformation toward more adaptive social-ecological systems?," *Ecology and Society*, 18(1): 13.

Merrell, W.J., Katsouros, M.H., and Bienski, J. (2001). "The Stratton Commission: The model for a sea change in national marine policy," *Oceanography*, 14(2): 11–16.

Millennium Ecosystem Assessment. (2005). *Ecosystems and Human Well-Being: Synthesis*. Washington, DC: Island Press.

Moser, S.C. and Ekstrom, J.A. (2010). "A framework to diagnose barriers to climate change adaptation," *Proceedings of the National Academies of Sciences*, 107(51): 22026–31.

Moser, S.C., Williams, S.J., and Boesch, D.F. (2012). "Wicked challenges at land's end: Managing coastal vulnerability under climate change," *Annual Review of Environment and Resources*, 37: 51–78.

Nelson, M., Ehrenfeucht, R., and Laska, S. (2007). "Planning, plans, and people: Professional expertise, local knowledge, and governmental action in post-Hurricane Katrina New Orleans," *Cityscape: A Journal of Policy Development and Research*, 9(3): 23–52.

Newig, J. and Fritsch, O. (2009). "Environmental governance: Participatory, multi-level and effective?," *Environmental Policy and Governance*, 19(3): 197–214.

Nicholls, R.J. and Kebede, A.S. (2012). "Indirect impacts of coastal climate change and sea-level rise: The UK example," *Climate Policy*, 12(suppl): S28–S52.

Nicholls, R.J., Wong, P.P., Burkett, V.R., Codignotto, J.O., Hay, J.E., McLean, R.F., Ragoonaden, S., and Woodroffe, C.D. (2007). "Coastal systems and low-lying areas," in M.L. Parry, O.F. Canziani, J.P. Palutikof, P.J. van der Linden and C.E. Hanson (eds.), *Climate Change 2007: Impacts, Adaptation and Vulnerability. Contribution of Working Group II to the Fourth Assessment Report of the Intergovernmental Panel on Climate Change*, Cambridge: Cambridge University Press. http://www.ipcc.ch/ (accessed August 2013).

Norris, F.H., Stevens, S.P., Pfefferbaum, B., Wyche, K.F., and Pfefferbaum, R.L. (2008). "Community resilience as a metaphor, theory, set of capabilities, and strategy for disaster readiness," *American Journal of Community Psychology*, 41: 127–50.

Olsen, S.B. (2002). "Assessing progress towards goals of coastal management," *Coastal Management*, 30: 325–45.

Olsen, S.B. (2003). "Frameworks and indicators for assessing progress in Integrated Coastal Management initiatives," *Ocean & Coastal Management*, 46(3/4): 347–61.

Olsen, S.B., Page, G.G., and Ochoa, E. (2009). *The Analysis of Governance Responses to Ecosystem Change: A Handbook for Assembling a Baseline*. Geesthacht, Germany: GKSS Research Centre.

Olsson, P., Folke, C., and Berkes, F. (2004). "Adaptive co-management for building resilience in social-ecological systems," *Environmental Management*, 34(1): 75–90.

Palmer, J. (2012). "Risk governance in an age of wicked problems: Lessons from the European approach to indirect land-use change," *Journal of Risk Research*, 15(5): 495–513.

Parry, M.L., Canziani, O.F., Palutikof, J.P., van der Linden, P.J., and Hanson, C.E. (eds.) (2007). *Climate Change 2007: Impacts, Adaptation and Vulnerability. Contribution of Working Group II to the Fourth Assessment Report of the Intergovernmental Panel on Climate Change*, Cambridge: Cambridge University Press. http://ipcc.ch/ (accessed August 2013).

Pelling, M. (2011). *Adaptation to Climate Change: From Resilience to Transformation.* London: Routledge.

Pelling, M. and Blackburn, S. (2013). *Megacities and the Coast: Risk, Resilience and Transformation.* London: Earthscan.

Peñalba, L.M., Elazegui, D.D., Pulhin, J.M., and Cruz, R.V.O. (2012). "Social and institutional dimensions of climate change adaptation," *International Journal of Climate Change Strategies and Management*, 4(3): 308–22.

Plummer, R. and Armitage, D.R. (2007). "Charting the new territory of adaptive co-management: A Delphi study," *Ecology and Society*, 12(2): 10.

Plummer, R., Armitage, D.R., and de Loë, R.C. (2013). "Adaptive comanagement and its relationship to environmental governance," *Ecology and Society*, 18(1): 21.

Plummer, R. and Baird, J. (2013). "Adaptive co-management for climate change adaptation: Considerations for the Barents region," *Sustainability*, 5: 629–42.

Preston, B.L., Westaway, R.M. and Yuen, E.J. (2011). "Climate adaptation planning in practice: An evaluation of adaptation plans from three developed nations," *Mitigation and Adaptation Strategies for Global Change*, 16: 407–38.

Renn, O. (2008). *Risk Governance: Coping with Uncertainty in a Complex World.* London: Earthscan.

Renn, O., Klinke, A., and van Asselt, M. (2011). "Coping with complexity, uncertainty and ambiguity in risk governance: A synthesis," *Ambio*, 40: 231–46.

Rittel, H.W. and Weber, J. (1973). "Dilemmas in a general theory of planning," *Policy Sciences*, 4: 155–69.

Rockström, J., Steffen, W., Noone, K., Persson, A., Chapin, F.S., Lambin, E.F., Lenton, T.M. et al. (2009). "A safe operating space for humanity," *Nature*, 461(24): 472–75.

Scarlett, L. (2013). "Collaborative adaptive management: Challenges and opportunities," *Ecology and Society*, 18(3): 26.

Small, C. and Nicholls, R.J. (2003). "A global analysis of human settlement in coastal zones," *Journal of Coastal Research*, 19(3): 584–99.

Smit, B., Pilifosova, O., Burton, I., Challenger, B., Huq, S., Klein, R.J.T., and Yohe, G. (2001). "Adaptation to climate change in the context of sustainable development and equity," in J.J. McCarthy, O. Canziani, N.A. Leary, D.J. Dokken, and K.S. White (eds.), *Climate Change 2001: Impacts, Adaptation and Vulnerability. Contribution of the Working Group II to the Third Assessment Report of the Intergovernmental Panel on Climate Change*, Cambridge: Cambridge University Press.

Sorensen, J. (1997). "National and international efforts at integrated coastal management: Definitions, achievements, and lessons," *Coastal Management*, 25(1): 3–41.

Stirling, A. (2010). "Keep it complex," *Nature*, 468: 1029–31.

Stojanovic, T., Ballinger, R.C., and Lalwani, C.S. (2004). "Successful integrated coastal management: Measuring it with research and contributing to wise practice," *Ocean & Coastal Management*, 47: 273–98.

Stokols, D., Perez Lejano, R., and Hipp, J. (2013). "Enhancing the resilience of human–environment systems: A social–ecological perspective," *Ecology and Society*, 18(1): 7.

Storbjörk, S. (2010). "'It takes more to get a ship to change course': Barriers for organizational learning and local climate adaptation in Sweden," *Journal of Environmental Policy and Planning*, 12(3): 235–54.

Taylor, B.M., Harman, B.P., and Inman, M. (2013). "Scaling-up, scaling-down, and scaling-out: Local planning strategies for sea-level rise in New South Wales, Australia," *Geographical Research*, 51(3): 292–303.

Termeer, C., Biesbroek, R., and van den Brink, M. (2012). "Institutions for adaptation to climate change: Comparing national adaptation strategies in Europe," *European Political Science*, 11: 41–53.

Tobey, J., Rubinoff, P., Robadue, Jr., D., Ricci, G., Volk, R., Furlow, J., and Anderson, G. (2010). "Practicing coastal adaptation to climate change: Lessons from integrated coastal management," *Coastal Management*, 38(3): 317–35.

Tompkins, E.L., Adger, W.N., Boyd, E., Nicholson-Cole, S., Weatherhead, K., and Arnell, N. (2010). "Observed adaptation to climate change: UK evidence of transition to a well-adapted society," *Global Environmental Change*, 20: 627–35.

Torell, E. (2000). "Adaptation and learning in coastal management: The experience of five east African initiatives," *Coastal Management*, 28(4): 353–63.

United Nations. (1992). "Results of the World Conference on Environment and Development: Agenda 21," Rio de Janeiro, Brazil; New York: United Nations Conference on Environment and Development.

USAID. (2009). *Adapting to Coastal Climate Change: A Guidebook for Development Planners*, Washington, DC: USAID.

van Asselt, M.B.A. and Renn, O. (2011). "Risk governance," *Journal of Risk Research*, 14(4): 431–49.

Walker, B.H., Carpenter, S.R., Anderies, J.M., Abel, N., Cumming, G.S., Janssen, M.A., Lebel, L., Norberg, J., Peterson, G.D., and Pritchard, L. (2002). "Resilience management in social–ecological systems: A working hypothesis for a participatory approach," *Conservation Ecology*, 6(1): 14.

Walker, B.H., Holling, C.S., Carpenter, S.C., and Kinzig, A.P. (2004). "Resilience, adaptability and transformability," *Ecology and Society*, 9(2): 5.

Walker, W.E., Haasnoot, M., and Kwakkel, J.H. (2013). "Adapt or perish: A review of planning approaches for adaptation under deep uncertainty," *Sustainability*, 5: 955–79.

Walters, C. (1997). "Challenges in adaptive management of riparian and coastal ecosystems," *Conservation Ecology*, 1(2): 1. http://www.consecol.org/vol1/iss2/art1/ (accessed August 2013).

Wilson, G.A. (2102). "Community resilience, globalization, and transitional pathways of decision-making," *Geoforum*, 43: 1218–31.

Wilson, P.A. (2009). "Deliberative planning for disaster recovery: Re-membering New Orleans," *Journal of Public Deliberation*, 5(1), Article 1. http://www.publicdeliberation.net/jpd/vol5/iss1/art1 (accessed August 2013).

World Bank. (2013). *Turn Down the Heat: Climate Extremes, Regional Impacts, and the Case for Resilience*. A report for the World Bank by the Potsdam Institute for Climate Impact Research and Climate Analytics. Washington, DC: World Bank.

World Commission on Environment and Development (WCED). (1987). *Our Common Future*. Oxford: Oxford University Press.

Yohe, G. and Tol, R.S.J. (2002). "Indicators for social and economic coping capacity—Moving toward a working definition of adaptive capacity," *Global Environmental Change—Human and Policy Dimensions*, 12(1): 25–40.

Zalasiewicz, J., Williams, M., Haywood, A., and Ellis, M. (2011). "The Anthropocene: A new epoch of geological time?," *Philosophical Transactions of the Royal Society A*, 369: 835–41.

Part II

Climate change and the coastal zone

North America

Chapter 4

Social-ecological change in Canada's Arctic

Coping, adapting, and learning for an uncertain future

Derek Armitage

Abstract: In Canada's Arctic, adaptive responses to global environmental change have become available with the emergence of new institutional arrangements and multilevel comanagement practices. Such arrangements have been shown, albeit with mixed results, to support knowledge building, learning, and conflict resolution; reduce vulnerability; and foster adaptive capacity. Experiences in Canada's Arctic may provide valuable lessons for other coastal contexts dealing with the implications of climate change, including: (1) the need to identify and act upon policy windows; (2) the role of bridging organizations in helping connect local, regional, and national actors in shared processes of learning; (3) the value of knowledge coproduction processes in understanding and responding to change; and (4) the use of interactive or participatory scenario processes to help communities explore vulnerabilities and build adaptive capacity.

INTRODUCTION

Implications of global climate change in Canada's Arctic will be profound (Arctic Climate Impact Assessment 2005). Johannessen and Miles (2010) suggest that changes in the physical environment of the Arctic will be *transformative*, pointing to projected amplifications in surface air temperature between 3°C and 6°C by 2080 and the exposure of significant areas of the Arctic Ocean because of sea ice loss. In particular, rapid changes to sea ice quality and extent will produce uncertain consequences for Arctic coastal communities (Figure 4.1) closely tied to their environments. Longstanding cultural adaptations of coastal communities (e.g., flexibility in resource use, detailed local environmental knowledge) and emerging coping mechanisms of Arctic residents in the face of change and uncertainty (e.g., modifying harvest activities and timing, minimizing risk, use of new technology) may be overwhelmed as tipping points or thresholds are crossed (Box 4.1). Moreover, climate and sea ice change will not act in isolation from other issues affecting the region, including oil, natural gas, and

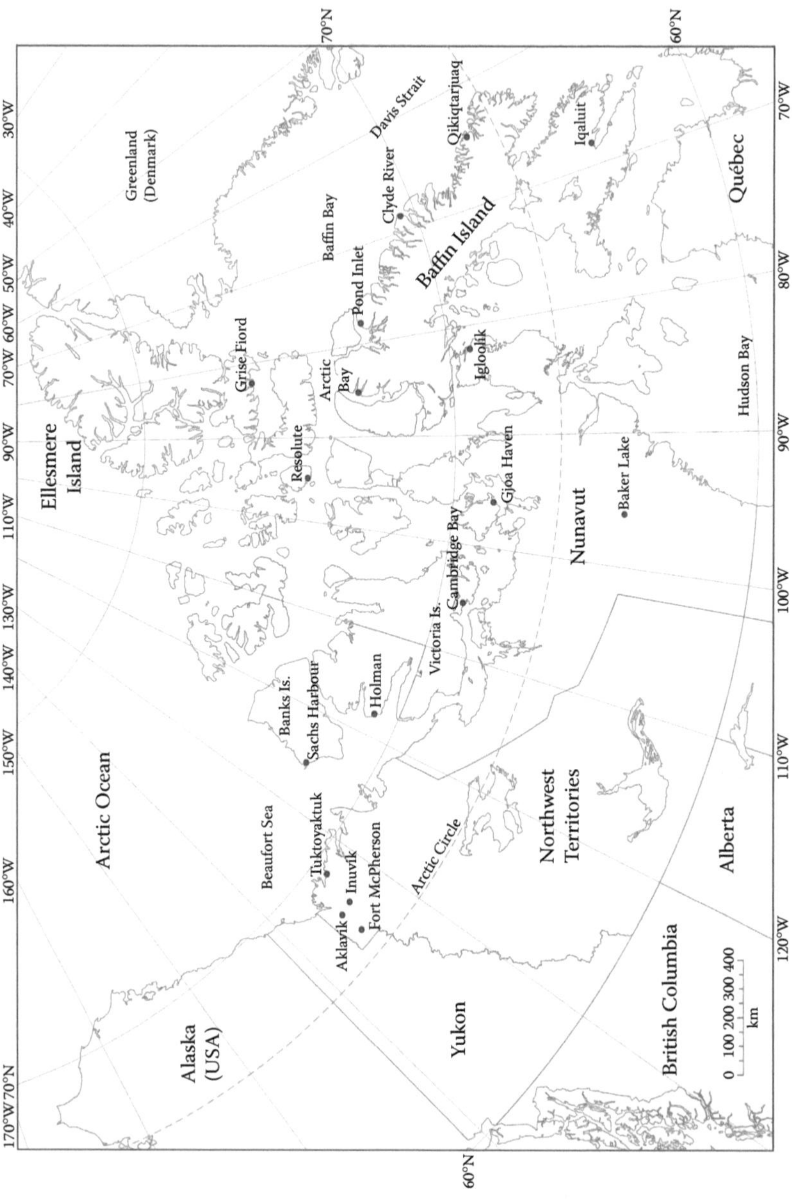

Figure 4.1 Canada's Arctic.

mineral developments; airborne pollutants; sovereignty claims and security concerns; the socioeconomic dimensions of renewable resource extraction; and demands of Northerners to have a greater role in decision-making.

In the context of rapid change, understanding the vulnerability and resilience of coastal communities is necessary, as are efforts to support coastal governance arrangements and emerging comanagement institutions that are flexible and adaptive. Internationally, sea ice change and its consequences will place additional pressure on the existing governance framework for cooperation developed by Arctic nations (Stokke and Honneland 2007). This governance context includes wildlife treaties, the establishment of the Conservation of Arctic Flora and Fauna Committee in 1991, the Arctic Council in 1996, and the 1982 United Nations Convention on the Law of the Sea (UNCLOS), which provides a legally binding framework to resolve state-based conflicts (Huebert and Yeager 2008). However, although the drivers of sea ice change occur largely outside the region, the effects of that change will be experienced most acutely at local and regional scales. Climate change adaptation efforts will necessitate, as a result, a move toward more process-based and multilevel coastal governance processes that account for change and uncertainty and that provide the flexibility for local communities, subregional, national, and international interests to respond quickly and collaboratively (Berkes et al. 2005; Keskitalo and Kulyasova 2009).

In Canada's Arctic, there is no overall framework for coastal governance. Canada's Oceans Act (Government of Canada 1996) consolidates existing federal responsibilities through an integrated legislative approach. In particular, the corresponding Oceans Strategy (Government of Canada 2002) seeks to foster the integrated management of coastal, marine, and estuarine activities and mandates the inclusion of land claims organizations, affected aboriginal organizations, and coastal communities. However, opportunities for more process-based approaches to coastal governance in Canada's Arctic are driven in large part by comprehensive land claims agreements between indigenous groups and the federal government and the institutions of public government they have subsequently established (Table 4.1). Specifically, settled land claims have helped to formalize comanagement institutional arrangements, defined here as the sharing of power and decision-making authority between national and territorial governments and local communities. These claims-based arrangements illustrate the importance of policy windows and are helping to establish bridging organizations in which diverse types of knowledge to facilitate climate change adaptation can be coproduced among indigenous and nonindigenous groups. Experiences with new institutional arrangements and the strategies of Arctic indigenous groups to respond to change in Canada's Arctic may offer lessons for building adaptive capacity in other coastal contexts.

Table 4.1 Summary of selected comprehensive land claims agreements in Canada's North (Canadian dollar)

Agreement	*Key features*	*Main comanagement institutions*
Inuvialuit Final Agreement (1984)	• Covers 91,000 km^2 • $152 million over 14 years, in addition to onetime payments of $7.5 million to a social development fund and $10 million to the Economic Enhancement Fund	• The Fisheries Joint Management Committee (FJMC) is the main institution responsible for fish and marine mammals. • Other boards were created to deal with environmental screening and impact assessment (i.e., Environmental Impact Review Board and the Environmental Impact Screening Committee). • Community-based Inuvialuit Hunters and Trappers Committees (HTCs) are the main comanagement partner at the local level.
Nunavut Comprehensive Land Claim Agreement (1993)	• Covers 1.9 million km^2 (one-fifth of the total land mass of Canada) • 23,000 Inuit beneficiaries • Capital transfer payments of $580 million (in 1989 dollars) with interest payable over 14 years, in addition to payments for specific funds and initiatives (e.g., Bowhead Whale Study)	• The Nunavut Wildlife Management Board (NWMB) is the primary institution responsible for wildlife and fisheries. • The Nunavut Impact Review Board (NIRB) addresses environmental impact and assessment concerns. • Hunters and Trappers Organizations (HTOs) are the local comanagement partners.

Source: Indian and Northern Affairs Canada (INAC), Final agreements and related implementation matters, 2010, http://www.ainc-inac.gc.ca/.

CONVERGING THREATS IN THE CANADIAN ARCTIC

The Canadian Arctic is socioculturally and ecologically unique. Majestic landscapes and seascapes (Figures 4.2 and 4.3), vulnerable ecologies, and a rich tapestry of Arctic cultures and knowledge make the region a special place (Armitage and Clark 2005).

Yet in a world of interconnected spaces and places, Canada's coastal Arctic communities are confronting a convergence of threats not previously experienced and impacts to ecosystems and people that are increasingly undesirable and inequitable. Globally, the largest temperature increases are projected to occur over the Polar region (ACIA 2005; Pachauri et al. 2007). The monthly sea ice extent for 1979–2010, for example, highlights a decline

Figure 4.2 Hamlet of Qikiqtarjuaq, Nunavut, Canada. (Courtesy of D. Armitage.)

Figure 4.3 Old whaling outpost, east coast of Baffin Island, Nunavut, Canada. (Courtesy of D. Armitage.)

of approximately 11.5% per decade (relative to the 1979–2010 average). The lowest recorded annual sea ice coverage as measured in September has occurred in 2007, 2008, and 2010 (Perovich et al. 2010) (see Figure 4.4). The implications of a projected seasonally ice-free Arctic within the next several decades are profound. Arctic marine environments are generally characterized by low species diversity with many species dependent on sea ice or seasonal sea ice transitions (Learmonth et al. 2006; Nuttall et al. 2005). Loss of sea ice will mean a loss of habitat for microorganisms, including ice algae, which will disturb Arctic food webs. Despite low species diversity, Arctic and sub-Arctic Seas are economically productive,

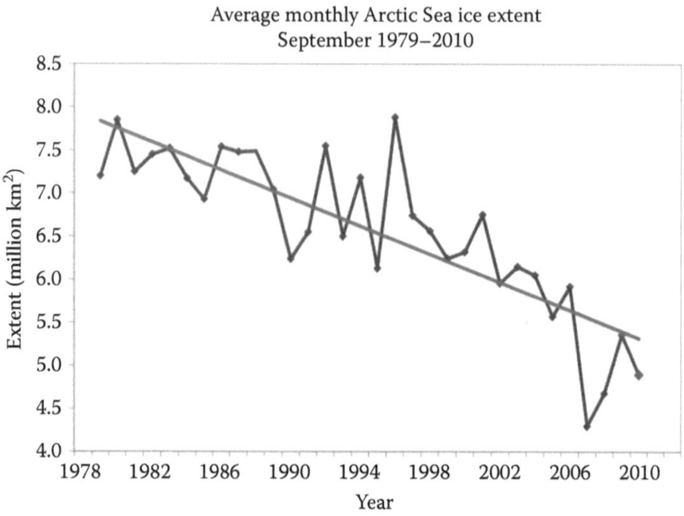

Figure 4.4 Monthly September ice extent for 1979–2010, showing a decline of 11.5% per decade. (Data from NSIDC, Weather and feedbacks lead to third-lowest extent, Arctic Sea Ice News & Analysis, http://nsidc.org/asina/2010/100410.html, 2010.)

yielding an estimated seven million tonnes of fish annually, with earnings of USD 15 billion and employment for approximately 0.6–1 million people (Anisimov et al. 2007).

Subsistence livelihoods and commercial resource extraction is and will be influenced by rapidly changing biophysical conditions (ARCUS 2009). Reid et al. (1994 cited in Hovelsrud et al. 2008) note, for example, that a majority of Inuit in the Canadian Arctic hunt, fish, and trap marine mammals and other wildlife, while Hovelsrud et al. (2008) highlighted significant economic revenues associated with resource harvesting (e.g., CAN500,000 in 2002 from seal by-products in Nunavut, Canada). Many Arctic people, including those in Canada, are moving from subsistence hunting to a cash-based or mixed wage economy, or experiencing shifts of subsistence hunting practices to industrial hunting/herding (Nuttall et al. 2005). Climate uncertainty and changing resource conditions may influence the economic viability of this transition, while more directly, the loss of sea ice will impact hunting and traveling. Decreased access to harvest populations and a greater risk of accidents is projected (Ford et al. 2007).

Biophysical changes and their implications (Box 4.1), however, will not act in isolation but will be tightly coupled with rapid economic expansion and sociocultural change. For instance, a loss of sea ice is projected to catalyze increased mineral, oil, and natural gas exploration in Canada's Arctic (Furgal and Prowse 2008). Areas of the Arctic are likely to become navigable

BOX 4.1 CROSSING PLANETARY SAFE BOUNDARIES? AN ARCTIC PERSPECTIVE

Rockström and colleagues (2009) highlight the global reach of anthropogenic pressures. To examine the challenge of global sustainability, they develop the notion of *planetary boundaries*, or the boundaries within which humanity can operate safely. Moving beyond one or more of these planetary boundaries, they argue, may have profoundly negative and uncertain outcomes for human well-being and ecosystem services and may trigger nonlinear, abrupt continental to planetary-scale environmental change. They suggest that the Earth System has approached the point at which abrupt global environmental change may not respond to further mitigation efforts.

Nine planetary boundaries have been identified, although they have tentatively quantified safe boundaries for only seven based on current understanding: (1) climate change (e.g., CO_2 concentration in the atmosphere at less than 350 ppm); (2) ocean acidification; (3) stratospheric ozone (e.g., less than 5% reduction in O_3 concentration from the preindustrial level of 290 Dobson units); (4) biogeochemical nitrogen and phosphorus cycles; (5) global freshwater use; (6) land system change; and (7) rate at which biological diversity is lost.

In the Arctic, the planetary boundaries concept highlights the importance of unsustainable trajectories, thresholds, and tipping points that (1) may not respond to current mitigation efforts and (2) may overwhelm historical adaptation strategies or current coping mechanisms of Arctic people. For example, Lenton et al. (2008) have identified trajectories of change in sea ice as one potential tipping point.

Responding to these thresholds and convergence of risks will require Arctic stakeholders to think carefully about a full range of adaptation options, learn from past experience and knowledge of Arctic people, and recognize that social learning processes leading to fundamental changes in institutions and governance may be the key to helping communities deal with the Arctic *triple squeeze* (see Figure 4.5).

that were previously not, both seasonally and permanently, creating greater accessibility to nonrenewable resources (Hovelsrud et al. 2008). Increased tourism, changes to travel, and subsequent changes to lifestyles in Arctic communities can be expected (Walsh et al. 2005).

One of the key findings of the Arctic Climate Impact Assessment (ACIA) (2005) was that reduced sea ice is very likely to increase marine transport and access to resources. The Arctic Marine Shipping Assessment (AMSA)

(2009) has projected that future marine activities will include more non-Arctic stakeholders and the potential for increased conflict among indigenous uses and commercial activities. There were 6,000 individual vessels traveling in the pan-Arctic region during the AMSA survey year and this number is expected to increase (AMSA 2009). The majority of the increase in vessel traffic will be destinational (as opposed to trans-Arctic) implying increased pressure to develop marine infrastructure, emergency response capacity given the threat of environmental impacts (oil spills), invasive species introduction, human–wildlife conflict, and waste and emissions (AMSA 2009; Furgal and Prowse 2008).

Indirect effects of climatic and environmental change are only now emerging as a seasonally ice-free Arctic will catalyze resource development and generate additional economic and sociocultural uncertainty (see Leichenko and O'Brien 2008). Over the past several decades, communities in Canada's Arctic have experienced significant cultural and social upheaval as a result of settlement and relocation schemes, negative experiences (abuse and neglect) with residential school systems, and the gradual erosion of land-based skills and knowledge with successive generations. Global environmental change, economic globalization, and cultural transition are resulting in a *triple squeeze* in a sensitive and distinct region (Figure 4.5).

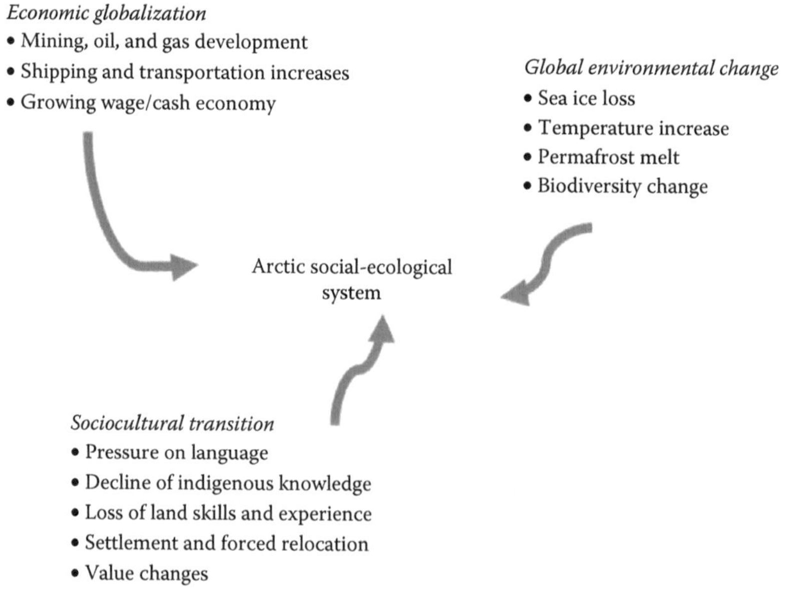

Figure 4.5 The Arctic *triple squeeze*, on sustainable development opportunities: (1) human population growth; (2) loss of ecosystem function; (3) climate change; (4) surprise. (Adapted from Rockström, J. and Karlberg, L., *Ambio*, 39(3): 257–65, 2010.)

COPING AND ADAPTING: THEN AND NOW

The triple squeeze on Arctic social-ecological systems is posing new challenges. Yet, a simple view of indigenous people in Canada's Arctic as vulnerable and passive victims in the context of change masks a more complex reality. Community-based research in the Canadian Arctic illustrates both sources of vulnerability and significant adaptive capacity of coastal communities (see Ford et al. 2007; Hovelsrud and Smit 2010; Krupnik and Jolly 2002; Laidler et al. 2009; Riedlinger and Berkes 2001). Change is not new to the Arctic, and the self-image of many Arctic people is one of flexibility and adaptability (Berkes and Armitage 2010; Berkes et al. 2005). Emerging vulnerabilities that stem from the gradual loss of traditional hunting and land-based skills, the historic imposition of top-down wildlife management regimes, and the erosion of social networks and culture are, therefore, framed by the experience and insights of Inuit and other northern people developing a range of strategies to cope and adapt to interacting biophysical, economic, and sociocultural stressors. For example, long-standing cultural adaptations to environmental change and uncertainty (Table 4.2) have enabled northern indigenous people to sustain themselves through generations. Some of these cultural adaptations include the development of a comprehensive knowledge of their environments, flexible group size and mobility, and social networks that involve sharing of risks and rewards.

Table 4.2 Cultural adaptations as sources of resilience to change

Adaptive strategy	*Description*
Group size flexibility and associated mobility on the land/sea ice	• Arctic ecosystems have generally low or patchy biological productivity, which means harvesting is uncertain. • Large social groups and permanent settlements are not historically feasible in an Arctic environment. • Small, mobile, and self-supporting groups are a more effective form of social organization.
Flexibility in response to seasonal harvest cycles	• Individuals and groups deal with unpredictability with opportunistic harvesting. • Seasonal cycles drive harvesting but with backup plans should target species not be available. • Resilience to unexpected events supported by traditional knowledge and cultural memory.
Comprehensive and detailed knowledge of land and development of related skills	• Diverse land/sea-based skills with accumulation of environmental knowledge over time. • Skill sets required to survive not necessarily gendered, with early knowledge and experience gained by children. • Critical survival skills developed by most individuals.

(Continued)

Table 4.2 (Continued) Cultural adaptations as sources of resilience to change

Adaptive strategy	*Description*
Social networks that foster sharing mechanisms	• Sharing of harvests occurs among multiple households, a particularly important mechanism in times of scarcity. • High social value attached to sharing with prestige for hunters and families that provide food. • Sharing through complex networks (not just direct family) and the basis for mutual support and minimization of risk.
Inter-community trade	• Regional differences in resource availability alleviated through trade. • Building of mutual support and trust among groups located in different geographic areas. • Extending social networks beyond immediate group an important activity.

Source: Berkes, F. and Armitage, D., *Études/Inuit/Studies*, 34(1): 109–31, 2010.

Such strategies have helped Arctic people maintain a diversity of options to deal with change, uncertainty, and unexpected events. The wider lessons associated with this environmental knowledge and capacity to adapt to change are invaluable as sources of resilience. Inuit and Inuvialuit (Box 4.2) lived experience, for instance, resonates with emerging climate change adaptation principles and best practices, including the importance

BOX 4.2 INUIT AND INUVIALUIT IN CANADA'S ARCTIC

In Canada's Arctic, the Inuvialuit and Inuit are descendents of the Thule people who migrated eastward from Alaska around AD 1000. The Inuvialiut, or Inuit, of the western Canadian Arctic reside primarily in the Mackenzie River Delta region, on Banks Island, and on parts of Victoria Island of the Northwest Territories (see Figure 4.1). Politically, the Inuvialuit are represented by the Inuvialuit Regional Corporation and their interests are reflected in the 1984 Inuvialuit Final Agreement (Table 4.1). The Inuit of the eastern Canadian Arctic reside primarily in Nunavut in numerous small communities scattered throughout the vast territory. Inuit are represented by Nunavut Tunngavik Incorporated, the main implementing body of the 1993 Nunavut Final Agreement. The territory of Nunavut itself was established in 1999. Inuit and Inuvialuit maintain a strong connection to the land and are close observers of environmental change. Although participation in the wage economy has increased, an important subsistence economy is characteristic of most communities and includes the harvest of a wide range of fish, caribou, and marine mammal species.

of maintaining diverse response options, drawing upon multiple adaptation strategies, and nurturing the human capacities and relational networks that enable adaptive capacity (see Chapin et al. 2010; Fazey et al. 2010).

There is, nevertheless, a danger these lessons will be lost. In the Canadian Arctic, for example, Inuit and Inuvialuit society has experienced dramatic change since the 1960s. Settlement schemes, residential schooling, and other interventions by the Canadian Government have rendered traditional adaptations increasingly less resilient. Increased variability in the Arctic environment and greater frequency of socioeconomic stresses further constrain long-term cultural adaptations because they make resource availability less predictable and interfere with the ability of people to access resources (Ford 2009; Krupnik and Jolly 2002). In response to increased uncertainty, communities in Canada's Arctic are making a range of adjustments or developing shorter-term coping strategies related to when, where, and how harvesting and travel activities take place (Table 4.3).

The loss of *Inuit Qaujimajatuqangit* (Inuit knowledge) accompanying demographic and social changes in the Canadian Arctic is, however, not absolute. Inuit and Inuvialuit experience and knowledge of their environments continue to evolve and provide important sources of understanding

Table 4.3 Coping strategies to deal with current change and uncertainty

Coping mechanism	*Description*
Modifying when harvesting is done	• Increased seasonal variability requires hunters to continuously adjust their seasonal calendar (e.g., shorter and warmer springs, increased rates of snow and ice melt). • Waiting is a common coping strategy creating challenges for others activities (participation in wage-based employment).
Modifying where harvesting is done	• Changing location or routes for harvesting or staying closer to community. • Difficulties in *reading* rapidly changing sea ice, permafrost thaw, snow conditions, and so on. • Hunters often need to find new routes to avoid unexpected and dangerous locations (slumps and mudslides) or to change hunting sites which may not coincide with locations of animals.
Adjusting how harvesting is done	• Use of all-terrain vehicles instead of snowmobiles when there is insufficient snow cover, more hunting effort from open water than increasingly dangerous ice edge. • Increased pressure to take all necessary supplies (at added cost and difficulty in transporting) because of uncertainty about accessing resources. • Use of new (and sometimes expensive) technologies to improve safety while on the land (e.g., global positioning system [GPS] units).

(Continued)

Table 4.3 (Continued) Coping strategies to deal with current change and uncertainty

Coping mechanism	*Description*
Adjusting the mix of species harvested	• Increasing unpredictability of hunts associated with changes in species mix and emergence of new species. • Reduced access to hunting areas has meant hunters have needed to switch target species and hunt locations, and use of fallback (but possibly less desirable) species.
Minimizing risk and uncertainty	• Experience when traveling on the sea ice now particularly important because of increased variability and unpredictability. • More careful monitoring of environment and weather conditions to avoid being caught in unsafe circumstances. • Greater use of technological solutions to offset risk (examples include GPS units, consulting satellite images, and adoption of very high frequency [VHF] radio).

Sources: Berkes, F. and Armitage, D., *Études/Inuit/Studies*, 34(1), 109–31, 2010; Berkes, F. and Jolly, D. *Conservation Ecology*, 5(2), 18, 2001; Ford, J. et al., *Arctic*, 60(2), 150–66, 2007; Laidler, G., *Climatic Change*, 78(2–4), 407–44, 2006.

(ACIA 2005; Laidler 2006; Riedlinger and Berkes 2001). There is much to learn from Arctic people by examining current and future vulnerability and adaptive capacity (Hovelsrud and Smit 2010).

MOVING FORWARD: ADAPTIVE CAPACITY AND LEARNING IN MULTILEVEL GOVERNANCE

How to respond effectively to rapid social-ecological change and uncertainty, while encouraging alternative trajectories that sustain ecosystem services and human well-being, is a central challenge in a rapidly changing Arctic. Historically evolved cultural adaptations (e.g., group mobility, resource sharing) and short-term adjustments (coping strategies) in response to change are of particular importance (see Tables 4.2 and 4.3). Some adaptation strategies required to address change in Arctic coastal communities will be technical in scope or involve engineered solutions. These strategies may include new building designs or building codes to address permafrost melt or coastal protection infrastructure in response to sea-level rise and storm surges. Engineered solutions and technically focused adaptation options, however, are only one dimension of a suite of strategies that will be required. Increasingly, those engaged in climate change adaptation are examining the links between adaptation, institutions, and broader governance processes (Collins and Ison 2009; Pelling et al. 2008).

In Canada's Arctic, new adaptive responses have become available with the emergence of institutional arrangements and multilevel comanagement practices (Berkes and Armitage 2010). Institutions and multilevel governance arrangements are particularly important because they can support knowledge building, learning, and conflict resolution and help to reduce vulnerability, build resilience, and increase adaptive capacity (Pahl-Wostl 2009). In Canada's Arctic, comanagement arrangements are gradually transforming how different actors (communities, managers) deal with change and uncertainty by encouraging the sharing of knowledge and facilitating social learning (Armitage 2005; Dale 2009; Kocho-Schellenberg 2010), although the limitations of comanagement arrangements are well documented (Nadasdy 2003).

Comanagement experiences and comprehensive land claims agreements provide an example of new institutional arrangements and governance practices that may be critical to future coping and adapting. Comprehensive land claims agreements (Table 4.1) have been instrumental in helping to gradually reshape relations between the Canadian Government and the Inuit and Inuvialuit by (1) providing significant capital transfer payments and dedicated funds for social and economic development; (2) a share of government royalties from oil, gas, and mineral development on Crown lands, along with greater control over resources on settlement lands; (3) clear rights to harvest wildlife on lands and waters; and (4) participation in wildlife management, conservation, and environmental protection activities. Specific provisions of the comprehensive claims agreements require communities have a direct role in decision-making about environment and natural resources, often through comanagement arrangements which require equal representation of government representatives, along with Inuit and Inuvialuit.

Comanagement arrangements, in particular, have emerged as a key component of land claims agreements in Canada's Arctic. Comanagement bodies are legally constituted, have formal mandates, and have a centrally located secretariat. These bodies are connected to local organizations (e.g., Hunters and Trappers Organizations [HTOs]) through which specific management actions are taken (enforcement, monitoring, some decisions on harvest quotas, etc.) and connected as well to higher-level authorities (e.g., federal departments). As a result, they provide an institutional mechanism through which comanagement actors can *learn to learn*, or learn to be adaptive (Pelling et al. 2008) and develop social and organizational networks that over time generate positive social and ecological outcomes in the face of environmental change. In all cases, final authority on key decisions rests with governments through the relevant federal minister, but claims-based comanagement institutions and partners still possess significant planning, management, and regulatory scope (White 2006).

The insights and practical strategies emerging in the Arctic in the context of these institutional arrangements provide valuable lessons for other coastal contexts grappling with climate change. Among these lessons are (1) the need to identify and act upon windows of opportunity; (2) the role of bridging organizations in helping connect local, regional, and national actors in shared processes of learning; (3) the value of knowledge coproduction processes in understanding and responding to change; and (4) the use of interactive or participatory scenario processes to help communities explore vulnerabilities and build adaptive capacity. Each of these lessons is discussed in brief.

Windows of opportunity

Experiences in Canada's Arctic point to the importance of recognizing and acting upon windows of opportunity for institutional and governance change that may help to reduce vulnerability of Arctic people and build their adaptive capacity. Such windows of opportunity to explore alternative pathways to change may come in diverse forms, including policy reform, new legislation, or crises (see Gunderson et al. 2009). Policy windows may also be triggered by social or ecological events and emerge at different scales, but the net effect is to allow actors in the system to experiment with novel approaches and shift practices. Several windows of opportunity are recognizable in the Canadian Arctic context. The comprehensive claims agreements and the provisions they contain (see Bankes 2005) are perhaps the clearest example of a policy window that may allow novel forms of governance to be tested. The comanagement arrangements that emerged from the comprehensive claims agreements have taken a significant time to develop, but they offer a means for sense-making and shared learning that is of critical importance. Despite their limitations, many comanagement arrangements have broadened how problems are defined and by whom (Berkes et al. 2005).

Rapid and unexpected change in Arctic sea ice has created another window of opportunity. The negative consequences of this trajectory of change are quickly becoming recognized outside of the research community. There is now increasing pressure on politicians to rethink international- and regional-level governance arrangements for shipping, exploration, and fisheries (among other economic sectors) in the context of a radically different Arctic environment (Huebert and Yeagar 2008). There is also recognition that any governance arrangement will need to link more effectively international- and local-level actors (Young 2009) and the institutional arrangements in Canada's Arctic may be helpful models in this regard.

Bridging organizations

Bridging organizations provide a forum in which different domains of practice (local, regional, national) are linked. Schultz (2009) has documented how bridging organizations create arenas for social processes and trust-building to support adaptation, coping, and transformation. Bridging organizations are particularly important in sustaining the relational networks that support social learning. Bridging organizations facilitate knowledge generation, preference formation, sense-making, and conflict resolution among actors in relation to specific environmental issues (Schultz 2009). Berkes (2009) has also identified the role of bridging organizations in supporting the vertical and horizontal linkages that improve information and resource flows and build trust. In the Canadian Arctic, the comanagement bodies (e.g., the Fisheries Joint Management Committee [FJMC] and the Nunavut Wildlife Management Board [NWMB]) are increasingly serving this function in ways that extend beyond their specific mandates by forming key nodes in multilevel institutional networks. Figure 4.6 illustrates how the regional NWMB, with members nominated the main Inuit organization (Nunavut Tunngavik Incorporated), the Government of Canada, and the Territorial Government of Nunavut, bridges the relevant national departments (Department of Fisheries and Oceans) with subregional and local resource management organizations. Although these organizations have been developed for certain functions (wildlife management), they are applicable when broader system stresses (climate, sea ice change) must be addressed and a more systems-based approach to problem understanding is required. Exactly how these processes may occur is not fully understood but the role of comanagement boards is not likely to diminish in the face of rapid change.

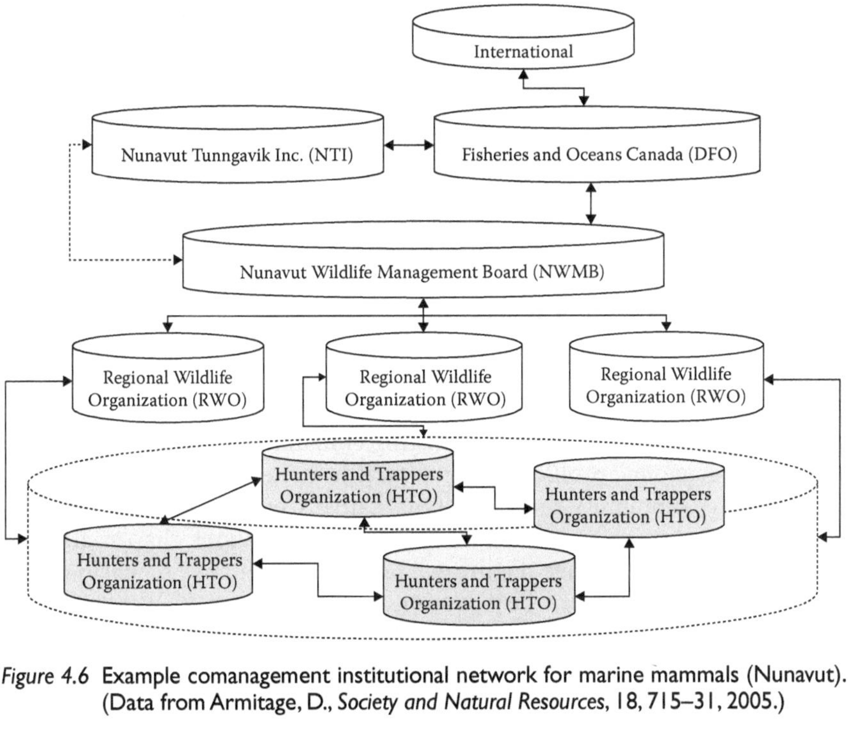

Figure 4.6 Example comanagement institutional network for marine mammals (Nunavut). (Data from Armitage, D., *Society and Natural Resources*, 18, 715–31, 2005.)

Knowledge coproduction

Knowledge is a key component of learning. A diversity of knowledge regarding complex social-ecological systems is fundamental to build adaptive capacity, particularly when explored through collaborative processes. With regard to climate change, Riedlinger and Berkes (2001) highlighted how bridging western science and indigenous knowledge produces complementarities in temporal and spatial scales (see also Laidler 2006) and thus contributes to a better understanding of impacts, adaptations, and monitoring needs. However, experiences in Canada's Arctic suggest that most effort is devoted to the idea of western scientific and indigenous knowledge *integration*. This may be too limited a view and may not be realistic given the fundamental differences in the epistemological roots of these knowledge systems. Rather, learning to support adaptation may depend upon a more holistic understanding of how knowledge can be *coproduced* through collaborative processes in which a plurality of knowledge sources and types are brought together to address a defined problem to build a systems-oriented understanding of that problem. This requires greater recognition of a multifaceted process of knowledge gathering, sharing, integration, interpretation, and application that contributes to learning and adaptation (Dale and Armitage 2011). In Canada's Arctic, experiences with knowledge coproduction are mixed, but there is evidence that the institutional context is creating the relational space for social interaction and trust building.

Participatory scenarios

Policy windows, bridging organizations, and knowledge coproduction reflect the institutional context for learning that can help coastal Arctic communities build adaptive capacity. For communities grappling with the triple squeeze, tangible methods are also required to help understand vulnerabilities and the sources of capacity to adapt. In the context of collaborative processes, qualitative and other visually based scenarios (Peterson 2007) can help make sense of the range of future socioeconomic, institutional, and biophysical conditions influencing community vulnerability (Wesche 2009). Scenarios and supporting narratives can be developed to reflect *possible futures* and can be based on existing climate data and impact projections, socioeconomic profiles, and potential future exposure-sensitivities as identified by community members. Qualitative scenarios are not meant to be predictive, but serve instead to create dialogue about change and the implications of that change. Figure 4.7 reflects a qualitative scenario developed for use in a northern First Nation Canadian context (Wesche 2009) where highly numerical or written outputs may not be appropriate in a predominately oral culture. Several scenarios were developed, each of which reflected different but linked dimensions of climate change and resource

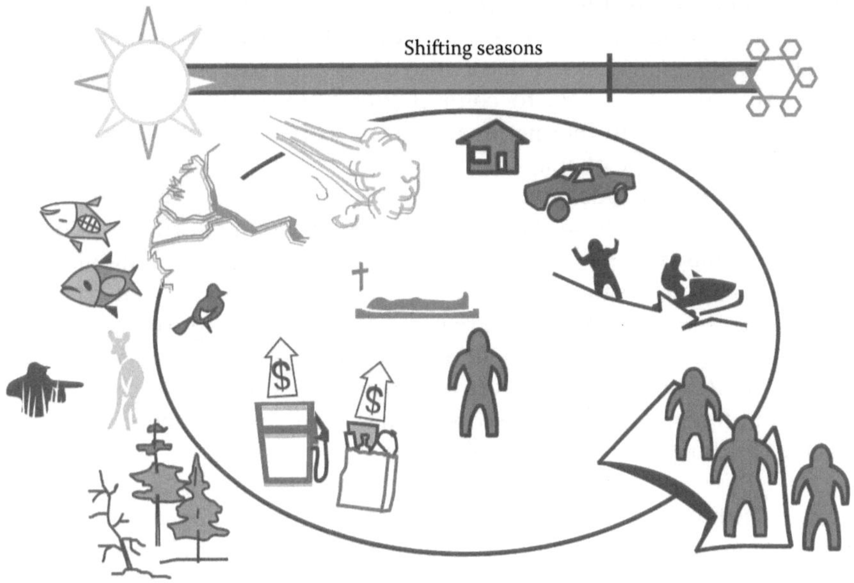

Figure 4.7 An example qualitative scenario used in a community context. (Data from Wesche, S. Responding to change in a northern aboriginal community [Fort Resolution, NWT, Canada]: Linking social and ecological perspectives, unpublished thesis, Wilfrid Laurier University, Waterloo, Ontario, Canada, 2009.)

development, as well as projected effects on livelihoods and economic opportunities, health and wellness, migration, and demographics. The most valuable outcomes of the scenario process include the dialogue generated on the nature of social and ecological changes faced by people, better understandings of exposure-sensitivities from an individual and community perspective, and identification of options to build capacity to address future exposure-sensitivities by reflecting on resource issues (e.g., human, financial), institutions, and governance (Wesche and Armitage 2010).

CONCLUSION

Climate change is but one of several interconnected drivers of change (Figure 4.5) that may lead to a fundamental transformation of the Arctic, with serious consequences for ecosystems and human well-being. Historical and contemporary responses to environmental change provide sources of resilience and reflect an impressive capacity of Arctic people to cope and adapt. There is much to be gained by understanding these experiences and the knowledge upon which they are based. At the same time, converging threats in the Arctic may overwhelm the ability of individuals and communities to respond proactively and equitably. Emerging strategies to adapt and transform also include institutional arrangements and comanagement practices that have evolved over the past two decades. Such governance processes help to connect groups in collaborative networks across levels (local to international) and provide opportunities for social learning—a key type of adaptation.

In the coming decades, the need to prepare for change and the lessons being learned now across the Arctic may be of vital importance in other coastal regions. Careful attention to questions about appropriate international governance regimes and the role for local, regional, and national governance arrangements will be required to enhance adaptive capacity. As outlined in this chapter, a bundle of strategies will be needed (see Berkes et al. 2005; Chapin et al. 2010) to augment the historical and present-day ability of Inuit and Inuvialuit to cope, adapt, and learn. Components of this bundle of strategies include the following:

- Identifying, encouraging, and building upon enabling conditions for adaptation and institutional change through policy and legislation. In the Arctic case, comprehensive land claims agreements have, despite many challenges with their implementation, played a catalytic role in creating institutional conditions that enhance collaboration and dialogue. Further integration of local–national (e.g., comanagement arrangement) and international (e.g., Arctic Council) governance mechanisms is required.
- Encouraging knowledge coproduction (linking indigenous knowledge and science) in which the multiple aspects of that process are carefully considered: knowledge gathering, sharing, integration, interpretation, and application (Dale and Armitage 2011). In collaborative coastal governance processes, knowledge coproduction can be an important mechanism for social learning which is itself a key type of adaptation.
- Maintaining the diversity of adaptation options that emerge with support for diverse cultures, languages, and knowledge systems. In a rapidly changing Arctic, sociocultural pressure and the loss of land-based skills and knowledge pose a challenge to future adaptation opportunities. Linking Inuit knowledge of environmental change with science will be possible only if the cultural foundations of that knowledge are maintained.
- Using scenarios (and other participatory simulations) to explore among diverse groups, interacting vulnerabilities associated with climatic and nonclimatic drivers of change, adaptive capacities, and resulting adaptation policy options.
- Making use of bridging organizations that can serve as arenas to link different actors, knowledge systems, and shared responsibilities for

action. In Canada's Arctic, a number of comanagement boards have a central role in linking actors vertically and horizontally and, therefore, can help facilitate multilevel and adaptive governance arrangements that over time serve to build trust.

ACKNOWLEDGMENTS

Many of the ideas and insights presented in this chapter are the result of productive collaboration with colleagues and graduate students. I would like to acknowledge in particular the contributions of Fikret Berkes, Ryan Plummer, Barry Smit, Sonia Wesche, and Aaron Dale. Special thanks to Pam Schaus for preparing the map. The chapter has benefited from the constructive feedback provided by chapter reviewers. Any limitations with the chapter are my responsibility. Research in the Canadian Arctic has been supported by ArcticNet, the International Polar Year (IPY-CAVIAR), and Canada's Social Sciences and Humanities Research Council (SSHRC).

REFERENCES

AMSA. (2009). "Arctic marine shipping assessment report 2009," Iceland: Arctic Council, The Protection of the Arctic Marine Environment (PAME) Working Group. http://arcticportal.org/uploads/4v/cb/4vcbFSnnKFT8AB5lXZ9_TQ/AMSA2009Report.pdf (accessed December 2011).

Anisimov, O.A., Vaughan, D.G., Callaghan, T.V., Furgal, C., Marchant, H., Prowse, T.D., Vilhjálmsson, H., and Walsh, J.E. (2007). "Polar regions (Arctic and Antarctic)," in Parry, M.L., Canziani, O.F., Palutikof, J.P., van der Linden, P.J., and Hanson, C.E. (eds.), *Climate Change 2007: Impacts, Adaptation and Vulnerability. Contribution of Working Group II to the Fourth Assessment Report of the Intergovernmental Panel on Climate Change*, Cambridge: Cambridge University Press. http://www.ipcc.ch/ (accessed July 2013).

Arctic Climate Impact Assessment (ACIA). (ed.) (2005). "Arctic climate impact assessment: ACIA scientific report," Cambridge; New York: Cambridge University Press. http://www.acia.uaf.edu/pages/scientific.html (accessed December 2011).

Arctic Research Consortium of the United States (ARCUS). (2009). "Sea Ice Outlook 2008—Summary report." http://www.arcus.org/ (accessed May 2009).

Armitage, D. (2005). "Community-based narwhal management in Nunavut, Canada: Change, uncertainty and adaptation," *Society and Natural Resources*, 18(8): 715–31.

Armitage, D. and Clark, D. (2005). "Issues, priorities and research directions for Oceans Management in Canada's North," in F. Berkes, R. Huebert, H. Fast, M. Manseau, and A. Diduck (eds.), *Breaking Ice: Renewable Resource and Ocean Management in the Canadian North*, Calgary, Ontario, Canada: Arctic Institute of North America and University of Calgary Press.

Bankes, N. (2005). "Exploring the roles of law and hierarchy in ideas of resilience," in F. Berkes, R. Huebert, H. Fast, M. Manseau, and A. Diduck (eds.), *Breaking Ice: Renewable Resource and Ocean Management in the Canadian North*, Calgary, Ontario, Canada: Arctic Institute of North America and University of Calgary Press.

Berkes, F. (2009). "Evolution of co-management: Role of knowledge generation, bridging organizations and social learning," *Journal of Environmental Management*, 90: 1692–702.

Berkes, F. and Armitage, D. (2010). "Co-management institutions, knowledge and learning: Adapting to change in the Arctic," *Études/Inuit/Studies*, 34(1): 109–31.

Berkes, F., Huebert, R., Fast, H., Manseau, M., and Diduck, A. (eds.) (2005). *Breaking Ice: Renewable Resource and Ocean Management in the Canadian North*, Calgary, Ontario, Canada: Arctic Institute of North America and University of Calgary Press.

Berkes, F. and Jolly, D. (2001). "Adapting to climate change: Social-ecological resilience in a Canadian western Arctic community," *Conservation Ecology*, 5(2): 18.

Chapin, F.S., III, Carpenter, S.R., Kofinas, G.P., Folke, C., Abel, N., Clark, W.C., Olsson, P. et al. (2010). "Ecosystem stewardship: Sustainability strategies for a rapidly changing planet," *Trends in Ecology and Evolution*, 25(4): 241–9.

Collins, K. and Ison, R. (2009). "Living with environmental change: Adaptation as social learning," *Environmental Policy and Governance*, 19: 351–7.

Dale, A. (2009). "Inuit qaujimajatuqangit and adaptive co-management: A case study of Narwhal Co-Management in Arctic Bay, Nunavut," unpublished thesis, Wilfrid Laurier University, Waterloo, Ontario, Canada.

Dale, A. and Armitage, D. (2011). "Marine mammal co-management in Canada's Arctic: Knowledge co-production for learning and adaptive capacity," *Marine Policy*, 35(4): 440–9.

Fazey, I., Gamarra, J., Fischer, J., Reed, M., Stringer, L., and Christie, M. (2010). "Adaptation strategies for reducing vulnerability to future environmental change," *Frontiers in Ecology and the Environment*, 8(8): 414–22.

Ford, J. (2009). "Sea ice change in Arctic Canada: Are there limits to Inuit adaptation?," in W.N. Adger, I. Lorenzoni, and K. O'Brien (eds.), *Adapting to Climate Change: Thresholds, Values, and Governance*, Cambridge: Cambridge University Press.

Ford, J., Pearce, T., Smit, B., Wandel, J., Allurut, M., Shappa, K., Ittusujurat, H., and Qrunnut, K. (2007). "Reducing vulnerability to climate change in the Arctic: The case of Nunavut, Canada," *Arctic*, 60(2): 150–66.

Furgal, C. and Prowse, T.D. (2008). "Northern Canada," in D.S. Lemmen, F.J. Warren, J. Lacroix, and E. Bush (eds.), *From Impacts to Adaptation: Canada in a Changing Climate 2008*, Ottawa, Ontario, Canada: Government of Canada.

Government of Canada. (1996). "Oceans Act," Ottawa: Department of Justice, Government of Canada. http://laws-lois.justice.gc.ca/ (accessed December 2011).

Government of Canada. (2002). "Oceans strategy," Ottawa: Fisheries and Oceans Canada, Government of Canada. http://www.dfo-mpo.gc.ca/ (accessed December 2011).

Gunderson L., Allen C., and Holling, C.S. (2009). *Fundamentals of Ecological Resilience*, Washington, DC: Island Press.

Hovelsrud, G. and Smit, B. (eds.) (2010). *Community Adaptation and Vulnerability in Arctic Regions*, Berlin, Germany: Springer.

Hovelsrud, G.K., McKenna, M., and Huntington, H.P. (2008). "Marine mammal harvests and other interactions with humans," *Ecological Applications*, 18(2): S135–S147.

Huebert, R. and Yeager, B. (2008). *A New Sea: The Need for a Regional Agreement on Management and Conservation of the Arctic Marine Environment*, Oslo, Norway: WWF International Arctic Programme.

Indian and Northern Affairs Canada (INAC). (2010). Final agreements and related implementation matters. http://www.ainc-inac.gc.ca/ (accessed October 2010).

Johannessen, O.M. and Miles, M.M. (2010). "Critical vulnerabilities of marine and sea ice–based ecosystems in the high Arctic," *Regional Environmental Change*, 11(S1): 239–48.

Keskitalo, C. and Kulyasova, K. (2009). "The role of governance in community adaptation to climate change," *Polar Research*, 28: 60–70.

Kocho-Schellenberg, J.-E. (2010). "A case study of the Husky Lakes beluga entrapment issue using network analysis: The structure of adaptive co-management, unpublished thesis," University of Manitoba, Winnipeg, Manitoba, Canada.

Krupnik, I. and Jolly, D. (eds.) (2002). *The Earth is Faster Now: Indigenous Observations of Arctic Environmental Change*, Fairbanks, AK: Arctic Research Consortium of the United States.

Laidler, G. (2006). "Inuit and scientific perspectives on the relationship between sea ice and climate change: The ideal complement?," *Climatic Change*, 78(2–4): 407–44.

Laidler, G., Ford, J., Gough, W.M., Ikummaq, T., Gagnon, A., Kowal, S., Qrunnut, K., and Irngaut, C. (2009). "Travelling and hunting in a changing Arctic: Assessing Inuit vulnerability to sea ice change in Igloolik, Nunavut," *Climatic Change*, 94(3/4): 363–97.

Learmonth, J.A., Macleod, C.D., Santos, M.B., Pearce, G.J., Crick, H.Q.P., and Robinson, R.A. (2006). "Potential effects of climate change on marine mammals," *Oceanography and Marine Biology: An Annual Review*, 44: 431–64.

Leichenko, R. and O'Brien, K. (2008). *Environmental Change and Globalization: Double Exposures*, Oxford: Oxford University Press.

Lenton, T.M., Held, H., Kriegler, E., Hall, J.W., Lucht, W., Rahmstorf, S., and Schellnhuber, H.J. (2008). "Tipping elements in the Earth's climate system," *Proceedings of the National Academy of Sciences of the United States of America*, 105(6): 1786–93.

Nadasdy, P. (2003). "Re-evaluating the co-management success story," *Arctic*, 56(4): 367–80.

NSIDC. (2010). "Weather and feedbacks lead to third-lowest extent," Arctic Sea Ice News & Analysis, October 4, 2010. http://nsidc.org/arcticseaicenews/2010/10/weather-and-feedbacks-lead-to-third-lowest-extent/ (accessed January 2012).

Nuttall, M., Berkes, F., Forbes, B., Kofinas, G., Vlassova, T., and Wenzel, G. (2005). "Hunting, herding, fishing, and gathering: Indigenous peoples and renewable resource use in the Arctic," in S. Symon, L. Arris, and B. Heal (eds.) *Arctic Climate Impact Assessment*, Cambridge; New York: Cambridge University Press. http://www.acia.uaf.edu/pages/scientific.html (accessed December 2011).

Pachauri, R.K. and Reisinger, A. (eds.); IPCC. (2007). *Climate Change 2007: Synthesis Report. Contribution of Working Groups I, II and III to the Fourth Assessment Report of the Intergovernmental Panel on Climate Change.* Geneva, Switzerland: IPCC. http://www.ipcc.ch/ (accessed July 2013).

Pahl-Wostl, C. (2009). "A conceptual framework for analysing adaptive capacity and multi-level learning processes in resource governance regimes," *Global Environmental Change*, 19(3): 354–65.

Pelling, M., High, C., Dearing, J., and Smith, D. (2008). "Shadow spaces for social learning: a relational understanding of adaptive capacity to climate change within organisations," *Environment and Planning A*, 40(4): 867–84.

Perovich, D., Meier W., Maslanik, J., and Richter-Menge, J. (2010). "Sea Ice Cover," Arctic Report Card 2010. http://www.arctic.noaa.gov/reportcard (accessed December 2011).

Peterson, D. (2007). "Using scenario planning to enable an adaptive co-management process in the Northern Highlands Lake District of Wisconsin," in D. Armitage, F. Berkes, and N. Doubleday (eds.), *Adaptive Co-Management: Collaboration, Learning and Multi-Level Governance*, Vancouver, British Columbia, Canada: University of British Columbia Press.

Riedlinger, D. and Berkes, F. (2001). "Contributions of traditional knowledge to understanding climate change in the Canadian Arctic," *Polar Record*, 37(203): 315–28.

Rockström, J. and Karlberg, L. (2010). "The quadruple squeeze: Defining the safe operating space for freshwater use to achieve a triply green revolution in the Anthropocene," *Ambio*, 39(3): 257–65.

Rockström, J., Steffen, W., Noone, K., Persson, A., Chapin, F.S., Lambin, E.F., Lenton, T.M. et al. (2009). "A safe operating space for humanity," *Nature*, 441(24): 427–75.

Schultz, L. (2009). "Nurturing resilience in social-ecological systems: Lessons learned from bridging organizations," PhD thesis, Stockholm University, Stockholm, Sweden.

Stokke, O.S. and Honneland, G. (2007). *International Cooperation and Arctic Governance: Regime Effectiveness and Northern Region Building*, London: Routledge.

Walsh, J.E., Anisimov, O., Hagen, J.O.M., Jakobsson, T., Oerlemans, J., Prowse, T.D., Romanovsky, V. et al. (2005). "Cryosphere and hydrology," in S. Symon, L. Arris, and B. Heal (eds.), *Arctic Climate Impact Assessment*, New York: Cambridge University Press. http://www.acia.uaf.edu/pages/scientific.html (accessed December 2011).

Wesche, S. (2009). "Responding to change in a northern aboriginal community (Fort Resolution, NWT, Canada): Linking social and ecological perspectives," unpublished thesis, Wilfrid Laurier University, Waterloo, Ontario, Canada.

Wesche, S. and Armitage, D. (2010). "'As long as the sun shines, the rivers flow and grass grows': Vulnerability, adaptation and environmental change in Deninu Kue Traditional Territory, Northwest Territories," in G. Hoveslrud and B. Smit (eds.), *Community Adaptation and Vulnerability in Arctic Regions*, Berlin, Germany: Springer.

White, G. (2006). "Cultures in collision: Traditional knowledge and Euro-Canadian governance processes in northern land-claim boards," *Arctic*, 59(4): 401–19.

Young, O. (2009). "Whither the Arctic? Conflict or cooperation in the circumpolar north," *Polar Record*, 45(232): 73–82.

Chapter 5

Climate change and infrastructure adaptation in coastal New York City

William Solecki, Cynthia Rosenzweig, Vivien Gornitz, Radley Horton, David C. Major, Lesley Patrick, and Rae Zimmerman

Abstract: A large fraction of New York City's four main infrastructure systems—energy, transportation, water and wastewater, and communications—lies less than three meters above mean sea level, making them vulnerable to coastal flooding during major storm events. The vulnerability of many of these systems was made evident during the storm surge associated with Hurricane Sandy in October 2012. By the end of the twenty-first century, sea-level rise will cause coastal floods to occur more frequently. In response, the city has considered a broad spectrum of flexible adaptation strategies, including operations and management measures as well as infrastructure investments and policy solutions that promote the co-benefits of providing protection against both existing and future climate risks. To sustain these climate adaptation strategies, climate indicators and climate-related coastal impact indicators should be regularly monitored and reassessed. The approaches, methods, and tools described here include a multijurisdictional stakeholder–scientist process, state-of-the-art scientific projections and mapping, and development of adaptation strategies based on an overarching risk-management approach.

INTRODUCTION

This chapter* describes approaches, methods, and tools that have been designed to help New York City and its agencies to respond to climate change and associated sea-level rise through a comprehensive climate risk and adaptation strategy identification process. Climate change adaptation planning in New York City is characterized by a multijurisdictional stakeholder-scientist process, state-of-the-art scientific projections and

* Elements of this chapter have been previously published in *Climate Change Adaptation in New York City: Building a Risk-Management Response* (Rosenzweig et al. 2010). We acknowledge the contributions made by the Boston Consulting Group in the formulation of the adaptation process and tools described herein.

mapping, and the development of adaptation strategies based on a risk-management approach. The chapter draws heavily from the authors' experience as members of the New York City Panel on Climate Change (NPCC) and contributors to its work (NPCC 2009, 2010; Rosenzweig et al. 2013). The NPCC provides expert advice to the City of New York and critical infrastructure stakeholders.

The discussion is organized into the following sections. First, we briefly review New York's coastal critical infrastructure. Next, projected climate changes and associated coastal zone hazards are presented. Impacts of past coastal storms and potential future extreme events conditions on critical infrastructure including Hurricane Sandy are discussed in the following section. Finally, adaptation opportunities and steps for moving forward are addressed.

Hurricane Sandy and its associated storm surge had a tremendous impact on New York City, killing 43 people and causing an estimated 19 billion dollars of losses (New York City 2013). The storm brought dramatic impacts to the city's critical infrastructure and its planning for future extreme events. The hurricane was associated with record storm surge that caused massive disruption of much of the city's urban systems. Much of the transportation was shut for a period following the storm, including the subway tunnels in the East River (connecting Manhattan with the outer boroughs) that were flooded. In response to Hurricane Sandy, Mayor Bloomberg commissioned the Special Initiative on Rebuilding and Resiliency (SIRR) to address the issue of what happened during Sandy, how climate change might alter the patterns and conditions of future events, and how to make the most vulnerable communities and systems in the city more resilient in the future. Whether or not Hurricane Sandy signals a tipping point in New York City policy and approach to extreme climate risks and planning, the events since its landfall have resulted in a dramatic increase in focus on these topics. This chapter is presented in this context—to serve both the discourse of the immediate Sandy aftermath and the longer-term narrative of climate change adaptation in coastal cities.

NEW YORK CITY AS A COASTAL CITY

New York City houses one of the densest infrastructure systems in the world. Because of its location, age, and composition, some of this infrastructure and materials may not be able to withstand projected strains and stresses from a changing climate. Regional infrastructure can be considered within the following main systems: Energy; transportation; water, wastewater, and other waste; and communications. These comprise the majority of the infrastructure of the New York City region, especially near the coast.

Much of the current infrastructure of the city was built during the twentieth century. Most of the available sites during that time were at the water's edge either on or near remnant wetlands or derelict industrial sites. The city's infrastructure profile includes some of the following characteristics (Zimmerman and Faris 2010):

- Three high-volume international airports.
- About two dozen major electricity generating facilities (Ascher 2005).
- 145,000 km of underground electrical distribution lines across about 55 distribution networks (O'Rourke et al. 2003).
- Multiple large-scale rail transit systems serve the region and three of the large ones involve almost 2,100 miles of track in the metropolitan area (including 659 track miles or 840 subway miles) which alone support about several billion trips by passengers annually in the region (Metropolitan Transportation Authority).
- 1.1 billion gallons of water supplied each day from a watershed of 1,972 square miles (New York City Department of Environmental Protection 2008).
- 6,700 miles of water distribution piping within the city's borders (New York City 2011).
- 14 wastewater treatment plants, 6,600 miles of sewer piping, and about 100 pumping stations (NYC-DEP 2008).
- Recycling or disposal of 15,500 tons per day of rubbish and other waste (New York City 2006).
- Vast networks of fixed and mobile telecommunication networks for personal, public (e.g., emergency), and business (e.g., Wall Street banking and investment companies).
- Approximately one million buildings within the five boroughs.

This complex set of infrastructure features is at the center of a sprawling metropolitan area that compromises an urban core city of 8.3 million (made up of five boroughs that are also counties) centered within another 28 counties and an additional 13 million residents. Together the 33 counties include territory of three states (Connecticut, New Jersey, and New York) and several thousand, often overlapping, jurisdictions including municipal, county, state, federal, and other agency/authority boundaries.

Severe coastal storms in the New York City region can cause extensive street, basement, and sewer floods and power outages in nearshore areas such as were experienced with Hurricane Sandy in October 2012. If extreme climate events become more frequent, there will be increased stress on these infrastructure systems as they play critical roles in emergency management. Furthermore, interdependencies, multiple owners, and complicated jurisdictions make coordination of adaptation planning especially challenging in the region (Zimmerman and Faris 2010).

CLIMATE RISK, SEA-LEVEL RISE, AND COASTAL FLOODING

Given New York's proximity to the coast and the nearly 1000 miles of shoreline (including tidal wetlands), sea-level rise and associated increased frequency, intensity, and duration of coastal flooding during storms present significant challenges to the city and its residents.

Observed sea-level rise in the New York City region

Prior to the industrial revolution, sea level had been rising along the East Coast of the United States at rates of 0.9–1.1 cm per decade. This was primarily due to regional subsidence as the earth's crust continues to slowly readjust to ice sheet melting since the end of the last ice age. Within the past 100–150 years, however, as global temperatures have increased, regional sea level has been rising more rapidly than during the preceding 1000 years (Engelhart and Horton 2012; Engelhart et al. 2009, 2011a, 2011b). Currently, rates of sea-level rise in the New York City metropolitan region range between 2.5 and 4.0 cm per decade, with a long-term rate for New York City since 1901–2006 averaging nearly 3 cm per decade (NOAA 2012). These sea level trends, as measured by tide gauges and satellites, include both the effects of recent global warming and the residual crustal adjustments to the removal of the ice sheets.

Coastal storms in the New York City region

Prevailing and dominant regional meteorological–oceanographic phenomena include hurricanes/tropical cyclones and extratropical storms characterized by strong northeast winds (known as nor'easters). Hurricanes strike New York relatively infrequently; when they do occur, they can produce large storm surges and wind damage. Nor'easters generally produce smaller surges and weaker winds than those rare hurricanes that do strike the region. Nevertheless, nor'easter effects can be large, in part because their long duration spanning several tidal cycles means an extended period of high winds, waves, and high water, often coinciding with high tides.

A large fraction of New York City and the surrounding infrastructure lies less than 3 m above mean sea level. The infrastructure in these areas is vulnerable to inundation during major storm events due to coastal surges and inland flooding caused by concurrent rainfall that is prevented from draining to the sea by the accompanying surge.* The current 1-in-100-year

* Surge is usually defined as the water level generated by a storm above that of the astronomical tide; flood level is the sum of the tide and the surge. National Oceanic and Atmospheric Administration (NOAA) tide gauges collect water level data at 6-min intervals and results are usually averaged hourly. For the NPCC study, the highest surge and flood levels per 24 h (i.e., daily) were used to calculate 1-in-100-year floods.

flood produces a water level (this includes stillwater + wave heights) approximately 4.57 m above the North American Vertical Datum of 1988 (Horton et al. 2010, 2011). The 1-in-100-year and 1-in-500-year coastal floods are generally produced by hurricanes, whereas the 1-in-10-year coastal flood event is generally associated with nor'easters. Documenting the occurrence rate of extreme storms over New York's history is challenging given reporting gaps and inconsistencies. Although no trend in observed storms is now evident, characterizing historical storms is a critical first step in understanding future storms and their impacts.

Sea-level rise projections

The NPCC sea-level rise projections for New York City are calculated as the sum of contributing components (Rosenzweig et al. 2013). These include changes in dynamic ocean height; thermal expansion; vertical land movements; loss of ice from glaciers, ice caps, and land-based ice sheets; gravitational, isostatic, and rotational effects resulting from recent ice mass loss; and land water storage. Others (e.g., Perrette et al. 2013; Slangen et al. 2012) have taken a similar regionalized approach to sea-level rise projections, with less specificity to the New York City region. For each of these components of sea level change, the 10th, 25th, 75th, and 90th percentiles of the distribution were estimated.[*] The sum of all components at each percentile is assumed to give the aggregate sea-level rise projection. This method does not take into account potential correlation among components. For example, freshening (infusion of nonsaltwater) of the North Atlantic due to Greenland ice sheet mass loss may cause changes in the Gulf Stream and North Atlantic Current, thereby affecting sea surface heights along the eastern seaboard.

Based on the above methods, by the 2020s sea level is projected to rise 10.2–20.3 cm (4–8 inches) according to the *middle range* 25th–75th percentile values[†] and 27.9 cm (11 inches) by the 90th percentile values. By the 2050s, the middle range estimate is 27.9–61 cm (11 to 24 inches) and 78.7 cm (31 inches) according to the 90th percentile estimate (Rosenzweig et al. 2013).

Future coastal floods and storms

As sea levels rise, coastal flooding associated with storms will very likely increase in intensity, frequency, and duration. The changes in coastal flood heights shown in Table 5.1 are solely due to changes in sea level through time. Any increase in the frequency or intensity of storms themselves would

* No uncertainty range was estimated for subsidence, because it is well-known for The Battery (Peltier 2012).

† In NPCC (Rosenzweig et al. 2010), the middle or central range was taken as the 17th–83rd percentiles.

Table 5.1 Coastal flood heights and recurrence for The Battery

	Baseline	*Low estimate (10th percentile)*	*Middle range (25th–75th percentile)*	*High estimate (90th percentile)*
2020s coastal flood heights				
Stillwater flood heights associated with 10-year flood	7.0 feet	7.2 feet	7.3–7.7 feet	7.9 feet
Flood heights associated with 100-year flood (stillwater + wave heights)	15.0 feet	15.2 feet	15.3–15.7 feet	15.8 feet
Stillwater flood heights associated with 100-year flood	10.8 feet	11.0 feet	11.1–11.5 feet	11.7 feet
Stillwater flood heights associated with 500-year flood	14.4 feet	14.6 feet	14.7–15.1 feet	15.3 feet
2020s flood recurrence				
Annual change of today's 10-year flood	10.0%	10.9%	11.8%–15.2%	16.9%
Annual change of today's 100-year flood	1.0%	1.1%	1.2%–1.5%	1.7%
Annual change of today's 500-year flood	0.2%	0.2%	0.2%–0.3%	0.3%
2050s coastal flood heights				
Stillwater flood heights associated with 10-year flood	7.0 feet	7.6 feet	7.9–9.0 feet	9.6 feet
Flood heights associated with 100-year flood (stillwater + wave heights)	15.0 feet	15.6 feet	15.9–17.0 feet	17.6 feet
Stillwater flood heights associated with 100-year flood	10.8 feet	11.4 feet	11.6–12.8 feet	13.4 feet
Stillwater flood heights associated with 500-year flood	14.4 feet	15.0 feet	15.3–16.4 feet	17.0 feet
2050s flood recurrence				
Annual change of today's 10-year flood	10.0%	14.3%	17.2%–31.3%	46.5%
Annual change of today's 100-year flood	1.0%	1.4%	1.7%–3.2%	5.0%
Annual change of today's 500-year flood	0.2%	0.3%	0.3%–0.4%	0.7%

Source: Rosenzweig, C. et al., *Climate Risk Information 2013: Observations, Climate Change Projections, and Maps*, New York City Panel on Climate Change, New York, 2013.

Note: The percentiles in the top row of 2020 and 2050 coastal flood heights and flood recurrence refer to the values for projected sea-level rise. Flood heights for the 2020s are derived by adding the sea-level rise projections for the corresponding percentiles to the baseline values. Baseline flood heights associated with the 10-year, 100-year, and 500-year floods are based on the stillwater elevation levels. For 100-year flood, height is also given for stillwater plus wave heights. Flood heights are referenced to the NAVD88 datum.

result in even more frequent future flood occurrences relative to the current 1-in-10- and 1-in-100-year coastal flood events. By the 2050s, the middle range sea-level rise projections alone suggests that coastal flood levels that currently occur on average once per decade may occur once every three to six years. The NPCC estimates that due to sea-level rise alone, today's 1-in-100-year flood may occur approximately five times more often by the 2050s with the high estimate for sea-level rise.

For some climatic variables, quantitative projections are unavailable due to insufficient data or incomplete knowledge of physical processes. Using instead qualitative indices of likelihood developed by the Intergovernmental Panel on Climate Change (IPCC), the frequency of intense hurricanes (Category 3 or greater on the Saffir–Simpson scale) and extreme winds was characterized as *more likely than not* to increase throughout the twenty-first century. However, this study did not consider changes in frequency and intensity of nor'easters, nor of hurricanes. (However, see Lin et al. 2012. They, on the other hand, did not fully examine the role of sea-level rise.)

COASTAL STORM IMPACTS ON CRITICAL INFRASTRUCTURE

The historical record of past storm surge and coastal flooding provides opportunities for current coastal zone managers to understand future risks. In this section, we examine the circumstances and impacts of past coastal storms on New York City's regional infrastructure. Although Sandy has become a crucial reference point for contemporary decision-making, it is in this longer historical context that a more complete understanding of infrastructure vulnerability in the New York's coastal zone can be best understood.

Lessons learned from past experiences

New York State has been frequently impacted by storm surge from tropical and extratropical storms. Though the observational record of storm events spans only the last 150 years (Table 5.2), contemporary newspapers, personal diaries, ship logs, and town histories have been used to extend this record back to the 1620s (Boose et al. 2001). In addition, large prehistoric storms have left telltale signs of their passing in the geological proxy record, and these signals have been used to extend the storm record over a few millennia (Donnelly and Woodruff 2007; Donnelly et al. 2001a, 2001b, 2004; Liu and Fearn 2000a, 2000b; Mann et al. 2009; Nyberg et al. 2007; Scileppi and Donnelly 2007). Using this paleotempestological approach to analyze local coastal sediment samples, Scileppi and Donnelly (2007) concluded that many strong hurricanes have impacted the New York City and Long Island region over the past ~3500 years.

Table 5.2 Major tropical and extratropical storms impacting New York City

Date	*Name*	*Category or level*[a]
September 3–5, 1815	Great September Gale of 1815	3
September 3, 1821	Norfolk and Long Island Hurricane	1–2
June 1825	Early June Hurricane	1
October 13, 1846	Great Hurricane of 1846	n/a
October 6–7, 1849	October Hurricane of 1849	n/a
September 1858	New England Storm	1
November 2, 1861	Expedition Hurricane	n/a
September 1869	September Gale of Eastern New England	1
March 11–14, 1888	Great Blizzard of 1888	n/a
August 23, 1893	Midnight Storm	1–2
September 21,1938	Long Island Express	3
September 15, 1944	Great Atlantic Hurricane	1
December 25–26, 1947	Major Blizzard	n/a
August 31, 1954	Hurricane Carol	3
September 12, 1960	Hurricane Donna	3
September 21, 1961	Hurricane Esther	1–2
March 3–8, 1962	Ash Wednesday Storm[b]	n/a
June 22, 1972	Hurricane Agnes	1
August 10, 1976	Hurricane Belle	1
February 6, 1978	Major Blizzard	n/a
September 27, 1985	Hurricane Gloria	2–3
August 28, 1991	Hurricane Bob	2
October 29–November 2, 1991	Halloween Storm, *Perfect Storm*[b]	3
December 11–12, 1992	Major Nor'easter[b]	3
March 13–14, 1993	Storm of the Century	3
September 16, 1999	Hurricane Floyd	2
September 21, 2003	Hurricane Isabel	TS
September 2, 2006	Hurricane Ernesto	TS
November 11–14, 2009	Nor'easter (*Nor'Ida*)	3
March 12–13, 2010	Nor'easter	n/a
August 28, 2011	Hurricane Irene	TS
September 8, 2012	Tropical Storm Lee	TS
October 29, 2012	Hurricane Sandy	TS

Source: Buonaiuto, F. et al., "Coastal zones," in C. Rosenzweig et al. (eds.), *NYS ClimAID: Integrated Assessment for Effective Climate Change Adaptation Strategies in New York State*, 2011; Rosenzweig, C. et al., *Climate Change Adaptation in New York City: Building a Risk Management Response*, Annals of the New York Academy of Sciences, New York, 2010, http://www.nyas.org/.

[a] Levels are based on intensity scale developed by Salmun et al. 2009.

[b] Extratropical storms with the highest surges since 1960.

Though the New York coastal area has been hit by six major (Category 3) hurricanes since 1851 (Blake et al. 2011; Ludlum 1963), nor'easters and winter storms are also responsible for major coastal flooding. For example, the 1962 Ash Wednesday winter storm (March 6–7) occurred at perigee (closest approach of the moon to the earth in its monthly orbit) and at new moon, close to the spring equinox that is associated with higher than average tides. The storm lasted for over five tidal cycles, resulting in 2.1 m flood elevations at The Battery—the southernmost tip of Manhattan Island. The December 1992 nor'easter, which arrived on a full moon and lasted for three tidal cycles, flooded The Battery with 2.4 m of water and led to the near complete shutdown of the metropolitan New York transit system with the Port Authority Trans-Hudson trains out of operation for 10 days (Gornitz et al. 2002). In comparison, the glancing blow to the City of the Great Hurricane of 1938 (also known as the *Long Island Express*), which made landfall approximately 100 km east of Manhattan, nevertheless caused flooding from the East River three blocks inland, yet did not linger long enough to cause impacts at high tide. (It was, however, the most disastrous twentieth century storm that affected Long Island and southern New England, producing surges of up to 10 m [including waves] and nearly 700 fatalities.) However, the flood height (3.38 m above NAVD88) from Hurricane Sandy that struck the New York City metropolitan area on October 29, 2012, was the highest in at least 200 years, resulting in 43 fatalities and an estimated USD 19 billion in damages.

Coastal storms, such as Sandy, in the New York City metropolitan region cause extensive damage to human-made structures and coastal ecosystems with the resulting loss of property and lives. Potential losses of almost USD 11 billion have been estimated for single-family residences (not including other buildings) under a realistic worst-case hurricane surge on Long Island, New York (the eastern portion of the New York City metropolitan region), which would rank 6th on a list of 13 most vulnerable US areas (FASS 2010). Sandy's destructiveness far exceeded this estimate with total damages reaching USD 19 billion (New York City 2013). Of the 43 deaths, 23 occurred in Staten Island, with the remainder spread throughout Queens, Brooklyn, and Manhattan. At least 800 buildings were destroyed or severely damaged, with thousands more impacted. Nearly two million people lost power during the storm. Most tunnels connecting Manhattan to Queens and Brooklyn were flooded; subway, bus, and ferry transportation were suspended (New York City 2013). The IPCC (2007) states that intense hurricanes and associated extreme wind events will more likely than not become more frequent due to expected warming of the upper ocean in the tropical cyclone genesis regions. Of the hurricanes that formed, the fraction of those that develop into intense hurricanes will likely increase (Horton et al. 2010). As sea-level rise alone will provide a higher baseline above which storm surges and wave action operate, any additional increase in storm intensity would magnify the risks to coastal communities.

Future challenges and opportunities

Climate change presents many challenges and some opportunities for managers of coastal infrastructure in and around New York City. As sea level rises, the areal extent of a current 1% flood area will penetrate much farther inland toward the later part of the century, especially in regions of relatively flat topography such as much of southern Brooklyn and Queens. Figure 5.1 highlights the dramatic landward progression of the 1-in-100-year flood zone, especially in the Greater Jamaica Bay area of Brooklyn and Queens,

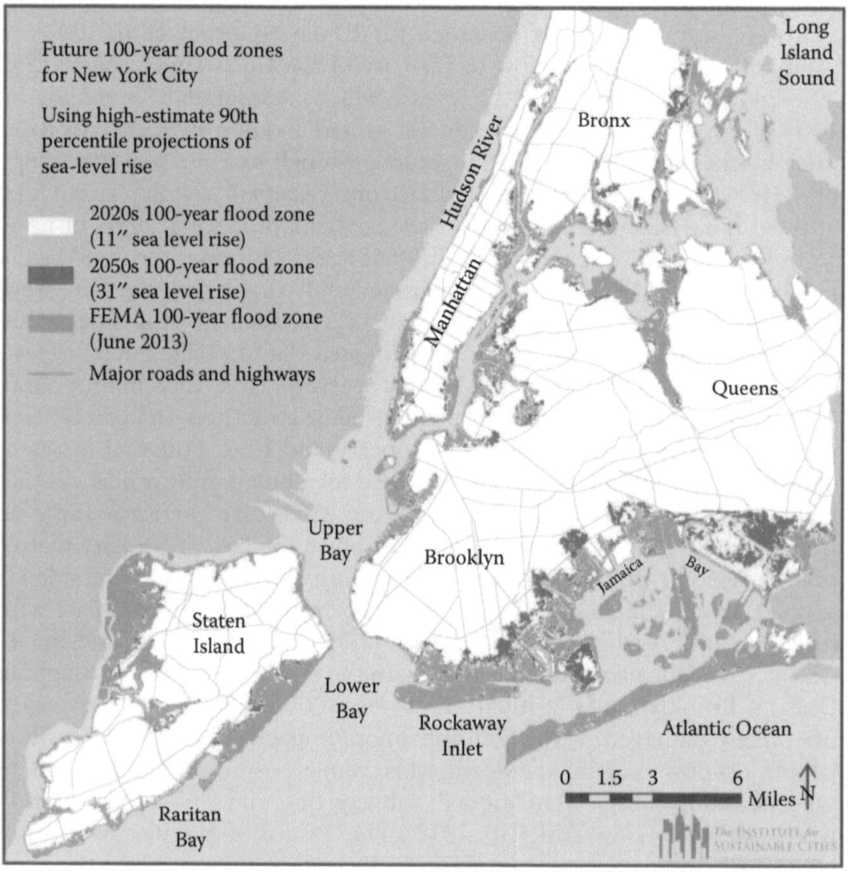

Figure 5.1 The potential areas that could be impacted by the 100-year flood in the 2020s and 2050s based on projections of the high-estimate 90th percentile sea-level rise scenario. (Data from Rosenzweig, C. et al., *Climate Risk Information 2013: Observations, Climate Change Projections, and Maps*, New York City Panel on Climate Change, New York, 2013; New York City, *PlaNYC A Stronger More Resilient New York* [*SIRR*], City of New York Press, New York, 2013; Map Authors: K. Grady, L. Patrick, W. Solecki, June 2013.)

for the 2020s and 2050s, for the high-estimate 90th percentile sea-level rise. The implications of including updated future sea level trends (as in this example) would be far reaching, including new communities for potential inclusion in the National Flood Insurance Program and changing the extent and base flood elevations of the New York City Flood Insurance Rate Maps.

Many components and facilities of the city's critical infrastructure would potentially become vulnerable to flooding from increased frequency of extreme precipitation events and sea-level rise (Table 5.3). Although it can be difficult and/or expensive to retrofit existing facilities, a number of very large new projects are being planned or are underway in New York City that provide an opportunity to incorporate climate change adaptations in the form of elevating, flood proofing, or providing heat-resistant materials for transportation structures. In the next section, we examine how these projects can be more effectively included in the city's emerging climate adaptation planning.

Table 5.3 Impacts of sea-level rise, coastal floods, and storms on critical coastal infrastructure by sector in the New York City region

Communications	*Energy*	*Transportation*	*Water and waste*
Higher average sea level			
• Increased saltwater encroachment and damage to low-lying communications infrastructure not built to withstand saltwater exposure • Increased rates of coastal erosion and/or permanent inundation of low-lying areas, causing increased maintenance costs and shortened replacement cycles	• Increased saltwater encroachment and damage to low-lying power plants not built to withstand saltwater exposure • Increased rates of coastal erosion and/or permanent inundation of low-lying areas, threatening power plants located there, resulting in higher maintenance costs and shorter replacement cycles	• Increased saltwater encroachment and damage to infrastructure not built to withstand saltwater exposure • Increased rates of coastal erosion and/or permanent inundation of low-lying areas, resulting in increased maintenance costs and shorter replacement cycles • Decrease clearance levels under bridges	• Increased saltwater encroachment and damage to freshwater and waste infrastructure • Increased pollution runoff from brownfields (e.g., solid and hazardous waste site and waste storage facilities) • Permanent inundation of low-lying areas, wetlands, piers, and marine transfer stations

(*Continued*)

Table 5.3 (Continued) Impacts of sea-level rise, coastal floods, and storms on critical coastal infrastructure by sector in the New York City region

Communications	*Energy*	*Transportation*	*Water and waste*
More frequent and intense coastal flooding			
• Increased need for emergency management actions with high demand on communications infrastructure; potential physical and chemical (saltwater corrosion) damage to communications infrastructure in low-lying areas	• Increased need for emergency management actions • Exacerbate flooding of streets and low-lying power plants and increase the use of energy to control floodwaters	• Increased need for emergency management actions • Exacerbate flooding of streets, subways, tunnel, and bridge entrances and cause structural damage to infrastructure due to wave action and saltwater inundation and associated corrosion	• Increased need for emergency management actions including reserve pressure sewer flooding • Exacerbate street, basement, and sewer flooding and will also lead to structural damage to infrastructure due to wave action • Episodic inundation of low-lying areas, wetlands, piers, and marine transfer stations

Source: Horton, R. and Rosenzweig, C., "Climate risk information," in C. Rosenzweig and W. Solecki (eds.), *Climate Change Adaptation in New York City: Building a Risk Management Response*, Annals of the New York Academy of Sciences, New York, 2010, http://www.nyas.org/; updated by authors.

MOVING FORWARD: ADAPTATION STRATEGIES

Discussion of adaptation strategies within the context of New York City critical infrastructure to date has focused on operations and management, investments in infrastructure, and policy solutions at the sector or regional or citywide scales. (For a recommended set of adaptation assessment steps, see Major and O'Grady 2010.) Increasingly, the attention has sharpened to the development of climate resiliency planning efforts that could provide adaptation opportunities which bring near-term benefits such as increased resource use efficiency as well as long-term benefits. In the following section, we discuss these elements in detail. In general, the approach of the city administration to climate change impacts has been increasingly framed as a flexible adaptation approach focused on promoting benefits from strategies that provide protection against both existing and future climate risks (New York City 2011). The Office of the Mayor's 2013 report *A Stronger More Resilient New York* describes incentives in all sectors of the city that emerged in response to the impacts of Hurricane Sandy. Scheduled capital upgrades of infrastructure and enhanced climate science understanding will provide opportunities for effective adaptation planning, strategies, and implementation in the future.

Operations and management

There is a great potential, at least in the near term, for adaptation measures related to current operations and management to deal with sea-level rise and storms. For the transportation sector with assets and operations near the coast, initiatives include elevating and installing barriers around assets to maintain systems operations during storms, developing resiliency planning exercises and improving communications about the restoration of services, and increasing system flexibility and redundancy through expanded bus and ferry networks (New York City 2013). For the water sector, which in New York City includes 14 wastewater pollution control plants that discharge into the New York/New Jersey Harbor Estuary, adaptation strategies related to operations and management include hardening pumping stations and wastewater treatment plants, developing cogeneration facilities at select plants, and reducing combined sewer overflows through green infrastructure and high-level storm sewers (New York City 2013).

Investments in infrastructure

Adaptation strategies for critical infrastructure in the coastal zone can include both *hard* and *soft* measures.

Hard measures

In response to projected rates of sea-level rise, especially if rates follow the high-end scenario, existing hard structures in the New York City region will need to be strengthened and elevated over time (Gornitz 2001). The SIRR report focused significant attention to the small- to medium-scale examples of hard measure storm surge protection features and strategies (New York City 2013).

In general, shoreline armoring is applied where substantial assets are at risk. Existing hard structures in the New York City region include seawalls, groins, jetties, breakwaters, bulkheads, and piers. Seawalls and bulkheads, a common form of shore protection in the region, often intercept wave energy, increasing erosion at their bases, which eventually undermines them. Erosion can be reduced by placing rubble at the toe of the seawall. Groins, often built in series, intercept littoral sand moved by longshore current, but may enhance beach erosion further downdrift, if improperly placed. Similarly, jetties designed to stabilize inlets or to protect harbors may lead to erosion. Individually engineered solutions can also be achieved by raising structures and systems or critical system components to higher elevations (Jacob et al. 2001). This may be done without moving them to higher ground.

One possible long-term *hard measure* that has been suggested for New York City and brought up in the SIRR is storm surge barriers designed to protect against high water levels, which would increase in height as sea level rises (and possibly also through increasing intensity of storms) (Hill 2011; Zimmerman and Faris 2010). Examples of such barriers are in place in Europe, including the Thames Barrier in London and the Maeslant Barrier in the Netherlands (Dircke et al. 2012). The risk of future casualties and storm surge damage from hurricanes and nor'easters might be reduced by barriers placed across vulnerable openings to the sea, but would not protect against river floods. Each barrier would require large open navigation channels for ships and a porous cross section allowing sufficient tidal exchange and river discharge from New York Harbor to maintain ship passage and water quality (Hill 2011).

At present, a large-scale storm surge barrier concept has not been accepted by the city government; Aerts et al. (2009, 2012) provide a balanced discussion of these proposed barriers. Storm surge barriers are seen as relevant as part of a long-term, staged response to rising sea levels and flooding, especially if rising sea levels and enhanced flooding proceed at the higher end of the projections. A key point is that those risks still need to be better characterized in regard to the efficacy of citywide measures. Such options, which would entail significant economic, environmental, and social costs, would require very extensive study before being regarded as appropriate for implementation, especially as alternative approaches to adaptation become discussed and debated. Within the SIRR, the city administration has proposed a number of local storm surge barriers to protect against *back door* flooding that occurred through the coastal zone of Brooklyn and Queens.

New York City could protect against some levels of surge over at least the next several decades with a combination of local measures (such as flood walls and reclaimed natural barriers), improved storm information and forecasting to help managers of power plants, airports, and wastewater treatment plants to prepare for future extreme events and better evacuation planning. In the aftermath of Sandy, a series of such hard measures have been put forth including the proposed installation of armor stone shoreline protection, repairing and raising bulkheads, complete tide and emergency floodgate repairs, and the installation of offshore breakwaters (New York City 2013).

Soft measures

Because the New York City coastline has extensive beaches and coastal wetlands as well as built-up areas, and because erosion problems are often associated with hard structures along the coast, *softer* approaches involving wetland and dune restoration and beach nourishment have emerged as a potential method of shoreline protection, where possible. The benefits of soft measures were extensively discussed in the post-Sandy context, and a limited number of new strategies including living shorelines, beach nourishment, and dune installation projects were put forward in the SIRR report (New York City 2013). Beaches, including Coney Island, Brighton Beach, and The Rockaways in the boroughs of Brooklyn and Queens, are maintained for public recreational use, while the nearby Gateway National Recreation Area, making up a large portion of Jamaica Bay, is an important nature reserve and bird migration stopover site. Beach nourishment or restoration consists of placing sand that has usually been dredged from offshore or other locations onto the upper part of the beach. Beach nourishment needs to be repeated over time because the erosional processes at work are continual. Under sea-level rise and associated enhanced coastal flooding, beaches will require additional sand replenishment to be maintained (Gornitz 2001).

Another *soft* approach is to enhance and expand the existing Bluebelt systems in the city (NYC-DEP 2008; Stalenberg 2011). The Bluebelt is a stormwater management system covering about one-third of Staten Island. The program preserves natural drainage corridors, including streams, ponds, and other wetland areas. Preserving these wetlands enables them to perform ecosystem functions of conveying, storing, and filtering stormwater, while providing open space and diverse wildlife habitats. The SIRR report included recommendations to dramatically expand the Bluebelt system because evaluation after Hurricane Sandy found it to be effective in providing temporary storage for storm surge waters and lessening the impacts of the storm-related flooding.

Creating a *soft edge* shoreline to reduce the seaward gradient also helps to dampen ship wakes and lessen wave energy and surge impacts, while providing new wildlife habitat (Nordenson et al. 2010). A replanted salt marsh along the waterfront of the Brooklyn Bridge Park in New York City is an example of a soft edge shore (NYC-DCP 2011). Salt marsh restoration underway at Jamaica Bay, New York City, will expand remaining wildlife habitat and help lessen storm surge flooding in surrounding Brooklyn and Queens neighborhoods (e.g., see Figure 5.1 above; USACE 2012). Other soft adaptation approaches are described in Nordenson et al. (2010). Green infrastructure strategies throughout New York City that incorporate vegetative areas and permeable pavement material to retard the movement of and absorb water have been extensively developed by New York City's sustainable stormwater management plan (2008) and by New York City's green infrastructure plan (2010).

Policy solutions

Adaptive land-use management and changes in zoning, design standards, and regulations are mechanisms by which coastal zone adaptation can proceed through policies. Titus et al. (2009) categorized policy options for dealing with sea-level rise as *protect, retreat, or abandon*. Adaptive land-use management could involve the development of erosion/flood setbacks; limiting new high-density construction in high hazard zones; and rezoning for low density and recreation uses. Creative land use, as is being considered in Rotterdam, could raise buildings on stilts, use ground floors for communal activities and parking, and design parks or open green spaces as water-absorbing areas (Aerts et al. 2009; Stalenberg 2011).

Potential adaptations related to land-use management being considered include developing plans allowing for coastal inundation in defined areas; strengthening building codes for construction of more *storm-proof* buildings (with the caveat that the public needs to know that no building can be made *fail-safe* indefinitely); and gradually retreating from the most at-risk areas or using these areas differently, such as for parkland that could be flooded with minimal damage (NYC-DEP 2008). This could entail obtaining vacant coastal land to act as buffers against flooding and storm damage and/or to allow for inland migration of coastal wetlands. Examples of policy solutions put forth in city's SIRR 2013 initiatives include adapting building and zoning codes to improve the flood resiliency of buildings in the 100-year floodplain, developing flood standards for the placement of telecommunication equipment in buildings, and adopting new design guidelines for wastewater facilities subject to storm surge and sea-level rise.

To effectively adapt to climate change, laws, regulations, and some basic legal frameworks that govern infrastructure must also be adopted. Sussman and Major (2010) considered the potential for zoning changes and limiting or even curtailing new construction in high hazard zones. They examined a wide range of current environmental laws and regulations at all levels relevant to New York City to determine their applicability to adaptation efforts. Laws applicable to New York City are enacted by legislative bodies, the United States Congress, the New York State Legislature, and the New York City Council. Regulations are issued by governmental agencies or authorities. They often have the force of law and may be issued in many forms including rules, orders, procedures, and administrative codes.

Indicators and monitoring

Another key recommendation of the NPCC is that climate change impacts and adaptation strategies should be regularly monitored and reassessed as part of any climate change adaptation strategy (Jacob and Blake 2010). This could be done taking into account changes in climate science, impacts, and adaptation strategies, as well as other factors such as population growth rates and technological advancements that will influence infrastructure in the region.

In order to successfully monitor future climate and impacts related to developing New York City's coastal adaptation strategy, two types of indicators to be tracked should be identified in advance. First, climate indicators such as global and regional sea-level rise can provide an indicative measure of overall climate trends and whether climate changes are occurring beyond the projected range. Given the large uncertainties in climate projections, continual monitoring of climate indicators can play a critical role in refining future projections and reducing uncertainties. Second, climate-related coastal impact indicators provide a way to identify consequences of sea-level rise and enhanced coastal flooding as they emerge. For example, more frequent transportation disruptions may be an indicator of climate-related changes in coastal storm frequency and/or strength.

Changes in these climate variables also may produce feedback effects, which in turn could exacerbate certain climate-related impacts. These would need to be monitored as well. Examples of such impacts include increased shoreline erosion and changes in the biological and chemical composition of coastal waters and hence in marine ecosystems.

In addition to monitoring climate and impacts, advances in scientific understanding, technology, and adaptation strategies should also be closely monitored. Technological advances, such as those in materials science and engineering, could influence design and planning, enabling *greener* solutions and potentially also resulting in cost savings.

One potential pitfall of monitoring over short timescales, especially for small regions, is that it is easy to mistake natural variability for a long-term trend. Creating an effective climate-monitoring program is a long-term commitment and requires different methods over a much longer timescale than more common short-term monitoring efforts. The NPCC has recommended that such a monitoring program be established and maintained. To accomplish this, it would be useful to establish federal and local partnerships between New York City and the National Oceanic and Atmospheric Administration's Regional Integrated Sciences and Assessments research groups and Regional Climate Centers.

CONCLUSION

The complex interaction between potential climate change and the management and construction of critical infrastructure in the New York City region illustrates the significant challenges confronting coastal cities throughout the world due to climate change. Hurricane Sandy was a devastating reminder of these risks and vulnerabilities. Coastal cities face a specific set of conditions that require a unique set of adaptation strategies due to their population densities and critical infrastructure in low-lying coastal zones, immobility, overlapping regulatory jurisdictions, and especially the variety and complex infrastructure networks, upon which the population is highly dependent. Although specifically designed for New York City, the comprehensive approaches, methods, and tools described here can be modified and applied to many urban coastal areas as well as noncoastal areas. These include a multijurisdictional stakeholder–scientist interaction process, state-of-the-art scientific projections and mapping, and development of a broad range of adaptation strategies based on an overarching risk-management approach.

Although climate change will exacerbate existing urban challenges and environmental stressors, it also provides opportunities for cities by encouraging safer and more strategic infrastructure investments and improving urban planning and regulation. Although most US cities are struggling to finance their required existing infrastructure investments without considering consequences of climate change, adaptation to climate change can provide additional incentives for increased cooperation among local, state, and federal agencies and new funding sources. If cities respond wisely with foresight, they will be better prepared to forestall any potential adverse impacts, improve climate management, and at the same time upgrade needed infrastructure that facilitates their citizens' comfort and movement.

REFERENCES

Aerts, J., Botzen, W., Bowman, M.J., Ward, P.J., and Dircke, P. (eds.) (2012). *Climate Adaptation and Flood Risk in Coastal Cities*, London; New York: Earthscan.

Aerts, J., Major, D.C., Bowman, M., Dircke, P., and Marfai, M.A. (2009). *Connecting Delta Cities: Coastal Cities, Flood Risk Management, and Adaptation to Climate Change*, Amsterdam, the Netherlands: Free University of Amsterdam Press.

Ascher, K. (2005) *Anatomy of a City*, London: Penguin Books.

Blake, E.S., Rappaport, E.N., and Landsea, C.W. (2007). "The deadliest, costliest, and most intense United States tropical cyclones from 1851 to 2006 (and other frequently requested hurricane facts)," Technical Memorandum NWS TPC-5, Miami, FL: National Oceanic and Atmospheric Administration.

Blake, E.S., Landsea, C.W., and Gibney, E.J. (2011). *The Deadliest, Costliest, and Most Intense United States Tropical Cyclones from 1851 to 2010 (and Other Frequently Requested Hurricane Facts)*. NOAA Technical Memorandum NWS NHC-6.

Boose, E.R., Chamberlin, K.E., and Foster, D.R. (2001). "Landscape and regional impacts of hurricanes in New England," *Ecological Monographs*, 71: 27–48.

Buonaiuto, F., Patrick, L., Hartig, E., Gornitz, V., Stedinger, J., Tanski, J., and Waldman, J. (2011). "Coastal zones," in C. Rosenzweig, W. Solecki, A. DeGaetano, S. Hassol, P. Grabhorn, and M. O'Grady (eds.), *NYS ClimAID:*

Integrated Assessment for Effective Climate Change Adaptation Strategies in New York State, New York: Annals of the New York Academy of Science, 1244: 121–62.

Dircke, P., Jongeling, T., and Jansen, P. (2012). "Navigable storm surge barriers for coastal cities: an overview and comparison," in J. Aerts, W. Botzen, M.J. Bowman, P.J. Ward, and P. Dircke (eds.), *Climate Adaptation and Flood Risk in Coastal Cities*, London; New York: Earthscan.

Donnelly, J.P., Butler, J., Roll, S., Wengren, M., and Webb, T., III (2004). "A backbarrier overwash record of intense storms from Brigantine, New Jersey," *Marine Geology*, 210: 107–21.

Donnelly, J.P., Roll, S., Wengren, M., Butler, J., Lederer, R., and Webb, T., III (2001a). "Sedimentary evidence of intense hurricane strikes from New Jersey," *Geology*, 29(7): 615–18.

Donnelly, J.P., Smith Bryant, S., Butler, J., Dowling, J., Fan, L., Hausmann, N., Newby, P. et al. (2001b). "700 yr sedimentary record of intense hurricane landfalls in Southern New England," *Geological Society of America Bulletin*, 113(6): 714–27.

Donnelly, J.P. and Woodruff, J.D. (2007). "Intense hurricane activity over the past 5,000 years controlled by El Nino and the West African monsoon," *Nature*, 447: 465–68.

Engelhart, S.E. and Horton, B.P. (2012). "Holocene sea level database for the Atlantic coast of the United States," *Quaternary Science Reviews*, 54: 12–25.

Engelhart, S.E., Horton, B.P., Douglas, B.C., Peltier, W.R., and Trnqvist, T. (2009). "Spatial variability of late Holocene and 20th century sea-level rise along the Atlantic coast of the United States," *Geology*, 37(12): 1115–18.

Engelhart, S.E., Horton, B.P., and Kemp, A.C. (2011a). "Holocene sea level changes along the United States' Atlantic Coast," *Oceanography*, 24(2):70–9.

Engelhart, S.E., Peltier, W.R., and Horton, B.P. (2011b). "Holocene relative sea-level changes and glacial isostatic adjustment of the U.S. Atlantic coast," *Geology*, 39(8): 751–54.

First American Spatial Solutions (FASS) (2010). "Corelogic finds more than $234 billion in residential storm surge exposure in 13 U.S. cities." http://www.faspatial.com/storm-surge (accessed August 2010).

Gornitz, V. (with contributions by S. Couch) (2001). "Sea-level rise and coasts," in C. Rosenzweig and W.E. Solecki (eds.), *Climate Change and a Global City: The Potential Consequences of Climate Variability and Change, Metro East Coast*, Report for the U.S. Global Change Research Program. New York: Columbia Earth Institute.

Gornitz, V., Couch, S., and Hartig, E.K. (2002). "Impacts of sea-level rise in the New York City metropolitan area," *Global and Planetary Change*, 32(1): 61–88.

Hill, D. (ed.) (2011). "Against the Deluge: storm surge barriers to protect New York City," *Conference Proceedings*, Polytechnic Institute of New York University, Brooklyn, NY, March 30–31, 2009.

Horton, R., Gornitz, V., Bowman, M., and Blake, R. (2010). "Climate observations and projections," in C. Rosenzweig, and W. Solecki (eds.), *Climate Change Adaptation in New York City: Building a Risk Management Response*, New York: Annals of the New York Academy of Sciences. http://www.nyas.org/ (accessed March 2012).

Horton, R. and Rosenzweig, C. (2010). "Climate risk information," in C. Rosenzweig and W. Solecki (eds.), *Climate Change Adaptation in New York City: Building a Risk Management Response*, New York: Annals of the New York Academy of Sciences. http://www.nyas.org/ (accessed March 2012).

Horton, R.M., Gornitz, V., Bader, D.A., Ruane, A.C., Goldberg, R., and Rosenzweig, C. (2011). Climate hazard assessment for stakeholder adaptation planning in New York City, *Journal of Applied Meteorology and Climatology*, 50: 2247–2266.

IPCC. (2007). "Summary for Policymakers," in S. Solomon, D. Qin, M. Manning, Z. Chen, M. Marquis, K.B. Averyt, M. Tignor, and H.L. Miller (eds.), *Climate Change 2007: The Physical Science Basis. Contribution of Working Group I to the Fourth Assessment Report of the Intergovernmental Panel on Climate Change*, Cambridge; New York: Cambridge University Press. http://www.ipcc.ch/ (accessed August 2013).

Jacob, K. and Blake, R. (2010). "Indicators and Monitoring," in C. Rosenzweig and W. Solecki (eds.), *Climate Change Adaptation in New York City: Building a Risk Management Response*, New York: Annals of the New York Academy of Sciences. http://www.nyas.org/ (accessed March 2012).

Jacob, K.H., Edelblum, N., and Arnold, J. (2001). "Infrastructure," in C. Rosenzweig and W. Solecki (eds.), *Climate Change and a Global City: The Potential Consequences of Climate Variability and Change, Metro East Coast*, New York: Columbia Earth Institute.

Lin, N., Emanuel, K., Oppenheimer, M., and Vanmarcke, E. (2012). "Physically based assessment of hurricane surge threat under climate change," *Nature Climate Change*, 2: 462–7.

Liu, K.-B. and Fearn, M.L. (2000a). "Holocene history of catastrophic hurricane landfalls along the Gulf of Mexico coast reconstructed from coastal lake and marsh sediments," in Z.H. Ning and K. Abdollahi (eds.), *Current Stresses and Potential Vulnerabilities: Implications of Global Change for the Gulf Coast Region of the United States*, Baton Rouge, LA: Franklin Press.

Liu, K.-B. and Fearn, M.L. (2000b). "Reconstruction of prehistoric landfall frequencies of catastrophic hurricanes in Northwestern Florida from lake sediment records," *Quaternary Research*, 54: 238–45.

Ludlum, D.M. (1963). *Early American Hurricanes 1492–1870*, Boston, MA: American Meteorological Society.

Major, D.C. and O'Grady, M. (2010). "Adaptation assessment guidebook," in C. Rosenzweig and W. Solecki (eds.), *Climate Change Adaptation in New York City: Building a Risk Management Response*, New York: Annals of the New York Academy of Sciences, Vol. 1196. http://www.nyas.org/ (accessed March 2012).

Mann, M.E., Woodruff, J.D., Donnelly, J.P., and Zhang, Z. (2009). "Atlantic hurricanes and climate over the past 1,500 years," *Nature*, 460: 880–5.

Metropolitan Transportation Authority, "The MTA network." http://web.mta.info/mta/network.htm.

New York City. (2006). Department of Sanitation, Solid Waste Management Plan. http://www.nyc.gov/html/dsny/html/swmp/swmp.shtml (accessed June 2012).

New York City. (2008). *PlaNYC 2030: Sustainable Stormwater Management Plan 2008*, New York: City of New York Press.

New York City. (2010). *NYC Green Infrastructure Plan*, New York: City of New York Press.

New York City. (2011). *PlaNYC Update April 2011*, New York: City of New York Press.

New York City. (2013). *PlaNYC: A Stronger More Resilient New York (SIRR)*, New York: City of New York Press.

New York City Department of City Planning (NYC-DCP). (2011). *New York City Comprehensive Waterfront Plan—Vision 2020*. www.nyc.gov/waterfront (accessed June 2012).

New York City Department of Environmental Protection (NYC-DEP). (2008). *Assessment and Action Plan: A Report Based on the Ongoing Work of the DEP Climate Change Task Force*, New York: New York City Department of Environmental Protection Climate Change Program.

NOAA. (2012). *Global Sea Level Rise Scenarios for the US National Climate Assessment*. Edited by Parris, A., P. Bromirski, V. Burkett, D. Cayan, M. Culver, J. Hall, R. Horton, K. Knuuti, R. Moss, J. Obeysekera, A. Sallenger, and J. Weiss, NOAA Tech Memo OAR CPO-1. 37 pp.

Nordenson, G., Seavitt, C., and Yarinsky, A. (2010). *On the Water: Palisade Bay*, New York: Metropolitan Museum of Art.

NPCC. (2009). Climate Risk Information: New York City Panel on Climate Change, New York: New York City.

Nyberg, J., Malmgren, B.A., Winter, A., Jury, M.R., Kilbourne, K.H., and Quinn, T.M. (2007). "Low Atlantic hurricane activity in the 1970s and 1980s compared to the past 270 years," *Nature*, 447: 698–702.

O'Rourke, T.D., Lembo, A., and Nozick, L. (2003). "Lessons learned from the World Trade Center disaster about critical utility systems," in *Beyond September 11th: An Account of Post-Disaster Research*, Natural Hazards Research & Applications Information Center, Public Entity Risk Institute and Institute for Civil Infrastructure Systems, Boulder, CO: University of Colorado.

Peltier, W.R. (2012). *GIA data sets*. http://www.psmsl.org/train_and_info/geo_signals/gia/peltier/.

Perrette, M., Landerer, F., Riva, R., Frieler, K., and Meinshausen, M. (2013). "A scaling approach to project regional sea-level rise and its uncertainties," *Earth System Dynamics*, 4: 11–29.

Rosenzweig, C. and Solecki, W. (eds.); New York City Panel on Climate Change (NPCC). (2010). *Climate Change Adaptation in New York City: Building a Risk Management Response*, New York: Annals of the New York Academy of Sciences. http://www.nyas.org/ (accessed March 2012).

Rosenzweig, C. and Solecki, W. (eds.); New York City Panel on Climate Change (NPCC). (2013). *Climate Risk Information 2013: Observations, Climate Change Projections, and Maps*, New York: New York City Panel on Climate Change.

Salmun, H., Molod, A., Buonaiuto, F.S., Wisniewska, K., and Clarke, K.C. (2009). "East Coast Cool-Weather Storms in the New York Metropolitan Region." *Journal of Applied Meteorology and Climatology*, 48(11): 2320–2330.

Scileppi, E. and Donnelly, J.P. (2007). "Sedimentary evidence of hurricane strikes in Western Long Island, New York," *Geochemistry, Geophysics, Geosystems*, 8(6): 25.

Slangen, A., Katsman, C., van de Wal, R., Vermeersen, L., and Riva, R. (2012). "Towards regional projections of twenty-first century sea level change based on IPCC SRES scenarios," *Climate Dynamics*, 38: 1191–209.

Stalenberg, B. (2011). "Innovative flood defences in highly urbanized water cities," in J. Aerts, W. Botzen, M.J. Bowman, P.J. Ward, and P. Dircke (eds.), *Climate Adaptation and Flood Risk in Coastal Cities*, London; New York: Earthscan.

Sussman, E. and Major, D.C. (2010). "Law and regulation," in C. Rosenzweig and W. Solecki (eds.), *Climate Change Adaptation in New York City: Building a Risk Management Response*, New York: Annals of the New York Academy of Sciences. http://www.nyas.org/ (accessed March 2012).

Titus, J.G., Hudgens, D.E., Trescott, D.L., Craghan, M., Nuckols, W.H., Hershner, C.H., Kassakian, J.M. et al. (2009). "State and local governments plan for development of most land vulnerable to rising sea-level along the US Atlantic coast," *Environmental Research Letters*, 4(4): 1–7.

U.S. Army Corps of Engineers (USACE). (2012). "Jamaica Bay Islands, Brooklyn, NY; Yellow Bar Hassock and Elders Point Marsh Islands" (last updated January 2012). http://www.nan.usace.army.mil/ (accessed August 2013).

Zimmerman, R. and Faris, C. (2010). "Infrastructure impacts and adaptation challenges," in C. Rosenzweig and W. Solecki (eds.), *Climate Change Adaptation in New York City: Building a Risk Management Response*, New York: Annals of the New York Academy of Sciences. http://www.nyas.org/ (accessed March 2012).

Chapter 6

Crisis on the delta

Emerging trajectories for New Orleans

Joshua A. Lewis, Ann M. Yoachim, and Douglas J. Meffert

> The justest notion you can form of it is, to imagine yourself two-hundred persons, who have been sent out to build a city, and who have settled on the banks of a great river, thinking nothing but upon putting themselves under cover from the injuries of the weather, and in the meantime waiting till a plan is laid out for them, and till they have built houses according to it.
>
> Pierre Charlevoix 1722 (1923: 273)

Abstract: The flooding of New Orleans and southeast Louisiana in 2005 initiated considerable attention and debate concerning the vulnerability of coastal cities to the effects of climate change. The current rate of relative sea-level rise along the Louisiana coast is one of the highest in the world, and the political struggles over flood protection and coastal wetland restoration are perhaps harbingers of looming debates in coastal regions worldwide. This chapter will elucidate how floods and hurricanes have served as triggering events that have spurred political initiatives and novel approaches to natural resources management, economic development, and, more recently, practice and policy for adapting to the effects of climate change. Further, the BP oil disaster in the Gulf of Mexico in 2010 presented new obstacles and opportunities for restoring the complex and valuable ecologies that underpin the resilience of this cultural and economic hub of the American South.

INTRODUCTION

Perhaps no event has served to draw attention to the issue of climate change more than the flooding of New Orleans in 2005. Journalists, commentators, and some scientists framed the flood as evidence for climate change, or at least a *canary in the coalmine*, a warning that coastal cities worldwide might suffer similar fates if ocean temperatures and sea level continued to rise. Al Gore's *An Inconvenient Truth* premiered less than a year later, offering images from

Katrina's aftermath for emotional effect, encouraging his viewers to consider "... how we react when we hear warnings from the leading scientists in the world" (Guggenheim 2006). Tying Katrina's intensity to climate change in a thorough manner is empirically difficult, making the association in the public consciousness a dubious, albeit rhetorically potent, simplification. As our opening quote suggests, adapting to climatic extremes has been a constant challenge for this society since the colonial era. Floods, hurricanes, epidemics, and land loss have plagued the region for centuries. Anthropogenic activities in the Mississippi's deltaic plain have accelerated land loss and promoted soil subsidence. Certainly though, sea-level rise projections for the Louisiana coast paint a dire picture. The prospect of stronger and possibly more frequent hurricanes is a bleak brushstroke in an increasingly harrowing outlook for the region. Not only is the sea rising but also much of the landscape itself is sinking. Utilizing conservative rates of both sea-level rise and coastal subsidence, Blum and Roberts (2009) project that terrain less than 1 m above current sea level will be submerged by the year 2100 unless significant coastal restoration is achieved (see Figures 6.1 and 6.2).

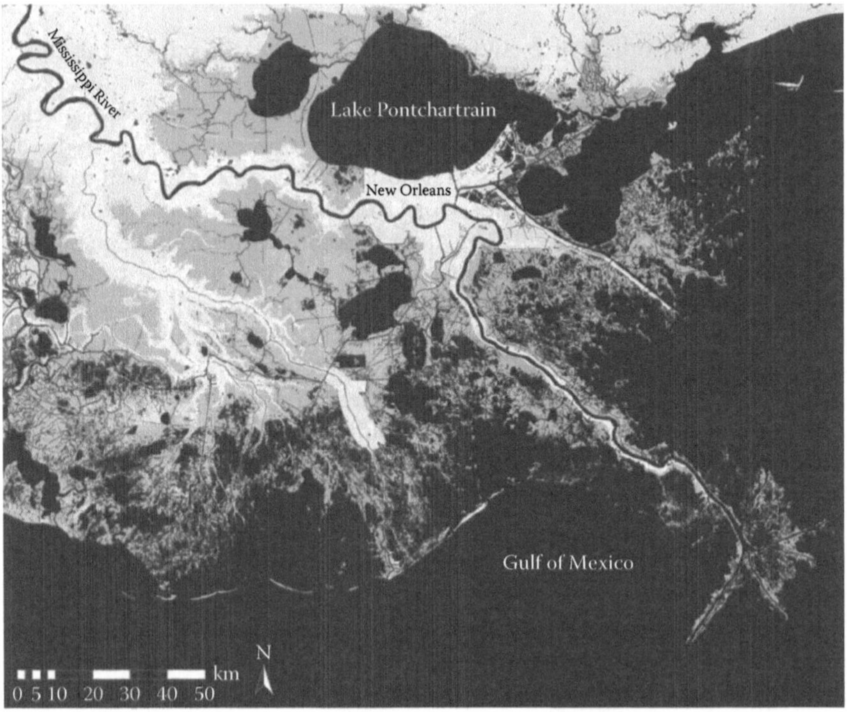

Figure 6.1 Louisiana's present day coastal footprint. (Data from National Elevation Data, US Geologic Survey, 1999; Aerial imagery from Landsat thematic mapper satellite image, 2005; Processing by Joshua A. Lewis and Jakob Rosenzweig.)

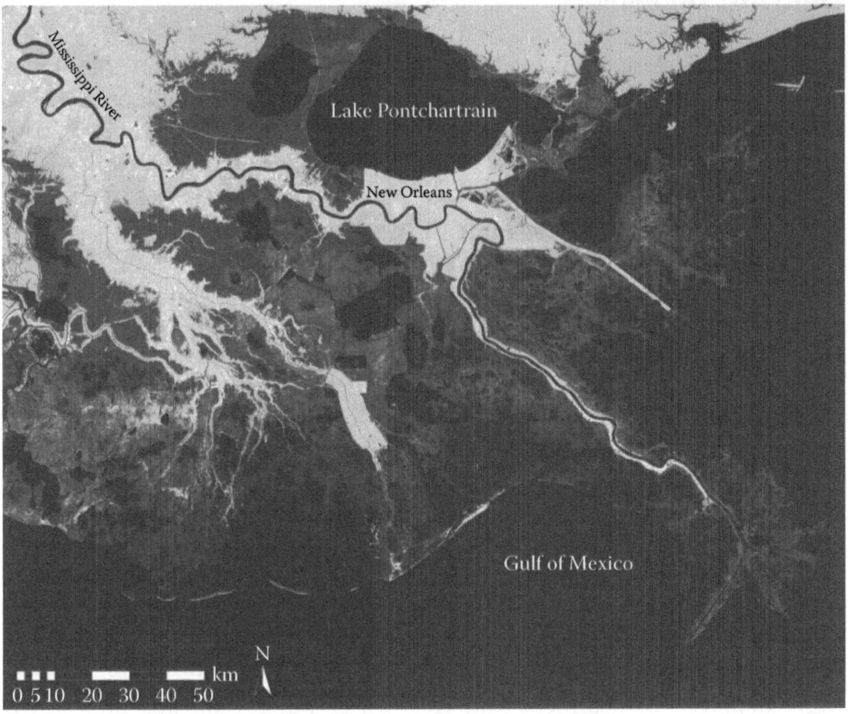

Figure 6.2 Projected land loss and submergence by 2100 under scenario. (Data from Blum, M.D. and Roberts, H.H., *Nature Geoscience*, 2, 488–91, 2009; National Elevation Data, US Geologic Survey, 1999; Aerial imagery from Landsat thematic mapper satellite image, 2005; Processing by Joshua A. Lewis and Jakob Rosenzweig.) (Note: Areas under one meter in elevation have been darkened.)

Although New Orleans serves as an effective symbol for Gore and others to communicate looming global challenges, it is important to remember that the dynamics of social-ecological systems are complex and most illuminatingly explored at multiple scales. As the saying goes, the governing (and the devil) is in the details. Even so, perhaps embedded in the dense narrative of this much-discussed delta there exist trends and decision points that other coastal communities might find instructive as they look toward the daunting task of climate change adaptation. In this chapter, we aim to articulate the roots of the contemporary social and ecological dilemmas facing planners and managers at both the urban (New Orleans) and coastal (deltaic plain) scales and offer some reflections on the emerging trends in climate adaptation measures we have observed in the post-Katrina era. It is, inevitably, an incomplete account. There is no shortage of excellent research and writing on the fate of the Louisiana coast, and we encourage curious readers to explore further some of the references in this chapter.

DELTAIC DILEMMAS

The peculiar geographic dilemmas of New Orleans and the lower Mississippi Delta were well rehearsed in journalistic and scholarly accounts long before the hurricanes of 2005 (Campanella et al. 2004; Colten 2005; Lewis 1976). Since the city's founding in the early eighteenth century, New Orleans has persistently been a place demarcated by extremes in temperature, humidity, and rainfall, not to mention riverine and coastal flooding (see Table 6.1). Key systemic variables like the timing and scope of freshwater pulses constitute a particular disturbance regime, which then upholds important ecological processes in both terrestrial and aquatic systems (Elmqvist et al. 2003; Walker and Salt 2006). The deltaic landscape is to a large degree sustained by these disturbances and the materials they disperse (see Table 6.1). Even hurricanes have historically had beneficial effects by pushing sediments suspended in coastal waters back into coastal wetland systems.

Geologically, the Mississippi's deltaic plain is in a constant state of collapse and regeneration. Land loss is as intrinsic to the delta system as land gain. Following the stabilization of eustatic sea level around 6,000–7,000 years ago, the Mississippi's deltaic plain began to gradually build new land, extending its coastline seaward (Day et al. 2007). New delta lobes extend their reach into the Gulf of Mexico; others, increasingly deprived of sediment and freshwater pulses, subside and slowly submerge beneath the sea. This process of delta lobe creation and abandonment plays out every ~500–1,500 years (Törnqvist et al. 1996). The fecund ecosystems that emerged in this rich environment are well known

Table 6.1 The disturbance regime of coastal Louisiana

Event	*Timescale*	*Typical effects*
Major changes in river channels	500–1000 years	New delta lobe formation, major sediment deposition
Major river floods	50–100 years	Major sediment deposition, land gain on existing delta lobe
Major storms	20–25 years	Major sediment deposition, enhanced biological production
Average river floods	Annual	Enhanced sediment deposition, freshening, nutrient input, enhanced vegetation growth
Normal storm events	Weekly	Enhanced sediment deposition, enhanced organism dispersal, enhanced sediment dispersal
Tides	Daily	Marsh drainage, stimulated marsh production, low net transport of water and material

Source: Day, J.W. Jr. et al., *Science*, 315(5819), 1679–84, 2007.

for their biodiversity, extensive migratory bird habitat, and abundant marine life.

Sustaining an urban place in a system so defined by flux and variation requires major human intervention. Up until the twentieth century, engineers and planners focused on mitigating the effects of events with short-term timescales like rainstorms and annual river flooding. Drainage and flood protection technology consisted largely of earthen levees along the Mississippi River and drainage canals. The promise of a port city that could act as a control point for the entire Mississippi Basin ultimately overrode concerns about environmental hazards (Campanella 2008). The same processes that built and sustained the sliver of land where the early settlers built their city also generated a complex matrix of interrelated hazards. As one coastal ecologist succinctly put it, "... coastal ecosystems provide goods and services that are prized by the public but are sustained by processes and pulses, such as river floods, which impinge negatively on human activities and structures" (Day et al. 2008: 486).

Although coastal fisherman, trappers, and hunters were well aware of the services the delta's ecosystems provide, urbanites in New Orleans took an understandably divergent stance. Indeed, public attitudes toward the city's remnant swamplands tended to focus on their lack of utility in economic development and the public health nuisances they generated. In the age of cholera and yellow fever, diseases affected the poor and immigrant communities most severely, as these residents often had no acquired immunity, or lived in low-lying areas with poor drainage. In 1893, the city's public officials commissioned a plan to drain most of the city's remaining swamplands, arguing that the swamps were an impediment to economic growth and promoted *unsanitary conditions* (Colten 2005: 83). Newly invented hydraulic pumps drained low-lying terrain for new subdivisions, drainage canals carried water out of the urban areas, and higher levees were raised along the river and the city's perimeter to prevent flooding. The machinery of urban development demanded conditions antithetical to the delta system: well-drained and stable soils not subject to frequent flooding. Although the swamps and marshes being drained sat slightly above sea level, the drained polders subsided quickly, in some places more than 3 m below sea level. In the early twentieth century, nearly all of the city neighborhoods and its 300,000 residents were above sea level. The scale and intensity of these drainage and levee interventions accelerated throughout the twentieth century such that, by 1960, half of the city's population had moved into areas below sea level (Campanella 2007). Producing a landscape resilient to major hurricanes and river floods would require even more intensive measures. Two key events, a major river flood in 1927 and a major hurricane in 1965, brought these matters to the fore.

A FUTURE FORGED IN CRISIS

In democratic societies, it has long been observed that opportunities for significant realignments in political systems often follow in the wake of disasters and crises. Some political scientists refer to such events as *triggering events* that generate *windows of opportunity* in the political sphere (Kingdon 1995; Olsson et al. 2006). Of course, not all actors in a political system will identify precisely the same set of problems underlying the crisis, but the singularity of the moment usually demands that some substantive measure in decreasing the system's vulnerability to similar catastrophes emerges. In the New Orleans area, two such catastrophes in the twentieth century have had expansive and far-reaching ramifications for regional hydrology and ecology, fundamentally altered the disturbance patterns endemic to the area, and further enabled human settlement in vulnerable areas.

In 1927, the lower Mississippi River Valley experienced extensive levee failures and flooding. Small towns along the river's lower reaches were all but wiped off the map. Prior to 1927, most of the earthen levees along the river's banks were maintained by state and local agencies, limiting their scale and effectiveness. River flooding was just a fact of life and, as we have described, part of the delta's natural disturbance regime. Hysteria gripped New Orleans as floodwaters moved southward. Economic elites in New Orleans, concerned about the city's economic reputation, lobbied state and federal officials to intentionally breach the river levee downstream from New Orleans as a means of relieving pressure on the city's defenses. Subsequently, the US Congress passed the Flood Control Act of 1928, which amounted to a federal takeover of the river's levee system. The US Army Corps of Engineers became the primary authority for flood control throughout the river valley (Barry 1997). Hence, the social-ecological governance of the region jumped scales, from local/state to primarily federal control. This scale of command and control management by the Corps of Engineers enabled not only the prevention of major river floods, but, as a result of their new jurisdiction, the agency created a series of spillways and barriers to prevent the Mississippi from abandoning its current delta lobe. Under the Army Corps' control and funded by the US Congress, the Mississippi Delta became a human-dominated system. Further, local economic actors were provided with an institutional framework to realize new coastal infrastructure projects, many of which ultimately undermined coastal ecosystems and accelerated land loss (Freudenburg et al. 2009).

With the Mississippi under the watchful eye of federal managers, only major hurricanes remained as a significant risk factor for the New Orleans region (Figure 6.3). In search of profits and prosperity, humans introduced a few disturbances of their own. With the river effectively tamed, New Orleanians continued their dispersal into low-lying terrain. Meanwhile, the Port of New Orleans embarked on an ambitious project to move many of

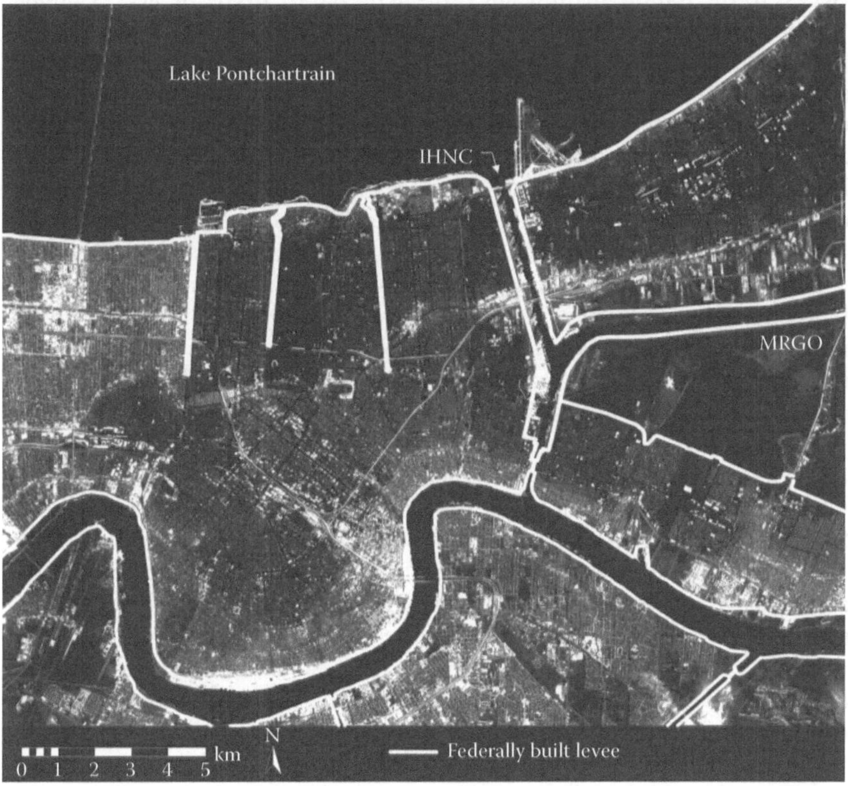

Figure 6.3 New Orleans levee system and flooding during Katrina. IHNC, Inner Harbor Navigation Canal; MRGO, Mississippi River Gulf Outlet. (Aerial imagery from EO-1 satellite, September 6, 2005.) (Note: Areas darkened on the image are partially submerged. In general, darkened areas are also at or below sea level.)

its facilities to an inner harbor, where water levels were stable and shipping channels not subject to the shifting sand bars in Mississippi's winding channel. The Inner Harbor Navigation Canal (IHNC), completed in the 1920s, bisected the city's sparsely populated ninth ward (see Figure 6.1). Next, Louisiana's powerful congressional delegation successfully secured major federal funding to excavate a 122-km shipping canal to connect the city's inner harbor facilities with the Gulf of Mexico. The Mississippi River Gulf Outlet (MRGO) broke ground in 1958. Ultimately, its touted economic impact never materialized. Instead, the MRGO disrupted salinity levels in the swamps and marshes east of the city and offered storm surges an unimpeded route directly into the IHNC. Freshwater wetland forests between the MRGO and the city began to reach critical salinity thresholds (Shaffer et al. 2009). Additionally, the oil and gas industry began extraction activities throughout the Louisiana coast in the twentieth century.

Access to the remote drilling sites was achieved by dredging access canals, which fragmented swamp and marsh ecosystems and accelerated their erosion. Further, scientists now believe that these extraction activities reduce subsurface pressure such that wetlands around drilling activities subside considerably faster than under undisturbed circumstances (Day et al. 2007). Kolker et al. (2011) found that subsidence rates in abandoned drilling areas are actually slowing down, leading to some discussion among coastal scientists to reconsider projected subsidence in such areas. About 1000 km upstream from New Orleans, damming and flood control activities in the Mississippi Basin significantly reduced the sediment load carried by the river by as much as half (Blum and Roberts 2009). With the river's natural distributaries blocked and its floods contained, this staggering drop in sediment loads was scarcely noticed, as the continent's bounty of nutrients and soil flowed uselessly into the Gulf of Mexico.

The summer of 1965 brought devastation to coastal communities and low-lying urban neighborhoods when Hurricane Betsy's storm surge overwhelmed the levee system along the city's eastern flank, flooding much of St. Bernard Parish and parts of New Orleans. At the time of Betsy's landfall, it just so happened that the US Congress was preparing to consider an extensive hurricane protection system for the region, which was passed swiftly in the wake of the storm. The levee system built not only protected developed areas but also placed much of the city's remnant eastern wetland landscapes behind concrete floodwalls, opening the most exposed portion of the city's footprint to urban development. Additionally, the US Army Corps of Engineers took control over levee walls within the City of New Orleans, taking responsibility for preventing the city's drainage infrastructure from flooding the city when storm surges entered the system from Lake Pontchartrain. Flood control and navigation projects have stimulated economic growth, urban expansion and protection from frequently occurring environmental disturbances and hazards. Paradoxically, these same projects have undermined essential deltaic processes, triggered massive coastal erosion, created rigid and unresponsive management systems, and compromised the region's social-ecological resilience to hurricane storm surges (Campanella et al. 2004; Ernstson et al. 2010; Kates et al. 2006).

In the two decades before Katrina, an increasingly large cadre of coastal residents, politicians, journalists, scientists, and environmental activists began to sound the alarm about the rapid pace of coastal erosion. The US Congress passed the Coastal Wetlands Planning, Protection and Restoration Act (CWPPRA) in 1990. CWPPRA-funded projects have included important incremental steps and experiments that will inform future measures. A number of small river diversions were created before Katrina and coastal planners should examine the data collected in these endeavors carefully. Despite these promising signs, the ragged coastline, open canals, and faulty floodwalls that were in place in 2005 proved woefully inadequate.

THE POST-KATRINA ERA

Hurricane Katrina made landfall in Louisiana on August 29, 2005, as a hurricane with sustained winds of 205 km h^{-1} spreading outward 190 km from the center. Coastal communities along the Gulf Coast of Mississippi and Louisiana were decimated by storm surges between 6 m and nearly 9 m, the latter being the highest ever recorded in the United States (IPET 2006). This immense wall of water overwhelmed the city's flood defenses in dozens of locations and was channeled deep into the city's inner harbor by the earlier mentioned MRGO canal. The performance of the levee system was hotly debated in the civil engineering community, though most analysts agree that there were significant flaws in the design, construction, and maintenance of the hurricane protection system—the system authorized in 1965 but never fully funded or completed. Some concrete levee walls failed catastrophically, while others were overtopped and suffered erosion and scouring around their base, causing them to fail in certain cases. The levee failures flooded 80% of the city and damaged 70% of the occupied housing units. Outside the city limits, the storm surge and increased wave action in the southeastern estuaries converted around 100 km^2 of Louisiana coastal wetlands into open water (Day et al. 2007). This equaled what was expected to be lost over the next 50 years and further increased the vulnerability of the region to future storms. In total, over 515,000 homes in Louisiana were damaged by Katrina and Hurricane Rita, which followed on September 23, 1.4 million people were displaced and over 1,100 people lost their lives (Jonkman et al. 2009; Liu et al. 2011). Despite the many daunting issues facing residents and officials, the region continues to recover, rebuild, and repopulate from the storms as evidenced by the data in Table 6.2. The region has recovered 89% of its pre-Katrina population or 1,167,764 residents and in New Orleans proper, 71% of its 2000 population of 484,674 have returned (GNOCDC 2011).

Table 6.2 Selected demographics for metropolitan New Orleans

Parish	*1950 Population*	*2000 Population*	*2010 Population*	*2010 Poverty rate (%)*
Orleans (City)	570,445	484,674	343,829	23.90
Jefferson	103,873	455,456	432,552	13.70
St. Bernard	11,087	67,229	35,897	21.30
Plaquemines	14,239	26,757	23,042	12.20
LaFourche	42,209	89,974	96,318	15.50
Terrebonne	43,328	104,503	111,860	15.70
St. Tammany	26,988	191,268	233,740	10.30

Source: US Census Bureau statistics.

An expansion of civic engagement and collective action among residents, neighborhood associations, and civil society organizations has emerged as an important feature of the recovery (Gotham and Campanella 2011). Yet the catalyst of much of this growth was involvement and response to initial land-use and rebuilding plans (Irazábal and Neville 2007). In a region so shaped by its relationship to water, one might expect adaptation strategies to be inherently intuitive; however, planning efforts immediately following Hurricane Katrina created an atmosphere of distrust and public discourse around *green space* became toxic (Olshansky and Johnson 2010). The *New York Times* profiled a noted New Orleans developer with ties to the Bush administration a month after Katrina struck. Referring to redevelopment in New Orleans post-Katrina, the developer was quoted as saying, "I think we have a clean sheet to start again … [a]nd with that clean sheet we have some very big opportunities" (Rivlin 2005). This brand of sweeping overtures by powerful figures did little to reassure displaced and returning residents that their homes and communities would maintain their pre-storm character. Strong opposition emerged among the public and local leadership to the notion that the city's most vulnerable neighborhoods should consider embracing a lower population density and repurpose vacant land as green space. The discourse surrounding these suggestions to *shrink the footprint* of the city became oriented not toward climate change adaptation or hazard mitigation, but to racial injustice, land grabs, and supposed selfish motives by planners and developers. The perception that the city's economic elite was conspiring to prevent certain neighborhoods from recovering was widespread during the recovery's early days. Subsequent plans avoided these contentious issues of large-scale spatial adaptation, taking the city's vast footprint as a given.

One such plan, the City of New Orleans first Master Plan, was adopted in 2010. The plan does highlight the need for 1-in-500-year flood protection, the city's responsibility in land-use and water management, the development of a climate change policy group, and the recognition that the city's survivability is directly tied to the health of the coastal ecosystem (Goody Clancy Associates 2010) (Figure 6.4). The comprehensive zoning ordinances currently being developed reflect the call for the city to learn to *live* with its abundance of water and support the adoption of passive surface and groundwater management strategies through provisions that allow for the development of green corridors on unused industrial zones, large-scale urban agriculture, and residential to open space conversion. The implementation of these strategies may reduce flooding from extreme weather events, but also are dependent on an effective structural protection system and the implementation of a comprehensive coastal restoration programme including using the Mississippi River to rebuild the coast through river diversions.

Reliance on structural engineering solutions for hurricane protection and long-term survivability for the City of New Orleans can be seen in both

Figure 6.4 A degraded swamp forest and the city's skyline. (Courtesy of Joshua A. Lewis.) (Note: Swamp forests provide the region with a number of ecosystem services, not the least of which is their capacity to buffer storm surges.)

planning processes and federal investment following Katrina and Rita. Although the majority of the federal spending (USD 75 billion) was devoted to immediate emergency response activities (GNOCDC 2011), USD 14.6 billion was directed to the US Army Corps of Engineers to construct the levees, floodgates, and storm surge barriers that comprise the New Orleans Risk Reduction System. The keystone in this system is the Inner Harbour Navigation Storm Surge Barrier (see Figure 6.5). The largest design-build project in the US Army Corps of Engineers' history, this barrier is designed to prevent storm surge from the Gulf of Mexico from entering the city's inner harbor port complex, which lies deep within the city itself. Dubbed by locals as the *Great Wall of St. Bernard*, an 8-m high extension of the surge barrier traverses 37 km of the vulnerable eastern flank of St. Bernard Parish (see Figure 6.6). Federal engineers also installed new floodgates and pumping stations where the city's main drainage and navigation canals meet Lake Ponchartrain, an important step in ensuring that storm surges have no obvious means of conveyance into the city's footprint. Once finished, costs to maintain this system will be borne by parish and state governments. By providing 1-in-100-year flood protection, the system does allow residents and businesses to gain access to the National Flood Insurance Program. Increasing attention by the private insurance industry to potential climate

Figure 6.5 A Colonel from the US Army Corps of Engineers discusses the Inner Harbor Navigation Canal Storm Surge Barrier with visitors. (Courtesy of Joshua A. Lewis.) (Note: Completed in 2011, the 3-km barrier is designed to prevent hurricane storm surges from entering the city's primary navigation canal.)

Figure 6.6 A group of visitors walks beneath a new levee wall designed to protect New Orleans and St. Bernard Parish from storm surges. (Courtesy of Joshua A. Lewis.)

change impacts including increased extreme weather events and sea-level rise scenarios makes this access essential to continued investment in the city, but reinforces the trend of continued development in low-lying areas close to flood hazards.

Katrina and Rita's wrath served as a major wake-up call for New Orleans residents and elected officials to the importance of flood protection through coastal wetland restoration. In the decades preceding the two storms, initial coastal restoration plans were called for and developed by a variety of stakeholders including state agencies with initial cost estimates at USD 14 billion. A regional integrated levee protection and coastal restoration plan designed to slow coastal wetland loss was just beginning to be implemented in 2005 (Constanza et al. 2006). It is widely acknowledged that the scale and funding of restoration activities must be ramped up significantly, and it should be mentioned that the vast majority of coastal scientists working in Louisiana are keenly aware of climate change and its likely effects on the region. The 2012 coastal master plan (being developed by the State of Louisiana) provides greater specificity on the locations and scale of large-scale freshwater and sediment river diversions that will provide for delta building, as well grappling with the contentious task of outlining areas of the coast that will be difficult to sustain due to the combined effects of relative sea-level rise and other anthropogenic causes of coastal wetland degradation. A well-regarded team of scientists and engineers officially advising and reviewing the project agree that uncertainty remains in the predictive model outcomes; however, the predictive models of erosion and risk on the Louisiana coast do include both sea-level rise and subsidence on a broad spatial scale as factors (e.g., the Mississippi Delta has a higher relative sea-level rise rate than the Chenier Plain of the southwest Louisiana coast). In a major step forward, Appendix C of the Master Plan document includes a detailed review of sea-level rise literature, concluding that all new ecosystem restoration projects should be engineered with the expectation that sea levels will rise between 0.5 and 1.5 m over the next 100 years. This report also acknowledges anthropogenic factors in climate change (CPRA 2012). It is encouraging that state officials are attentive to the latest climate science and that the Master Plan itself includes bold measures to avert even more catastrophic land loss. The state plan, among others, endorses the multiple lines of defense strategy, which recognizes the need for a hurricane protection system that includes healthy coastal wetlands, man-made levees and barriers, as well as the adoption of other measures such as elevation and flood proofing in light of increasing regional vulnerability due to climate change (CPRA 2007; Goody Clancy Associates 2010; Lopez 2006; USACE 2007).

Representative of this increased awareness of the importance of wetlands and the need for integration of structural and nonstructural protection to reduce the vulnerability of New Orleans is a collaborative effort and engagement of multiple government agencies, local and national nonprofits and academia, to restore the degraded wetland ecosystem along the city's vulnerable eastern flank. This coordinated effort highlights the complex legal, political, and funding challenges of implementing large-scale ecosystem restoration projects. Bowing to pressure from coastal residents and environmental activists, Congress officially decommissioned and closed the MRGO channel in 2008 and a rock dam was installed to prevent saltwater intrusion. The nearby Lower Ninth Ward and neighboring St. Bernard Parish are slated to benefit from an authorized, but not yet funded, ecosystem restoration plan designed to restore wetlands destroyed by the MRGO. The plan's aquatics models, however, do not take climate change scenarios into account (USACE 2010) and in general tend to treat climate change as one of many *uncertainties* where adaptive strategies will be triggered when greater clarity is achieved in the science. The plan includes a small freshwater diversion project, but it has emerged as the most controversial element of the project for a variety of reasons. Other pilot projects in the area include a USD 10 million wastewater wetland assimilation project funded by the State Coastal Impact Assistance Program, a program now administered under the Bureau of Ocean Energy, Management and Enforcement that was formerly the Mineral Management Service, the agency that was responsible for regulating oil and gas production in the Gulf of Mexico when the BP Deepwater Horizon Disaster occurred in April of 2010. Ironically, this event may be the most significant catalyst for funding coastal adaptation and restoration moving forward. The prospects for extensive federal coastal restoration funding seemed to be waning along with national attention to Katrina recovery, but the BP disaster served to further underscore the urgency of these projects.

MOVING FORWARD

The recovery and rebuilding efforts following hurricanes Katrina and Rita served as a *window of opportunity* to address long-standing ecosystem degradation and burgeoning climate change impacts. Yet, as this chapter has shown, these opportunities cannot be divorced from the realities of the cultural, economic, and political drivers in the region, including the dependency on the oil and gas industry, fishing, tourism and navigation, and the connection between these livelihoods and the unique culture of the region. Moving forward, any successful adaptation measures must address the complexity of these social, ecological, and economic considerations. We recommend at least the four measures outlined subsequently.

Divert the river, restore the coast

A primary focus of this chapter has been the adoption or lack thereof of climate change adaptation strategies in New Orleans following Hurricane Katrina, but the city cannot exist under even the most conservative sea-level rise estimates without a comprehensive coastal restoration effort that reintroduces the flow of the Mississippi River to rebuild land through diverting sediment. At the same time, large-scale coastal habitat building and enhancement through engineering efforts (e.g., dredging, pumping, and pipeline conveyance of sediment) is required to build coastal habitats quickly, albeit at a higher initial cost. More conventional engineering protection measures (levees, floodwalls, storm surge barriers, etc.) are also necessary for the protection of the culture, economy, and people of New Orleans. More specifically, the 2012 Master Plan attempts to balance restoration and protection goals including (1) maximizing land creation; (2) sustaining land; (3) capitalizing on natural processes; (4) favoring navigation interests; (5) maintaining ecological productivity; and (6) the impact of more small diversions compared to fewer, larger river diversions. The cost estimate of constructing restoration projects on this scale continues to evolve; the 2012 Master Planning process has focused primarily on both USD 50 billion and USD 20 billion funding scenarios. Given the time horizon of the projects in this plan (i.e., both 20-year near-term projects and longer-term projects on the scale of a half century), the complexity of stakeholders and evolving retrospective and predictive modeling that will occur, the ultimate projects that are completed and operated will undoubtedly be an iterative process as the Master Plan is revisited and revised over time.

Relocate exposed coastal communities

As noted above, to protect New Orleans' critical coastal infrastructure and to respond to ongoing coastal erosion, ultimately communities are going to need to relocate. This will require the support of diverse state and federal agencies. The adaptation of communities to date, in terms of relocating to less vulnerable areas, has been market-driven more than policy-driven. More specifically, repetitive losses of flooding and the collective relocation of amenities necessary for a vibrant community (schools, stores, post offices, etc.) to less vulnerable areas are the drivers for individuals and families to decide to move their residences to less vulnerable areas. To date, the state has not proactively engaged with the communities that will have to relocate in the next 50 years to provide them with the knowledge and economic resources to relocate before their property losses occur. The state must acknowledge the need to encourage thousands of rural inhabitants to relocate in the next 10–20 years—particularly in vulnerable areas outside of existing and proposed levee protection systems and work with and compensate these individuals in a fair, transparent, and equitable manner.

Strong institutions to manage diverse interests (oil and gas)

The challenge moving forward is to be aware of the limited adaptive capacity associated with oil and gas dependency and enhance our focus on restoring (on a large-scale) natural coastal ecosystems that provide for our fisheries, tourism, and other related economies. In terms of process, unfortunately, Louisiana is an example of a state with multiple institutions that, to date, have been unable to coordinate and implement adaptive measures like large-scale freshwater/sediment diversions necessary to achieve the goals stated above and balance those efforts related to energy economies. The difficult choices that lie ahead require adaptive governance, which can respond to increased vulnerability; the balancing of short and long-term needs and goals; and leadership and engagement at the local, state, and federal levels.

Develop economic sectors beyond those heavily dependent on natural resource extraction

Dependence on the oil and gas industry for restoration and adaptation funds cannot be overlooked nor can the fact that the sea-level rise and increased hurricane intensity faced by the region is in part due to burning of fossil fuels nationally and internationally. It was only after Hurricane Katrina, through the Gulf of Mexico Energy Security Act (GOMESA), that a portion of the federal oil and gas revenues were directed to the state to help support Louisiana's coastal restoration and protection efforts described previously. Pending legislation in both houses of Congress (the proposed *RESTORE* Act) could result in 80% of the Clean Water Act fines (USD 5.4 billion to USD 21.4 billion) from the BP Deepwater Horizon Disaster going to the affected Gulf States—instead of the federal treasury. Fifty percent of this amount could be allocated toward Louisiana with the majority of it going toward coastal restoration—but as is the case with recovery and rebuilding after Hurricanes Katrina and Rita, even if awarded, these funds will require the reconciling of federal, multiple states, and local interests and the acceptance of human-induced climate change.

Even with funds for coastal restoration, the long-term impacts of the Deepwater Horizon oil spill and other coastal degradation phenomena put the economic viability of our fisheries in question as we look to the twenty-second century. These uncertainties along with the certainty regarding the finite abundance of oil and gas are issues that support Louisiana's investing in economies that don't depend on these two primary practices of resources extraction. Investments in renewable energy (hydrokinetic, algae, etc.) economies and others (e.g., those that capitalize on Louisiana's abundance of water) should be explored on local, regional, state, and national levels.

CONCLUSION

It remains to be seen for how long New Orleans will remain one of the world's most potent allegories for the travails of climate change adaptation and disaster recovery. New crises emerge throughout the world each year, each with its own special tragedies, dilemmas, and insights. Undoubtedly, the dynamics of global trade and demand for fossil fuels will generate a compelling economic justification for the region's continued habitation, not to mention the region's rich cultural landscape and a population singularly dedicated to staying put. As we've elucidated here, human habitation in the deltaic plain has never been, and never will be, an easy or straightforward enterprise. Throughout the world, the search for a more sustainable balance between natural deltaic processes, economic growth, and contemporary urbanism will continue as long as people call deltaic plains home. For the Mississippi River delta, if current sea-level rise projections prove true, the next century will witness a transformation in this human-dominated delta. It remains to be seen whether this transformation will be driven by farsighted leadership or by crisis driven myopia. As we've demonstrated, historical trends suggest the latter, but some emerging characteristics of the Katrina and BP oil disaster recovery processes provide some compelling, if small-scale, examples of how this urbanized river delta might be sustained into the future.

REFERENCES

Barry, J. (1997). *Rising Tide: The Great Mississippi Flood and How it Changed America*, New York: Touchstone.

Blum, M.D. and Roberts, H.H. (2009). "Drowning of the Mississippi Delta due to insufficient sediment supply and global sea-level rise," *Nature Geoscience*, 2(7), 488–91.

Campanella, R. (2007). "Above-sea-level New Orleans: The residential capacity of Orleans Parish's higher ground," A whitepaper prepared for the Tulane/Xavier Center for Bioenvironmental Research. http://www.richcampanella.com/ (accessed February 2012).

Campanella, R. (2008). *Bienville's Dilemma: A Historical Geography of New Orleans*, Baton Rouge, LA: University of Louisiana Press.

Campanella, R., Etheridge, D., and Meffert, D.J. (2004). "Sustainability, survivability, and the paradox of New Orleans," in C. Alfsen-Norodom, B.D. Lane, and M. Corry (eds.), *Urban Biosphere and Society: Partnership of Cities*, Annals of the New York Academy of Sciences, volume 1023, June 2004.

Charlevoix, F. (1923). *Journal of a Voyage to North America, Vol. II*, trans. Louise Phelps Kellogg, Chicago, IL: The Caxton Club.

Coastal Protection and Restoration Authority (CPRA). (2007). *Louisiana's 2007 Comprehensive Master Plan for a Sustainable Coast.* http://coastal.louisiana.gov/ (accessed December 2011).

Coastal Protection and Restoration Authority (CPRA). (2012). *Louisiana's 2012 Coastal Master Plan.* http://coastalmasterplan.louisiana.gov/ (accessed February 2012).

Colten, C.E. (2005). *An Unnatural Metropolis*, Baton Rouge, LA: LSU Press.

Constanza, R., Mitsch, W., and Day, J.W. Jr. (2006). "A new vision for New Orleans and the Mississippi Delta: Applying ecological economics and ecological engineering," *Frontiers in Ecology and the Environment*, 4(9), 465–72.

Day, J.W. Jr., Boesch, D.F., Clairain, E.J., Kemp, G.P., Laska, S.B., Mitsch, W.J., Orth, K. et al. (2007). "Restoration of the Mississippi Delta: Lessons from Hurricanes Katrina and Rita," *Science*, 315(5819), 1679–84.

Day, J.W., Christian, R.R., Boesch, D.M., Yáñez-Arancibia, A., Morris, J., Twilley, R.R., Naylor, L., Shaffner, L., and Stevenson, C. (2008). "Consequences of climate change on the ecogeomorphology of coastal wetlands," *Estuaries and Coasts*, 31(3), 477–91.

Elmqvist, T., Folke, C., Nyström, M., Peterson, G., Bengtsson, J., Walker, B., and Norberg, J. (2003). "Response diversity, ecosystem change, and resilience," *Frontiers in Ecology and the Environment*, 1(9), 488–94.

Ernstson, H., Leeuw, S.E., Redman, C.L., Meffert, D.J., Davis, G., Alfsen, C., and Elmqvist, T. (2010). "Urban transitions: On urban resilience and human-dominated ecosystems," *Ambio*, 39(8), 531–45.

Freudenburg, W.R., Barbara, S., and Erikson, K.T. (2009). "Disproportionality and disaster: Hurricane Katrina and the Mississippi River-Gulf Outlet," *Social Science Quarterly*, 90(3): 497–515.

Goody Clancy Associates. (2010). "Resilience: Living with water and natural hazards," in *Plan for the 21st century: New Orleans 2030, Vol. 2 Strategies and Actions (Ch. 12)*, New Orleans, LA: City Planning Commission. http://www.nola.gov/ (accessed November 2011).

Gotham, K.F. and Campanella, R. (2011). "Coupled vulnerability and resilience: the dynamics of cross-scale interactions in post-Katrina New Orleans," *Ecology and Society*, 16(3): 12.

Greater New Orleans Community Data Center (GNOCDC). (2011). *The New Orleans Index at 6.* http://www.gnocdc.org/ (accessed November 2011).

Guggenheim, D. (Director) (2006). *An Inconvenient Truth: A Global Warning* [DVD], Hollywood, CA: Paramount.

Interagency Performance Evaluation Task Force (IPET). (2006). "Performance evaluation of the New Orleans and Southeast Louisiana hurricane Protection System," MMTF 00038-06, Vicksburg, MS: U.S. Army Corps of Engineers.

Irazábal, C. and Neville, J. (2007). "Neighbourhoods in the lead: Grassroots planning for social transformation in post-Katrina New Orleans?," *Planning Practice and Research*, 22(2), 131–53.

Jonkman, S.N., Maaskant, B., Boyd, E., and Levitan, M.L. (2009). "Loss of life caused by the flooding of New Orleans after Hurricane Katrina: Analysis of the relationship between flood characteristics and mortality," *Risk Analysis*, 29(5), 676–98.

Kates, R.W., Colten, C.E., Laska, S., and Leatherman, S.P. (2006). "Reconstruction of New Orleans after Hurricane Katrina: A research perspective," *Proceedings of the National Academy of Sciences of the United States of America*, 103(40): 14653–60.

Kingdon, J.W. (1995). *Agendas, Alternatives, and Public Policies*, New York: Harper Collins.

Kolker, A.S., Allison, M.A., and Hameed, S. (2011). "An evaluation of subsidence rates and sea-level variability in the northern Gulf of Mexico," *Geophysical Research Letters*, 38(21): 1–6.

Lewis, P. (1976). *New Orleans: The Making of an Urban Landscape*, Cambridge, MA: Ballinger.

Liu, A., Anglin, R., Mizelle, R., and Plyer, A. (eds.) (2011). *Resilience and Opportunity: Lessons from the US Gulf Coast after Katrina and Rita*, Washington, DC: The Brookings Institution Press.

Lopez, J. (2006). "The multiple lines of defense strategy to sustain coastal Louisiana," Lake Pontchartrain Basin Foundation. http://www.saveourlake.org/PDF-documents/MLODSfullpt2-06.pdf (accessed December 2011).

Olshansky, R. and Johnson, L. (2010). *Clear as Mud*, Chicago, IL: APA Planners Press.

Olsson, P., Gunderson, L.H., Carpenter, S.R., Ryan, P., Lebel, L., Folke, C., and Holling, C.S. (2006). "Shooting the rapids: Navigating transitions to adaptive governance of social-ecological systems," *Ecology and Society*, 11(1): 18.

Rivlin, G. (2005). "A Mogul who would rebuild New Orleans," *The New York Times* September 29. http://www.nytimes.com/ (accessed December 2011).

Shaffer, G.P., Day, J.W. Jr., Mack, S., Kemp, G.P., van Heerden, I., Poirrier, M.A., Westphal, K.A. et al. (2009). "The MRGO navigation project: A massive human-induced environmental, economic, and storm disaster," *Journal of Coastal Research*, SI 54: 206–24.

Törnqvist, T.E., Kidder, T.R., Autin, W.J., van der Borg, K., de Jong, A.F.M., Klerks, C.J.W., Snijders, E.M.A., Storms, J.E.A., van Dam, R.L., and Wiemann, M.C. (1996). "A revised chronology for Mississippi River subdeltas," *Science*, 273(5282): 1693–6.

United States Army Corps of Engineers. (2007). Louisiana Coastal Protection and Restoration (LACPR): Plan Formulation Atlas. New Orleans, LA: U.S. Army Corps of Engineers.

United States Army Corps of Engineers (USACE). (2010). MRGO ecosystem restoration plan draft feasibility report and draft environmental impact statement. Appendix I: Aquatics model report. http://mrgo.gov/ProductList.aspx?ProdType=study&folder=1368 (accessed February 2012).

Walker, B. and Salt, D. (2006). *Resilience Thinking: Sustaining Ecosystems and People in a Changing World*, Washington, DC: Island Press.

Part III

Climate change and the coastal zone

South and Southeast Asia

Chapter 7

Building resilient coastal communities by enabling participatory action

A case study from India

R. Ramesh, Ahana Lakshmi, Annie George, and R. Purvaja

Abstract: India is among the many countries that are likely to be extensively affected by climate change events, especially sea-level rise and changes in the precipitation patterns. Agriculture is the most important source of livelihood and is particularly vulnerable. This chapter provides a brief overview of the status of climate change-related activities in India at the national level and describes a case study of a bottom-up approach to empower a coastal deltaic agricultural community in Tamil Nadu state through participatory irrigation management as an adaptive response to the threat of climate change and sea-level rise. It is clear that, in coastal areas, planning for integrated management needs to take into account issues such as irrigation requirements for agriculture to ensure that the vulnerability of coastal communities to climate change-related factors is reduced. To achieve this, there is a need for a holistic approach, at the appropriate spatial scale, and greater interaction between the different institutional structures concerned.

INTRODUCTION

Recent research has shown that uneven rise in sea level in the Indian Ocean threatens coastal areas along the Bay of Bengal and the Arabian Sea (Han et al. 2010). The authors note that, if future anthropogenic warming effects in the Indo-Pacific warm pool dominate natural variability, the northern Indian Ocean may experience significantly greater sea-level rise than the global average. The Intergovernmental Panel on Climate Change (IPCC) has predicted with high confidence that projected sea-level rise could result in coastal flooding and loss of coastal ecosystems, saline intrusion, and threaten the stability of coastal wetlands (Cruz et al. 2007). Asia is the most populous continent in the world, with increasing urbanization and industrialization along the coast likely to exacerbate the effects of climate change. This chapter provides a brief overview of the status of climate change-related activities in India at the national level and describes a specific case study pertaining to the potential impact of climate change on a coastal deltaic agricultural community and an adaptive response.

THE INDIAN COAST

India is the second most populous country in the world with a population of 1.21 billion (Census 2011 provisional data). India's coastline is about 7,500 km long and supports around 30% of the nation's total population. A variety of natural ecosystems, including dense mangroves (such as Sundarbans, Coringa, Pichavaram, and Muthupet) and coastal lagoons (Pulicat, Chilika, and Vembanad), deltaic systems (Mahanadi–Brahmani, Godavari–Krishna, and Cauvery), coral reefs (Gulf of Mannar and Gulf of Kachchh), and tidal flats (Vedaranyam, Rann of Kachchh) provide millions with livelihoods in the primary sectors of fisheries and agriculture. Agriculture remains the principal source of livelihood for more than 58% of the population. Compared to other countries, India faces a substantial challenge, because food security has to be ensured for approximately 17.5% of the world's population (Ministry of Agriculture 2011) with only a 2.3% share in world's total land area. Indian deltas are areas of intensive agriculture, with rice the dominant crop.

According to the IPCC, Asia will be one of the most severely affected regions of the world as a result of *business-as-usual* global warming (Cruz et al. 2007). India is likely to experience increased exposure to extreme events, including cyclones and tropical storms, floods, and severe vector-borne diseases. Sea-level rise may cause large-scale inundation along the coastline and recession of flat sandy beaches. The ecological stability of mangroves and coral reefs may be at risk. The 2010 report of the Indian Network for Climate Change Assessment (INCCA 2010) provides an assessment of the impact of climate change in the decade of the 2030s on four key sectors of the Indian economy: agriculture, water, natural ecosystems and biodiversity, and health. The assessment was undertaken in four climate-sensitive regions of India, namely, the Himalayan region, the Western Ghats, the Coastal Areas, and the Northeast Region. Climate change impacts of highest relevance in coastal areas are with respect to sea-level rise and changes in the occurrence and frequency of storm surges.

Estimates of sea-level rise for the Indian coast over recent decades lie between 1.06 and 1.75 mm yr^{-1}, with a regional average of 1.29 mm yr^{-1}, after glacial isostatic corrections are applied (Unnikrishnan and Shankar 2007). These estimates are consistent with the 1–2 mm yr^{-1} global sea-level rise trend reported by the IPCC (Bindoff et al. 2007). The east coast is considered more vulnerable than the west due to its flat terrain and the numerous deltas. Areas with a large number of creeks and backwaters are likely to be at a higher risk of inundation (Shetye et al. 1990). According to country-level estimates of urban, rural, and total population and land area in the low elevation coastal zone (10 m or less elevation) generated by the Global Rural-Urban Mapping Project for India, large patches of the east coast and some areas in the west coast are particularly vulnerable to sea-level rise (McGranahan et al. 2007).

India's climate regime is dominated by the southwest and northeast monsoons and their onset is also accompanied, in the Arabian Sea as well as in the Bay of Bengal, by a number of low pressure systems, some of which turn into deep depressions and cyclones. The frequency of formation of cyclones in the Bay of Bengal is higher than in the Arabian Sea, and the shallow depth of the Bay of Bengal and the low flat coastal terrain produce much larger storm surges on the east coast. A decrease in the frequency of cyclones has been observed over the past four to five decades, but cyclone intensity has increased (Ramesh Kumar and Shankar 2010). As global warming develops, storm surges and flooding due to intense events are likely to increase over time, especially in the southern part of the east coast where tidal ranges are low (Unnikrishnan et al. 2010).

Agricultural crops are grown in close proximity to the coast, especially in the deltaic plains. According to the INCCA report, there is likely to be a marginal increase in the production of rice in many coastal areas: It has been estimated that rain-fed rice yields may increase on the east coast, while a decline in rice yields is expected on the west coast of India (INCCA 2010). Most agricultural operations in India are directly or indirectly controlled by the monsoon. In addition, there are many other abiotic and biotic factors that have both direct and indirect relevance to climate impacts and are likely to affect rice production. Box 7.1 outlines aspects of the Indian government's response.

BOX 7.1 INDIA TAKES ACTION ON CLIMATE CHANGE AND COASTAL ZONE MANAGEMENT

India is the world's fourth largest economy. It is the fifth largest greenhouse gas emitter, accounting for about 5% of global emissions, although on a *per capita* basis India's emissions are 70% below the world average (Centre for Climate Change and Energy Solutions 2008). As in many other countries, India has a number of policies that, while not primarily driven by climate concerns, contribute to climate mitigation either by reducing or by avoiding greenhouse gas emissions. The National Action Plan for Climate Change outlines a number of steps to simultaneously advance development and the climate change-related objectives of adaptation and mitigation. The plan identifies eight core

(*Continued*)

national missions, which includes the National Water Mission. There are also other initiatives, for example, in the area of disaster management response to extreme climate events where the approach to disaster management has moved from relief to prevention, mitigation, and preparedness.

The main objective of the National Water Mission, which is of particular significance for agricultural activities in coastal areas, is "conservation of water, minimizing wastage and ensuring its more equitable distribution both across and within States through integrated water resources development and management" (Ministry of Water Resources 2009). Five goals have been identified, including (Goal 2) "Promotion of citizen and state actions for water conservation, augmentation and preservation" and (Goal 5) "Promotion of basin level integrated water resources management." Strategy II.2 under Goal 2 is "Expeditious implementation of programme for repair, renovation and restoration of waterbodies in areas/situations sensitive to climate change by (1) Increasing capacity of minor tanks, and (2) Rehabilitating water bodies, with changed focus." The promotion of participatory irrigation management through Command Area Development and Water Management Programme is covered in Strategy II.6.

Mainstreaming climate change into various activities is being actively followed, especially in the protection of coastal areas. Two important developments have been the publication of the CRZ Notification 2011 and the initiation of the Integrated Coastal Zone Management (ICZM) Project. India enacted the framework Environment (Protection) Act in 1986. Subsequently, under this act, a CRZ Notification was issued in 1991. A stretch of the coastline, 500 m from the high tide line and the space between the high tide line and low tide line, was declared as the CRZ and activities in this area were regulated, restricted, or prohibited. This notification has undergone a number of changes over the past twenty years (MoEF 2005, 2009; Ramesh et al. 2010). The new CRZ Notification, published in January 2011, now includes the seaward side, extending twelve nautical miles into the sea (territorial waters) and 500 m from the high tide line on the landward side. This is to be further extended landward up to the *hazard line*, which is being delineated by the Survey of India.

The Government of India intends to apply efficient ICZM approaches for the management of the coastal and marine areas in India. According to Ministry of Environment and Forests (MoEF 2005), the objectives of ICZM are to help achieve (1) security of lives and property in disaster-prone coastal

(*Continued*)

zones; (2) conservation, preservation, restoration, and development of coastal resources and ecosystems; (3) security of livelihood of coastal communities and overall food security; (4) security of cultural and heritage sites; and (5) goals of national development and growth in such ways that the development is sustainable. Accordingly, one of the primary components will include mapping, delineation, and demarcation of the hazard line for the coast of India. The maximum upper limits of the hazard line would be 7 km inland and the lower limits would be 500 m from the high tide line. This exercise will define the boundaries of the coastal zone in mainland India (which, in turn, will establish planning boundaries of the state/local ICZM plans) and will incorporate the effects of recurrent coastal hazards, including potential incremental effects induced by climate change (most notably sea-level rise) within ICZM plans. The hazard line for the mainland coast will be mapped and delineated as the landward composite of the coastal 100-year flood lines (including sea-level rise impacts) and the 100-year predicted erosion lines.

BUILDING A RESILIENT COASTAL COMMUNITY: A CASE STUDY

The availability of irrigation facilities is an important factor in determining the vulnerability of agriculture to climate change and reducing that vulnerability by improving irrigation infrastructure can be an effective precautionary response, with immediate as well as long-term benefits. Agricultural lands in coastal areas are under tremendous pressure with escalating demands from industry (especially to establish power plants and special economic zones) and for housing. In addition, dams constructed across rivers also play a vital role in controlling water availability for agriculture. This situation is complicated in the coastal deltaic areas because of the proximity to the ocean and issues such as tidal ingress through backwaters and estuaries, which can cause salinization of water and land. And, of course, when the area is also low lying, the effects of sea-level rise due to climate change are likely to be greater. Anticipating future impacts and taking suitable adaptive action is advisable. However, identifying vulnerable communities and, more importantly, identifying precise actions to be taken may not happen unless there is a trigger. One such trigger, with the follow-up action that occurred, is described in this case study of Nagapattinam district, Tamil Nadu, India.

The case study site

The coast of Tamil Nadu, a state in southern peninsular India, was the worst affected nationwide by the 2004 Indian Ocean tsunami. Within Tamil Nadu state, Nagapattinam district was declared as the worst affected. Initially, it was believed that the fishing community was most affected as most fishing hamlets are located within 500 m of the shoreline. During a more systematic analysis of the communities affected by the tsunami in the district of Nagapattinam, it was found that the agricultural community was also affected by the flooding of their fields and the destruction of their crops by the tsunami waves. A study undertaken by the nongovernmental organization, NGO Coordination and Resource Centre (NCRC), to examine methods for disaster proofing agriculture in Nagapattinam found that there was a decreasing trend in the yield from agricultural land over the years, even during normal years, irrespective of disasters (NCRC 2006). The salinity in the soil and lower fertility of the soil seemed to be the main factors contributing toward this decreasing trend. It was also found that the pattern of landholding was skewed, with a large proportion of marginal farmers owning a very small proportion of agricultural lands. Multiple disasters faced by the community have affected them in ways much more than just loss to physical or tangible assets such as land. The result of these disasters has meant reduced food security for families, in some cases even starvation, forced labor, migration, and increased debts.

The Nagapattinam district comprises a narrow strip of land with a 189-km-long coastline, located in the low elevation coastal zone and highly vulnerable to sea-level rise (Figure 7.1). The coastal town is also located at the tail end of the Cauvery delta. Eighty-five percent of the population of the district is dependent on agriculture and rice is the main crop produced. The delta is known for floods and droughts within the same year as the quantum of water released from the upstream Mettur Dam depends on the southwest monsoon as well as the release of water from the reservoir in the upper reaches of the river from the neighboring state of Karnataka. In addition, the region is affected by storms originating from the Bay of Bengal. During the northeast monsoon, flooding in Nagapattinam district is common as this low-lying delta region is vulnerable to storm surges generated by cyclones.

There are fourteen river systems entering Nagapattinam district from which emerge 1,141 irrigation canals, totaling 1,324 km in length, and 180 drainage canals, totaling a length of 346 km. About 30,000 hectares of agricultural land is irrigated. Tail-end regulators (TERs) are placed at appropriate points, just before which the river is channelized into irrigation canals (Figure 7.2). After the tail-end regulator, the river acts as a drain for the excess water from the fields as well as floodwater.

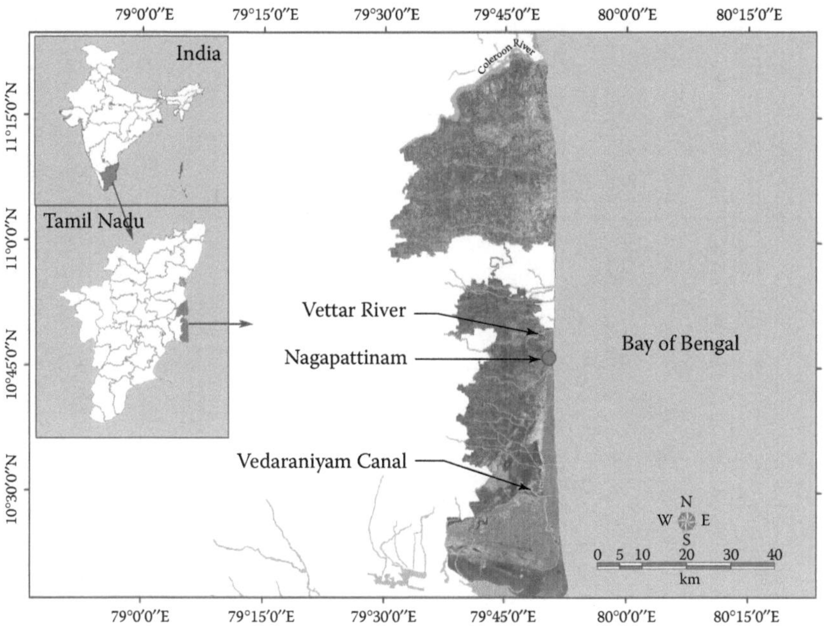

Figure 7.1 Map of the case study area: Nagapattinam, Tamil Nadu, India. (Data from Sivanappan, R.K., *Mapping and Study of Coastal Water Bodies in Nagapattinam District*, NGO Coordination and Resource Centre, Nagapattinam, Tamil Nadu, India, 2007.)

Each network of irrigation and drainage channels is commonly referred to by the name of the river on which it is located. Thus, Kaduvaiaru TER (TER-9) refers to the irrigation–drainage network of the Kaduvaiaru River, which enters through Nagapattinam taluk (Figure 7.3). The tail-end regulator is located at a place called Vadugacherry, about 9 km from the Bay of Bengal. There is a high level of saltwater ingress through the backwaters of this area.

A survey of the status of the irrigation system, structured around 14 tail-end regulators, was carried out in 2006 (Sivanappan 2007). It was observed that both drainage and irrigation channels were extensively damaged, causing flooding and salinization of the farm lands. The silted and damaged structures and systems could not contain or effectively hold the heavy flow of water, resulting in flooding of the adjoining cultivable lands. Alternatively, when the base flow of water was nonexistent due to the closure of the shutters at the upstream Mettur Dam, sea water entered the fields during the periodic high tides, resulting in saline water intrusion and consequent salinization of the soil, making the land progressively less productive.

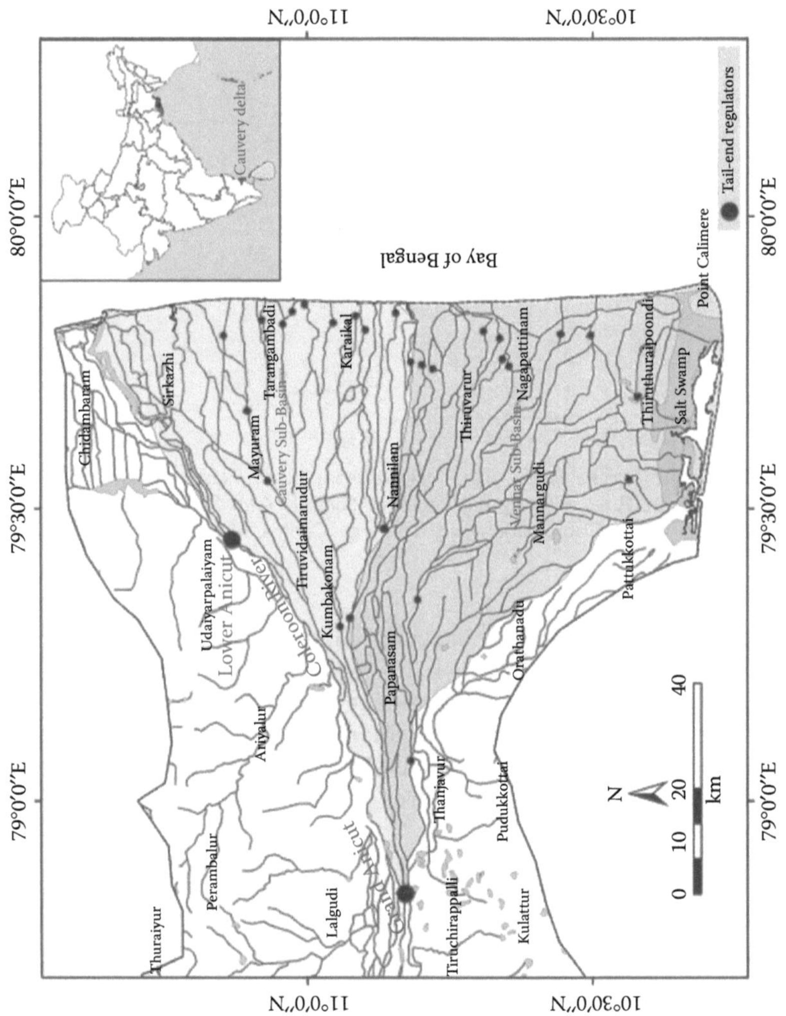

Figure 7.2 The Cauvery delta with tail-end regulators. (Data from Sivanappan, R.K., *Mapping and Study of Coastal Water Bodies in Nagapattinam District*, NGO Coordination and Resource Centre, Nagapattinam, Tamil Nadu, India, 2007.)

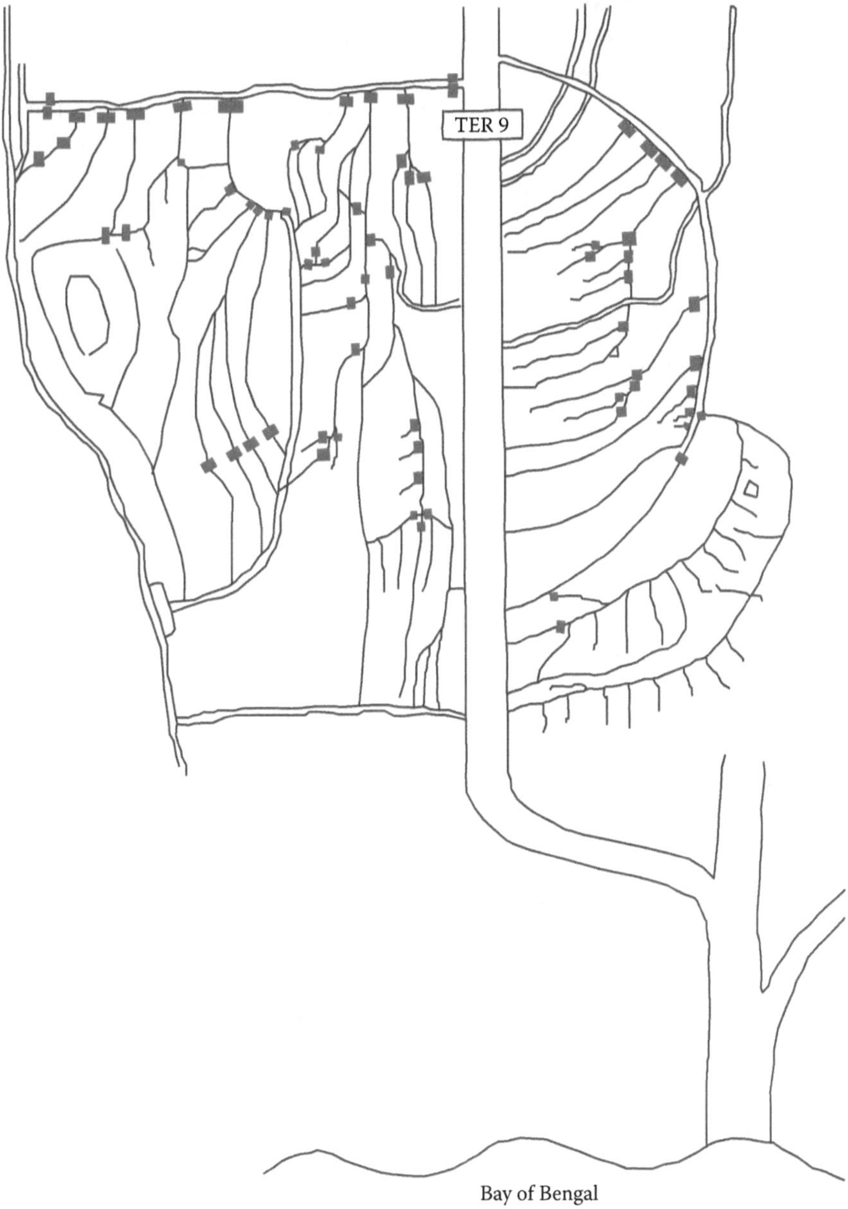

Figure 7.3 Example of a tail-end regulator-9 at Kaduvaiyaru, Nagapattinam, Tamil Nadu, India. (Data from Sivanappan, R.K., *Mapping and Study of Coastal Water Bodies in Nagapattinam District*, NGO Coordination and Resource Centre, Nagapattinam, Tamil Nadu, India, 2007.)

Participatory irrigation management

Taking into consideration the fact that it was not only repairs of the defective agricultural infrastructure but also long-term maintenance that would be a sustainable solution to the frequent flooding or droughts that the farming communities faced, a set of interventions was planned and executed as a *participatory agri-infrastructure maintenance* model, including both the local *panchayat* (village administration) and the actual users as active participants. This bottom-up approach involved extensive social mobilization and bringing together stakeholders at various levels, including some of the key departments of the Government of Tamil Nadu. The specific objectives of the project were as follows: safe disposal of excess floodwater to mitigate water logging/inundation; to increase the availability of water through harvesting rain/floodwater as well as improving recharge potential of the groundwater; to contain backwater salinization by improving the efficiency of the drainage channels; and to increase the intensity of cropping through improved irrigation systems.

Three river systems were chosen in Phase I: the South Rajan River system under TER-1; the Manjalaru River system under TER-4; and the Kaduvaiyaru River system under TER-9. The project was carried out during 2007–2008. The interventions are summarized in Table 7.1. Although physical interventions were important, the project was also committed to building up the capacities of the local communities in operating and maintaining their resources. Twenty-two *panchayat*-level wise water user associations were formed during this phase and these were later federated at the TER level for holistic planning.

Because the areas where interventions took place are located close to the Bay of Bengal, the effects were quickly noticed in the year following the repairs. When water was let into the repaired channels, the flow was maintained and reached the cropped areas on time. Discussions with

Table 7.1 Types of interventions to river systems

Type of intervention	*Activity*
Technical	Canal desilting Strengthening of bunds Repair of canal structure
Institutional	Social mobilization Water user groups Capacity building Micro plans development Linkages with government at various levels
Financial	Grants for repair Linkages for government funding
Legal	Registration of associations Legal literacy/statutory compliance

farmers also brought out the information that the monsoonal rains had been unpredictable in recent times, a fact that agrees with the climate change predictions by INCCA (2010) of more intense rainfall and a reduction in number of rain days. The high water table in these areas means that water stagnation used to be high. After the repairs were carried out to the irrigation and drainage channels, the extent of stagnation during periods of intense rainfall was reduced and has resulted in the saving of crops. Thus, at the end of the first phase, 250 additional acres were brought under cultivation. In actual monetary terms, 250 acres with 1.2 MT yield per acre will bring in an income of INR 31,50,000 in one year and INR 94,50,000 in three years. Three thousand two hundred acres were saved from flooding. Even at a conservative 50% saving of the crops, in monetary terms, this translates to INR 20,16,000 annually and INR 60,48,000 at the end of three years. The income generated and saved jointly has more than paid back the investment costs that the project had incurred over the last three years of the project period (Figure 7.4).

What helped and what hindered—Barriers to mainstreaming

The Tamil Nadu Farmers' Management of Irrigation Systems Act (2000; http://www.ielrc.org/content/e0001.pdf) enables sections of major and minor irrigation canals to be handed over to the community through water users' groups or associations for operation and maintenance, including water budgeting, regulating of water use, levying water user fees, and settlement of local disputes. To create viable water user associations was, however, a major challenge as it required a shift in mindset from government-dependant action to users working in a participatory mode. This change in mindset required extensive social mobilization, more time consuming than that envisaged. Eventually, intensive stakeholder exercises and participatory resource mapping, along with continuous field visits and interactions with the users by the project staff, achieved the desired level of community participation.

Strategies that proved effective included approaching and involving leaders (political, traditional, and informal); ensuring universal inclusion of different water users by meeting them at locations and times convenient to them (e.g., meeting agricultural laborers in the field); conducting meetings at the site of the work, which helped in getting the community together and in dissemination of the project information; ensuring that all planning was bottom-up and building the capacity of the water user associations, including financial capabilities; and ensuring transparency. What hindered the project was when a community would prefer action to systematic planning. In some cases, there were attempts to do away with democratic processes and social mobilization. Overall, coordination between people and agencies required time, patience, and prudence.

Figure 7.4 A water-regulating siphon system before (top), during (middle), and after (bottom) repair. (Courtesy of Bedroc, http://bedroc.in/.)

It was found that one of the essential requirements for success was that project staff members were able to identify the key person, a champion, who, once convinced about the efficacy of a project, would pull out all the stops to enable its implementation. Also, the activities had to be carried out over a large area for consistent and long-term results. It is not enough to de-silt just one or two channels, but the irrigation system complex has to be examined holistically. Networking with and the approval of various line departments such as the Public Works Department, in charge of structures, and the Agriculture Department also proved crucial. Networking also enabled pooling of scarce funds and resources for optimal utilization and ensured that there was no duplication of effort (George and Lakshmi 2011).

CONCLUSION

India is a country vulnerable to the impacts of climate change. Overall, the Government of India has recognized the challenges and is working toward various adaptation and mitigation strategies. The vulnerability of coastal areas to sea-level rise and flooding/erosion due to intensified cyclones has been recognized in CRZ Notification 2011 (Ministry of Environment and Forests 2005, 2009) in the demarcation of a composite hazard line and in restricting activities along the coast. ICZM is being promoted to ensure the well-being of coastal communities. It is clear that, in coastal areas, planning for integrated management needs to take into account issues such as irrigation requirements for agriculture to ensure that, for example, should there be an increase in sea-level rise-related events, there is no saline ingress into the irrigation systems and that coastal communities are not affected. This calls for greater interaction between the different institutional structures concerned.

India's National Water Mission (Ministry of Water Resources 2009), which is a part of the country's National Action Plan for Climate Change (Government of India 2008), has defined various strategies to achieve specific goals with respect to a changing climate regime. One approach is to promote participatory irrigation management, a form of community-based adaptation, mobilizing and empowering the community. This method is likely to ensure better sustainability of any system developed, apart from troubleshooting at the grassroots level, without waiting for a problem to magnify before action is taken. The case study described here, of participatory irrigation management in a low elevation coastal zone that is likely to be strongly affected by climate change-related factors, predates the National Water Mission, but it indicates that the mission is on the right track in calling for citizen's action in water conservation and management. It also emphasizes the need for promotion of basin-level integrated water resources management, another major aspect of the mission's goals. Such activities can have far-reaching effects in building local community resilience. Climate change is a slow process with effects felt over a long stretch of time. Building adaptive resilience among coastal communities can also be a long drawn-out process, but the consequences of such interventions will have an impact over a greatly extended period of time.

REFERENCES

Bindoff, N.L., Willebrand, J., Artale,V., Cazenave, A., Gregory, J., Gulev, S., Hanawa, K. et al. (2007). "Observations: Oceanic climate change and sea level," in S. Solomon, D. Qin, M. Manning, Z. Chen, M. Marquis, K.B. Averyt, M. Tignor, and H.L. Miller (eds.), *Climate Change 2007: The Physical Science Basis. Contribution of Working Group I to the Fourth Assessment Report of the Intergovernmental Panel on Climate Change*, Cambridge: Cambridge University Press. http://www.ipcc.ch/ (accessed August 2013).

Centre for Climate Change and Energy Solutions. (2008). *Climate Mitigation Measures in India*. http://www.c2es.org/ (accessed November 2011).

Cruz, R.V., Harasawa, H., Lal, M., Wu, S., Anokhin, Y., Punsalmaa, B., Honda, Y., Jafari, M., Li, C., and Huu Ninh, N. (2007). "Asia," in M.L. Parry, O.F. Canziani, J.P. Palutikof, C.E. Hanson, and P.J. van der Linden (eds.), *Climate Change 2007: Impacts, Adaptation and Vulnerability. Contribution of Working Group II to the Fourth Assessment Report of the Intergovernmental Panel on Climate Change*, Cambridge: Cambridge University Press. http://www.ipcc.ch/ (accessed August 2013).

George, A. and Lakshmi, A. (2011). *Participatory water resource management in select areas of Nagapattinam*, Nagapattinam, Tamil Nadu, India: Bedroc. http://bedroc.in/ (accessed December 2011).

Government of India. (2008). *National Action Plan on Climate Change*, New Delhi, India: Prime Minister's Council on Climate Change. http://pmindia.nic.in/climate_change.htm (accessed December 2011).

Han, W., Meehl, G.A., Rajagopalan, B., Fasullo, J.T., Hu, A., Lin, J., Large, W.G. et al. (2010). "Patterns of Indian Ocean sea-level change in a warming climate," *Nature Geoscience*, 3, 546–50.

Indian Network for Climate Change Assessment (INCCA). (2010). *Climate Change and India: A 4x4 Assessment—A Sectoral and Regional Analysis for 2030s*, presentations as on November 16, 2010, New Delhi, India: Ministry of Environment and Forests. http://moef.nic.in/ (accessed December 2011).

McGranahan, G., Balk, D., and Anderson, B. (2007). "The rising tide: Assessing the risks of climate change and human settlements in low elevation coastal zones," *Environment & Urbanization*, 19(1): 17–37.

Ministry of Agriculture. (2011). *Annual Report 2010–11*, New Delhi, India: Department of Agriculture and Cooperation, Ministry of Agriculture. http://agricoop.nic.in/ (accessed December 2011).

Ministry of Environment and Forests (MoEF). (2005). *Report of the Expert Committee Chaired by Prof. M.S. Swaminathan to Review the Coastal Regulation Zone Notification 1991*, New Delhi, India: Ministry of Environment and Forests. http://moef.nic.in/mef/crz_report.pdf (accessed December 2011).

Ministry of Environment and Forests (MoEF). (2009). *Final Frontier: Agenda to Protect the Ecosystem and Habitat of India's Coast for Conservation and Livelihood Security*, Report of the Expert Committee on the draft Coastal Management Zone (CMZ) Notification, New Delhi, India: Ministry of Environment and Forests. http://www.indiaenvironmentportal.org.in/files/cmz_report.pdf (accessed December 2011).

Ministry of Water Resources. (2009). *National Water Mission*, New Delhi: Ministry of Water Resources. http://mowr.gov.in/ (accessed December 2011).

NGO Coordination and Resource Centre (NCRC). (2006). *Understanding Vulnerabilities of Agricultural Communities to Frequent Disasters and Coping Mechanisms: A Sample Study of Tsunami Affected Agricultural Villages in Nagapattinam District*, Nagapattinam, Tamil Nadu, India: NGO Coordination and Resource Centre. http://www.ncrc.in/publications.php (accessed October 2011).

Niyas, N.T., Srivastava, A.K., and Hatwar, H.R. (2009). *Variability and Trend in the Cyclonic Storms over the North Indian Ocean, Meteorological Monograph No. 3/2009*. Pune, India: Indian Meteorological Department.

Ramesh, R., Purvaja, R., Senthil Vel, A., Lakshmi, A., and Bhat, J.R. (2010). "Coastal regulation zone notification: A review of the chronology of amendments," in J.R. Bhat (ed.), *Coastal Environment*, Gland, Switzerland: IUCN-Mangroves for the Future and Ministry for Environment and Forests.

Ramesh Kumar, M.R. and Sankar, S. (2010). "Impact of global warming on cyclonic storms over north Indian Ocean," *Indian Journal of Marine Sciences*, 39(4): 516–20.

Shetye, S.R., Gouveia, A.D., and Pathak, M.C. (1990). "Vulnerability of the Indian coastal region to damage from sea level rise," *Current Science*, 59: 152–6.

Sivanappan, R.K. (2007). *Mapping and Study of Coastal Water Bodies in Nagapattinam District*, Nagapattinam, Tamil Nadu, India: NGO Coordination and Resource Centre.

Unnikrishnan, A.S., Manimurali, M., and Ramesh Kumar, M.R. (2010). "Sea-level changes along the Indian coast—Impacts & vulnerability," Presentation to the *Ministry of Environment and Forests, National Workshop*, November 16, 2010. http://moef.nic.in/ (accessed September 2011).

Unnikrishnan, A.S. and Shankar, D. (2007). "Are sea level rise trends along the coasts of the north Indian Ocean consistent with global estimates?," *Global and Planetary Change*, 57(3/4): 301–7.

Chapter 8

Climate adaptation technologies in agriculture and water supply and sanitation practice in the coastal region of Bangladesh

Saleemul Huq and M. Golam Rabbani

Abstract: The coastal zone of Bangladesh, covering 32% of the land area and home to 30% of the population, is one of the regions that is most vulnerable to climate change and sea-level rise. The fertile land of the Ganges–Brahmaputra delta provides a productive base for agriculture, and the coast offers a diversity of natural resources, such as marine fisheries and shrimps, forest, salt, and minerals. The high level of physical vulnerability is made worse by factors such as an increasing population density, poverty, and limited access to services, especially water supply and sanitation, energy, and health services. Examples of current adaptation technologies and practices in agriculture and the water supply and sanitation sectors are presented and relevant components of the National Adaptation Programme of Action (NAPA) discussed. Three potential barriers to the effective implementation of adaptation projects have been identified and need to be addressed: lack of awareness of the seriousness of the climate threat; lack of integration of the climate issue in the development of policies, plans, and programs in climate-sensitive sectors; and lack of adequate tools, knowledge, and methodologies for guidance and advice in decision-making.

INTRODUCTION

The countries on the coasts of the Indian Ocean and Bay of Bengal are highly vulnerable to climate change. The people of Bangladesh are reported to be among the most at risk worldwide from sea-level rise (World Bank 2007). Even at current rates of sea-level rise, more than one million people could be directly affected by sea-level rise in 2050 in the Ganges–Brahmaputra–Meghna delta (Ericson et al. 2006). As a result of saline intrusion in both soil and freshwater, millions of people may face serious challenges in accessing safe water for drinking and other domestic uses. Factors already affecting Bangladesh include variations in temperature and rainfall, increased intensity of floods and recurrent flooding, the frequent incidence of cyclone and storm surges, drought, and salinity intrusion. Climate change is one of the greatest challenges to the development process, and the adversity of climate change impacts is greater when they combine with factors such as increasing density of population, poverty, and reduced access to services, especially water supply and sanitation, energy, and health services. A large proportion of the south Asian population lacks access to water services and more than 27% of the south Asian population are already without adequate food (Rahman et al. 2007a).

BACKGROUND AND CONTEXT

The coastal region of Bangladesh

The coastal zone of Bangladesh consists of 19 out of the nation's 64 districts. Twelve districts are classified as exposed coast and seven are defined as interior coast. The coastal zone covers 32% of the country in terms of land area and accommodates about 30% of the total population of the country (Islam 2004). The coast offers a diversity of natural resources, such as marine fisheries and shrimps, forest, salt, and minerals, and a location for high potential exploitation of both onshore and offshore natural gas. Export processing zones, harbors, airports, tourism complexes, and industrial units are located in this area. Some of the districts, including Chittagong, Cox's Bazar, and Patuakhali, have significant investment for tourism and associated business and trade.

Agriculture is the prime and economically most important sector of the country. The contribution of the broad agriculture sector to GDP was 20.29% in 2009–2010 (Ministry of Finance 2011), though this is down from around 26% in 1999–2000. There is a substantial indirect contribution to overall GDP, especially through the growth of the broad service sector including the wholesale and retail trade, hotel and restaurants, transport, and communication. In addition, the agriculture sector currently includes 43.6% of the total labor force in the country. The exact contribution of the agriculture sector from the coastal zone to GDP is not known, but it is reported that crop production, especially rice production per hectare, in the coastal zone has decreased slightly over the past several years. Total rice production (*Aus*, *Aman*, and *Boro*) over the nation increased to 28.9 million tons in 2008 from 27.3 million tons in 2007. Ten out of thirty-one agro-ecological zones are located in the coastal zone of Bangladesh (Islam 2004).

Bangladesh is well known as the land of the rivers (Box 8.1). Rivers, streams, and canals occupy about 7% of the country's surface. Surface water resources are dominated by the three major rivers: the Ganges, the Brahmaputra, and the Meghna. The Ganges–Brahmaputra–Meghna basins discharge about 43,000 m^3/s averaged over the year into the Bay of Bengal, draining over 1.65 million km^2 (Chowdhury and Bhuiya 2000). The country shares 57 trans-boundary rivers, 54 with India and 3 with Myanmar.

BOX 8.1 THE MEGADELTAS AND MEGACITIES OF ASIA

The large deltaic plains of the world are highly populated and subject to a range of natural and anthropogenic pressures. These low-lying areas are threatened from both land and sea, with the present-day impact of flooding, loss of sediment, erosion, subsidence, salinization, and storm surges aggravated by high population densities, poverty, and limited infrastructure and services. Climate change and sea-level rise will render an already precarious situation that much worse. According to Ericson et al. (2006), current rates of global sea-level rise, sediment trapping, flow diversion, and subsidence could, by 2050, result in more than one million people being affected in each of the three megadeltas (deltas with an area greater than 10,000 km^2), the Ganges–Brahmaputra–Meghna in Bangladesh, the Mekong in Vietnam, and the Nile in Egypt. Three quarters of the population affected worldwide by these trends live in Asian deltas or megadeltas (Ericson et al. 2006).

There are 11 megadeltas within Asia. The Intergovernmental Panel on Climate Change (Cruz et al. 2007) has concluded that Asian deltas are highly threatened by climate change and that responding to this threat will present important challenges. "The sustainability of megadeltas in Asia in a warmer climate will rest heavily on policies and programmes that promote integrated and co-ordinated development of the megadeltas and upstream areas, balanced use and development of megadeltas for conservation and production goals, and comprehensive protection against erosion from river flow anomalies and sea-water actions that combines structural with human and institutional capability building measures," according to Cruz et al. (2007).

Thirteen of the world's 20 largest cities are located in the coastal zone and seven of these are to be found in the Asian megadeltas. In 2005, Mumbai, Guangzhou, Shanghai, Ho Chi Minh City, and Kolkata were among the 10 most populated cities in the world (World Bank 2010). Based on case studies of climate risks and adaptation prospects for three Asia megacities, Bangkok, Ho Chi Minh City, and Manila, the World Bank (2010) concluded that, as climate change develops to 2050, the frequency of extreme events experienced in these cities will increase, as will the flood-prone area and the proportion of the population exposed to flooding. The costs of damage could range from 2% to 6% of the regional GDP, with damage to buildings an important component of the costs (land subsidence proved a major factor in the damage cost estimates). Impacts on the poor and vulnerable will be substantial, but even better-off communities would be affected. Three recommendations were advanced by the World

(*Continued*)

Bank study: (1) better management of urban environment and infrastructure; (2) climate-related risks as an integral part of city and regional planning; and (3) targeted, city-specific solutions, combining infrastructure investments, zoning, and ecosystem-based strategies. "Cutting edge approaches to urban adaptation are needed," the study concludes (World Bank 2010).

Mick Kelly

Nearly 1,000 billion m^3 of water enters into Bangladesh every year from these trans-boundary rivers but a very small portion of this total volume is available during the dry season, especially the month of February. Small, isolated wetlands, including ponds, lakes, *haors* (backswamps), and canals, serve the demand for water during the dry season. Access to improved drinking water source was 89% in 2008, while sanitation coverage was 63% in the same year (WHO/UNICEF n.d.). In 2004, access to safe water was 3% lower in the coastal region than at the national level, but sanitation coverage in the coastal region was 9% higher (Islam 2004).

Climate sensitivity, vulnerability, and impacts in the coastal zone

Over Bangladesh as a whole, the temperature is predicted to increase by 1.1°C in the monsoon season (May–September) and by 1.8°C in winter (December–February) by the year 2050 (World Bank 2000). These estimates are in good agreement with those used in Bangladesh's National Adaptation Plan of Action (Ministry of Environment and Forests 2005). Recently, temperatures have been generally increasing in the monsoon period of June–August, with maximum and minimum temperatures increasing at the rate of 0.05°C and 0.03°C a year, respectively (Rahman and Alam 2003). Average winter time (December–February) maximum and minimum temperatures show contrasting decreasing and increasing trends at a rate of –0.001°C and 0.016°C a year, respectively (Rahman and Alam 2003). The perception of local communities regarding temperature trends shows a similar picture in most of the regions of the country. That increased temperature, especially during the pre-monsoon period (March–May), is a major problem has emerged from various studies conducted in the coastal zone (Bangladesh Centre for Advanced Studies 2010).

It is predicted that rainfall levels on average could become greater and more irregular overcoming decades, though winter rainfall levels may decrease (Ministry of Environment and Forests 2005). Annual rainfall rose at about 4 mm a year over the period 1978–2008. The northwestern districts of the country usually face drought to some extent almost every year during the

March–May period, with severe drought occurring at least nine times since 1973. Some coastal districts also suffer from drought problems. Coastal flooding and water logging due to excessive rainfall is frequent. The recent floods in 1998, 2004, and 2007 affected most of the coastal districts. In future, increased snow melt from the Himalayan permafrost due to increasing temperature may force more water to flow through the Ganges, Meghna, and Brahmaputra river systems and their river networks resulting in additional flooding extending over the central flood plain of Bangladesh (Rahman et al. 2007b). Increased flooding due to climate change may affect large areas, with high potential for substantial casualties in coastal areas.

Temperature and rainfall variations are already affecting both rice and nonrice crops in the coastal zone. For example, erratic rainfall behavior (late onset, excessive rainfall over a short period, lack of rainfall in particular time of the season, etc.) is causing a reduction in agricultural yields (Bangladesh Bureau of Statistics 2010). Around a third of a million tons of rice (*Aus* and *Aman* variety) was damaged recently by excessive rainfall and flooding in Jessore, one of the major rice-producing coastal districts of the country (Bangladesh Bureau of Statistics 2010). Total production of *Aus* (local and high-yield variety) decreased from 3,606 tons in 2007 to 2,955 tons in 2008 in Satkhira, another vulnerable coastal district. A similar trend has been observed in the total production of other rice varieties (*Aman* and *Boro*) in Satkhira (Bangladesh Bureau of Statistics 2009).

According to the estimates used in developing the NAPA (Ministry of Environment and Forests 2005), Bangladesh could face a sea-level rise of 32 cm by the year 2050. Sea level at Hiron Point near Sundarban has been rising at 5.3 mm a year over the period 1977–2002 (Centre for Environment and Geographic Information Services 2006). Other stations along the Bangladesh coastline also show an increasing trend (SAARC Meteorological Research Centre 2003). In the future, low-lying coastal lands might be gradually inundated, affecting all agricultural activities, water supply and sanitation systems, and other infrastructure unless these are adequately protected. The primary physical effects of sea-level rise on the Bangladesh coast have been identified as the intrusion of saline water, drainage congestion, changes in coastal morphology as erosion rates and sediment flows alter, and amplification of the impact of extreme events such as storm surges (World Bank 2000).

Saline water intrusion represents one of the major physical effects of sea-level rise on the coast of Bangladesh. Saltwater is already intruding into fresh water and increasing the level of salinity in many coastal districts in the southern part of the country, including Patuakhali, Pirojpur, Satkhira, Bhola, Khulna, Feni, and Noakhali (Islam 2004). According to Rabbani et al. (2010), salinity intrusion is affecting about 0.83 million hectares of land, resulting in reduction in crop yields. The supply of clean water to the domestic, industrial, and agricultural sectors, and for business purposes, is also being affected.

It has been reported that at least 11 out of 19 coastal districts have a higher extent of severe malnutrition of children than that of the national average (Islam 2004). In addition, incidences of diarrhoea and dysentery are higher in coastal regions where salinization is affecting both water and soil. Saltwater will increasingly intrude through different channels (rivers, canals, etc.) as sea-level rise progresses. Higher sea level may occupy low-lying areas and push the saline water front further into the water channel and there could be upward pressure on the saline/freshwater interface in the groundwater aquifers. A higher level of storm surge would carry saline water further, reducing the quality of surface water. Any reduction in freshwater discharge or flow in rivers during the winter months would also increase saltwater intrusion. It is feared that, in the future, salinization could render the whole coast unsuitable for some rice varieties (i.e., *Boro*) and wheat production (Huq et al. 1999).

Drainage would be hampered by the combined effect of sea-level rise, subsidence, siltation, higher riverbed levels, and reduced sedimentation in areas that are flood-protected. Drainage congestion will result in increased waterlogging. Infrastructure development could also adversely affect the natural drainage capacity of the coastal zone. Morphological processes along the coast of Bangladesh are extremely dynamic. As climate change develops, there is concern that increased riverflow might heighten rates of bank erosion and disruption of the balance between sediment transport and deposition could lead to even greater rates of change in coastal morphology.

The geographical structure of the Bangladesh coast makes it particularly vulnerable to recurrent cyclones and storm surges, and this vulnerability would be exacerbated by sea-level rise. The coast of the country has been affected by at least two super cyclones (with winds greater than 220 km h^{-1}) and 19 very severe cyclones (119–220 km h^{-1}) over the past 40 years. Model results indicate an increase in precipitation intensity, and possibly wind intensity, in a warmer climate. Unnikrishnan et al. (2006) project no significant change in cyclone frequency in the Bay of Bengal, but large increases in the frequency of storm surges. Cyclones and storm surges affect most severely life, livelihoods, and coastal ecosystems. For example, in 2007, cyclone Sidr killed more than 3,000 people and affected the lives of about six million others (Rabbani et al. 2010). The storm damaged around 80% of the Sundarban area and destroyed crops over about 0.35 million hectares. In addition, embankments and roads, power supply networks, and sources of safe water, such as tube wells, ponds, and reservoirs, were seriously affected. Over 6,000 small ponds were contaminated with saline water. The problems of the local communities were aggravated when cyclone Aila hit the coast just one year after Sidr. At the time of writing, some coastal districts, notably Satkhira, Patuakhali, and Barguna, are still suffering from lack of safe water supply and proper sanitation practices.

Table 8.1 presents the assessment of the sectoral impacts of climate change and sea-level rise for Bangladesh developed by the NAPA team (Ministry of Environment and Forests 2005).

Table 8.1 Intensity of impacts on different sectors due to climate change

	Physical vulnerability							
		Sea-level rise			Flood			
Sectoral vulnerability	*Extreme temperature*	*Coastal inundation*	*Salinity intrusion*	*Drought*	*River flood*	*Flash flood*	*Cyclone and storm surges*	*Erosion and accretion*
Crop agriculture	+++	++	+++	+++	+	++	+++	
Fisheries	++	+	+	++	++	+	+	
Livestock	++	++	+++			+	+++	
Infrastructure	+	++			++	+	+	+++
Industries	++	+++	++		++	+	+	
Biodiversity	++	+++	+++		++		+	
Health	+++	+	+++		++		++	
Human settlement							+++	+++
Energy	++	+			+		+	

Note: +++ high, ++ moderate, + lower impact.

ADAPTATION IN AGRICULTURE AND WATER SUPPLY AND SANITATION

Governmental and nongovernmental organizations have introduced a number of adaptation technologies that are being practiced at the community level, not only in the coastal zone but also in other climate-prone areas in Bangladesh. Some of the current adaptation technologies in crop agriculture and water supply and sanitation are spontaneous in nature, but others are reactive (Rabbani 2010). Most of these available adaptation technologies in terms of both crops and water supply and sanitation are likely to address present climate variability. Communities may well, however, need improved technologies in every sector that needs to be resilient to the changes of climate discussed earlier. Examples of current adaptation technologies and practices in agriculture (particularly in rice production) and the water supply and sanitation sectors are presented here.

The Ministry of Agriculture and associated agencies of the Government of Bangladesh have introduced a number of crop varieties that are resilient to climate-induced hazards such as flooding and salinization (Table 8.2). Three rice varieties are grown in three different seasons: *Aus*, *Aman*, and *Boro* (Bangladesh Centre for Advanced Studies 2010). The season of *Boro* refers to the cultivation that takes place in the months of December–May. In the case of the varieties that are farmed during this season, the seeds are sown first and then transplanted, and the production has to be irrigated. The season of *Aus* starts in April and continues till August. In Bangladesh, these months are known for the monsoon and the crops grown at *Aus* are rain fed. These varieties are low yield and threatened by heavy monsoons. Agriculturally, June–December is known as the *Aman* season. This is the time when most natural climate-related hazards hit the country and harvests are often affected by natural calamities. Some varieties of *Aman* are scattered and raised, while others are transplanted. The yield of *Aman* is greater than those grown in *Aus* and lower than those grown in *Boro*.

The farmers in the coastal zone are using a number of stress-tolerant rice varieties to adapt to changing conditions (Figure 8.1). That some of the varieties are growing rapidly in popularity among farmers may be due to higher production and net profit. A recent study conducted in the coastal zone reveals that BR 47, which is a saline-tolerant variety, provides the highest production among all the varieties being practiced in different seasons in the study area (Bangladesh Centre for Advanced Studies 2010). Although this saline-tolerant variety takes 30–50 days more than any drought-tolerant variety to reach harvest, it gives 30%–40% more yield than the latter per unit (Table 8.2). Most of the farmers, however, in the study area still use traditional varieties of rice (Figure 8.1). Figure 8.2 spells out the case on the cost and net profit from the cultivation

Table 8.2 Climate-related stress-tolerant rice varieties in the coastal region and Bangladesh

Climate-related stress	*Climate-tolerant rice variety*	*Growing season*	*Growth duration (days)*	*Average yield (ton/ha)*
Flood (submergence)	BRRI dhan 51	*Aman*	142–154	4.0
	BRRI dhan 52	*Aman*	145–155	4.5
Salinity in soil, surface, and groundwater	BRRI dhan 40	*Aman*	145	4.5
	BRRI dhan 41	*Aman*	148	4.5
	BR 10	*Aman*	150	5.5
	BR 23	*Aman*	150	5.5
	BRRI dhan 27	*Aus*	115	4.0
	BR 47	*Boro*	152	6.0
Drought	BRRI dhan 42	*Aus*	100	3.5
	BRRI dhan 43	*Aus*	100	3.5
	BRRI dhan 33	*Aman*	118	4.5
	BRRI dhan 39	*Aman*	122	4.5

Sources: *Daily Star*, "Two salinity tolerant varieties of T-Aman paddy soon," July 18, 2010, http://www.thedailystar.net/; *Financial Express*, "BRRI releases 2 new stress-tolerant rice varieties," April 10, 2011, http://www.thefinancialexpress-bd.com/; Bangladesh Rice Research Institute (BRRI), "Achievement of BRRI-modern varieties," n.d., http://www.brri.gov.bd/; Huq, S. and Rabbani, G., "Adaptation technologies in agriculture: The economics of rice-farming technology in climate—Vulnerable areas of Bangladesh," in L. Christiansen et al. (eds.), *Technologies for Adaptation: Perspectives and Practical Experiences*, UNEP, Denmark, 2011; Mazumdar, M.L.H., "Adapting agriculture to climate change," *Daily Star*, January 1, 2011, http://www.thedailystar.net/; Salam, M.A. et al., *International Rice Research Newsletter*, 32(1), 42–3, 2011.

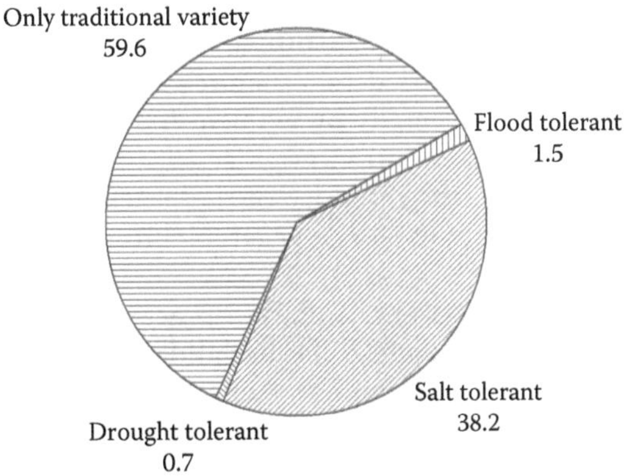

Figure 8.1 Number of households practicing different types of stress-resistant rice varieties in coastal areas (percentage of sample of 401 households).

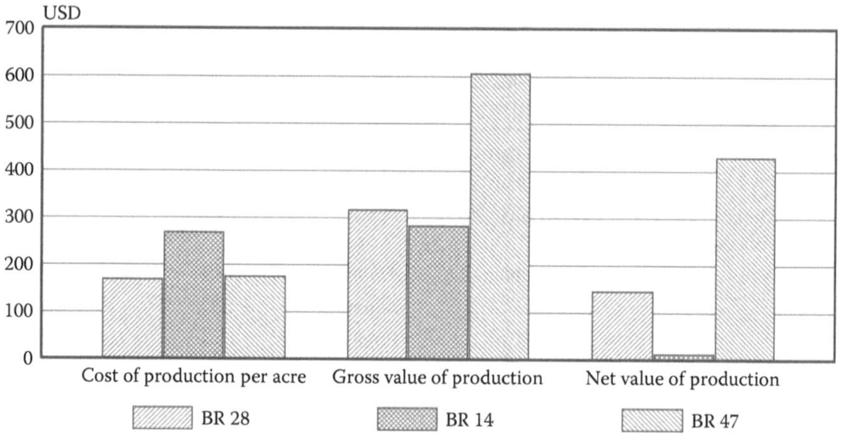

Figure 8.2 Estimates of cost and profit of traditional and salt-tolerant *Boro* variety (BR 47) in the coastal region.

of traditional and saline-tolerant BR 47. The salt-tolerant variety requires 20% less resources than the average cost of production for traditional varieties. And the net profit from the salt-tolerant variety is 81% greater than the average profit of the traditional varieties.

The *floating garden*, locally known as *baira* cultivation, is an adaptation technology in the agricultural sector that can easily be replicated in flood-prone areas. A floating garden is built using water hyacinth with some amount of soft soil or cow dung as a base on which vegetables are grown. Many villages in the coastal region in Bangladesh exposed to flood or inundation practice *baira* cultivation for basic or alternative income. Floating agriculture is being promoted by the governmental and nongovernmental organizations through training and cross visits (Ministry of Environment and Forests 2005). Communities that are facing climate-related hazards have long tried to generate additional income through alternative livelihoods, especially by women. Income diversification is an absolutely spontaneous adaptation practice that has been happening for many years. But, of late, the intensity of the practices has increased and diversified from a single crop (single vegetable) to multiple crops (various types of vegetables).

Safe water supply and proper sanitation practice is still a challenge for many communities in Bangladesh. Households and communities deploy a number of coping and adaptation technologies. For example, at the household level, one indigenous water treatment technology consists of three of four containers or pots, each covered by a clean cloth, arranged vertically and separately in a bamboo or wooden enclosure. All of the containers, except the one at the base, have small leaks at the bottom to pass on the water. Some communities use pond sand filter technology, especially for the purpose of safe drinking water. Pond sand filter technologies are mainly provided by the governmental and nongovernmental organizations to ensure safe water for vulnerable communities. Many communities protect small isolated water bodies and ponds for domestic use. It is evident that a large number of households in Shyamnagar, Upazila subdistrict, under the district of Satkhira, still use pond water for drinking purposes. Many of these ponds on which poor communities depend on for domestic water needs and small-scale livelihoods (e.g., vegetable farming, home gardening) were submerged as Cyclone Aila hit in May 2009; most parts of Upazila (subdistrict) are already exposed to saline intrusion at different levels of concentration. Many communities are using water collected by rainwater harvesting technology. This practice is again increasing, especially in the water-scarce coastal areas of the country. A recent initiative taken by the Bangladesh Centre for Advanced Studies (BCAS) in association with the NGO Forum for Public Health with support from Swiss Agency for Development and Cooperation is working on an action research project to identify context-specific water supply and sanitation technology in five ecospecific areas of the country, including the coastal zone. A set of criteria (i.e., hydrologically feasible, locally repairable, locally acceptable, user-friendly, gender-friendly, locally affordable, seasonally durable, functionality during disaster, and sustainability) is being used to identify the context-specific WatSan technology to adapt with future changing conditions.

THE NATIONAL ADAPTATION PROGRAMME OF ACTION

The NAPA for Bangladesh was published in 2005 (Ministry of Environment and Forests 2005). It presents a set of adaptation measures that complement national goals, including poverty reduction, and the objectives of multilateral environmental agreements (MEAs) and that, if not implemented without delay, would increase vulnerability or increase adaptation costs at a later date. The priority actions or projects were developed on the basis of a series of consultative workshops and sectoral reports. The final workshop involved over 100 stakeholders from the government and civil sectors. The NAPA report concluded that, in the Bangladesh setting, logic and justice compel focused attention on the vulnerabilities and multiple stresses affecting the lives and livelihoods of the poor (Ministry of Environment and Forests 2005). In prioritizing adaptation needs and activities, poverty reduction and security of livelihoods, with a gender perspective, were the most important criteria. The formal ranking order was based on the following main criteria:

1. Impact of climate change on the lives and livelihoods of the communities
2. Poverty reduction and sustainable income generation of communities
3. Enhancement of adaptive capacity in terms of skills and capabilities at community and national levels
4. Gender equality (as a cross-cutting criteria)
5. Enhancement of environmental sustainability
6. Complementarity and synergy with national and sectoral plans and programs and other MEAs.

The resulting list of priority activities or projects is summarized in Table 8.3. Details of selected projects that are particularly relevant to the themes of this chapter, agriculture and water supply and sanitation practice in the coastal zone, are as follow (see Ministry of Environment and Forests [2005] for additional information).

NAPA Project No. 1 concerns the reduction in climate change hazards through coastal afforestation. The justification for this action is that coastal forests play a vital role in stabilizing shorelines and providing protection against cyclones and other extreme events, as supported by recent experiences with tsunamis. The proposal is for a community-based afforestation program, based on deep-rooted, salt-tolerant species. Community participation will be a key element of the project; it is considered that the involvement of the local people, especially women, will enhance adaptive capacities and livelihoods in general. There will be synergy with the National Biodiversity Strategy and Action Plan, where afforestation is one of the critical working components.

The focus of Project No. 2 is the provision of drinking water to coastal communities to combat enhanced salinity due to sea-level rise. Given the threat to water supplies posed by climate change and sea-level rise, finding alternative sources of safe drinking water, such as rainwater harvesting and surface and groundwater treatment, is considered essential for the safety of the present and future generations.

The theme of Project No. 3 is capacity building, specifically, in integrating climate change in planning, designing of infrastructure, conflict management, and land-water zoning for water management institutions. It is observed in the NAPA document (Ministry of Environment and Forests 2005) that the National Water Management Plan (2001) is based on the current situation and does not take into account sustainability issues such as climate change, yet consideration of climate change issues and adaptive measures needs to be a regular part of the activities of water sector managers. There is a need for the knowledge of engineers and water sector managers to be more contextualized with climate change science and adaptation options.

Table 8.3 Projects advanced under the Bangladesh National Adaptation Programme of Action

No.	*Project*	*Type of project*
1	Reduction in climate change hazards through coastal afforestation with community participation	Intervention
2	Providing drinking water to coastal communities to combat enhanced salinity due to sea-level rise	Intervention
3	Capacity building for integrating climate change in planning, designing of infrastructure, conflict management, and land–water zoning for water management institutions	Capacity building
4	Climate change and adaptation information dissemination to vulnerable community for emergency preparedness measures and awareness raising on enhanced climatic disasters	Awareness/ capacity building
5	Construction of flood shelter and information and assistance center to cope with enhanced recurrent floods in major floodplains	Intervention
6	Mainstreaming adaptation to climate change into policies and programs in different sectors (focusing on disaster management, water, agriculture, health, and industry)	Capacity building
7	Inclusion of climate change issues in curriculum at secondary and tertiary educational institutions	
8	Enhancing resilience of urban infrastructure and industries to impacts of climate change	Capacity building
9	Development of eco-specific adaptive knowledge (including indigenous knowledge) on adaptation to climate variability to enhance adaptive capacity for future climate change	Intervention
10	Promotion of research on drought, flood and saline tolerant varieties of crops to facilitate adaptation in future	Research
11	Promoting adaptation to coastal crop agriculture to combat increased salinity	Intervention
12	Adaptation to agriculture systems in areas prone to enhanced flash flooding—North East and Central Region	Intervention
13	Adaptation to fisheries in areas prone to enhanced flooding in North East and Central Region through adaptive and diversified fish culture practices	Intervention
14	Promoting adaptation to coastal fisheries through culture of salt tolerant fish special in coastal areas of Bangladesh	Intervention
15	Exploring options for insurance to cope with enhanced climatic disasters	Research

Source: Ministry of Environment and Forests, *National Adaptation Programme of Action (NAPA): Final Report*, Ministry of Environment and Forests, Government of the People's Republic of Bangladesh, Dhaka, Bangladesh, 2005, http://www.moef.gov.bd/.

Institutional and policy development is essential to assist water sector managers in the development of multiobjective projects involving all stakeholders. It is noted that conflict situations often arise in the water resources sector and that climate stress is likely to aggravate tension. Capacity for negotiating sustainable conflict management, particularly in the water sector, needs to be strengthened.

Project Nos. 10 and 11 respond to the threat of increased salinization in the coastal zone through research on drought, flood, and saline-tolerant varieties of crops and by promoting adaptation in coastal crop agriculture. It is accepted that crop agriculture needs new approaches and technologies to deal with salinization in the coastal area and that improved varieties of all types of crops must be developed to withstand potential climate impacts. Suggested adaptations include, for communities affected by tidal surge flooding, the use of wet bed, no-tillage methods for maize production after loss of the *Aman* rice crop, and before the next *Boro* rice crop; this would also help to meet fuel and fodder needs. Selected vegetables and fruits could be produced on raised beds to meet the day-to-day demands of affected households and, as well as providing for the household, this could also generate some income. Project No. 11 would motivate affected communities to adopt the technologies such as these.

The Bangladesh Climate Change Strategy Action Plan (Ministry of Environment and Forests 2009) also addresses both the agriculture and water supply and sanitation sectors. More than 50 projects recommended by the Action Plan in the areas of food security, social protection and health, comprehensive disaster management, infrastructure, research and knowledge management, mitigation and low carbon development, and capacity building and institutional strengthening are being implemented by different government institutions Many of these projects are related to agriculture and water supply and sanitation and some projects are working directly with climate-resistant crop varieties.

CONCLUSION

Climate change and sea-level rise pose a substantial threat to a developing nation such as Bangladesh. By virtue of their circumstances, though, the people of Bangladesh have a long history of responding to natural hazards and coping strategies exist at the household level and beyond that can provide a strong foundation for robust adaptation measures. Moreover, the Bangladesh NAPA provides an effective platform for longer-term adaptation planning, and its pro-poor ethos could be considered a model for other countries. In building on these foundations, difficulties will undoubtedly have to be faced. Resourcing is a critical issue. Beyond that, the NAPA report identifies three potential barriers to the effective implementation of adaptation projects (Ministry of Environment and Forests 2005): (1) lack of awareness of both the seriousness of the climate threat and what actions can be taken in response; (2) lack of integration of climate change impacts in the development of policies, plans, and programmes in climate-sensitive sectors; and (3) lack of adequate tools, knowledge, and methodologies for guidance and advice in decision-making.

Ensuring that these deficits—in awareness, integration, and understanding—do not adversely affect the process of adaptation must be a high priority.

REFERENCES

Bangladesh Bureau of Statistics. (2009). *2008 Yearbook of Agricultural Statistics of Bangladesh*, Dhaka, Bangladesh: Statistics Division, Ministry of Planning, Government of Bangladesh.

Bangladesh Bureau of Statistics. (2010). *2009 Yearbook of Agricultural Statistics of Bangladesh*, Dhaka, Bangladesh: Statistics Division, Ministry of Planning, Government of Bangladesh.

Bangladesh Centre for Advanced Studies. (2010). "Economics of adaptation to climate change In Bangladesh," A report prepared by Bangladesh Centre for Advanced Studies (BCAS), Dhaka, Bangladesh.

Bangladesh Rice Research Institute (BRRI). (n.d.). "Achievement of BRRI-modern varieties." http://www.brri.gov.bd/ (accessed February 2011).

Centre for Environment and Geographic Information Services. (2006). "Impact of sea level rise on land use suitability and adaptation options," A report prepared for the Ministry of Environment and Forests, Government of Bangladesh.

Chowdhury, K.R. and Bhuiya, A. (2000). "Environmental processes: Flooding, river erosion, siltation, and accretion-physical impacts," in A.A. Rahman, S. Huq, and G.R. Conway (eds.), *Environmental Aspects of Surface Water Systems of Bangladesh*, Dhaka, Bangladesh: The University Press Limited.

Cruz, R.V., Harasawa, H., Lal, M., Wu, S., Anokhin, Y., Punsalmaa, B., Honda, Y., Jafari, M., Li, C., and Huu Ninh, N. (2007). "Asia," in M.L. Parry, O.F. Canziani, J.P. Palutikof, P.J. van der Linden, and C.E. Hanson (eds.), *Climate Change 2007: Impacts, Adaptation and Vulnerability. Contribution of Working Group II to the Fourth Assessment Report of the Intergovernmental Panel on Climate Change*, Cambridge: Cambridge University Press. http://www.ipcc.ch/ (accessed August 2013).

Daily Star. (2010). "Two salinity tolerant varieties of T-Aman paddy soon," July 18. http://www.thedailystar.net/ (accessed August 2011).

Ericson, J.P., Vorosmarty, C.J., Dingman, S.L., Ward, L.G., and Meybeck, M. (2006). "Effective sea-level rise and deltas: Causes of change and human dimension implications," *Global and Planetary Change*, 50: 63–82.

Financial Express. (2011). "BRRI releases 2 new stress-tolerant rice varieties," April 10. http://www.thefinancialexpress-bd.com/ (accessed August 2011).

Huq, S., Karim, Z., Asaduzzaman, M., and Mahtab, F. (eds.) (1999). *Vulnerability and Adaptation to Climate Change for Bangladesh,* Dordrecht, the Netherlands: Kluwer Academic Publishers.

Huq, S. and Rabbani, G. (2011). "Adaptation technologies in agriculture: The economics of rice-farming technology in climate—Vulnerable areas of Bangladesh," in L. Christiansen, A. Olhoff, and S. Traerup (eds.), *Technologies for Adaptation: Perspectives and Practical Experiences*, Denmark: UNEP.

Islam, M.R. (2004). *Where Land Meets the Sea: A Profile of the Coastal Zone of Bangladesh*, Dhaka, Bangladesh: The University Press Limited.

Mazumdar, M.L.H. (2011). "Adapting agriculture to climate change," *Daily Star*, January 1. http://www.thedailystar.net/ (accessed February 2011).

Ministry of Environment and Forests. (2005). *National Adaptation Programme of Action (NAPA): Final Report*, Dhaka, Bangladesh: Ministry of Environment and Forests, Government of the People's Republic of Bangladesh. http://www.moef.gov.bd/ (accessed January 2012).

Ministry of Environment and Forests. (2009). *Bangladesh Climate Change Strategy Action Plan 2009*, Dhaka, Bangladesh: Ministry of Environment and Forests, Government of the People's Republic of Bangladesh. http://www.moef.gov.bd/ (accessed January 2012).

Ministry of Finance. (2011). *Bangladesh Economic Review 2011*, Dhaka, Bangladesh: Economic Division, Ministry of Finance, Government of the People's Republic of Bangladesh. http://www.mof.gov.bd/ (accessed January 2011).

Rabbani, M.G. (2010). "Community based adaptation technology: current practices in Bangladesh," *City Voices CITYNET: The Regional Network of Local Authorities for the Management of Human Settlement*, 1(1): 6–7.

Rabbani, M.G., Rahman, A.A., and Islam, N. (2010). "Climate change and sea level rise: Issues and challenges for coastal communities in the Indian Ocean region," in D. Michel and A. Pandya (eds.), *Coastal Zone and Climate Change*, Washington, DC: The Henry L. Stimson Center.

Rahman, A. and Alam, M. (2003). *Mainstreaming Adaptation to Climate Change in Least Developed Countries (LDCs): Working Paper 2: Bangladesh Country Case Study*, London: International Institute of Environment and Development (IIED).

Rahman, A., Alam, M., Alam, S., Uzzaman, M.R., Rashid, M., and Rabbani, M.G. (2007a). *Risks, Vulnerability and Adaptation in Bangladesh*, A background paper prepared for UNDP Human Development Report 2007. Dhaka, Bangladesh: Bangladesh Centre for Advanced Studies.

Rahman, A.A., Huda, A.S., and Rabbani, M.G. (2007b). *Situation Analysis of Capacity Building Needs for IWRM in South Asia*, Dhaka, Bangladesh: Bangladesh Centre for Advanced Studies.

SAARC Meteorological Research Centre (SMRC). (2003). *The Vulnerability Assessment of the SAARC Coastal Region due to Sea Level Rise: Bangladesh Case*, Dhaka, Bangladesh: SMRC-No.3, SMRC Publications.

Salam, M.A., Rahman, M.A., Bhuiyan, M.A.R., Uddin, K., Sarker, M.R.A., Yasmeen, R., and Rahman, M.S. (2011). "BRRI dhan 47: A salt-tolerant variety for the boro season," *International Rice Research Newsletter*, 32(1): 42–3.

Unnikrishnan, A.S., Rupa Kumar, K., Fernandes, S.E., Michael, G.S., and Patwardhan, S.K. (2006). "Sea level changes along the Indian coast: observations and projections," *Current Science (Bangalore)*, 90(3): 362–8.

WHO/UNICEF. (n.d.). Joint Monitoring Programme (JMP) For water supply and sanitation. http://www.wssinfo.org/ (accessed February 2011).

World Bank. (2000). *Bangladesh: Climate Change and Sustainable Development*, Report no. 21104-BD, Dhaka, Bangladesh: World Bank Rural Development Unit, South Asia Region.

World Bank. (2007). *The Impact of Sea Level Rise on Developing Countries: A Comparative Analysis*, World Bank Policy Research Working Paper 4136, Washington, DC: World Bank.

World Bank. (2010). *Climate Risks and Adaptation in Asian Coastal Megacities: A Synthesis Report*, Washington, DC: World Bank.

Chapter 9

Coastal zone management and climate policy in Vietnam

P. Mick Kelly

Abstract: The coast of Vietnam is undergoing rapid transformation as a result of socioeconomic reform and environmental change. Increasing environmental stress and conflict over resources stemming from accelerating rates of development and the emerging threat of climate change and sea-level rise converge in the coastal zone to create a substantial challenge to individual livelihoods and community well-being, a challenge that coastal managers and planners must confront. This chapter reviews recent trends in coastal zone management and climate policy in Vietnam, identifying opportunities and weaknesses. Over the past two decades, significant progress has been made in developing a framework for integrated coastal zone management and, in parallel, a national strategy on climate change. Ensuring effective coordination between these two processes, hence capturing synergistic benefits, would clearly be advantageous. Disaster risk reduction would provide a common focus of immediate relevance. Alongside the availability of resources, three particular issues that will determine the success of the process of policy development and implementation are highlighted: the importance of coordination, vertically and horizontally within society; the need for full participation; and the desirability of a flexible, step-by-step approach (so that lessons can be learnt from experience). Models are available within Vietnamese experience suggesting that these issues can be tackled effectively.

INTRODUCTION

The physical characteristics of the Socialist Republic of Vietnam and the socioeconomic situation of its people render this Southeast Asian country particularly vulnerable to the impact of climate change and sea-level rise. The coastline of Vietnam extends 3,440 km and the coastal zone is frequently affected by extreme weather events, most notably, the incidence of tropical cyclones and typhoons. The Red River delta in the north and the Mekong River delta in the south are densely populated and home to most of the nation's agriculture, commerce, and industry. A dynamic environment, the coastal zone of Vietnam has been actively managed for centuries providing a fertile, but at times precarious, basis for the well-being of the nation's people (Khanh 1984). The natural dynamism of the environment has been matched in recent decades by rapid social change as the process of economic renovation has led to mixed outcomes for the inhabitants of the coastal zone (Adger et al. 2001). Although, overall, economic health has improved and there has been a marked diversification of economic opportunities, inequity has risen and there has been a loss of social capital.

This chapter considers the response of the Vietnamese government to the multiple threats faced by inhabitants of the nation's coastal zone. Over the past 20 years, a series of initiatives has laid a framework for integrated coastal zone management (ICZM) in Vietnam in an attempt to resolve resource conflicts and environmental degradation. At the same time, national policy on climate change, which, given the vulnerability of the coastal zone, has major implications for coastal management, has been developed. The challenges and opportunities afforded by these synergistic developments are discussed, drawing on the experience of the Red River delta and its environs as an example.

THE RED RIVER DELTA: PAST, PRESENT, AND FUTURE

The Red River delta in northern Vietnam covers an area of approximately 17,000 km^2. It is triangular in shape extending 120 km along the coast and around 240 km inland (Sekhar 2005; Thanh et al. 2004). It is one of the most densely populated regions in the world. Seventeen million people live in the Red River delta, with population density exceeding 3,000/km^2 in places (Thang et al. 2011). The capital of Vietnam, Hanoi, with a population of around six million, is located in the delta, which was originally an inlet of the Gulf of Tonkin, being formed over millennia by deposition. The Red River (*Sông Hông*) rises in Yunnan Province, People's Republic of China, and has two major tributaries, the Lo River (*Sông Lô*) and Black River (*Sông Da*). The silt of the Red River is rich in iron oxide, hence its name, and the land of the delta is very fertile. The entire region is less than 3 m above sea level, with much of it standing at 1 m or less. It is prone to flooding due to high river flow, high tides, storm surges, and strong wind-induced waves (Vietnam Ministry of Water Resources et al. 1994). There are around 3,000 km of river dykes and 1,500 km of sea dykes (Thanh et al. 2004) (Figure 9.1).

The Red River delta is the ancestral home of the Vietnamese people. Dyke-building, for water retention in rice agriculture and to withstand river flooding, can be traced back to the Bronze Age (Benson 1997; Khanh 1984). Sea dykes were built for land reclamation, rather than coastal defense, in earlier times. Over time, as the network of river and sea dykes evolved, social structures developed to ensure maintenance of the system and a communal view of coastal

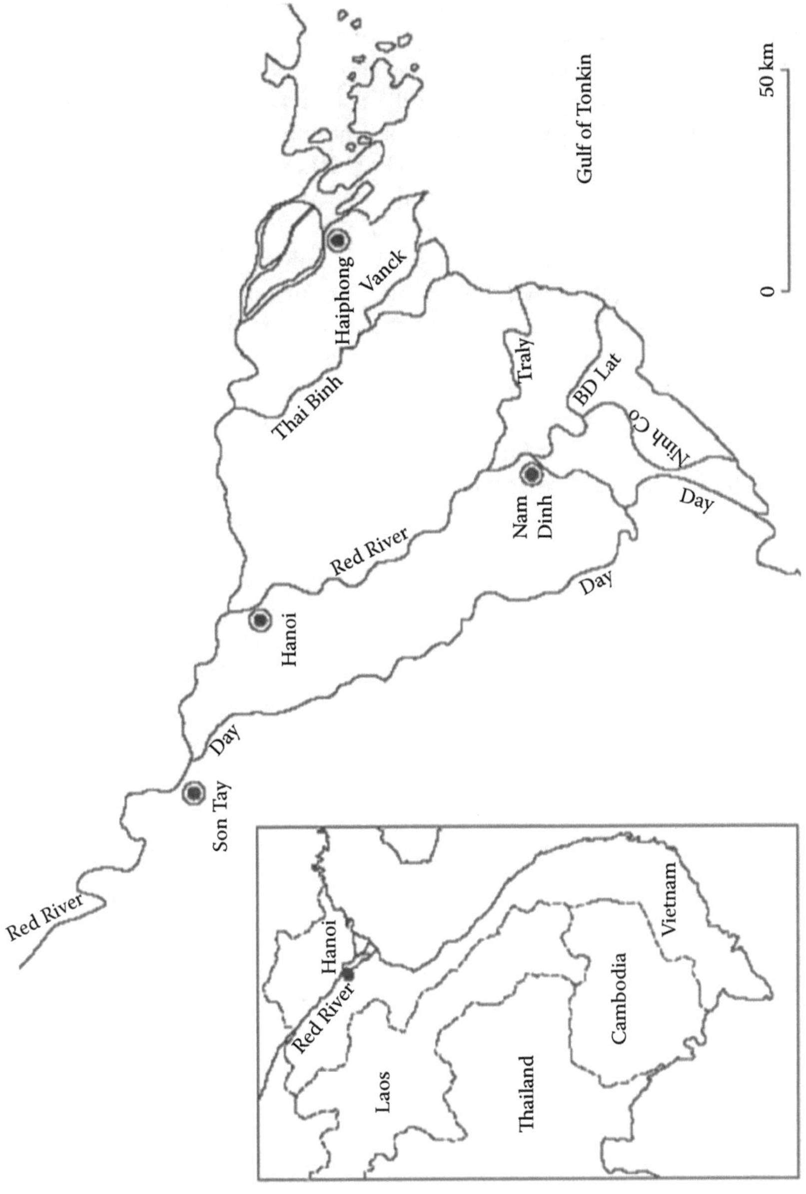

Figure 9.1 The Red River delta. (Data from Mai, C.V. et al. *Journal of Coastal Research*, 25, 105–16, 2009.)

protection emerged with all levels of society accepting a moral responsibility to engage with the process of defense against flooding. A complex set of institutions evolved to ensure compliance. Protection against natural hazards was seen as a continual process that would anticipate impacts rather than solely provide a response in the immediate aftermath of an event. The legacy of the traditional system can be seen in aspects of the late twentieth century flood protection and cyclone response system (Kelly et al. 2001a). The system of shoreline protection, dyke management, and flood control is now being redeveloped, based on modern design technologies and techniques (Mai et al. 2009).

Over the past 20 years, the coastal zone of Vietnam has come under increasing pressure as a result of a growing population, economic development, and environmental degradation (Sekhar 2005; Thanh et al. 2004). As a major generator of national income, through the agriculture, fishing, marine transportation, and tourism industries and the increasing development of heavy industries around the major population centers, the coastal zone has become a magnet for internal migration. Land continues to be reclaimed from the sea for industrial expansion, ports, and tourism. Aquaculture has become a major focus for investment and, as farms often replace mangrove forests, the services afforded by the coastal ecosystem are being degraded (Tri et al. 2001). Tourist centers in the coastal zone are experiencing rapid growth with local infrastructure, such as waste disposal systems, under pressure. Deforestation in the hills above the delta is increasing the risks associated with heavy rainfall, and urbanization and industrialization are decreasing water quality in the delta and the coastal waters. Ericson et al. (2006) identify sediment loss as the major contributor to the contemporary rate of effective sea-level rise in the coastal zone of the Red River delta, estimating a current rate of 3–5 mm per year.

Vietnam is among the nations most threatened by the rise in sea level associated with global warming. According to Pham Khoi Nguyen, Minister of Natural Resources and Environment,

> the impacts of climate change, in particular sea-level rise, will heavily affect the country's economic production, livelihood, environment, infrastructure, public health, and threaten the achievements of poverty reduction, food and energy security, sustainable development, as well as the fulfillment of the Millennium Development Goals.
>
> Ministry of Natural Resources and Environment 2010: 12

In a survey of low-lying areas in 84 coastal developing countries (Dasgupta et al. 2007, 2009), Vietnam ranks first in terms of gross domestic product (GDP), urban area and wetlands are at risk in the event of a 1-m rise in sea level. A 1-m rise in sea level could inundate 5,000 km^2 of the Red River delta (Tran et al. 2005). Sectors considered vulnerable to sea-level rise include agriculture and food security, aquaculture, sea and coastal ecological systems, water resources (surface and groundwater), energy, tourism,

Figure 9.2 Damage resulting from Typhoon Damrey in September 2005. (Courtesy of L.Q. Huy.)

residential space, infrastructure and industrial zones; poor farmers and fishermen, senior citizens, children, and women are particularly at risk (Ministry of Natural Resources and Environment 2010).

It may be that the changing frequency and character of extreme weather and climate events (Ho et al. 2011) could have more significant consequences for the coastal zone in the immediate future than the long-term trend in sea level. Any increase in the frequency or strength of tropical cyclones striking the coast of Vietnam would seriously impact lives and livelihoods (Kelly and Adger 2000; Kelly et al. 2001a). In 1997, Tropical Storm Linda was responsible for the deaths of over 1,000 people in the south of Vietnam, with economic losses approaching USD 600 million. In September 2005, Typhoon Damrey, one of the most powerful storms of recent decades, caused extensive overtopping of sea dykes in the Red River delta, and an 800-m length of sea dykes was completely washed out in Nam Dinh Province (Mai et al. 2009). Drought and flooding already affect the productivity of agriculture, and the Ministry of Natural Resources and Environment projects reduced outputs for both the spring and summer rice crops in the Red River delta as climate changes (Ministry of Natural Resources and Environment 2008) (Figure 9.2).

Although the physical threat to the coastal zone of the Red River delta is clear, assessing the full implications of the risks associated with climate change and sea-level rise requires assessment of the human dimension, the vulnerability of the local population (see Box 9.1). Gioa Thuy district

BOX 9.1 DEFINING VULNERABILITY: IMPLICATIONS FOR POLICY

It is useful to distinguish between the biophysical threat faced by a population, which can be termed exposure, and the capacity of that population to respond to stress, vulnerability (Adger 1999; Kelly and Adger 2000; Adger and Kelly 2001). Vulnerability, in this sense, is determined by social and economic conditions and is socially differentiated. Following Sen (1990), the key to understanding what shapes levels and patterns of vulnerability is the extent to which individuals, groups, or communities are *entitled* to make use of resources, in the most general sense of the word. Analyzing vulnerability, therefore, is a matter of assessing the *architecture of entitlements*, the availability and distribution of entitlements, and how this changes over time as a result of social or biophysical trends. Various factors determine access to resources. At the individual level, diversity of income sources and social status within the community play a part. At the community level, levels of informal and formal social security and insurance and institutional and market structures are important determinants. Social capital, the existence of networks of social connections and shared norms and values, is a key factor.

Focusing on the present-day social and economic determinants of vulnerability can be an effective starting point in determining sustainable adaptive strategies (Adger and Kelly 2001). By identifying *win–win* solutions, this approach can reduce the conflict between immediate development needs and longer-term environmental protection. Drawing on case studies conducted in northern Vietnam, Kelly and Adger (2000) identify four measures that might, as a starting point, reduce the vulnerability to climate change of the poorer members of coastal communities and enhance their adaptive capacity: (1) poverty reduction; (2) risk-spreading through income diversification; (3) redressing the impact of the loss of common property management rights; and (4) compensating for the erosion of forms of collective action or investment. They note that the underlying causes of vulnerability, the maldistribution of resources, must also be addressed.

lies in Nam Dinh Province on the coast of the Red River delta, bordering the Red River (Figure 9.3). The district includes Xuan Thuy National Park, a coastal wetland environment with extensive mangrove forests, declared as Vietnam's first Ramsar site in January 1989. The district is located in one of the fastest developing economic zones in Vietnam. Agriculture is the main economic activity, but local livelihoods are relatively diverse compared to elsewhere in rural Vietnam. In a case study of vulnerability to climate

Figure 9.3 Giao Thuy district. (Data from Mai, C.V. et al. *Journal of Coastal Research*, 25, 105–16, 2009.)

change in the area, then known as Xuan Thuy district, Adger and Kelly (1999) made use of three indicators—poverty, inequality, and institutional adaptation—to define the manner in which socioeconomic trends were increasing or decreasing levels of vulnerability, particularly in the case of exposure to cyclone impacts (Table 9.1). A key factor proved to be the myriad social and economic changes accompanying the economic renovation process of the previous 10–20 years.

Table 9.1 Characteristics of vulnerability in Xuan Thuy district

Vulnerability indicator	*Proxy for*	*Observations*	*Mechanism for translation into vulnerability*
Poverty	Marginalization	One-fifth of population below poverty line on the basis of basic needs Poverty and income stratified: salt-making communities—low incomes with low variance; agriculture and rice-based communities—higher income with higher variance	Perception of marginalization in land allocation and access to resource on part of salt-making communities Landlessness—higher rate of permanent migration among poor
Inequality	Degree of collective responsibility, informal and formal insurance, and underlying social welfare function	Estimated income equality higher than for rural Vietnam Aquaculture contributes most to inequality, followed by wages and remittances, agriculture, and salt-making	Concentration of productive capital and entitlements in fewer hands
Institutional adaptation	Architecture of entitlements determines exposure, institutions as conduits for collective perceptions of vulnerability, endogenous political institutions constrain or enable adaptation	Formal: powerful communes (including those in aquaculture) divert district resources for individual benefit Informal: re-emergence of post-collective credit and rural institutions	Increased impact of climate extremes exacerbated through neglect of collective action Offset by adaptation and policy learning and by informal civil society mechanisms

Source: Adger, W.N. and Kelly, P.M., "Social vulnerability and resilience," in W.N. Adger et al. (eds.), *Living with Environmental Change: Social Vulnerability, Adaptation and Resilience in Vietnam*, Routledge, London, 2001.

The system of reforms that become known as *doi moi* was introduced in 1986, accelerating and broadening an existing reformist tendency (Fforde 2009). The collectivized agricultural system had already been amended in the early 1980s through the creation of household contracts that devolved control of some aspects of agricultural production from the collective to the household-level. Since the late 1980s, the major priorities of the *doi moi* program have been industrialization, rural development, and poverty reduction. State-owned enterprises in the manufacturing and other sectors have been opened up to market forces and land tenure and control has been reformed. Over the past three decades, there has been significant progress in reducing inflation and poverty levels, coupled with strong economic growth averaging close to 7% *per annum* during the 1990s (Fforde 2009).

At the time of the case study, some 10 years after the introduction of *doi moi*, Xuan Thuy was, by rural Vietnamese standards, a relatively wealthy district. As a result of the economic renovation process, aquaculture, fishing, sea-food processing, and other trading and services related to agriculture and aquaculture sectors supplemented the incomes of many households. The poorer households remained, however, dependent on a limited range of resources and income sources. They had reduced access to credit and were more reliant on livelihoods, such as salt-making, exposed to extreme events. The rapid development of aquaculture since the introduction of *doi moi* had had mixed effects on levels of vulnerability in the area. All sectors of the community should have benefited from the trickle-down effects of increased wealth generation. By increasing levels of inequality, though, the expansion of aquaculture had skewed access to resources and the loss of a common access resource, the mangrove forests, had adversely affected social capital. Aquaculture was also tying up considerable amounts of capital in an activity inherently exposed to extreme weather events, particularly cyclone impacts.

Recent research in Gioa Thuy district has demonstrated that these trends have continued over the past 10 years (Huy 2010, 2011). The community has become wealthier overall, and the development of economic groups whose livelihoods are largely independent of the environment, such as traders, has contributed significantly to the local economy, assisting a large number of households to diversify their livelihoods. Nevertheless, a large part of the district's population remains poor and the majority of the population is still highly dependent on the environment. Income inequality increased sharply during the early years of the present century. Vulnerability is also being affected, adversely in many circumstances, by changes in market policies and the dismantling of the agricultural cooperatives.

Changing land rights have also affected levels of vulnerability. The Bac Cua Luc wetland lies in Hoanh Bo district in Quang Ninh Province, to the immediate north of the Red River delta. Historically, this mangrove wetland has been managed collectively. During the 1990s, two areas were enclosed and, with government subsidy, allocated as reclaimed agricultural land to settlers from more highly populated areas or leased or sold for aquaculture. Adger et al. (1997, 2001) discuss the effects of this privatization of a common resource on the two local communes, Le Loi and Thong Nhat. The conversion enhanced inequality in income distribution as the poorer members of the communities were more dependent on the converted mangrove forests, whereas the wealthy could take advantage of new investment opportunities. It also increased the chances of dissent among the local people regarding management of the remaining, more limited mangrove resource, undermining the social capital inherent in forms of collective action, common property management in this case. The individual vulnerability of the poorer members of the community was heightened, as was the collective vulnerability of the community as a whole.

MANAGING THE COASTAL ZONE

According to Vietnam's Ministry of Natural Resources and Environment (2006), the main environmental threats to coastal sustainability are unsustainable resource use, ecosystem degradation, environmental pollution and natural and environmental disasters, population pressure, and conflict of interest between stakeholders. Capacity to manage these pressures, which reflect the classical growth/environment dichotomy, is constrained. Although a national framework for environmental protection and associated infrastructure has been developed over the past 30 years, implementation remains patchy (Kelly et al. 2001b) and legal enforcement is lax (Sekhar 2005).

A number of projects and programs intended to address the issue of coastal resource management have been undertaken. The Vietnam Coastal Zone Vulnerability Assessment was conducted in the mid-1990s (Thang et al. 2011; Vietnam Vulnerability Assessment 1996). It was concluded that the overall vulnerability of Vietnam to a 1-m rise in sea level was high given the potential damage, lives that could be lost, and the cost of adaptive measures. ICZM was recommended as an effective adaptive response. Following this study, three ICZM pilot projects were conducted at the provincial level* under the Vietnam–Netherlands Integrated Coastal Zone Management Program (VNICZM) (Sekhar 2005; Thang et al. 2011). The VNICZM project was informed by the Vietnamese vision of an effective integrated coastal program, with an ICZM program being simultaneously carried out in Hanoi and the coastal provinces in a manner consistent with Vietnam's constructive, holistic, and gradual approach to decentralization (Thang et al. 2011). The reference to gradual decentralization is significant, reflecting a general commitment by central government to this process; the

* The administrative structure in Vietnam consists of the national, provincial, district, and commune levels.

emphasis on simultaneous execution is also noteworthy. The district level will act as the focus of the ICZM process in order to base strategy on local problems. As Eucker (2008) notes, effective coordination between the center and the provincial, district, and commune levels will be essential to the success of any such program. Sekhar (2005) argues that the decentralization process should extend to granting legal authority at the commune level to resolve local resource conflicts.

The ICZM strategy for Nam Dinh Province was developed as one of the initial pilot projects (People's Committee of Nam Dinh Province 2003; Eucker 2008). The goal of the initiative was to solve the present and potential problems and challenges at its coastal zone regarding the pollution of environment, degradation of the natural resources and biodiversity, coastal erosion, and natural hazards, aiming at hunger elimination and poverty alleviation, improvement of the material and spiritual living conditions of the local communities, and achievement of the sustainable development (People's Committee of Nam Dinh Province 2003). On the physical side, the main threats to the coastal zone in this area were identified as short-term hazards, such as cyclone impacts, and longer-term problems, such as coastal erosion. In discussing coastal protection strategies in the Red River delta, Mai et al. (2009: 107) describe Nam Dinh Province as "the most dynamic part of the coastal zone." On the human side, the main threats included low awareness of the value of coastal resources, lack of understanding of common resources and the need to safeguard values, difficult living conditions and population growth, and the tendency to exploit resources for immediate benefits. Cited as presenting difficulties in effective management were an inappropriate framework of resources and environment management, including the sectoral and territory-oriented structure of resources management, which tended to lead to multisectoral conflicts and cross-boundary issues, and a lack of linkage between policy and management actions related to resources and environment. Stakeholder input was a major aspect of development of the strategy, with workshops and participatory rural appraisals conducted at the commune level.

The resulting ICZM strategy for Nam Dinh Province consisted of six components:

- Public awareness and education on natural resources, environment, and sustainable development
- Training, mobilizing human resources for ICZM
- Protection and conservation of coastal natural resources and values
- Prevention and minimization of natural hazards
- Rehabilitation and improvement of habitats and polluted and degraded objects/environmental components
- Development of coastal potentials and values

Although the Nam Dinh ICZM strategy covered several decades, the action programs focused on urgent issues, mainly for the period to 2010 covered by most provincial and national plans and strategies. A number of difficulties in implementing ICZM locally were identified, including the low level of knowledge and high levels of unemployment and poverty on the part of local communities and limitations in local government management capacity (human resources, finance, and technology). Immediate actions to be undertaken included awareness-raising among coastal inhabitants and capacity-building for local officials. In Gioa Thuy district, sustainable development of the Ramsar site, the Xuan Thuy National Park, was identified as a major concern. A provincial steering committee, including representatives from government departments and agencies, was established to ensure multisectoral participation. Eucker (2008) considers that improved coordination between and within relevant sectors and more effective use of institutional and human resources would be among the most promising outcomes of the strategy.

There is strong support at the ministerial level for a coastal management approach based on ICZM. According to Minister Pham Khoi Nguyen, "the present and the future coastal challenges are complex and need to be addressed in a holistic way" (Nguyen 2011: 107). A coastal management center has been established within the Vietnam Administration of Sea and Islands in the Ministry of Natural Resources and Environment. It is recognized that participation of the coastal inhabitants, nongovernmental organizations, and stakeholders at the local level is an essential component of any ICZM program and, as a means of triggering participation, an awareness-raising project has been conducted with children as the target group (Vahtar et al. 2011). The Ministry of Natural Resources and Environment document, *Vietnam's ICZM Strategy 2020 and Orientation up to 2030* (2006), provides a draft framework for ICZM in Vietnam and a basis for the intended development of the national ICZM strategy. The overall objective of Strategy 2020 is "To achieve sustainable development through integrated management of the Vietnam Coastal Zone by creating inter-sectoral, interagency and inter-governmental coordination and cooperation mechanisms" (Ministry of Natural Resources and Environment 2006: 30). An extremely detailed blueprint is advanced for achieving this goal. A national integrated coastal management work program for the period 2008–2013 has been developed, and, building on the lessons of the pilot projects, ICZM programs are underway in a number of coastal provinces (Thang et al. 2011).

As well as being responsible for ICZM development, the Ministry of Natural Resources and Environment has a central role in the development of domestic policy on climate change. Vietnam issued its first National Communication under the United Nations Framework Convention on Climate Change (UNFCCC) in 2003 (Ministry of Natural Resources and Environment 2003), having ratified the treaty in 1994. In 2008, the

Ministry of Agriculture and Rural Development published an Action Plan Framework for Adaptation to Climate Change in the Agriculture and Rural Sector (Ministry of Agriculture and Rural Development 2008) and, that same year, the National Target Programme in Response to Climate Change (Ministry of Natural Resources and Environment 2008) received governmental approval. The National Target Programme was largely a research, planning, and communication exercise (Fortier 2010), laying the foundations for an implementation program to follow.

Vietnam's Second National Communication was submitted to the UNFCCC in 2010 (Ministry of Natural Resources and Environment 2010), and a National Climate Change Strategy, concerned with the implementation of mitigation and adaptation measures, was approved the following year (Government of Vietnam 2011). All of the Strategic Tasks of the 2011 National Climate Change Strategy are relevant to management of the coastal zone. Strategic Task 1 covers the development of an effective climate change and sea-level rise monitoring network and disaster risk reduction measures. The focus on disaster risk reduction reflects the priorities of the national government. Over the 15-year period to 2009, over half of the funding for climate impact and adaptation efforts, all from the national government, was devoted to disaster risk reduction (Fortier 2010), consistent with a precautionary *win–win* approach focused on immediate problems.*

Strategic Task 3 concerns appropriate adaptation to sea-level rise in vulnerable areas. Plans include development of a socioeconomic master plan suitable for a changing climate that will take account of the increased risk of storm, flooding, inundation, drought, land loss, and environmental degradation in vital or sensitive areas, including the Red River delta. Developing infrastructure and planning for residential areas is also a goal, with vital coastal and riparian embankments to be reinforced and upgraded. New livelihoods and production practices suitable to climate change and sea-level rise impacts will be developed. Strategic Task 6 covers increasing the role of the government in the response to climate change, adjusting and integrating climate change issues into strategies, schemes and planning, including the development of climate-resilient economic areas, and completing institutional arrangements such as a climate change law. The engagement of the entire political system will be enhanced to ensure sufficient governance and coordination between sectors. A uniform system of coordination mechanisms will be developed between government agencies and businesses, as well as a mechanism to involve communities and nongovernmental organizations. Increased public consultation and participation in implementation activities is a notable feature of the

* The largest proportion of climate funding, entirely contributed by foreign donors, was devoted to mitigation projects, over double that allocated to the impact and adaptation sector (Fortier 2010).

strategy. Community capacity development to cope with climate change is the theme of Strategic Task 7. Increased community involvement in climate change adaptation activities is a major goal, with an emphasis on local coping experience and the role of the authorities and local communities. Local livelihoods for flexible adaptation to different levels of vulnerability will be developed.

The National Climate Change Strategy recognizes that an effective response to climate change requires coordination not only between government agencies but with stakeholders in the broader public sector and the private sector. Here, experience in the development of ICZM should provide useful lessons given the clear government commitment to gradual decentralization as a means of dealing effectively with local problems in the coastal zone. A parallel here, perhaps, is the significance currently attached to community-based adaptation in the climate context in view of the necessity of taking full account of local circumstances (Huq et al. 2008). In the case of disaster risk reduction, there has been a largely effective model of coordination in the coastal zone that spans relevant sectors and administrative levels (Figure 9.4; Kelly et al. 2001a).

CHALLENGES AND OPPORTUNITIES

That sea-level rise and the myriad changes in local climate generated by global warming pose a serious threat to a coastal developing nation such as Vietnam is beyond doubt. The question is how to respond to the long-term and, to some extent, uncertain development of anthropogenic climate change and its impacts, particularly when faced with other, arguably more pressing, demands. One inescapable conclusion is that, on the grounds of both efficiency and effectiveness, coordination between, if not integrated development of, policy and planning in the intrinsically related areas of coastal zone management and climate adaptation must be a priority. There are many commonalities, but disaster risk reduction provides a cross-cutting theme of immediate relevance. Lessons can be drawn from experience in all three of these areas, coastal zone management, climate change, and disaster risk reduction, highlighting challenges to be met and opportunities to be taken.

Resourcing, in the broad sense of financial, technical, and human capacity, must always be a critical issue in a developing country such as Vietnam. Mai et al. (2009: 113), for example, in a survey of the coastal protection system in Vietnam, conclude that, as a result of budget constraints, lack of information, and other factors, the dyke system has usually been "designed and constructed for low design conditions with low quality." There has been a high occurrence of heavy wave overtopping, with a failure occurrence of about once in every seven years for dykes constructed in the early 1990s (the design return period was 20 years). Since the early 1990s, following

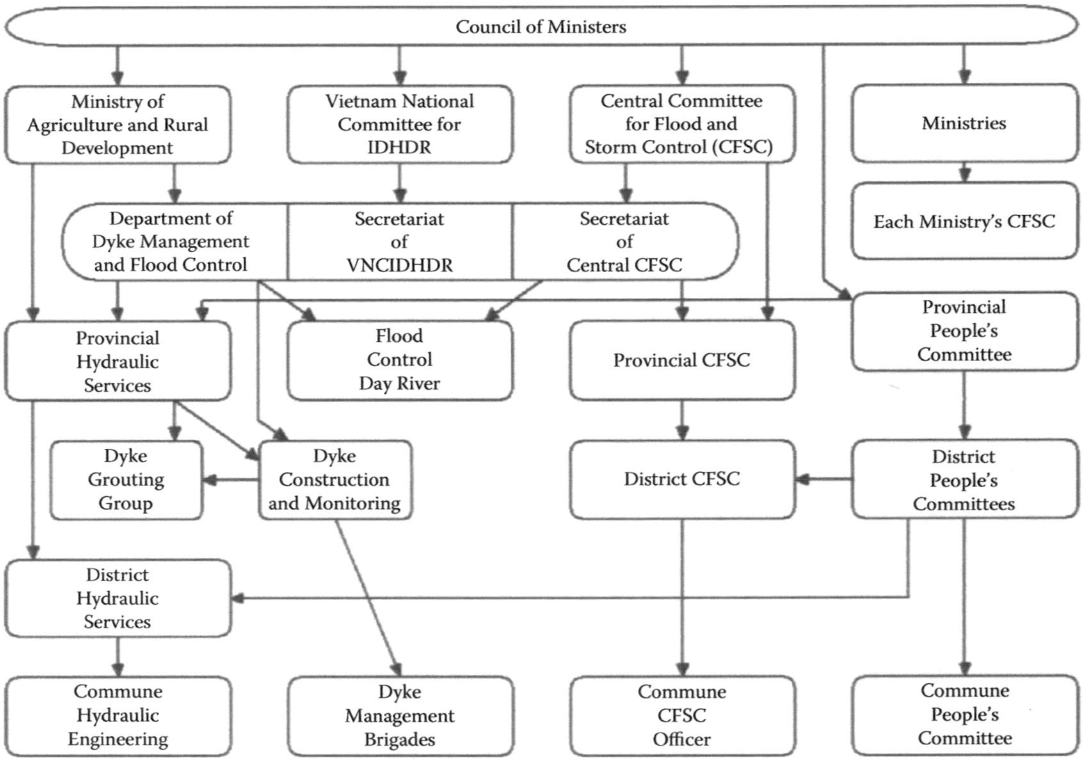

Figure 9.4 Institutional structure of the flood and storm protection system in Vietnam in the late 1990s. (Data from Kelly, P.M. et al., "Managing environmental change in Vietnam," in W.N. Adger et al., eds., *Living with Environmental Change: Social Vulnerability, Adaptation and Resilience in Vietnam*, Routledge, London, 2001.)

the lifting of the trade embargo, the Vietnamese government has received greater foreign assistance, with the advantages and disadvantages that this brings. Sekhar (2005: 819) observes that, in the context of the development of IZCM in Vietnam, "foreign donors have brought a significant change in thinking and behaviour in the national administration with an emphasis on broader vision, integrated action, and stronger ownership." On the other hand, the national agenda has been swayed by the availability of finance for particular projects and there has also been a tendency for overseas consultants to undermine local expertise, of which Vietnam has a considerable resource. For example, with respect to one statement of environmental strategy, Sekhar (2005: 819) dryly comments, "much of the document has been written by external experts whose understanding of the local problems and conflicts is minimal." It is not difficult to find other examples. Capacity strengthening and empowerment of domestic experts is needed to ensure a sound foundation for the construction of sustainable strategies.

Coordination is central to ICZM, essential in the response to climate change, the impacts of which will reach into all sectors of society, and of proven benefit in the case of disaster risk reduction. Yet, as Sekhar (2005) observes, the traditional planning approach in Vietnam is top-down and sectoral in nature, rendering difficult the level of coordination necessary for a truly integrated coastal management process. The centralized nature of Vietnamese decision-making and planning also militates against participation and an inclusive process of reaching decisions (Fortier 2010; Sekhar 2005), an important goal in order to ensure engagement across society. It is clear that the government recognizes the need for greater coordination. Stakeholder consultation is a notable aspect of the 2011 National Climate Change Strategy and has been a strong feature of the ICZM pilot projects conducted over the past decade. Eucker (2008: 69) concludes that IZCM has "a vast potential to improve the management process in the coastal zone and to provide an impact on general governance aspects." The historic strength of the storm and flood warning and response system demonstrates that coordination within the administrative structure and between sectors can be achieved. Nevertheless, much will depend on how successful the government is in renovating the existing planning paradigm.

In a related vein, Fortier (2010: 241) criticizes the strong technocratic bias in the development of climate policy in Vietnam to 2009, arguing that the strategic response to climate change must broaden public participation in climate policy-making in order to overcome its intrinsic limitations and the potential for abuse created by the fact that only a few political and technocratic actors have contributed to its development. He regards the technocratic focus on climate scenarios as a basis for planning and, in particular, the focus on a single scenario (cf. Ministry of Natural Resources and Environment 2009) as a *privileged representation of Vietnam's future*, as a serious risk. This approach, he argues, tends to prioritize biophysical at the expense of sociopolitical processes, may exclude actual outcomes to the detriment of the vulnerable, and can unduly benefit more powerful sectors of society. Fortier (2010) also regrets the failure on the part of the Vietnamese government to question the country's economic model in responding to the climate threat, relying instead on *climate-proofing* traditional development plans. This is, of course, a criticism that could be directed at most nations. Aspects of the 2011 National Climate Change Strategy, such as the emphasis placed on broader societal engagement and the commitment to development of resilient livelihoods, go some way, at least in principle, toward addressing Fortier's concerns. Nonetheless, the technocratic bias remains.

The overarching challenge is that of planning for an uncertain future at a time when both the biophysical environment and the socioeconomic context are undergoing rapid change. It is here that the technocratic and top-down tendency in Vietnamese planning and the rigidity that can ensue may prove particularly limiting, mitigating against the flexibility that Kelly et al. (1994) and Secrett (1996) argue is critical if sustainable adaptation is to be ensured. Kelly (2000) defines three elements of a sustainable response to the climate problem. A sustainable response must take full account of the range of technical, social, economic, cultural, and political processes that will determine effectiveness in a particular situation and guarantee equity, fairness, both within and between generations. There must be a careful assessment of priorities, placing long-term concerns in the context of the full range of present-day demands and aspirations. Finally, sustainable adaptation is characterized by caution and staged implementation,* based on flexibility, diversity, continual evaluation of performance, and an informed and engaged community.

CONCLUSION

Vietnam has a strong potential to develop an integrated approach to coastal zone management, climate change, and disaster risk reduction. Alongside strengthening capacity, that is, the availability of financial, information, technological, and human resources, three particular issues warrant highlighting. In all cases, there are examples or models available within Vietnamese experience that suggest these issues can be tackled successfully. First, there is the need for effective coordination throughout society. In the context of disaster

* Boateng (2012), reviewing adaptation options in Vietnam, proposes a coastal adaptation framework that provides holistic and multistage processes to develop sustainable coastal adaptation policy. He suggests an integrated shoreline management planning approach (DEFRA 2006), based on a littoral cell and sediment budget concept, and recommends that consideration be given to nonstructural approaches to adapting to the flood risk in the Red River delta.

risk reduction, coordination across sectors and levels has been an essential element of the response to emergencies for many centuries. Second, full participation, inclusion, and engagement must be ensured. As Adger et al. (2001) observe, Vietnam has a long-standing commitment to equity, in terms of power and access to resources, and this can provide an effective basis for limiting social exclusion. Finally, flexibility, a step-by-step approach, is desirable so that lessons can be learnt from experience and plans adapted accordingly.

The people of Vietnam have battled adversity through their history. Managing the dynamic coastal environment has proved a constant backdrop and has provided a striking demonstration of the ingenuity and commitment the Vietnamese people have brought to bear on whatever challenge has confronted them. For the twenty-first century, ensuring the security of the vulnerable people and resources of the coastal zone in the face of the combined impact of environmental degradation, conflicting demands, and climate change calls, once more, for urgent mobilization of the nation's resources.

ACKNOWLEDGMENTS

I thank Luong Quang Huy for his advice and assistance and my coeditors for their insights and inspiration.

REFERENCES

Adger, W.N. (1999). "Social vulnerability to climate change and extremes in coastal Vietnam," *World Development*, 27(2): 249–69.

Adger, W.N. and Kelly, P.M. (1999). "Social vulnerability to climate change and the architecture of entitlements," *Mitigation and Adaptation Strategies for Global Change*, 4: 253–6.

Adger, W.N. and Kelly, P.M. (2001). "Social vulnerability and resilience," in W.N. Adger, P.M. Kelly, and N.H. Ninh (eds.), *Living with Environmental Change: Social Vulnerability, Adaptation and Resilience in Vietnam*, London: Routledge.

Adger, W.N., Kelly, P.M., and Ninh, N.H. (eds.) (2001). "Environment, society and precipitous change," in *Living with Environmental Change: Social Vulnerability, Adaptation and Resilience in Vietnam*, London: Routledge.

Adger, W.N., Kelly, P.M., Ninh, N.H., and Thanh, N.C. (1997). "Property rights and the social incidence of mangrove conversion in Vietnam," Global Environmental Change Working Paper 97-21, Norwich; London: Centre for Social and Economic Research on the Global Environment, University of East Anglia; University College London.

Benson, C. (1997). *The Economic Impact of Natural Disasters in Vietnam*, London: Overseas Development Institute.

Boateng, I. (2012). "GIS assessment of coastal vulnerability to climate change and coastal adaption planning in Vietnam," *Journal of Coastal Conservation*, 16: 25–36.

Dasgupta, S., Laplante, B., Meisner, C., Wheeler, D., and Jianping Y. (2007). *The Impact of Sea Level Rise on Developing Countries: A Comparative Analysis*, Policy Research Working Paper Series 4136, Washington, DC: World Bank.

Dasgupta, S., Laplante, B., Meisner, C., Wheeler, D., and Yan, J. (2009). "The impact of sea level rise on developing countries: a comparative analysis," *Climatic Change*, 93(3): 379–88.

Department for Environment, Food and Rural Affairs (DEFRA). (2006). *Shoreline Management Plan Guidance. Volume 2: Procedures*, London: DEFRA, Government of the United Kingdom.

Ericson, J.P., Vorosmarty, C.J., Dingman, S.L., Ward, L.G., and Meybeck, M. (2006). "Effective sea-level rise and deltas: causes of change and human dimension implications," *Global and Planetary Change*, 50: 63–82.

Eucker, D.M. (2008). "Governance in Vietnam: Implications for integrated coastal zone management," in R.R. Krishnamurthy, A. Kannen, A.L. Ramanathan, S. Tinti, B.C. Glavovic, D.R. Green, Z. Han, and T.S. Agardy (eds.), *Integrated Coastal Zone Management*, Singapore: Research Publishing.

Fforde, A. (2009). "Economics, history, and the origins of Vietnam's post-war economic success," *Asian Survey*, 49(3): 484–504.

Fortier, F. (2010). "Taking a climate chance: A procedural critique of Vietnam's climate change strategy," *Asia Pacific Viewpoint*, 51(3): 229–47.

Government of Vietnam. (2011). *National Climate Change Strategy*, Hanoi, Vietnam: Ministry of Natural Resources and Environment, Government of the Socialist Republic of Vietnam.

Ho, T.M.H., Phan, V.T., Nhu-Quan Le, N.Q., and Nguyen, Q.T. (2011). "Extreme climatic events over Vietnam from observational data and RegCM3 projections," *Climate Research*, 49: 87–100.

Huq, S., Reid, H., Granich, S., Kelly, M., and Kuylenstierna, J. (eds.) (2008). "Special issue on community-based adaptation," *Tiempo*, 68. http://www.tiempocyberclimate.org/newswatch/ (accessed January 2012).

Huy, L.Q. (2010). *Climate Change Adaptation: Engaging Local Society in the Research Process*, PhD thesis, University of East Anglia, United Kingdom.

Huy, L.Q. (2011). "Engaging local society," *Tiempo*, 78: 3–8. http://www.tiempocyberclimate.org/newswatch/ (accessed January 2012).

Kelly, P.M. (2000). "Towards a sustainable response to climate change," in M. Huxham, and D. Sumner (eds.), *Science and Environmental Decision Making*, Harlow: Pearson Education.

Kelly, P.M. and Adger, W.N. (2000). "Theory and practice in assessing vulnerability to climate change and facilitating adaptation," *Climatic Change*, 47(4): 325–52.

Kelly, P.M., Granich, S.L.V., and Secrett, C.M. (1994). "Global warming: Responding to an uncertain future," *Asia Pacific Journal on Environment and Development*, 1(1): 28–45.

Kelly, P.M., Hien, H.M., and Lien, T.V. (2001a). "Responding to El Niño and La Niña—Averting tropical cyclone impacts," in W.N. Adger, P.M. Kelly, and N.H. Ninh (eds.), *Living with Environmental Change: Social Vulnerability, Adaptation and Resilience in Vietnam*, London: Routledge.

Kelly, P.M., Lien, T.V., Hien, H.M., Ninh, N.H., and Adger, W.N. (2001b). "Managing environmental change in Vietnam," in W.N. Adger, P.M. Kelly, and N.H. Ninh (eds.), *Living with Environmental Change: Social Vulnerability, Adaptation and Resilience in Vietnam*, London: Routledge.

Khanh, P. (1984). "An age-old undertaking," in *Water Control in Vietnam*, Hanoi, Vietnam: Foreign Languages Publishing House.

Mai, C.V., Stive, M.J.F., and Van Gelder, P.H.A.J.M. (2009). "Coastal protection strategies for the Red River Delta," *Journal of Coastal Research*, 25(1): 105–16.

Ministry of Agriculture and Rural Development. (2008). *Action Plan Framework for Adaptation and Mitigation of Climate Change of the Agriculture and Rural Development Sector, Period 2008–2020*, Hanoi, Vietnam: Government of the Socialist Republic of Vietnam.

Ministry of Natural Resources and Environment. (2003). *Initial National Communication under the United Nations Framework Convention on Climate Change*, Hanoi, Vietnam: Government of the Socialist Republic of Vietnam.

Ministry of Natural Resources and Environment. (2006). *Vietnam's ICZM Strategy 2020 and Orientation up to 2030*, Hanoi, Vietnam: Government of the Socialist Republic of Vietnam.

Ministry of Natural Resources and Environment. (2008). *National Target Program to Respond to Climate Change*, Hanoi, Vietnam: Government of the Socialist Republic of Vietnam.

Ministry of Natural Resources and Environment. (2009). *Climate Change: Sea Level Rise Scenarios for Vietnam*, Hanoi, Vietnam: Government of the Socialist Republic of Vietnam.

Ministry of Natural Resources and Environment. (2010). *Viet Nam's Second National Communication to the United Nations Framework Convention on Climate Change*, Hanoi, Vietnam: Government of the Socialist Republic of Vietnam.

Nguyen, P.K. (2011). "Introductory statement—Vietnam," in R. Misdorp (ed.), *Climate of Coastal Cooperation*, Leiden, the Netherlands: Coastal & Marine Union. http://www.coastalcooperation.net/ (accessed January 2012).

People's Committee of Nam Dinh Province. (2003). *ICZM Strategy for Nam Dinh Province*, Nam Dinh, Vietnam: People's Committee of Nam Dinh Province.

Secrett, C.M. (1996). *Adapting to Climate Change in Forest Based Land Use Systems: A Guide to Strategy*, Atmospheric Environment Issues in Developing Country Series—No. 2, Stockholm, Sweden: Stockholm Environment Institute.

Sekhar, N.U. (2005). "Integrated coastal zone management in Vietnam: Present potentials and future challenges," *Ocean & Coastal Management*, 48(9/10): 813–27.

Sen, A.K. (1990). "Food economics and entitlements," in J. Drèze and A.K. Sen (eds.), *The Political Economy of Hunger Vol. 1*, Oxford: Clarendon.

Thang, H.C., Misdorp, R., Laboyrie, H., Pos, H., van Zetten, R., and Huan, N.N. (2011). "Vietnam: A decade of coastal cooperation," in R. Misdorp (ed.), *Climate of Coastal Cooperation*, Leiden, the Netherlands: Coastal & Marine Union. http://www.coastalcooperation.net/ (accessed January 2012).

Thanh, T.D., Yoshiki, S., Huy, D.V., Nguyen, V.L., Ta, T.K.O., and Tateishi, M. (2004). "Regimes of human and climate impacts on coastal changes in Vietnam," *Regional Environmental Change*, 4: 49–62.

Tran, V.L., Hoang, D.C., and Tran, T.T. (2005). "Building of climate change scenario for Red River catchments for sustainable development and environmental protection," in *Preprints, Science Workshop on Hydrometeorological Change in Vietnam and Sustainable Development*, Hanoi, Vietnam: Ministry of Natural Resources and Environment, Government of the Socialist Republic of Vietnam.

Tri, N.H., Hong, P.N., Adger, W.N. and Kelly, P.M. (2001). "Mangrove conservation and restoration for enhanced resilience," in W.N. Adger, P.M. Kelly, and N.H. Ninh (eds.), *Living with Environmental Change: Social Vulnerability, Adaptation and Resilience in Vietnam*, London: Routledge.

Vahtar, M., Thu, L.V., Dao, L.T.A., Toan, P., and Misdorp, R. (2011). "Awareness raising through an educational programme," in R. Misdorp (ed.), *Climate of Coastal Cooperation*, Leiden, the Netherlands: Coastal & Marine Union. http://www.coastalcooperation.net/ (accessed January 2012).

Vietnam Ministry of Water Resources; United Nations Development Programme; and United Nations Department of Humanitarian Affairs. (1994). *Strategy and Action Plan for Mitigating Water Disasters in Viet Nam*, New York; Geneva, Switzerland: United Nations.

Vietnam Vulnerability Assessment. (1996). *Vietnam Coastal Zone Vulnerability Assessment and First Steps towards Integrated Coastal Zone Management*, Report No. 7, Final Report, April 1996, The Hague, the Netherlands: WL | Delft Hydraulics/Deltares, International Coastal Zone Management Centre and Ministry of Transport, Public Works and Water Management.

Chapter 10

A climate for change

A comparative analysis of climate change adaptation in rapidly urbanizing Australian and Chinese city regions

Darryl Low Choy, Chen Wen, and Silvia Serrao-Neumann

Abstract: The chapter compares the current approaches to incorporating climate change adaptation considerations into growth management planning initiatives for the most rapidly urbanizing metropolitan regions in the People's Republic of China and Australia—namely, the Yangtze delta region and the South East Queensland region, respectively. In contrasting these initiatives against contemporary practices and accepted principles for climate change adaptation at regional scale, the chapter considers how these growth management initiatives require their respective communities and institutions to address climate adaptation at regional and local spatial scales, within the context of the longer-term temporal scale that these strategic plans traditionally cover. In examining current strategic planning practices, the chapter has investigated how they are attempting to deal with the uncertainty of evolving climate change science and the extended time frames that strategic adaptation policies need to address. The chapter concludes with a comparison of the challenges, lessons, and potential adaptive pathways for the contrasting settings of Australian and Chinese cities drawn from the case studies.

INTRODUCTION

Developed and developing countries are witnessing widespread urbanization, in many cases at quite rapid rates. This is especially the case for China and Australia where urbanization has reached 50% and 86%, respectively, in 2010. In response to projected continued rapid urbanization, governments at a number of levels in both countries have instigated formal planning processes in order to manage the anticipated urban growth, to protect important resources, to guide infrastructure investments, and to deliver a higher quality of life for their populations.

Against this background of growth management, a raft of global, national, regional, and local drivers of change are predicted to influence and impact these rapidly urbanizing city regions. One significant influence that is anticipated to impact at all levels in the short to long term in both countries is climate change. This will be potentially acute in urbanizing coastal regions.

The coastal regions of the Yangtze River delta (YRD) region and South East Queensland (SEQ) are People's Republic of China and Australia's most rapidly growing metropolitan regions, respectively. Urbanization in both regions has occurred at unprecedented rates in the past two decades, 57% in the YRD (Chinese National Development and Reform Commission 2010b) and 64% in SEQ (Department of Transport and Main Roads 2010). Urbanization in both regions has occurred with an intense concentration on their respective coastal zones. Governments in both countries have recently undertaken regional-scale, growth management planning initiatives for these coastal regions. Of particular interest is the degree of consideration given to climate change adaptation in these strategic planning processes, especially in light of the evolving but still uncertain climate science. Adaptation in climate change terms is seen as the "adjustment in natural or human systems in response to actual or expected climatic stimuli or their effects, which moderates harm or exploits beneficial opportunities" (IPCC 2007: 869). These regional strategies could play a crucial leadership role in guiding or directing their respective communities and institutions to address climate adaptation especially within the uncertain and long-term context that these plans traditionally address.

The government-led growth management initiatives that attempt to respond to this intense and rapid urbanization, largely focused on the coastal zones of both regions, provide an opportunity to examine how climate change challenges are being addressed in the strategic planning instruments of both regions.

REGIONAL STRATEGIC PLANNING

The nature and purpose of regional planning will depend largely on the political and economic context in which it occurs (Glasson and Marshall 2007). Hence, it can be expected that there will be differences between the forms of regional planning undertaken in a socialist economy and society to that practiced in a free-market economy of a democratic society. Nevertheless, the primary purpose remains essentially the same—that is, deciding on the general distribution of new activities and developments (Glasson and Marshall 2007). Likewise, there will be a number of similar functions including its foundation on forward thinking leading to policies and programs aimed at delivering a desired future state for the region being planned in accordance with a vision which may or may not have included input from the community at large.

As a process to manage a highly uncertain future, strategic planning has been embedded into many public and community planning systems. Kaufman and Jacobs (1998) remind us that the principal features of strategic planning include action and results orientation, environmental scanning, broad participation opportunities, and assessment of community strengths and weaknesses. These features are also critical elements in any planning process undertaking for climate change adaptation.

In recent time, growth management has emerged as the dominant planning paradigm with respect to rapidly urbanizing regions. It is argued that growth management issues are best addressed at the regional level especially in the case of a large metropolitan area with multiple local governments (Nelson et al. 1995). City-regional planning in these instances has come to represent a dominant intent—that of urban containment (Taylor 1998). Growth management involves the employment of "government regulations (to) ... control the type, location, quality, scale, rate, sequence or timing of development" (Nelson et al. 1995: xi). It is seen as "part of a larger effort to shape the desired community of tomorrow, based on a vision of the future that recognizes global resource limits and our responsibilities to future generations" (Nelson et al. 1995: xii).

As essentially a spatial planning exercise, regional strategic planning is ideally positioned to address issues of settlement patterns, spatial configuration of urban centers, urban form, the use and development of land and natural resources, and landscape protection. In this manner, both climate change mitigation and adaptation considerations can be taken into account in this place-based, problem-solving undertaking aimed at achieving sustainable development (Davoudi et al. 2009).

In terms of linking climate change to regional strategic planning, Hallegatte (2009: 241) identified a number of sectors in which climate change should be taken into account, because they typically involve long-term planning, long-lived investment timescales, and some irreversibility in choices, and are exposed to changes in climate conditions. They include water infrastructure (e.g., dams, reservoirs), 30–200 years; land-use planning (e.g., in flood plain or coastal areas), >100 years; coastline and flood defenses (e.g., dikes, seawalls), >50 years; building and housing (e.g., insulation, windows), 30–150 years; transportation infrastructure (e.g., port, bridges), 30–200 years; urbanism (e.g., urban density, parks), >100 years; and energy production (e.g., nuclear plant cooling systems), 20–70 years.

Clearly, there are obvious imperatives for addressing climate change adaptation in rapidly growing coastal regions within regional strategic planning processes. These processes would, however, benefit from being informed by a set of overarching and guiding principles.

PRACTICES AND PRINCIPLES FOR STRATEGIC CLIMATE CHANGE ADAPTATION

With increasing understanding of climate change, especially at local and regional scales, coupled with the advent of improved climate science at these scales (as opposed to global data and projections that have dominated to date), there are increasing opportunities to address the specific nature of climate change, particularly adaptation matters, through the regional strategic planning process than hitherto.

In the first instance, however, the generic matter of the relationship between climate change and other more traditional policy areas needs to be considered. Should climate change be addressed in a discrete or integrated fashion, that is, as a stand-alone policy area or embedded across the full spectrum of policy areas. This and similar considerations have also caused some to inquire if climate policies (mitigation and adaptation) should be mainstreamed into overall sustainable development policies at all levels of governance (Davoudi et al. 2009).

A further generic consideration deals specifically with the plan's starting point—its vision statement. To this end, has climate change been acknowledged in the vision statement that presumably deals with future time horizons of similar proportions to that of climate change?

In a procedural sense, the strategic planning process could address the community's adaptive capacity for climate change through a strengths, weaknesses, opportunities, and threats (SWOT) analysis, thus highlighting what Kaufman and Jacobs (1998) see as a key part of the process—the assessment of a community's strengths and weaknesses. This assessment should in turn underpin a range of associated policies and programs.

Given the uncertainty associated with planning for the future, Courtney et al. (1997) identified three distinct strategic postures, that is, the intent of the strategy relative to the current and future state of the region. The postures are as follows: (1) shape the future; (2) adapt to the future; and (3) reserve the right to play. The second type (adapt to the future) is where planners take the existing situation and its future evolutions as given and adapt strategies to react to the opportunities presented (Dimitriou 2007).

Implementation of a strategy in a climate of uncertainty requires the selection of one or a number of strategic *moves* involving a portfolio of actions. Courtney et al. (1997) identified three such moves involving the following:

1. *Big bet moves*—include major investment or acquisition commitments, usually in the form of *shaping* strategies involving making *big bets*. These focused strategies may result in positive outcome in one or more scenarios but would have negative effects in others.
2. *Option moves*—designed to capture the main benefits of the best-case scenarios while minimizing loses from the worst-case scenarios. Typically included in *reserving the right* postures, they can also be found in *shaping* strategies.
3. *No-regret moves*—include strategic and pre-emptive initiatives that will pay dividends in any scenario and even without the occurrence of climate change, particularly from a cost/benefit perspective. Typically, they aim to improve climate robustness. An example is land-use policies that limit

urbanization and development in areas subject to natural hazards with the intention of reducing disaster losses in the present climate, but where climate change makes them even more desirable (Hallegatte 2009).

In addition to the *no-regret* strategy, Hallegatte (2009) argues that in the highly uncertain (climatic) environment where decisions and (infrastructure) investments have to be made, more reliance should be placed on uncertainty-management methods, including the following:

- Favoring reversible and flexible options
- Buying *safety margins* in new investments
- Promoting soft adaptation strategies, including long-term prospective
- Reducing decision time horizons

Previously, regional strategic planning has considered a range of issues that until recently were not necessarily associated with climate change. An example being natural hazards such as flooding, storm surge, and a range of other disasters. Traditionally, governments acted in the public good through undertakings such as land-use planning that sought to regulate and prevent development in, for example, flood plains, that is, acting to minimize or eliminate the vulnerability of its community. This also involved the development of disaster preparedness policies and disaster management plans. Hence, many existing regional strategic plans may include relevant policies dealing with such events/hazards but not necessarily under a climate change heading.

Furthermore, the policy outcomes from a regional strategic planning process need to be innovative and consider the issue of adaptation from a range of different perspectives. To this end, a number of pertinent questions could include the following:

- Does the plan contain innovative initiatives such as low carbon *climate-proof* settlements or protection of green space to maximize livable microclimates?
- Does the plan acknowledge that community-based adaptation needs to be locally driven and involve those most at risk (Bicknell et al. 2009: 362)?
- Does the plan recognize the distinct differences between urban and rural areas and communities, in terms of different adaptation measures that acknowledge their different spatial concentrations of people, hazards, and stressors, the number of hazards and their potential for exacerbating each other, income, livelihoods, land-use management, and infrastructure (Bicknell et al. 2009: 34–5)?
- Does the plan attempt to build community resilience to offset or counter the impacts of future climate change (directly or indirectly)?
- Does the plan initiate studies of likely impacts of climate change and do they have a social dimension in addition to traditional biophysical science foci?

In terms of what should be considered best practice for high growth coastal areas, the planning process should acknowledge a fundamental set of principles (framework) that should be imbedded into regional (growth management) strategic plans to address climate change adaptation. Guidance as to the nature of these principles can be derived from the recently released New South Wales Coastal Planning Guidelines (NSW Department of Planning 2010), which are structured around six *coastal planning principles* that must be applied in decision-making processes in coastal areas with respect to sea-level rise.

- *Principle 1:* Assess and evaluate coastal risks taking into account declared sea-level rise planning benchmarks.
- *Principle 2:* Advise the public of coastal risks to ensure that informed land-use planning and development decision-making can occur.
- *Principle 3:* Avoid intensifying land use in coastal risk areas through appropriate strategic and land-use planning.
- *Principle 4:* Consider options to reduce land-use intensity in coastal risk areas where feasible.
- *Principle 5:* Minimize the exposure of development to coastal risks.
- *Principle 6:* Implement appropriate management responses and adaptation strategies, with consideration for the environmental, social, and economic impacts of each option.

The above considerations and principles provide a basis and framework upon which the existing regional strategic plans for the two case study areas can be assessed in terms of their overt or inferred address of climate change adaptation in the face of continuing rapid urbanization.

CURRENT PLANNING PRACTICE FOR GROWTH MANAGEMENT IN COASTAL AREAS

In the last three decades, the SEQ region (22,890 km^2) has experienced unprecedented growth which intensified the development of its narrow coastal zone and lower reaches and estuaries of its river systems (Figure 10.1). During the past 25 years, SEQ's population has increased from approximately 1.5 million to nearly 3 million. This growth, which represented over one-twentieth of the country's population, was largely due to interstate migration. The region's population is currently projected to grow to 4.4 million by 2031, still retaining the State's highest concentration and density of population (Department of Infrastructure and Planning 2009).

This ongoing growth in the region prompted the Queensland State Government, in the early 1990s, to implement sustained urban growth strategies to manage development and change in SEQ through a collaborative

Figure 10.1 South East Queensland (SEQ) region—Australia.

voluntary approach to regional planning (Low Choy 2008; Low Choy et al. 2007). However, as SEQ continued to be one of the fastest growing regions in Australia (Australian Bureau of Statistics 2010), the state government was forced to shift from the initial collaborative voluntary approach and institute a statutory plan to manage the ongoing growth. The first statutory Regional Plan was released in 2005 followed by a revised version in 2009 (Department of Infrastructure and Planning 2009; Office of Urban Management 2005).

The state's planning legislation designates the Regional Plan as an instrument superior to state planning policies and all other plans, policies, and strategies of state agencies and local governments in the region (Department of Infrastructure and Planning 2009). Hence, it is both a planning instrument and a statutory instrument that guides and directs all statutory and nonstatutory planning initiatives in SEQ, including state and local government level as well as the nongovernment sector (McDonald et al. 2010). The Regional Plan establishes a performance-based planning system for SEQ under which its policy intentions are guided and directed by a series of 12 desired regional outcomes. The current Regional Plan, the SEQ Regional Plan 2009–2031, puts climate change adaptation for the first time into the regional planning agenda as it nominates climate change as one of its regional policies as well as a key issue permeating all other policies throughout the plan (Department of Infrastructure and Planning 2009).

In parallel, the YRD boasts a strong position for development and growth among other regions in the People's Republic of China. Spanning an area of 210,700 km^2, centered on Shanghai, Jiangsu Province, and Zhejiang Province, the region holds a strategic position and is the driving force in the country's modernization (see Figure 10.2). Because of the country's latest economic reform and subsequent opening up to international markets, the YRD has been leading the nation's development through a number of reforms. This region has become "one of the most important engines to improve the nation's comprehensive strength, international competitiveness, while it provides an impetus to the sound and fast development of the national economy" (Chinese National Development and Reform Commission 2010b: (i) More recently, the rate of urbanization in this region has exceeded 60% leading to a population in excess of 97 million. Looking forward toward 2020, and an anticipated future rate of urbanization of 75% in the core region, has led the conclusion that "Shanghai is marching towards its goal of becoming an international metropolis ... (and the region) to be listed among those world-class urban agglomerations" (Chinese National Development and Reform Commission 2010b: 2–4).

The reform and development of the YRD is guided by the YRD Regional Plan 2010–2015 (Chinese National Development and Reform Commission 2010b). The plan was released in 2010 by the national government under its Guiding Opinions of the State Council on the further Reform and Opening

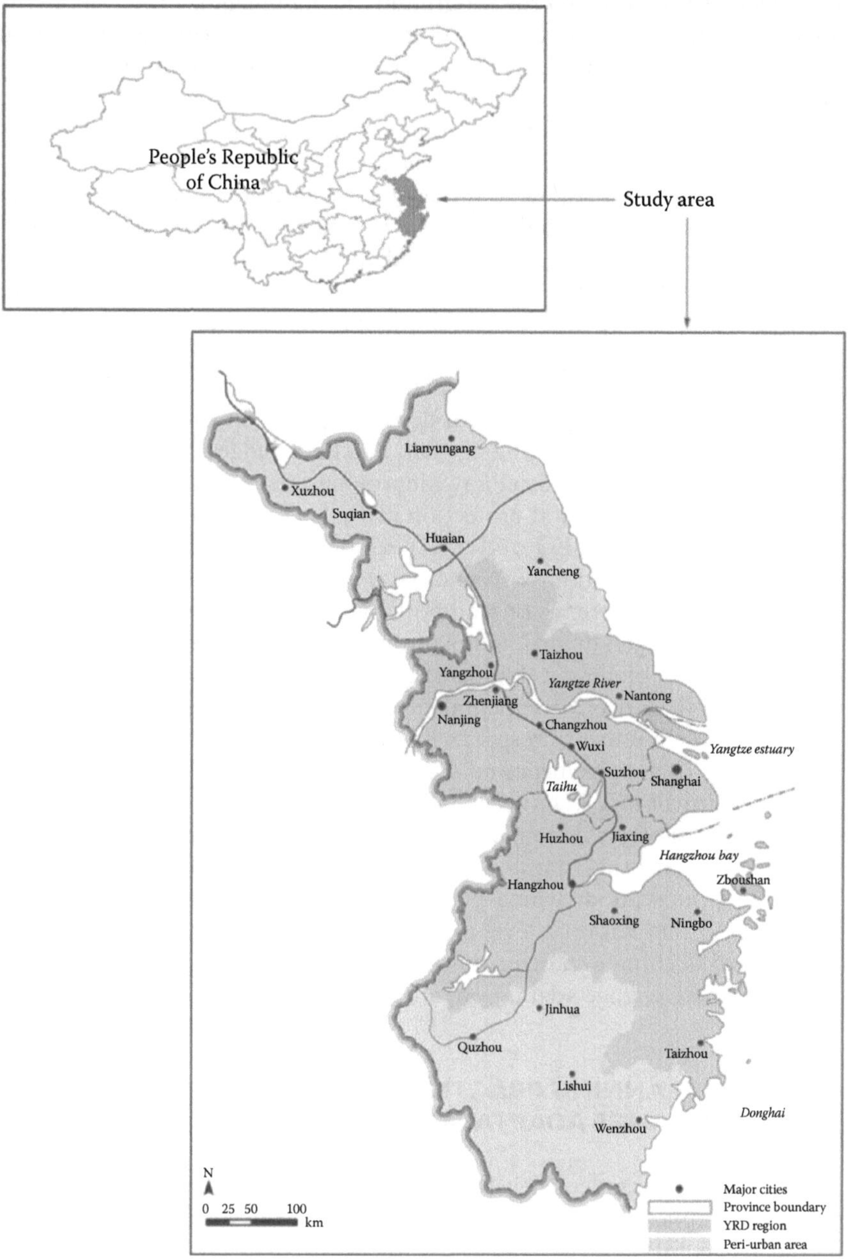

Figure 10.2 Yangtze River delta (YRD) region—People's Republic of China.

Up policy. It is considered the programmatic document set to promote the YRD's competitiveness in the global economy through regional integration and economic transformation. To these ends, it has both an internal and an external focus and establishes development goals that consider social, economic, and environmental outcomes. It attempts to guide development within the region addressing issues such as population and growth distribution; industrial development; physical infrastructure requirements; social services; natural resource utilization; environmental protection; institutional reform; and planning organization and implementation. Consistent with national policy, the plan's external focus is addressed through a section dedicated to "Opening up and Cooperation with the Outside World" (Chinese National Development and Reform Commission 2010b). The plan also may serve as the basis for other relative plans.

Although the current YRD Plan does not place climate change adaptation into the regional planning agenda, it acknowledges the importance of promoting a low-carbon economy, improving environmental quality, and encouraging environmental protection as well as land-use control.

Comparatively, the SEQ and YRD regions present a number of similarities, including rapid growth in coastal areas that leads to challenges in managing land-use expansion and potential climate change impacts. Both regions are managed by regional planning initiatives which include similar aims seeking to guide development, use of natural resources, infrastructure provision, and environmental protection. Nevertheless, these initiatives differ in a number of respects. Key differences include the focus of the plans—growth management and control in SEQ and growth management and development in YRD; their planning horizon—2031 in SEQ and 2020 with an even shorter economic focus in YRD; their experience in developing and implementing statutory plans—from 2005 in SEQ and 2010 in YRD; and the scale at which planning directives are set—regional/state level in SEQ and national level in the People's Republic of China. Lastly, with regard to climate change, only the SEQ Regional Plan clearly advances toward climate change adaptation, albeit in an incipient form.

CURRENT PLANNING PRACTICE FOR INCORPORATING CLIMATE CHANGE ADAPTATION INITIATIVES

At the Australian federal level, there are currently three main initiatives that have implications for climate adaptation at the state and local levels. These are as follows: the National Climate Change Adaptation Program (Department of Climate Change 2009a,b); the National Climate Change Adaptation Framework (Council of Australian Governments n.d.); and the Adapting to Climate Change in Australia Position Paper (Commonwealth of Australia 2010). A recurrent theme across all three initiatives is the identification of the need to develop adaptation strategies and to integrate climate adaptation into the decision-making and policy development processes. Ultimately, these initiatives aim at reducing the risks and vulnerabilities of key sectors and regions to the impacts of climate change. The Position Paper, in particular, highlights the need for vertical and horizontal integration across scales and sectors and outlines the roles and responsibilities of federal, state, territory, and local governments in order to create conducive policy environments for climate adaptation (Low Choy et al. 2010).

At the state level, the Queensland Government currently has two climate policies that have implications for climate change adaptation in the SEQ region, namely, ClimateQ: Toward a Greener Queensland 2009 (Department of Environment and Resource Management 2009) and ClimateSmart Adaptation 2007–2012 (Queensland Climate Change Centre of Excellence 2007). ClimateQ outlines the key themes and sectors for investments and policy development as part of the State's response to climate change impacts. One key theme specifically emphasizes the need to adapt to climate change impacts. The strategy also considers planning as one of the key sectors that requires the development of policy initiatives to adapt to climate change. The ClimateSmart five-year plan provides the foundation for increasing the State's resilience to climate change impacts through better understanding of climate change risks and vulnerabilities as well as embedding climate change impacts into decision-making. The plan is in alignment with the National Climate Change Adaptation Framework and creates practical steps toward climate change adaptation by identifying 62 actions for the priority sectors, including human settlement.

In the Chinese case, climate change-related policies are set at the national level through the National Climate Change Program (The State Council of the People's Republic of China 2007). The program is in alignment with the overall national economic and social development plan and regional plan and aims to address climate change by implementing a sustainable development strategy. It comprises actions in several key areas, including greenhouse gas mitigation, climate change adaptation, climate change science and technology, public awareness on climate change, and institutions and mechanisms. Ultimately, the program seeks to place the People's Republic of China in an innovative position worldwide by establishing a resource-conserving and environmentally friendly society which controls and mitigates its greenhouse gas emissions; enhances its capacity to adapt to climate change; encourages the harmonious development between economy, population, resources, and the environment; and contributes to improving global climate conditions.

More recently, the Chinese National Development and Reform Commission (2010a) released the National Policies and Actions Adapting to Climate Change—2010 Annual Report, which puts forward several actions and policies to mitigate climate change. Hence, the focus of these actions and policies is energy conservation as part of a major national economic and social development strategy. However, they also take into consideration synergies between mitigation and adaptation to ensure greater flexibility and effectiveness of related policies, institutions, specific measures, and managing systems through continuous adjustment and systematic operation. Key actions included in the report are the launching of a national low-carbon provinces and cities pilot program and continuation of the Eleventh Five-Year Plan (the nation's previous plan for economic and social development from 2006–2010) initiatives. The latter aimed at reducing energy consumption per unit of domestic gross product by 20% by 2010, followed by a continuous 20% decrease between 2010 and 2015. Additionally, the provinces and cities involved in the pilot program should fully integrate climate change into their Twelfth Five-Year Plan (the nation's current overall economic and social development plan for the period 2011–2015) as well as developing a low-carbon development plan.

In summary, both countries have overarching initiatives that provide the foundation steps to climate adaptation at the regional level. The distinction lies on the scale at which those initiatives are set—national and state level in Australia and national level in People's Republic of China. Despite the existence of these overarching initiatives, climate change adaptation in both countries remains a key challenge. For instance, in the Australian context, the initiatives lack statutory power therefore hindering their ability of influencing climate adaptation policies at state, regional, and local levels as well as the nongovernment sector. This limitation is demonstrated, for example, in the current SEQ Regional Plan that includes climate change as a key policy theme; however, in practice, policies have not yet been further developed to a degree that can effectively guide climate adaptation at the local government level. This represents a crucial challenge as local governments are critical players in addressing climate change impacts as they are ultimately responsible for land-use planning and development control (Low Choy et al. 2010). In the Chinese case, the initiatives have limited influence upon other regional and local policies, including the nongovernment sector, due to the lack of integration and consideration of climate change at the provincial level. For example, although the current YRD Plan acknowledges the importance of promoting a low-carbon economy, improving environmental quality, and encouraging environmental protection as well as land-use control, it does not place climate change adaptation into the regional planning agenda. This creates a critical challenge for the region to adapt to climate change as it not only compromises future regional sustainability but also hinders the region's ability to incorporate climate adaptation in land-use planning and growth control.

INCORPORATION OF STRATEGIC CLIMATE CHANGE ADAPTATION PRINCIPLES

This case study review will be contrasted against contemporary practices and accepted principles for climate change adaptation at regional scale. These include strategic planning and its long-term perspective and vision statement, existing planning initiatives dealing with natural hazards and coastal management, strategic measures for dealing with uncertainty, coastal planning principles, and understanding of adaptive capacity. Table 10.1 presents a summary of inferred and tangible examples, which illustrate how these practices and principles are currently addressed by the SEQ and YRD Regional Plans.

One of the key issues involved in dealing with climate change is related to the long-term nature of its impacts compared to the relative short-term nature of the policy/political cycle, particularly in Western democracies (Compston and Bailey 2008). In addressing the climate change challenge, therefore, it is essential that policies are coherent with the long-term nature of climate change impacts. Strategic planning facilitates the development of long-term policies; hence, these plans can play a critical role in tackling the climate change challenge (Hallegatte 2009). In the SEQ context, for instance, the 20-year time period established by the Regional Plan places the region in an advantageous position to manage climate change impacts, that is, it allows the region to deal with future impacts on, but not limited to, its coastline and flood defenses, buildings and housing, land-use planning, and infrastructure. This advantage is reinforced by existing planning instruments that do not necessarily deal specifically with climate change hazards but have direct links with them. This is the case of the State and Regional Coastal Management Plans, which provide guidance and direction to coastal management including land-use planning in coastal areas, and the State Planning Policy 1/03, which addresses natural hazards such as flooding, landslide, and bushfire. Additionally, infrastructure provision in SEQ is in alignment with the Regional Plan and subject to a yearly review process securing the necessary mechanisms for the implementation of long-term policies set by the plan (Department of Infrastructure and Planning 2010). These instruments certainly create scope for addressing future climate change impacts that are forecasted to affect SEQ's coastal areas.

In the YRD context, although strategic planning has a relatively shorter time frame, reaching up to 2020, it can still position its plan in an advantageous position to manage sustainable growth and climate change impacts. In addition, the overarching national policy on climate change sets ambitious targets to be met by 2010 and 2015, including a 20% reduction on energy consumption as well as the implementation of the low-carbon settlements pilot program. Although mostly focused on climate mitigation, these strong policies could set precedents in the Chinese response to climate change and influence other future policies related specifically to climate adaptation at regional

Table 10.1 Assessment of contemporary practice and accepted principles in the SEQ and YRD regional plans

Contemporary practices and accepted principles		*Regional strategies*	
		South East Queensland Regional Plan 2009–2031	*Yangtze River Delta Regional Plan 2010–2015*
Long-term strategic planning		Planning horizon 2031.[a]	Planning horizon 2020.[a]
Acknowledgment of climate change in vision statement		"… SEQ … resilient to climate change"[a]	[b]
Existing planning initiatives—natural hazards and coastal management		State Planning Policy 1/03; State and Regional Coastal Management Plans; South East Queensland Infrastructure Plan and Program 2009–2031.[a]	The plan may serve as a guide to other plans.[b]
Existing planning initiatives—climate change uncertainty		[b]	[b]
Coastal planning principles	1. Assess and evaluate coastal risks taking into account declared sea-level rise planning benchmarks.	Planning period incorporates long-term issues such as climate change (p. 7).[c] Plan's vision aims for a resilient region to climate change (p. 10).[c] Future development on islands will be guided by flooding and storm surge data.[c] Desired Regional Outcome 1.4—local planning schemes should be in accordance with the Queensland coastal plan and potential sea-level rise; new developments to avoid projected sea-level rise of 0.8 m by 2100 (p. 44–45).[a]	Future development will be guided by flooding, ground sediment, and storm surge data (p. 32).[c] Planning considers the ecological land use and enhances the protection of woodland and grassland (p. 32).[c]
	2. Advise the public of coastal risks to ensure that informed land-use planning and development decision-making can occur.	Development of local planning schemes is to include public engagement.[b]	Planning document is open to the public.[b]
	3. Avoid intensifying land use in coastal risk areas through appropriate strategic and land-use planning.	Nominated future urban growth corridor is outside the coastal area (p. 8).[a]	There are other eight development corridors outside the coastal area (p. 8–9).[a]

	4. Consider options to reduce land-use intensity in coastal risk areas where feasible.	Plan identifies the need for adaptation strategies to protect areas at risk from higher sea levels (p. 11).[a]	The plan encourages the rational exploitation and utilization of the resources from shallow tidal-flat areas (p. 33).[b]
	5. Minimize the exposure of development to coastal risks.	Desired Regional Outcome 1.4—areas with high exposure to storm tide, sea-level rise inundation, and coastal erosion are to be avoided (p. 44).[a]	Plan lays strict protection on coastal wetlands and establishes new natural reserves and special marine reserves with pertinence (p. 32).[a]
	6. Implement appropriate management responses and adaptation strategies, with consideration for the environmental, social, and economic impacts of each option.	Desired Regional Outcome 1.4—establishment of adaptation strategies to minimise vulnerability to climate change impacts (p. 44).[a] Implementation of State Planning Policies and State Coastal Management Plan to address disaster management (including sea-level rise, storm surge and coastal erosion) at local level (p. 45).[a] Implementation of SEQ Climate Change Management Plan to guide regional policies and support climate adaptation responses (p. 45).[a]	Plan reinforces the establishment of a marine environment monitoring system and reinforces the ecological conservation and restoration in the coastal zones (p. 34).[a]
Understanding of adaptive capacity		[b]	[b]

[a] Tangible example
[b] Not evident
[c] Inferred example

levels. In that case, these policies could comprise a strategic posture of adapting to the future as suggested by Courtney et al. (1997). Potentially, this strategic posture could enable regional planning in both regions to incorporate coastal planning principles (NSW Department of Planning 2010), therefore taking into account climate change uncertainties. This is evident in the policies of both regional plans which deal with future population growth through land-use planning. These policies control the location of new developments in areas at risk by identifying future growth corridors outside the coastal area.

In SEQ, guidelines set in the State and Regional Coastal Management Plans also assist land-use planning in avoiding locating new developments in areas at risk. Conversely, other policies promote the increase in population density through infill development in a number of urban centers located along the coast, thereby prospectively placing a larger number of people at an increased risk of harm from sea-level rise, coastal inundation, and storm surges. In the YRD case, the plan's urban and industrial footprint has been designated in areas of high development potential and low ecological value. Its land-use pattern will be guided by considerations of flooding, ground sediment, and storm surge, as well as guaranteeing protection of ecologically important water areas, woodland, grassland, and forests. The plan contains strict protective measures for coastal wetlands, including restoration in the coastal zone, creates new natural and special marine reserves, and makes provision for a marine environment monitoring system. Future regional development is set in accordance with a spatial pattern characterized by *1 core and 9 belts*, with Shanghai being the core of development as well as the pilot city for launching the low-carbon experimental economy. Besides the coastal development corridor, there are another eight development corridors outside the coastal area (Shanghai–Nanjing and Shanghai–Hangzhou–Ningbo corridor, Yangtze River corridor, Bay corridor, Nanjing–Taihu–Hangzhou corridor, Taihu corridor, eastern Longhai corridor, Grand Canal corridor, and Wenzhou–Lishui–Jinshan–Quzhou corridor).

There is no doubt that climate change uncertainties pose a critical challenge for the decision-making process (Hallegatte 2009). As highlighted earlier in the chapter, planning for climate change will require a number of strategic *moves* that is, big bet moves, option moves, and no-regret moves (Courtney et al. 1997). These moves appear to be at an embryonic stage in the SEQ Regional Plan. Although there are large investments allocated to infrastructure provision in the region, climate change impacts are yet to be considered in these decisions. Furthermore, due to political and legal barriers (England 2007; Hennessy et al. 2007), strategic planning in SEQ still struggles with embracing more robust policies involving, for example, no-regret moves (Courtney et al. 1997) and/or uncertainty-management methods (Hallegatte 2009). These may include policies involving planned retreat of coastal communities and associated mechanisms for potential compensation costs as well as the consideration of dynamic buffer zones in the development of coastal areas. The evidence for such moves is even less conspicuous in the YRD as the regional plan does not address climate adaptation. However, although strong policies are not only the prerogative of the state, as argued by Giddens (2009), the strong central direction set by the Chinese state would certainly influence the development of those strategic moves should they be considered.

It is important to have an understanding of the regional adaptive capacity when developing and implementing strategic policies/moves. For instance, a high adaptive capacity could facilitate their implementation and, therefore, contribute to reduce the regions' vulnerability to climate change. Nonetheless, as Adger et al. (2007) have argued, high adaptive capacity does not necessarily result in the implementation of actions to reduce climate change vulnerability, including through climate adaptation strategies. The regional planning process in SEQ reflects this assumption. Despite the overall high adaptive capacity observed in the SEQ region (Low Choy et al. 2010), its current policies and programs lack the inclusion/discussion of its community's adaptive capacity. This is also the case of the YRD Regional Plan. Kaufman and Jacobs (1998) acknowledge that the assessment of the community's strengths and weaknesses, which could assist in understanding its adaptive capacity, constitutes a key part of strategic planning. This, in turn, should inform proposed policies and programs. The exclusion of aspects related to adaptive capacity in both strategic planning processes influences the development of effective adaptation measures in two ways: (1) it restricts its capability of promoting the development of locally driven community-based adaptation (Satterthwaite et al. 2009a), and (2) it reduces its ability to deal with complex interactions between climate hazards and other stressors that are inherent to the social, ecological, and economic characteristics of its communities (Satterthwaite et al. 2009b).

In summary, the SEQ and YRD regions have existing mechanisms through their regional planning processes that could lead to climate adaptation at regional scale, particularly along their coastal areas. These include, for example, the existence of strategic policies that certainly contribute to reduce the regions' vulnerability to climate change impacts. However, the existence of conflicting policies, as well as the lack of consideration of climate change in others, also shows that climate change adaptation still remains a key challenge yet to be effectively addressed by the current planning practice for growth management in both regions.

DISCUSSION

Growth management has dominated the planning agenda in both the YRD and SEQ regions, with cursory consideration of climate change adaptation issues and policies. The review has revealed that, while climate change adaptation measures have not yet come to the fore in the current regional planning processes in both high growth regions, their planning endeavors, especially the regional strategic planning, comprise processes that can facilitate more detailed climate change considerations. It has also been noted that both regions have existing planning processes that could facilitate greater attention and uptake of climate change science and adaptation considerations into their respective regional plans and policies.

The cycle of formal plan review should incorporate an adaptive management framework into the regional strategic planning process. If this is achieved, especially in an institutional sense, then this will present an opportunity to upgrade the plan in light of experience from existing policy implementation and, importantly, the incorporation of new (climate) science as it becomes available or is enhanced. Hence, the existence of such processes will assist future efforts to incorporate climate change adaptation measures into planning and decision-making. In the SEQ case, there is a statutory requirement to review its plan every five years, while in the Chinese case this is unclear.

Both plans strongly demonstrate that each could adopt a stronger *no-regrets* approach that could maximize benefits from a raft of strategic and preemptive initiatives that will pay dividends regardless of the occurrence and impacts of climate change. Both plans already incorporate consideration of natural hazards in the more traditional policy and program sense. Hence, both regions have previous experience in dealing with natural hazards that could inform the incorporation of robust strategies in new or reviewed plans aimed at dealing with climate change impacts, including forecasted extreme weather events and sea-level rise.

A key challenge remains for both regions in terms of improving the understanding of their adaptive capacity for the purpose of regional planning strategies. Although regional planning in SEQ entails a public consultation process that could facilitate such understanding, this is absent in the Chinese case and possibly constitutes a critical distinction in regional planning processes undertaken in a socialist economy and society, to that practiced in a free-market economy of a democratic society. Nevertheless, the complexity involved in dealing with climate change demands that public consultation processes become more thorough as to provide information that can guide the development and implementation of community-focused policies to, in turn, contribute to enhance both individual and collective adaptive capacity. Additionally, it is important to highlight that adaptive capacity also includes institutional mechanisms/instruments that can be used to deal with climate change impacts. Strong institutional adaptive capacity ensures that communities are better prepared for and facilitates their recovery process from extreme weather events.

Furthermore, there is also the need to undertake a critical review of current institutional arrangements and climate change governance related to strategic planning and growth management in both regions. This is particularly important to ensure better vertical and horizontal coordination in the development and implementation of policies by harnessing potential synergistic policies as well as avoiding conflicting ones that can hinder the adaptation process. Better coordination and integration of climate policies will certainly reduce the regions' risk of harm from potential extreme weather events. Spatial planning can also serve as a center of this effort, and supported by good climate governance, it can contribute to enhance the response capacity to climate change impacts at both local and regional scales, thereby promoting effective adaptation.

A WAY AHEAD

This review has identified a number of practical steps that can be taken to build resilience, adaptive capacity, and sustainability of rapidly urbanizing regions in the face of the climate change challenge. First, the critical role that strategic planning can play should not be underestimated. It can facilitate an adaptive management approach which can address the uncertainty surrounding the evolving climate change science and thus begin to reduce regional vulnerabilities to climate change impacts. A system of triggers and thresholds, for example, could be adopted to guide such management initiatives instead of continuing to base them solely on robust and absolute science. It is also important that those initiatives enable a mix of strategies ranging from enforcement through to providing incentives to address the adaptation challenges. Strategic planning also offers appropriate mechanisms to assist regional planning in incorporating long-term strategies that are consistent with the long-term scale of climate change impacts.

Nonetheless, addressing climate change through effective adaptation also requires other critical challenges to be overcome. Hence, climate governance needs to be improved to foster strong leadership and the implementation of more contentious strategies, especially those available through a stronger *no-regrets* approach. Addressing the likelihood of the occurrence of natural hazards has been a long-standing aspect of conventional planning approaches. Likewise, existing institutional arrangements will need to be enhanced to enable better climate policy coordination and integration across all tiers of government.

Lastly, the success of climate change adaptation in both regions will ultimately depend on the uptake of appropriate climate adaptation policies at appropriate community levels. Clearly, this must be accompanied by an assessment of the community's adaptive capacity. Strong higher level leadership from state and regional perspectives can play a critical role in direction setting for local and community levels. The resulting cascading effects can facilitate implementation at lower levels with a higher degree of responsibility being taken by the community, which is essential for achieving long-term climate adaptation. This in turn can make a major contribution to building community resilience, adaptive capacity, and, ultimately, sustainability.

REFERENCES

Adger, W., Agrawala, S., Mirza, M., Conde, C., O'Brien, K., Pulhin, J., Pulwarty, R., Smit, B., and Takahashi, K. (2007). "Assessment of adaptation practices, options, constraints and capacity," in M.L. Parry, O.F. Canziani, J.P. Palutikof, P.J. Van Der Linden, and C.E. Hanson (eds.), *Climate Change 2007: Impacts, Adaptation and Vulnerability. Contribution of Working Group II to the Fourth Assessment Report of the Intergovernmental Panel on Climate Change*, Cambridge: Cambridge University Press. http://www.ipcc.ch/ (accessed August 2013).

Australian Bureau of Statistics. (2010). *Regional Population Growth, Australia, 2008–09*. http://www.abs.gov.au/ (accessed October 2010).

Bicknell, J., Dodman, D., and Satterthwaite, D. (eds.) (2009), *Adapting Cities to Climate Change: Understanding and Addressing the Development Challenges*, London: Earthscan.

Chinese National Development and Reform Commission. (2010a). *National Policies and Actions Adapting to Climate Change—2010 Annual Report*, Beijing, People's Republic of China: NRDC.

Chinese National Development and Reform Commission. (2010b). *Yangtze River Delta Regional Plan 2010–2015*, Beijing, People's Republic of China: NRDC.

Commonwealth of Australia. (2010). *Adapting to Climate Change in Australia—An Australian Government Position Paper*, Canberra, Australia: Commonwealth of Australia. http://www.climatechange.gov.au/ (accessed March 2012).

Compston, H. and Bailey, I. (eds.) (2008). *Turning Down the Heat: The Politics of Climate Policy in Affluent Democracies*, Basingstoke: Palgrave Macmillan.

Council of Australian Governments. (n.d.). *National Climate Change Adaptation Framework*, A report commissioned on February 10, 2006, Canberra, Australian: Council of Australian Governments.

Courtney, H., Kirkland, J., and Viguerie, P. (1997). "Strategy under uncertainty," *Harvard Business Review*, 75(6): 66–79.

Davoudi, S., Crawford, J., and Mehmood, A. (eds.) (2009). *Planning for Climate Change: Strategies for Mitigation and Adaptation for Spatial Planners*, London: Earthscan.

Department of Climate Change and Energy Efficiency. (2009a). *Local Adaptation Pathways Program*, Canberra, Australia: Commonwealth of Australia. http://www.climatechange.gov.au/ (accessed March 2012).

Department of Climate Change and Energy Efficiency. (2009b). *National Climate Change Adaptation Program*, Canberra, Australia: Commonwealth of Australia. http://www.climatechange.gov.au/ (accessed March 2012).

Department of Environment and Resource Management. (2009). *ClimateQ: Toward a Greener Queensland*, Brisbane, Queensland, Australia: Queensland Government. http://rti.cabinet.qld.gov.au/documents/2009/may/climateq%20toward%20a%20greener%20qld/Attachments/ClimateQ_Report_web_FINAL_20090715.pdf (accessed March 2012).

Department of Infrastructure and Planning. (2009). *South East Queensland Regional Plan 2009–2031*, Brisbane, Queensland, Australia: Queensland Government. http://dlgp.qld.gov.au/ (accessed March 2012).

Department of Infrastructure and Planning. (2010). *South East Queensland Infrastructure Plan and Program 2009–2031 (SEQIPP)*, Brisbane, Queensland, Australia: Queensland Government. http://dlgp.qld.gov.au/ (accessed March 2012).

Department of Transport and Main Roads. (2010). *Draft Connecting SEQ 2031: An Integrated Regional Transport Plan for South East Queensland*, Brisbane, Queensland, Australia: Queensland Government. http://www.connectingseq.qld.gov.au/ (accessed March 2012).

Dimitriou, H.T. (2007). "Strategic planning thought: Lessons from elsewhere," in H.T. Dimitriou and R. Thompson (eds.), *Strategic Planning for Regional Development in the UK*, London: Routledge.

England, P. (2007). *Climate Change: What Are Local Governments Liable For?* Brisbane, Queensland, Australia: Griffith University Urban Research Program. http://www.griffith.edu.au/ (accessed October 2010).

Giddens, A. (2009). *The Politics of Climate Change*, Cambridge: Polity Press.

Glasson, J. and Marshall, T. (2007). *Regional Planning*, London: Routledge.

Hallegatte, S. (2009). "Strategies to adapt to an uncertain climate change," *Global Environmental Change*, 19: 240–7.

Hennessy, K., Fitzharris, B., Bates, B.C., Harvey, N., Howden, M., and Hughes, L. (2007). "Australia and New Zealand," in M.L. Parry, O.F. Canziani, J.P. Palutikof, P.J. Van Der Linden, and C.E. Hanson (eds.), *Climate Change 2007: Impacts, Adaptation and Vulnerability. Contribution of Working Group II to the Fourth Assessment Report of the Intergovernmental Panel on Climate Change*, Cambridge: Cambridge University Press. http://www.ipcc.ch/ (accessed August 2013).

Intergovernmental Panel on Climate Change (IPCC). (2007). *Climate Change 2007: Impacts, Adaptation and Vulnerability, Contribution of Working Group II to the Fourth Assessment Report of the Intergovernmental Panel on Climate Change*, Cambridge: Cambridge University Press. http://www.ipcc.ch/ (accessed August 2013).

Kaufman, J.L. and Jacobs, H.M. (1998). "A public planning perspective on strategic planning," in S. Campbell and S. Fainstein (eds.), *Readings in Planning Theory*, Malden, MA: Blackwell.

Low Choy, D. (2008). "The SEQ regional landscape framework: is practice ahead of theory?," *Urban Policy and Research*, 26: 111–24.

Low Choy, D., Baum, S., Serrao-Neumann, S., Crick, F., Sanò, M., and Harman, B. (2010). *Climate Change Vulnerability in South East Queensland: A Spatial and Sectoral Assessment*, Unpublished report for the South East Queensland Climate Adaptation Research Initiative, Brisbane, Queensland, Australia: Griffith University.

Low Choy, D., Sutherland, C., Gleeson, B., Dodson, J., and Sipe, N. (2007). *Change and Continuity in Peri-Urban Australia. Peri-Urban Structures and Sustainable Development*, Brisbane, Queensland, Australia: Griffith University Urban Research Program, Monograph 4. http://www.griffith.edu.au/ (accessed January 2010).

McDonald, J., Baum, S., Crick, F., Czarnecki, J., Field, G., Low Choy, D., Mustelin, J., Sanò, M., and Serrao-Neumann, S. (2010). *Climate Change Adaptation in South East Queensland Human Settlements: Issues and Context*, Unpublished report for the South East Queensland Climate Adaptation Research Initiative, Brisbane, Queensland, Australia: Griffith University.

Nelson, A.C., Duncan, J.B., Mullen, C.J., and Bishop, K.R. (1995). *Growth Management Principles and Practices*, Chicago, IL: Planners Press.

NSW Department of Planning. (2010). *NSW Coastal Planning Guidelines: Adapting to Sea Level Rise*, Sydney, Australia: NSW Government. http://www.planning.nsw.gov.au/ (accessed March 2012).

Office of Urban Management. (2005). *South East Queensland Regional Plan 2005–2026*, Brisbane, Queensland, Australia, Queensland Government. http://www.dlgp.qld.gov.au/ (accessed March 2012).

Queensland Climate Change Centre of Excellence. (2007). *ClimateSmart Adaptation 2007–2012. An Action Plan for Managing the Impacts of Climate Change*, Brisbane, Queensland, Australia: Queensland Government. http://www.climatechange.qld.gov.au/ (accessed March 2012).

Satterthwaite, D., Dodman, D., and Bicknell, J. (2009a). "Conclusions: Local development and adaptation," in J. Bicknell, D. Dodman, and D. Satterthwaite (eds.), *Adapting Cities to Climate Change: Understanding and Addressing the Development Challenges*, London: Earthscan.

Satterthwaite, D., Huq, S., Reid, H., Pelling, M., and Lankao, P. (2009b). "Adapting to climate change in urban areas: The possibilities and constraints in low- and middle-income nations," in J. Bicknell, D. Dodman, and D. Satterthwaite (eds.), *Adapting Cities to Climate Change: Understanding and Addressing the Development Challenges*, London: Earthscan.

Taylor, N. (1998). *Urban Planning Theory Since 1945*, London: Sage.

The State Council of the People's Republic of China. (2007). *National Climate Change Program*, Beijing, People's Republic of China. http://www.china.org.cn/english/environment/213624.htm (accessed March 2012).

Part IV

Climate change and the coastal zone

Australasia

Chapter 11

The evolution of coastal vulnerability assessments to support adaptive decision-making in Australia

A review

Robert Kay, Ailbhe Travers, and Luke Dalton

Abstract: Australia has been assessing the vulnerability of its coastal zone to potential climate change impacts for over 20 years. The Australian coast is long, physically and biologically diverse, and its effective management is critical to the economic, social, and cultural well-being of the nation. Developing effective responses to the threats posed by climate change on coastal resources is a national priority. As a result, a broad variety of assessment approaches and tools have been used over the years at different geographic scales attempting to address a wide range of scientific, technical, policy, and planning questions. Reflections on the mosaic of coastal vulnerability and adaptation studies and policy responses summarized in this chapter provide useful insights into key successes, challenges, and opportunities to inform the ongoing adaptive journeys of Australian coastal managers. This may also provide helpful information for those in other countries who are at different stages of their coastal adaptive journey than Australia's and/or have different levels of capacity and capability.

INTRODUCTION

Australia is essentially a coastal nation. It is an island continent, and despite its vast interior, the great majority of its population and economic activities are concentrated in its coastal zones. The coast of Australia, like coastlines around the world, is dynamic and evolving in response to both natural and anthropogenic change. Given the disproportionately large percentage of the population located in this area—85% of Australians live within 50 km of the coast (DCCEE 2009)—it is unsurprising that coastal impact assessments have been underway in Australia for a number of years. Of particular interest here are assessments concerned with the vulnerability of the coastal zone to potential climate change impacts. A wide range of these assessments has been undertaken at various temporal and spatial scales over the past 20 years. A broad variety of assessment approaches and tools have been used at different geographic scales attempting to address a range of policy and planning questions.

This chapter provides an overview of the status of coastal climate change vulnerability and adaptation (V&A) assessment in Australia. The evolution of the current practice of coastal adaptation is discussed with reference to completed, ongoing, and planned coastal adaptation initiatives around the country. An important focus of this chapter is the multiscalar approach to V&A assessment adopted within Australia given the complexity of the policy landscape at a local, state, and federal level. An overview of the mosaic of coastal V&A studies and policy responses provides useful insights into key successes, challenges, and opportunities for the ongoing adaptive journeys of Australian coastal managers. The reflections presented here may also be helpful for those in other countries who have embarked on their coastal adaptive journey later than Australia or have different levels of capacity and capability. In particular, lessons that can be learned from Australian experience could assist coastal nations to optimize the allocation of skills, capacities, and resources to improve decision support within the broad range of coastal adaptation decision contexts, allowing them to move forward faster and at a lower expense.

WHAT MAKES THIS PLACE SPECIAL?

The Australian coastline is one of the longest of any nation, with a mainland extent of more than 30,000 km but a total length that may be as much as 120,000 km if all estuaries, indentations, islands, and island territories are included (Short and Woodroffe 2008; Thom and Short 2006). The shores of Australia include open coasts with rocky headlands, cliffs, and sandy beaches, and sheltered coasts, bays, and estuaries with muddy and sandy tidal flats (Australia State of the Environment Committee 2001). More than 10,000 beaches stretch for more than 50% of the shoreline, with the remainder being rocky or muddy. There are more than 700 coastal waterways, primarily estuaries each with a series of associated low-lying shorelines and wetlands.

The Australian coast is often considered as four broad coastal regions (Thom and Short 2006) each with distinctive ecosystems and natural assets, and each is vulnerable to a different degree to extreme weather and to the potential impacts of climate change:

- *Region 1:* The Muddy North—highly tidal, cyclone influenced and muddy
- *Region 2:* The Limestone South and West—small tides, carbonate rocks, high wave and wind energy

- *Region 3:* Eastern Headlands and Bays—small tides, quartz sands, moderate wave energy, many bays
- *Region 4:* The Barrier Reef—northern Queensland, including low-lying rocky mainland coasts and the Great Barrier Reef and its islands

The coast is a central part of Australian life with all major Australian cities (with the exception of the capital Canberra) located along the coast (see Figures 11.1 and 11.2). The nonmetropolitan coastal population has grown rapidly either through in-migration to Australia or through internal migration from capital cities to smaller coastal communities (Gurran et al. 2005). The coastal zone also holds special cultural and social significance for indigenous Australians (Smyth 1993). The country's *coastal culture* of *sand, sea, and surf* is also central to the psyche of indigenous and nonindigenous Australians alike (Kay and Alder 2005; Kay and McKellar 2000).

Australia's coastal zones are also critical to key national economic sectors, in particular, tourism/domestic recreation and the increasing importance of export industries. The rapid growth of resource industries in

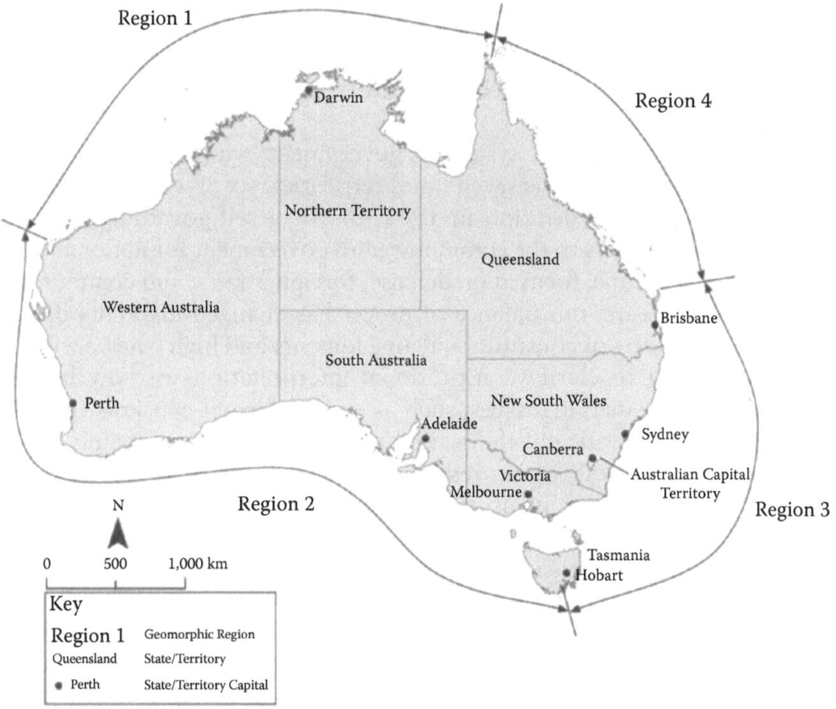

Figure 11.1 Australian coastal regions.

Figure 11.2 Images of the diversity of the Australian coast. Locations (clockwise from top left): London Arch, Port Campbell National Park, Victoria (Courtesy of L. Dalton); Cairns, Queensland (Courtesy of R. Kay); Kakadu, Northern Territory (Courtesy of R. Kay); Tweed Heads, New South Wales (Courtesy of R. Kay); Mandurah, Western Australia (Courtesy of R. Kay).

recent years (mainly mining and oil/gas) has required access to specialized port facilities. This in turn has flow-on effects for the development of regional service centers in coastal locations, the majority of which are found in Australia's north.

Australia has a federated system of government, with the federal (commonwealth) government and seven states/territories (six of which are coastal, see Figure 11.4). At federation in 1901, the then self-governing colonies granted limited powers to the commonwealth government for national-level issues that, at the time, focused on defense, foreign affairs, and communications. Over the years, this balance of power has changed markedly due to the combined effects of constitutional amendments and high court decisions that have sought to clarify constitutional interpretations of how best to address new and emerging issues, such as environmental concerns (Harvey and Caton 2003). Critically, the main powers for land-use planning (originally held by the states) remain vested with them. These powers are exerted through land-use policies, plans, strategies, and regulations—including those covering the use and management of coastal areas (Kay and Alder 2005). The state-based governance systems have evolved in their own way, shaped by national requirements and various inter-governmental coordinating processes. Importantly, a variety of state-based coastal management systems have resulted, with significant consequences for climate change adaptation governance.

The complexity of the coastal management landscape in Australia is further convoluted by the different degrees, styles, and detailed mechanics of the delegation of powers from states to municipal-level governments (Gibbs and Hill 2011). Consequently, there are very different opportunities for local governments to adapt to climate change impacts depending on which state they are in. Also, depending on the state system, there are differing degrees to which local governments can interpret the variety of adaptive options available to them. This has resulted in a range of technical standards, policies, plans, strategies, and regulations applied to address climate change in Australia. This is seen by many as inefficient and is currently the subject of a national inquiry by the Productivity Commission to specifically address barriers to effective adaptation action.*

Significantly, although there is considerable public and private infrastructure located in coastal zones that is exposed to current coastal hazards and potential future climate change, there are two key factors that have served to reduce these exposure levels when compared to other developed countries in Europe and North America. The first factor is the adoption, since European colonization, of public foreshore reserves during land subdivision (although with marked differences between states in the mechanics of this process) (Gibbs and Hill 2011; Harvey and Caton 2003). The widths of these foreshore reserves have, however, historically been set on the assumption of a stable set of climatic factors, including a stable sea level. There has been an understanding that these climatic factors are highly variable but it was assumed that they fluctuate around a stable mean, an assumption that is no longer valid in the face of changing climate and rising sea level. The second, related factor is that most significant coastal development has occurred since the late 1960s when coastal management systems were becoming established. This has allowed the conscious design of flexible hazard buffers, particularly in coastal areas where high value assets were located. Although these public reserve buffers have largely served to reduce exposure (to different degrees in different states), their contemporary condition is generally poor, in particular, in highly urbanized areas. In recent years, consideration of climate change has been included in buffer design. Given the pace of growth in areas of the country driven by the mining and oil/gas industries, there is significant opportunity to reduce future exposure to climate change impacts through well-conceived and effectively implemented initiatives. Indeed, this has been a goal of federal, state, and local governments to greater or lesser degrees, over the past 20 years, aided by specific assessments of vulnerability, as discussed in the subsequent section.

* See the inquiry website for more details: http://www.pc.gov.au/projects/inquiry/climate-change-adaptation.

ASSESSING AUSTRALIAN COASTAL VULNERABILITY

Studies to assess the vulnerability of the Australian coastline to climate change impacts date from early scientific assessments in the late 1980s* and early tests of impact assessment methodologies in the early 1990s (Kay and Hay 1993; Kay et al. 1992, 1996a; Woodroffe and McLean 1993). During the mid-1990s, a national program of case-study assessments was developed (Kay et al. 1996a; Waterman 1996) within the context of drafting a National Coastal Policy (as reported on by Kay and Lester 1997). The program developed a number of approaches to the assessment of potential climate change impacts that were specific both to context information availability and researcher perspectives. However, this early period of work in Australia is now largely forgotten.

The momentum built during the mid-1990s came to a rapid halt on election of the Federal Liberal Government in 1996, whose views on climate change differed markedly from the previous government (Hamilton 2007; Pearse 2007). The result was that there was effectively no national-level effort with respect to climate change impact assessment from 1996 until the mid-2000s. Although the sponsorship of coastal vulnerability assessments by the federal government was in hiatus during this period, individual researchers continued to undertake various impact assessments that subsequently influenced the evolution of the Australian V&A landscape (Table 11.1) (Kay et al. 2005). In addition, the active professional coastal management community in Australia continued to discuss impact assessments and their methods, including at a regular set of two-yearly national Coast-to-Coast conferences† and state-based conferences in three of the six coastal states (Queensland, New South Wales, and Western Australia). Many of these conferences included specific sessions on climate change, sometimes multiple sessions. As such, there is an extensive *gray* literature that incorporates both conceptual and analytical papers as well as case studies on specific climate change V&A assessments on which to draw experiences (Preston and Kay 2010).

A gradual change of view within the federal government occurred in the mid-2000s, indicated by the release of a report recommending assessment of potential climate change risks (Allen Consulting Group 2005) and the commissioning by the federal government of various explorations of coastal vulnerability approaches (Abuodha and Woodroffe 2006) and potential coastal adaptation planning responses (Kay et al. 2005). During this period, there was an apparent reluctance to develop national climate change policy (for both mitigation and adaptation) (Pearse 2007) with a focus instead on undertaking further research into impacts and adaptation. Important events in this respect were the Australian Federal Government's 2007 election

* Ten case studies of potential coastal impacts published in Pearman (1988).

† See website http://coast2coast.org.au/.

Table 11.1 Principal methods adopted to assess vulnerability of the Australian coast to climate change

Approach	*Geographical application*	*Principal methods*	*References*
Wetland mapping	Northern and northwestern coasts	Wetland mapping in Kakadu and elsewhere in the NT, in line with Ramsar wetland assessments	Finlayson et al. (2002); Eliot et al. (2005)
Landform mapping	South Australia	Holocene landform mapping as a guide to vulnerability	Bryant et al. (2001); Harvey et al. (1999)
Storm surge zones	Queensland	Queensland Climate Change and Community Vulnerability to Tropical Cyclones project	Queensland Government (2004)
Beach vulnerability	New South Wales	Fuzzy and probabilistic modeling	Cowell et al. (1995, 1996)
Beach vulnerability	Tasmania	Mapping beaches for Bruun rule and assessing inundation risk	Sharples (2004)

Source: Kay, R.C. et al., *Assessing, Mapping and Communicating Australia's Coastal Vulnerability to Climate Change*, A background reference paper for the Australian Greenhouse Office, Canberra, Australia, 2005.

commitment of AUD25 million for coastal adaptation, commitment to a National Adaptation Research Program and subsequently the creation of the National Climate Change Adaptation Research Facility (NCCARF). The establishment of NCCARF was particularly significant in that a continuing adaptation research program was instigated that includes a set of specific coastal adaptation research activities under the Settlements and Infrastructure and Marine Biodiversity and Resources Themes (Cox et al. 2010).*

On the one hand, the lack of policy developments within this time frame can be viewed as remiss, especially given adaptation analyses undertaken during the mid-1990s. On the other hand, it could also be argued that the federal government's sequencing of vulnerability assessments ahead of policy development (rather than in parallel) was prudent. Whatever the interpretation of coastal adaptation activities during the period from mid-1996 to 2007, their significance is perhaps not so much as what occurred (or didn't occur) during that period, but rather the reaction to them by subsequent federal governments.

The climate change landscape changed rapidly on the election of the Rudd Federal Government in late 2007. Symbolically, the first act of the new government that signaled its commitment to addressing climate change issues was to sign the Kyoto Protocol. Within Australia, this triggered a

* Funded research projects under the Settlements and Infrastructure Theme are listed on the NCCARF website at: http://www.nccarf.edu.au/settlements-infrastructure/.

substantial growth in the number of climate change risk assessments, with a particular emphasis on climate change risk in Australian population centers. To a large degree, this surge in interest is indicated by the various programs and initiatives that have been established at all levels of government, including (Preston and Kay 2010) the following:

- A series of five projects which all included significant biophysical impact assessments as well as trialing a range of methodological approaches to assessment; the National Coastal Vulnerability Assessment (NCVA) Case Studies—a series of six pilot studies to illustrate a range of policy issues, risks, and threshold sensitivities related to climate change impacts for selected coastal locations; and the Local Adaptation Pathways Program (LAPP) Phases I and II. LAPP 1 consisted of 32 projects covering 60 local governments, mostly in coastal and urban areas. LAPP 2 was made up of seven projects in 35 councils, mostly in regional and remote areas (Turnbull 2011).
- Various projects and studies on climate change impacts and adaptation assessment, either directly funded by the federal government or by local government or individual land developers (in some cases as part of development approval processes). For example, in Western Australia, the Vulnerability of the Cottesloe Foreshore to the Potential Impacts of Climate Change project was funded by the federal department of Emergency Management Australia (CZM Pty and Damara Ltd 2008).

In addition to the suite of assessments outlined above, an Australia-wide strategic level assessment of the magnitude of the problem Australia faces in its coastal zone—for coastal communities, infrastructure, ecosystems, and industries—has been undertaken through the *Climate Change Risks to Australia's Coasts: A First Pass National Assessment* (DCCEE 2009, 2011a). This study found that climate change risks to Australia's coastal assets are large and will increase significantly into the future, and in some highly vulnerable areas, climate change impacts will be felt in the near term.

Key conclusions from these assessments are that "Climate change risks to Australia's coastal areas are one of the biggest issues Australia's coastal society is facing over the long term" (ACCC 2011: 2) with greater than AUD226 billion in commercial, industrial, road, and rail, and residential assets potentially exposed to inundation and erosion hazards at a sea-level rise of 1.1 m (high end scenario for 2100) (DCCEE 2011a). In response to these risks, the government's specialist advisory body, the Australian Coasts and Climate Change Council, concluded "significant government investment and leadership to better understand the risks, and start building resilience and capability to facilitate a calm and rational transition will be critical" (ACCC 2011: 2).

POLICY RESPONSES

A relatively long history of adaptive responses to potential climate change impacts on Australia's coast has evolved in tandem with the ethos of vulnerability assessment (Bonyhardy 2010). For example, at a local government level, a limited number of *early adaptors* emerged during the late 1980s and early 1990s (Bonyhardy 2010). For example, in 1991, South Australia was the first state to integrate climate change factors into its coastal planning policy, by incorporating sea-level rise factors of 0.3 m to the year 2050 and 0.7 m rise to 2100. The impetus for this was attributed to the publication of the IPCC First Assessment Report at that time, in conjunction with expert advice from technical advisors and the hosting in South Australia of the Greenhouse '88 conference (Huppatz and Caton 2010). Since then there have been very different policy responses designed to promote adaptation of coastal zones by state/territory governments, and shaped by these responses, at the local level (Good 2011). An analysis of the benchmarks used across the Australian states and territories to determine the magnitude of potential future sea-level rise to be incorporated into coastal planning decisions provides a means for analyzing adaptive responses, albeit in response to one climate change driver only. These benchmarks for sea-level rise are largely incorporated through policy statements and recommendations nested within state coastal planning policies (Gibbs and Hill 2011; Good 2011).

The Western Australia State Coastal Planning Policy states that benchmarks for sea-level rise must be taken into account when planning authorities consider the *setback and elevation* for new developments to allow for the impact of coastal processes over a 100-year planning time frame (to 2110). In 2012, a new draft of the policy increased the benchmark for sea-level rise from 0.38 to 0.9 m for a 100-year planning horizon, reflecting the evolution of sea-level rise scenarios from the IPCC Third Assessment Report to those within the Fourth Assessment Report (Western Australian Planning Commission 2010). In Queensland, in considering the impacts of coastal hazards on existing and proposed development (particularly urban development) in the coastal zone, sea-level rise benchmarks based on the IPCC Fourth Assessment Report are to be considered as outlined in the Queensland Coastal Plan—Coastal Hazards Guideline (Section 4.5 Risk Assessment Factors) (Bell 2012).

In New South Wales, South Australia, and Victoria, the application of sea-level rise benchmarks is slightly more flexible. The Draft New South Wales Sea Level Rise Policy Statement (2009) emphasizes the need to undertake coastal risk assessment incorporating the sea-level rise benchmarks so that both current and future hazards can be evaluated; however, it is noted that the sea-level rise planning benchmarks are not intended to be used as a blanket prohibition on the development of land projected to be affected by sea-level rise. New local environment plans (LEPs) and development applications will continue to be assessed on their merits using a risk-based approach to determine whether the impacts of sea-level rise and other coastal processes can be mitigated and managed over time. In Victoria, the Victorian Coastal Strategy (2008) recommends a policy of planning for sea-level rise of not less than 0.8 m by 2100 should be implemented with the policy being generally applied for planning and risk management purposes. The South Australian Coast Protection Board Policy Document (endorsed in 2002) recommends that new developments should consider sea-level rise benchmarks for both 2050 and 2100. Interestingly, this benchmark remains the same as that recommended in 1991 as a result of the release of the IPCC First Assessment Report in 1990 (Huppatz and Caton 2010) (Figure 11.3).

The remaining Australian states and territories of Northern Territory and Tasmania currently have no sea-level rise planning benchmarks in place. Interestingly, the federal government's NCVA, published in 2009, uses a sea-level rise figure of 1.1 m to 2100 (DCCEE 2009). This does not reflect federal policy rather 1.1 m was selected as a *plausible value* for sea-level rise based on the prevailing climate change science (DCCEE 2009).

Although there has been a clear adaptive policy response at the state/territory level—albeit inconsistent between them—the focus of these policies has been on providing limited guidance for undertaking site-specific vulnerability assessments for new coastal developments. There has been

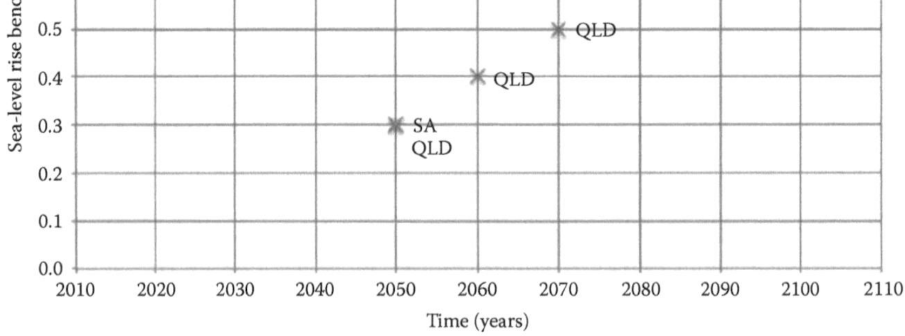

Figure 11.3 Sea-level rise planning benchmarks in Australia. QLD, Queensland; SA, South Australia; VIC, Victoria; WA, Western Australia. (Adapted from Good, M. *Government Coastal Planning Responses to Sea Levels, Australia and Overseas: Technical Report*, Antarctic Climate and Ecosystems Cooperative Research Centre, Hobart, Australia, 2011.)

less emphasis on guiding the methods and tools for undertaking such assessments for new developments, or using vulnerability assessments to guide adaptation responses for existing coastal developments, as outlined in the next section.

LESSONS LEARNED FROM AUSTRALIAN EXPERIENCE IN COASTAL V&A ASSESSMENT

The V&A assessment studies outlined in the previous section represent a considerable body of knowledge. A review by Preston and Kay (2010: 187) of the lessons emerging from the federally funded risk-based approaches to climate change impact and adaptation assessments funded in the 2007–2009 period concluded there were two sets of key challenges emerging from operational practice:

Methodological challenges related to

- The estimation of the likelihood of adverse consequences in risk assessment studies
- Identifying critical thresholds for climate change in specified systems and, more generally, quantifying relationships between climate change drivers, and system processes and responses
- Identifying thresholds for action—setting management thresholds for implementation of adaptation actions
- Overcoming data gaps and inconsistencies
- The general lack of comparability across different assessments and scales

Communication challenges related to

- Aligning stakeholder expectations with methodological and resource limitations
- Aligning stakeholder terminology to describe risk and vulnerability, as well as methodological approaches to risk assessment, with those of researchers and technical experts
- Clearly communicating the trade-offs between strategic, but generic, adaptation and specific adaptation options for site-specific problems and issues
- Linking risk to adaptation—translating information about risk at the local scale into actions that build capacity and reduce vulnerability

These challenges were subsequently confirmed by the formal review of the LAPP (Turnbull 2011). This review stressed the benefits of the LAPP in raising awareness of the need for climate change adaptation within the participating local governments (and other participating organizations) while stressing that the significant barriers to implementation of adaptive actions include the following:

- Inadequate funding
- Limited staff capacity due to limited numbers of staff or lack of skills and experience
- Insufficient direction and guidelines from state and territory governments
- Limited technical guidance on adaptation measures
- Having to wait for state management plan(s) or statutory land-use plan(s)
- Uncertainty about how action implementation measures

The above conclusions were echoed by participants in a LAPP review workshop held in June 2011, coordinated by ICLEI—Local Governments for Sustainability (ICLEI 2011). In addition, the ICLEI workshop stressed the importance of ongoing support for networking and knowledge exchange within and between local governments to assist in ongoing adaptation activities.

The outcomes of the various evaluations, discussion sessions, and feedback provided to the Australian Government combined to prompt a recognition that risk-based approaches may not be well suited to coping with uncertainty, an issue that must be dealt with when developing adaptation plans. This resulted in a change in direction through the Coastal Adaptation Decision Pathways (CADP) program* that explored investment and decision pathways that build resilience in the face of increasing climate risk, focusing on pathways that facilitate transformation of business practices attuned to long-term climate change projections (DCCEE 2011b). Competitive applications to receive funding from a pool of AUD4.5 million were sought with 13 projects subsequently initiated in June 2011 around the Australian coast (see Figure 11.4) with a scheduled completion in mid-2012. Through the CADP program, the department sought to identify and invest in a small number of detailed and nationally relevant projects over a short project life. One selection criterion used was a prerequisite detailed understanding of coastal vulnerability issues in the study area or vulnerability assessments proposed in parallel to the CADP using resources drawn from other sources.

It is intended that the CADP program develops flexible adaptation pathways as a set of strategies that can evolve over time in an iterative manner as new information about climate risks and adaptation options becomes

* See detail at http://www.climatechange.gov.au/government/initiatives/coastal-adaptation-decision-pathways.aspx.

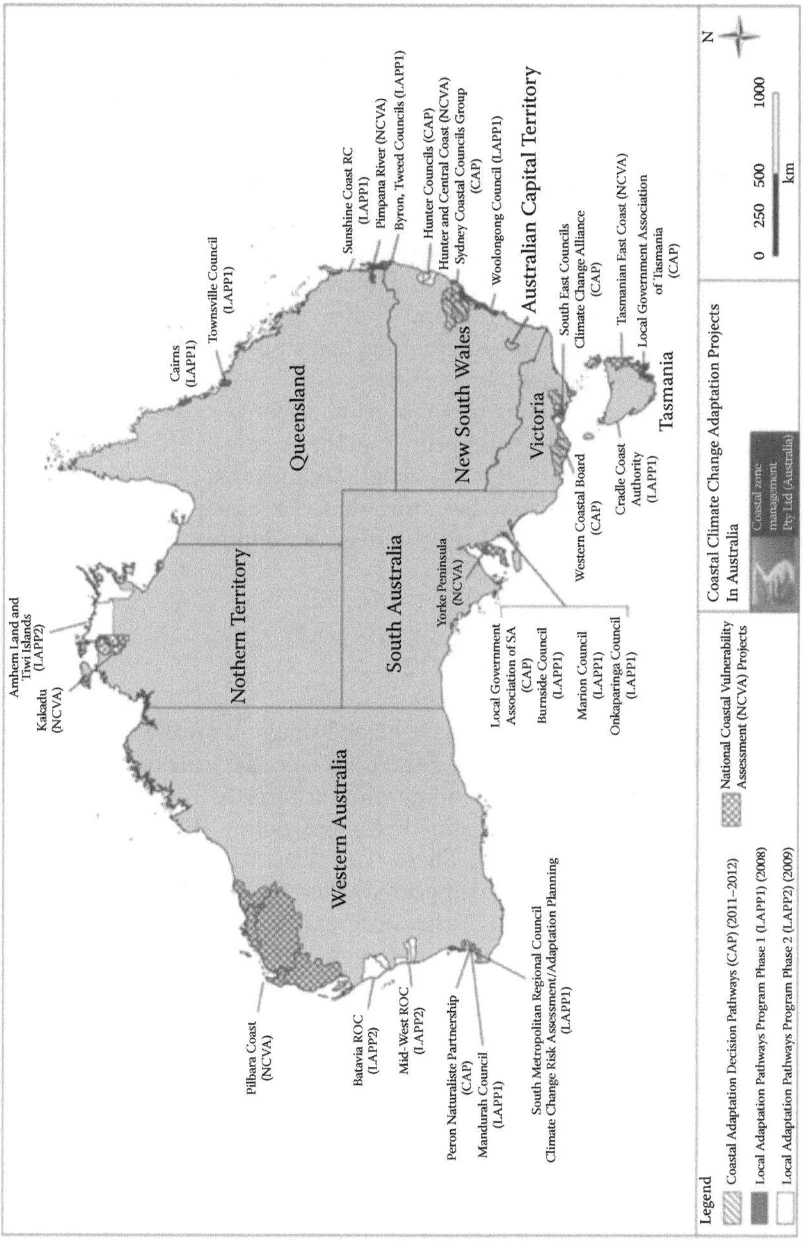

Figure 11.4 Coastal climate change adaptation projects.

available. In doing so, this approach seeks to avoid the high cost of early *over-engineering* (DCCEE 2011b)—in other words maladaptation. Additionally, the CADP projects encompass an economic component that is concerned with more broadly investigating how economics can better inform adaptation decision-making. The programs seek to use such assessment techniques to identify the nature and complexity of potential climate change impacts on the coast as well as their degree of knowledge and confidence (certainty) to help guide decision-makers in the short term and to support the broader stakeholder community in the longer term (DCCEE 2011b).

Given that these projects are ongoing, it is not possible, as yet, to evaluate the CADPs *flexible adaptation pathway* approach in supporting adaptation in coastal zones. In this regard, an independent study being undertaken by the Commonwealth Scientific and Industrial Research Organisation (CSIRO) through the Climate Adaptation Flagship (Wise, personal communication.) will be valuable, especially to assist in evaluating the ongoing barriers and opportunities to supporting effective coastal adaptation in Australia. It does, however, appear that the Australian Government is seeking to develop vulnerability assessment methods that explicitly support adaptive decisions, rather than assuming that risk-based assessment tools would (implicitly or explicitly) lead to effective adaptive decision-making at the coast.

BARRIERS AND OPPORTUNITIES

Coastal vulnerability assessments have helped shape Australian adaptation policy at strategic national/state level at the site-specific scale for new greenfield developments using specific adaptation criteria (with a strong focus on sea-level rise benchmarks) defined in state policy (Bonyhardy 2010; McDonald 2010; Preston 2010). There is evidence that as a result adaptive decision-making is actively supported, thereby reducing future exposure to climate change. Although this progress is encouraging, a number of challenges remain. These challenges largely relate to adaptive responses for coastal land that is already developed and as such, in its existing form, may lack the opportunity for developing an adequate coastal hazard buffer. Overall, the key to addressing these challenges will be to develop strategies that allow for use of coastal land in conjunction with a consideration of the relative risk that society is willing to accept. For example, multiple land-uses might be seen in one lifetime for the same piece of land as its risk profile changes and society's appetite to subsidize private/public risk changes.

The aforementioned challenges appear to be understood by the Australian Government and are reflected in the impetus for developing the CADP program of case studies around the Australian coastline. The long journey to the development of the program in Australia provides a useful frame of reference for drawing lessons on how national-level coastal adaptation assessments have evolved and the continuing challenges faced in supporting effective adaptive decision-making. In addition, a report from the Coasts and Climate Change Council has recently been provided to the federal government recommending a coordinated national response to the pivotal issues that coastal decision-makers face given projected climate change impacts (Thom 2012). The government's response to this report is expected in the second quarter of 2012 (Thom, personal communication). Finally, it is important to recognize that the initiatives described above have been undertaken in a highly charged political atmosphere (Speck 2010). Although these tensions are played out at all levels of government, it is at the federal level, within the context of greenhouse gas mitigation, and in particular the transition in Australia to placing a national price on carbon, that these tensions have been (and continue to be) most visible.

PRACTICAL RECOMMENDATIONS FOR BUILDING COMMUNITY RESILIENCE, ADAPTIVE CAPACITY, AND SUSTAINABILITY

The short review of Australian coastal vulnerability assessment and adaptation policy presented in this chapter enables a number of valuable conclusions to be drawn.

The first, and perhaps most obvious conclusion, is that the translation of outcomes of vulnerability assessments into adaptive policy and ultimately to reducing the exposure of coastal communities to climate change is a long-term (decadal scale) endeavor. When considered over a decadal time frame, there have clearly been multiple contributing factors (both positive and negative) that have shaped the evolution of thinking and practice in seeking to address the future impacts of climate change on the coast of Australia. Although this makes tracking causality in the evolution of practice from a vulnerability focus to one that more explicitly focuses on supporting adaptive decision-making, it also points to a future where such multiple inputs will continue to influence both the assessment and implementation of adaptive actions—especially within a contested political landscape. Although there has been considerable progress in this regard, for example, through the development of the National Position Statement for adaptation (DCCEE 2010) and various responses at state government level, there remains considerable opportunity to improve a truly national coastal adaptive policy approach. There have been repeated calls such a coordinated and transparent approach that seeks to harness the considerable knowledge and experience of Australian coastal managers and policy makers. The Australian Government, states/territories, local government, private sector stakeholders, indigenous people, and other coastal residents all have a role to play. The new Select Council on Climate Change (SCCC), constituted under the Council of Australian Governments, may play a significant role in this regard. One of the terms of reference of the SCCC is to develop national adaptation priorities and work plans. Its first meeting was held on May 4, 2012, and at the time of writing it is too early to comment on its specific role in coastal adaptation policy development.

Second, there is an urgent need to translate the considerable scientific and technical knowledge gained from over 20 years of vulnerability assessment into a nationally consistent and coherent set of methods, tools, and technical standards. There is a case for the states/territories to continue to tailor approaches and standards, such as sea-level rise benchmarks, to suit their own legislative and policy contexts. Moreover, the states and territories provide a useful test bed for evaluating alternative approaches. Nevertheless, there are clear efficiencies in developing a nationally consistent framework. Ensuring the legal robustness of methods, tools, and standards will be an expensive and complex process that would benefit from a national view. This is likely to become compelling, especially in the face of a likely increase in the number and scope of legal challenges regarding the development of, or restriction of, coastal land development. In turn, ensuring that such a transition to a nationally consistent approach is undertaken openly and transparently with key stakeholder input should facilitate the development of consistent frameworks, methods, and tools for understanding vulnerability within specific social, cultural, biophysical, and decision contexts.

Lessons learnt in implementing the CADP program should provide useful guidance of wide applicability regarding effective means of developing adaptive policy in the face of uncertainty. The focus on the development of flexible adaptation pathways that can evolve over time in light of new information addresses one of the critical challenges in adapting to climate change: how to plot a course when the destination remains unclear.

In conclusion, a consideration of coastal vulnerability assessments undertaken in Australia through the past 20 years proffers useful reflections on the Australian coastal adaptive journey to date and potential trends into the future. To paraphrase the folklore historian Henry Glassie,* the history of coastal vulnerability assessments in Australia is useful in that it helps provide a map from the past, through the present day to the future—and, in so doing, helps coastal managers navigate the vital adaptive journey. That this adaptive journey should proceed in a systematized, coherent fashion, through the adoption of nationally relevant policy that facilitates locally applicable outcomes, would appear self-evident. By sharing the journey between all organizations with coastal management responsibilities, a destination of resilient, well-adapted Australian coastal communities would be

* http://www.quotegarden.com/history.html.

more easily reached. As this review has shown, however, there remain considerable challenges in this regard, despite the investments in research, vulnerability assessments, adaptation planning, and policy development. The Australian experience demonstrates that the route from vulnerability assessment to meaningful coastal adaptation policy and practice is not a straightforward one. Nevertheless, it is a journey in the right direction and the travelogues that result should provide an invaluable guide to future generations of coastal planners and inhabitants and to other nations following a similar road.

ACKNOWLEDGMENT

The comments of Bruce Thom on earlier drafts of this chapter are very much appreciated.

REFERENCES

Abuodha, P.A. and Woodroffe, C.D. (2006). *International Assessments of the Vulnerability of the Coastal Zone to Climate Change, Including an Australian Perspective*, A report to the Department of Climate Change, Canberra, Australia, http://ro.uow.edu.au/cgi/viewcontent.cgi?article=1188&context=scipapers.

Allen Consulting Group. (2005). *Climate Risk and Vulnerability; Promoting an Efficient Adaptation Response in Australia*, Final report to the Australian Greenhouse Office, Commonwealth of Australia, Canberra, Australia: Department of Environment and Heritage.

Australian Coasts and Climate Change Council (ACCC). (2011). "Report to Minister Combet." http://www.climatechange.gov.au/ (accessed January 2012).

Australia State of the Environment Committee. (2001). *Australia State of the Environment 2001*, Independent report to the Commonwealth Minister for the Environment and Heritage, Canberra, Australia: CSIRO Publishing on behalf of the Department of the Environment and Heritage.

Bell, J. (2012). "Planning for climate change and sea-level rise—Queensland's new coastal plan," *Environmental and Planning Law Journal*, 29: 61–74.

Bonyhardy, T. (2010). "Swimming in the streets: The beginnings of planning for sea level rise," in T. Bonyhady, J. McDonald, and A. Macintosh (eds.), *Adaptation to Climate Change: Law and Policy*, Sydney, New South Wales, Australia: Federation Press.

Bryant, B., Harvey, N., Belperio, A., and Bourman, B. (2001). "Distributed process modeling for regional assessment of coastal vulnerability to sea-level rise," *Environmental Modeling and Assessment*, 6: 57–65.

Coastal Zone Management (CZM) Pty and Damara Ltd. (2008). *Vulnerability of the Cottesloe Foreshore to the Potential Impacts of Climate Change*, A report prepared for the Town of Cottesloe. http://www.cottesloe.wa.gov.au/ (accessed January 2012).

Cowell, P.J., Roy, P.S., and Jones, R.A. (1995). "Simulation of large-scale coastal change using a morphological behaviour model," *Marine Geology*, 126: 45–61.

Cowell, P.J., Roy, P.S., Zeng, T.Q., and Thom, B.G. (1996). "Relationships for predicting coastal geomorphic impacts of climate change," *Proceedings of the International Conference on the Ocean and Atmosphere, Pacific*, Adelaide, Australia: National Tidal Facility, pp. 16–21.

Cox, R., Cane, J., Farrell, C., Hayes, P., Kay, R., Kearns, A., Low Choy, D., McAneney, J., McDonald, J., Nolan, M., Norman, B., Nott, J., and Smith, T. (2010). *National Climate Change Adaptation Research Plan for Settlements and Infrastructure*, National Climate Change Adaptation Research Facility. http://www.nccarf.edu.au/ (accessed December 2011).

Department of Climate Change and Energy Efficiency (DCCEE). (2009). *Climate Change Risks to Australia's Coasts: A First Pass National Assessment*, Canberra, Australia: Commonwealth of Australia. http://climatechange.gov.au/ (accessed January 2012).

Department of Climate Change and Energy Efficiency (DCCEE). (2010). *Adapting to Climate Change in Australia—An Australian Government Position Paper*, Canberra, Australia: Commonwealth of Australia. http://www.climatechange.gov.au/ (accessed January 2012).

Department of Climate Change and Energy Efficiency (DCCEE). (2011a). *Climate Change Risks to Coastal Buildings and Infrastructure—A Supplement to the First Pass National Assessment*, Canberra, Australia: Commonwealth of Australia. http://climatechange.gov.au/ (accessed January 2012).

Department of Climate Change and Energy Efficiency (DCCEE). (2011b). *Coastal Adaptation Decision Pathways Projects*. http://www.climatechange.gov.au/ (accessed February 2012).

Eliot, I.G., Saynor, M.J., Eliot, M., Pfitzner, K., Waterman, P., and Woodward, E. (2005). *Assessment and development of tools for assessing the vulnerability of wetlands and rivers to climate change in the Gulf of Carpentaria, Australia*, Report prepared by the Environmental Research Institute of the Supervising Scientist and the National Centre for Tropical Wetland Research for the Australian Greenhouse Office, Canberra, Australia.

Finlayson, C.M., Begg, G.W., Howes, J., Davies, J., Tagi, K., and Lowry, J. (2002). *A manual for an inventory of Asian wetlands: version 1.0*, Wetlands International Global Series 10, Kuala Lumpur, Malaysia.

Gibbs, M. and Hill, T. (2011). *Coastal Climate Change Risk—Legal and Policy Responses in Australia*, A report to the Department of Climate Change and Energy Efficiency. http://www.climatechange.gov.au/ (accessed April 2012).

Good, M. (2011). *Government Coastal Planning Responses to Sea Levels, Australia and Overseas: Technical Report*, Hobart, Tasmania, Australia: Antarctic Climate and Ecosystems Cooperative Research Centre.

Gurran, N., Squires, C., and Blakely, E. (2005). *Meeting the Sea Change Challenge: Sea Change Communities in Coastal Australia*, A report for the National Sea Change Task Force, Sydney, New South Wales, Australia: Planning Research Centre, University of Sydney.

Hamilton, C. (2007). *Scorcher: The Dirty Politics of Climate Change*, Melbourne, Victoria, Australia: Black Inc. Agenda.

Harvey, N., Belperio, T., Bourman, B., and Bryan, B. (1999). "A GIS-based approach to regional coastal vulnerability assessment using Holocene geological mapping of the northern Spencer Gulf, South Australia," *Asia Pacific Journal on Environment and Development*, 6: 1–25.

Harvey, N. and Caton, B. (2003) *Coastal Management in Australia*, Melbourne, Victoria, Australia: Oxford University Press.

Huppatz, T. and Caton, B. (2010). "Taking sea level rise into account: A history of coast protection policy in South Australia and its translation to planning controls," in *Proceedings of Coast to Coast 2010*, September 20–24, Adelaide, Australia. http://conference2010.coast2coast.org.au/ (accessed January 2012).

ICLEI Oceania. (2011). *From Risk to Action, Looking Backward: Evaluation—Looking Forward: Implementation*, A local adaptation pathways program forum Report. http://www.pc.gov.au/__data/assets/pdf_file/0019/114526/sub038-attachment1.pdf.

Kay, R.C. and Alder, J. (2005). *Coastal Planning and Management (Second Edition)*, London: E&F Spon (Routledge).

Kay, R.C., Crossland, C., Gardner, S., Waterman, P., and Woodroffe, C.D. (2005). *Assessing and Mapping Australia's Coastal Vulnerability to Climate Change: Expert Technical Workshop*, December 13–14, Australian Greenhouse Office, Canberra, Australia (unpublished).

Kay, R.C., Eliot, I.E., Caton, B., and Waterman, P. (1996a). "A review of the Intergovernmental Panel on Climate Change's common methodology for assessing the vulnerability of coastal areas to sea-level rise," *Coastal Management*, 24(1): 165–88.

Kay, R.C., Eliot, I.E., and Klem, G. (1992). *Analysis of the IPCC Sea-Level Rise Vulnerability Assessment Methodology Using Geographe Bay, SW Western Australia as a Case Study*, A report to the Department of Arts, Sports, Environment and Territories, Canberra, Australia: Coastal Risk Management International Ltd.

Kay, R.C. and Hay, J.E. (1993). "A decision support approach to coastal vulnerability and resilience assessment: A tool for Integrated Coastal Zone Management," in R. McLean, and N. Mimura (eds.), *Vulnerability Assessment to Sea-Level Rise and Coastal Zone Management*, Intergovernmental Panel on Climate Change, Coastal Zone Management Sub-Group, Eastern Hemisphere Workshop, Tsukuba, Japan, August 3–6.

Kay, R.C. and Lester, C. (1997). "Benchmarking Australian coastal management," *Coastal Management*, 25: 265–92.

Kay, R.C. and McKellar, R. (2000). "What matters for your coast? Moving forward in coastal management by looking inwards," *Proceedings of the 10th New South Wales Coastal Conference*, Yamba, New South Wales, Australia, November 20–24.

McDonald, J. (2010). "Mapping the legal landscape of climate change adaptation," in T. Bonyhady, J. McDonald, and A. Macintosh (eds.), *Adaptation to Climate Change: Law and Policy*, Sydney, New South Wales, Australia: Federation Press.

New South Wales Department of Environment and Climate Change. (2009). Draft Sea Level Rise Policy Statement. http://www.environment.nsw.gov.au/resources/climatechange/09125draftslrpolicy.pdf.

Pearman, G. (ed.) (1988). *Greenhouse: Planning for Climate Change, Proceedings of the Greenhouse 1987 Conference*, Melbourne, Victoria, Australia: CSIRO Publishing.

Pearse, G. (2007). *High & Dry: John Howard, Climate Change and the Selling of Australia's Future*, Camberwell, Victoria, Australia: Penguin Group (Australia).

Preston, B.J. (2010). "The role of courts in relation to adaptation to climate change," in T. Bonyhady, J. McDonald, and A. Macintosh (eds.), *Adaptation to Climate Change: Law and Policy*, Sydney, New South Wales, Australia: Federation Press.

Preston, B.L. and Kay, R.C. (2010). *Managing Climate Risk in Human Settlements, Proceedings of the Greenhouse 2009 Conference*, Perth, Australia: CSIRO Press.

Queensland Government. (2004). "Queensland climate change and community vulnerability to tropical cyclones," *Ocean Hazards Assessment, Synthesis Report: An Overview of Discussion of Results from Project Stages 2, 3 and 4*, Queensland Government, Australia.

Sharples, C. (2004). *Indicative Mapping of Tasmanian Coastal Vulnerability to Climate Change and Sea Level Rise: Explanatory Report*, Tasmanian Department of Primary Industries, Water and Environment, Australia, 126pp.

Short, A. and Woodroffe, C.D. (2008). *The Coast of Australia*, Melbourne, Victoria, Australia: Cambridge University Press.

Smyth, D. (1993). *A Voice in All Places: Aboriginal and Torres Strait Islander Interests in Australia's Coastal Zone*, Canberra, Australia: Commonwealth of Australia.

Speck, D.L. (2010). "A hot topic? Climate change mitigation policies, politics, and the media in Australia," *Human Ecology Review*, 17(2): 125–34.

Thom, B. (2012). "Caring for our coasts?," Keynote address—*Australian Coastal Councils Conference*, March 6–7, Hobart, Tasmania, Australia. http://www.seachangetaskforce.org.au/Conference.html (accessed September 2012).

Thom, B.G. and Short, A.D. (2006). "Introduction: Australian Coastal Geomorphology 1984–2004," *Journal of Coastal Research*, 22: 1–10.

Turnbull (2011). *Review of the Local Adaptation Pathways Program*, A report prepared for the Department of Climate Change and Energy Efficiency, Canberra, Australia.

Victorian Coastal Council. (2008). Victorian Coastal Strategy, State of Victoria, Melbourne, Australia. http://www.vcc.vic.gov.au/resources/VCS2008/home.htm.

Waterman, P. (1996). *Australian Coastal Vulnerability Assessment Project Report*, Canberra, Australia: Department of the Environment, Sport and Territories.

Western Australian Planning Commission. (2010). *Position Statement—State Planning Policy No. 2.6 State Coastal Planning Policy Schedule 1 Sea Level Rise.* http://www.planning.wa.gov.au/ (accessed April 2012).

Woodroffe, C.D. and McLean, R.F. (1993). *Cocos (Keeling) Islands: Vulnerability to Sea Level Rise*, Canberra, Australia: Climate Change and Environmental Liaison Branch, Department of Environment, Sports and Territories.

Chapter 12

Adapting Australian coastal regions to climate change

A case study of South East Queensland

Timothy F. Smith, Darryl Low Choy, Dana C. Thomsen, Silvia Serrao-Neumann, Florence Crick, Marcello Sano, Russell Richards, Ben Harman, Scott Baum, Stephen Myers, Vigya Sharma, Marcus Bussey, Julie Matthews, Anne Roiko, and R.W. (Bill) Carter

Abstract: An increasing proportion of Australians are living in the coastal zone, which is also becoming increasingly prone to sea-level rise, storm surge, and flooding. The likely severity of the potential impacts of climate change on Australia's coastal communities has also been recognized by the Australian government through the recent national coastal inquiry. Similarly, some coastal local governments have become key advocates for progressive adaptation policies, although many continue to face significant capacity constraints in terms of resources, skills, and support from other tiers of government. South East Queensland has been identified as being particularly vulnerable to the impacts of climate change; local governments experience adaptive capacity issues primarily relating to infrastructure provision, emergency response capacity, and the changing socioeconomic characteristics of the region. Biophysical, socioeconomic, and political drivers of change are also likely to influence the effectiveness of adaptation options in the region into the future.

INTRODUCTION

Climate change is of particular significance to Australian coastal regions as almost 250,000 existing dwellings may be at risk from projections of a 1.1 m rise in sea level by 2100 (Department of Climate Change and Energy Efficiency 2010). Although there have been multiple studies addressing the potential *exposure* of Australian coasts to climate change effects (e.g., sea-level rise, storm surge, and flooding), there have been few that address climate change adaptation comprehensively by also determining levels of *sensitivity* and *adaptive capacity* within exposed communities. Furthermore, several authors have identified that the impacts of climate change will be context specific and may affect different sectors and places in different ways (e.g., Barnett 2002; Smit and Wandel 2006; Smith and Thomsen 2008; Smith et al. 2010; Tompkins and Adger 2004; Vincent 2007). Hence, charting effective adaptation pathways will require assessments of exposure, sensitivity, and adaptive capacity combined with an understanding of how these may differ across sectors and locations. This chapter provides an introduction to the Australian coastal zone in relation to climate change before describing a novel and mixed-method approach to identifying potential adaptation pathways and key issues for coastal adaptation. The chapter focuses on South East Queensland (SEQ), where Australia's largest single investment in climate change adaptation research at the regional scale is currently supporting such studies.

BACKGROUND

The three key threats to the Australian coastal zone have been broadly categorized by Smith et al. (2011) as population growth, the unsustainable use of coastal resources, and rapid environmental change (including climate change). For example, over 85% of Australia's population currently resides within 50 km of the coastline, and projections indicate that the trend of significant growth in coastal regions, both metropolitan and nonmetropolitan, is set to continue (Australian Bureau of Statistics 2001, 2004). Urbanization of the coastal zone has the potential to place unsustainable demands on coastal resources and infrastructure (Mosadeghi et al. 2009; Smith et al. 2011). Negative impacts of population growth, and associated unsustainable resource use, are likely to result in decreased quantity and quality of coastal ecosystem services and amenities. Climate change will exacerbate the effects of existing stresses on coastal zones. For example, climate change is likely to result in some habitat change and modification of freshwater flows (Tobey et al. 2010). Management of the combined stressors anticipated for the coastal zone presents a significant challenge for policy and the Australian government (Norman 2009). This challenge is compounded by the institutional fragmentation (including lack of clarity on roles and responsibilities in relation to climate change adaptation) that exists for the management of the Australian coastal zone (Lazarow et al. 2008). Despite these shortcomings, an urgent response is required. According to the United Nations Intergovernmental Panel on Climate Change (IPCC), SEQ has been identified as one of the six *hot spots* in Australia where the vulnerability to climate change is likely to be high (Parry et al. 2007). As recently as the beginning of 2011, over two-thirds of the state of Queensland was inundated by floodwaters, with the largest number of fatalities occurring in SEQ.

SOUTH EAST QUEENSLAND

SEQ is one of the fastest growing coastal regions in Australia with its 2009 population of just over 2.5 million estimated to reach 4.4 million people by 2031 (Department of Infrastructure and Planning 2009). Although there has been significant growth across the region, there has been particularly high and rapid growth in the coastal areas of the Gold Coast to the south and the Sunshine Coast to the north (Minnery 2001; O'Connor et al. 1998). The trend of population growth in the region (including coastal areas) is projected to continue (Roiko et al. in press). The focus of this chapter is on three coastal local government areas in SEQ—the Sunshine Coast, Moreton Bay, and the Gold Coast (Figure 12.1). In the Queensland context, as in most other states, the statutory planning functions have largely been delegated by respective state governments to local government. Hence, along with mandatory corporate and operational planning functions, local government has increasingly assumed greater responsibilities for environmental and natural resource management planning in the coastal zone (Low Choy 2006).

All three of the focus areas have begun climate change adaptation planning either through the local government authorities or through other stakeholder groups within the region. The Sunshine Coast Regional Council* has recognized that its climate change risk is exacerbated by the region's population growth, its coastal location, development pressures, dispersed settlement pattern, and reliance of climate-sensitive economies (Sunshine Coast Regional Council 2010). In response to these threats, the council has prepared the *Sunshine Coast Climate Change and Peak Oil Strategy 2010–2020* and related action plan to provide support for addressing environmental, social, and economic issues associated with climate change. The framework for the strategy consists of four key policy approaches that relate to leadership (through partnership and advocacy), mitigation, adaptation, and energy transition (to reduce oil dependency). Similarly, the Gold Coast City Council is taking a multifaceted approach and envisages that the city's leaders and communities will work collectively to achieve climate change resilience for the future (Gold Coast City Council 2009). The council's response to the risk of climate change targets five key areas for action: governance and leadership, advocacy and awareness, resilient infrastructure, planning and regulation, and research. Moreton Bay Regional Council, in beginning its adaptation planning, has commissioned a scoping study, *Scoping Climate Change Risk for Moreton Bay Regional Council*, which highlights several risks stemming from primary (e.g., physical effects), secondary (e.g., regulatory), and tertiary (e.g., social response) factors (Burton et al. 2009).

METHODS

This section describes the case study approach taken to examine the key challenges and opportunities facing coastal adaptation initiatives in SEQ. Specifically, the study focused on the Moreton Bay, Sunshine Coast, and Gold Coast City local government areas to assess (1) the capacity of coastal

* The Sunshine Coast Regional Council has recently been de-amalgamated to form two new local government areas, namely, the Sunshine Coast Council and the Noosa Council.

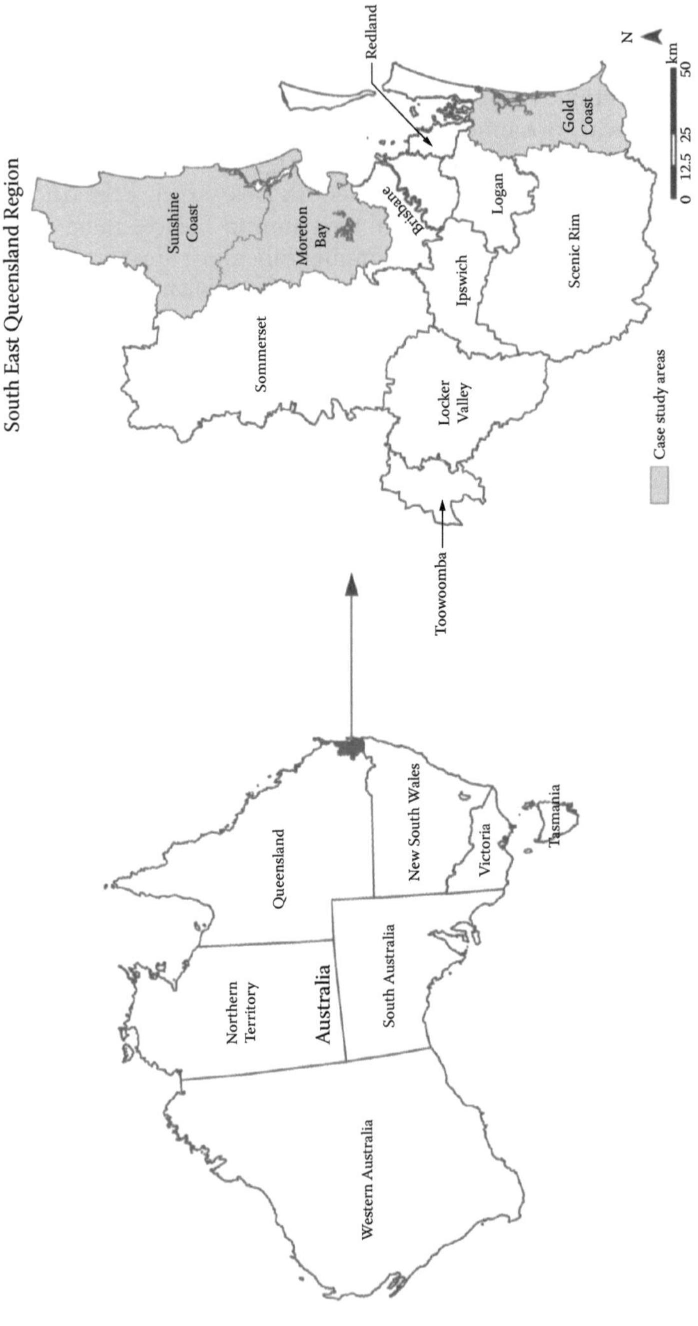

Figure 12.1 South East Queensland.

communities to adapt to climatic changes and (2) the identification of adaptation options for coastal communities under future climate scenarios. Data for the study were collected through three workshops within each of the local government areas. Participants included representatives from local and state governments and other organizations involved in urban planning and management, coastal management, infrastructure, emergency management, and human health.

The workshops adopted a multimethod approach to combine a set of participatory techniques including (1) systems thinking, (2) Bayesian belief network (BBN) modeling, and (3) scenario planning (Figure 12.2).

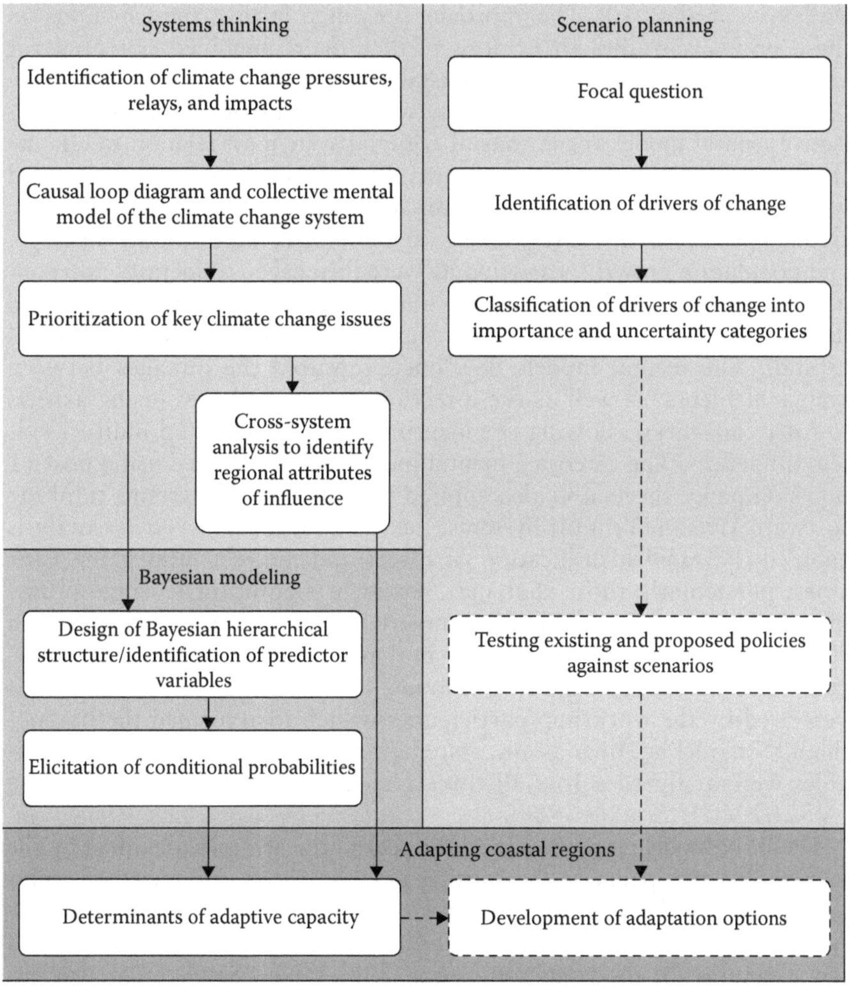

Figure 12.2 Multimethod approach used to inform effective adaptation to climate change in coastal regions.

These established methods are suited to exploring the key challenges and opportunities facing adaptation in coastal regions because they facilitate the direct involvement of stakeholders and build collective understanding of complex systems. Although the integration of these three methods has been considered previously within the fields of business and management (Millet 2009; Schriefer and Sales 2006), to date, it remains an innovative approach for the analysis of adaptive capacity and adaptation options in climate change and coastal management. Nevertheless, recent examples show their individual potential to improve understanding and engage stakeholders in coastal regional systems under the influence of climate change (see, e.g., Catenacci and Giupponi 2010; Smith et al. 2009; Willekens et al. 2010). The approach presented in this chapter builds on these previous studies to explore further their combined potential for adaptation to climate change in coastal regions.

Workshops began using the systems thinking approach to generate a collective mental model of the coastal regional system (in relation to climate change adaptation) among participants. Participant responses were framed with a set of predetermined pressures related to climatic and nonclimatic changes predicted for the region (e.g., sea-level rise, increasing storm surge, and population growth). Participants were then asked to identify and connect these issues to other associated influences and impacts they considered important to the system (e.g., water supply, government policy, and social capital). The mental models developed captured the linkages between system elements, as well as the direction of such linkages (using arrows to infer causality), allowing the identification of direct and indirect system influences. The emerging mental models were recorded using post-IT notes on paper sheets and also entered directly into the systems thinking software *Vensim* (Ventana Systems, Inc., Harvard, MA). *Vensim* analysis enabled the rapid identification of direct and indirect influencing variables, particularly those that may represent significant leverage points in the systems—as identified by numerous connections to other system elements. *In situ* capturing of the mental model also enabled real-time discussion and enquiry of the system model to validate critical system issues perceived by the workshop participants, which then formed the basis of the BBN modeling. In addition, system pressures, relays, and impact variables were analyzed across all three systems models using the software *INFLUENCE* (Walker 1988).

The BBN modeling process used the set of critical issues identified by the workshop participants as the starting point in the development of multilevel hierarchical trees. The participants then refined the BBNs by identifying adaptive capacity determinants for each critical issue, which created the structures for the conditional probability tables. Similar to Castelletti and Soncini-Sessa (2007), expert elicitation (though individual interviews) was the method by which the conditional probability tables were populated (mostly post-workshop). The BBNs were then used to model the probability of influence of adaptive capacity determinants based on a range of conditions (e.g., the probability of influence of a condition of high social capital, low resources, and low education on asset protection—where social capital, resources, and education were considered to be the adaptive capacity determinants of asset protection).

Following the systems thinking and BBN processes, scenario planning was also used to assist the identification of adaption pathways for SEQ. Scenario planning investigates unexpected elements that might appear in the future (Inayatullah 2009). In this study, the scenario planning was used to complement the systems thinking and BBN approaches that focused on current contexts, by identifying the key drivers of change necessary to inform future climate change adaptation options. To frame responses for this stage, workshop participants were presented with the following definition of drivers of change: "environmental, social, economic, political and technological factors (natural or human-induced) that directly or indirectly cause a change in a system and affect several temporal and spatial levels, internal or external to the system." Using this definition, participants identified and ranked (from low to high) current and future drivers of change for the region and its coastal settlements in terms of importance and uncertainty. These drivers of change were then used to construct scenarios (i.e., structured accounts of possible futures) against which existing and proposed climate change adaptation strategies could be evaluated.

KEY ISSUES FOR COASTAL ADAPTATION

The systems thinking approach generated a shared mental model for each of the workshops that included a suite of interconnected biophysical, socio-economic, technological, and political issues. Using *Vensim*, participants were able to explore complex webs of causal relationships between issues before identifying particular issues of importance to their sectors. In addition, the cross-sectoral analysis of systems models allowed the identification of regional *influential*, *relay*, or *outcome* variables (Table 12.1). Based on Godet (1994), we define these as

- *Influential variables* are variables that influence the system but remain unaffected by the system.
- *Relay variables* are variables that both influence and are influenced by the system.
- *Outcome variables* are variables that result from processes within the system and do not influence the system.

Table 12.1 Regional cross-sectoral pressures, relays, and outcome variables relating to climate change

Pressures	*Relays*	*Outcomes*
Heat	Government policy	Health
Air conditioning	Energy demand	Housing
Growing population	Infrastructure	Politics
Sea-level rise		Growing population
Extreme weather		Emergency services
Drought		Water supply
Storm surge		Self-help
Terrorism		Property value
Peak oil		Hardship
Price of energy		Money and cost of living
Intense rain		Stress
Reduced investment		Support services
Rates		

Influential variables typically included the climatic and nonclimatic issues that were used to stimulate discussion for the model building process (e.g., intense storms, sea-level rise, and growing population) and reflect the nominated starting points of each mental model. Nevertheless, these influential variables are consistent with the key stressors identified for the Australian coastal zone (e.g., Department of Climate Change and Energy Efficiency 2010). Only three *relay variables* emerged from the regional systems model: infrastructure, energy demand, and government policy. These are important to note as relay variables are often key leverage points in a system and provide a potential focus for adaptation strategies. And, finally, social issues such as adverse health impacts, housing impacts, hardship, and mental stress dominated *outcome variables*.

The system conceptualization process also provided the basis for the selection of priority management issues for further analysis using BBNs through the exploration of system linkages by workshop participants. Overall, BBNs were developed around 15 priority management issues (five for each workshop representing one key issue for each sector: urban planning and management, coastal management, infrastructure, emergency management, and human health). From a broad cross-sectoral perspective, what emerged from this process as determinants of, or influences on, adaptive capacity in coastal areas was the relative importance of community support/resilience, adequate funding, and proactive policy. There was also some overlap in the priority issues that were selected across the three coastal regions. For example, key adaptive capacity variables such as the capacity to respond to emergencies and the status of well-being in the community were identified consistently within the emergency management and health sector groups. Conversely, there was more diversity in the focus across the three coastal areas for the infrastructure sector, encompassing variables such as health, economics, policy, and design. There was also a range of perceptions of adaptive capacity determinants between individual stakeholders within each sector that was reflected in the sensitivity analysis of the BBNs.

The key current and future drivers of change for the three coastal local government areas were also identified using a scenario planning approach. Table 12.2 provides examples of the significant drivers of change identified by stakeholders across different levels of uncertainty. The consideration of the different levels of uncertainty is important for two reasons. First, the dominant drivers of change (i.e., those with low uncertainty) provide the context within which decisions will need to be made and to which policies will need to adjust. Second, the highly uncertain drivers may shape the future direction of the region and its coastal settlements, even if the direction of this change is unknown. Based on a scenario planning approach, current and proposed climate change adaptation policies should consider both the certain and uncertain drivers of change, including the different futures that these drivers can lead to or shape (Schoemaker 1991).

The drivers of change identified in the workshops were classified into the following categories: (1) biophysical (climate change, natural hazards, and environmental considerations); (2) socioeconomic (economic, social, and demographic shifts); (3) governance/political; (4) built environment; and (5) technological/scientific (see Table 12.2). The two most prominent and recurring categories of drivers in all workshops and across all levels of uncertainty were the socioeconomic and biophysical categories, representing 42% and 28% of identified drivers, respectively. Although the workshops involved distinct groups of stakeholders, it is important to highlight that the relevance given to these categories was consistent across all workshops. This result may not be entirely surprising considering the ongoing growth and development of coastal areas across SEQ and recurrence of extreme weather events, including the recent drought that affected the region during the last decade and placed enormous stress on water supply (Queensland Water Commission 2010) as well as intense storms leading to extensive beach erosion and coastal flooding (Department of Climate Change and Energy Efficiency 2010).

The socioeconomic and biophysical drivers for the region will lead to future challenges that will need to be addressed in order to achieve effective adaptation. They also reveal some opposing influences and pressures that require management by decision-makers. For example, the pressure of accommodating SEQ's growing population, especially along the coastal zone, will need to be balanced against environmental considerations, including climate change impacts and resource constraints. Economic, social, and demographic shifts, such as an ageing population or a decline in volunteer

Table 12.2 Significant drivers of change identified and ranked into levels of uncertainty by stakeholders at the three coastal workshops

Categories/types of drivers		*Examples of specific drivers identified by stakeholders ranked by level of uncertainty*		
		Low uncertainty	*Medium uncertainty*	*High uncertainty*
Biophysical	Climate change and natural hazards	• Storm surge • Flooding	• Cyclones (sequence) • Extreme heat (heatwaves)	• Increase in intensity of natural disasters • Increased likelihood of bushfires
	Environmental (e.g., biophysical aspects, natural resources, environmental impacts, and degradation)	• Declining water availability • Deteriorating natural resources	• Water availability • Increasing biosecurity threats	• Loss of ecological and social resilience • Food/produce shortage—reliance on imported
Socioeconomic	Economic shifts (e.g., globalization, peak oil, recession, economic activity)	• Increased tourism • Greater levels of affluence • Increased energy costs	• Increased cost of food production • Mining industry crashes	• Oil vulnerability • International economic recession • Change in employment (different jobs)
	Social shifts (e.g., population growth, social resilience)	• Population growth • Declining voluntarism	• Decline in social resilience • Increased reliance on recovery needs	• Vulnerability of communities to climate change • Societal collapse
	Demographic shifts (e.g., mobility, demographic profile)	• Ageing population • Population change	• Migration	

Governance/ political	Changing policies Public consultation and engagement Changes in governance	• Government policy reform (tax, health) • Transport paradigm shift	• Decreased number of resources in departments • Loss of trust in community	• Failed state • New paradigm for community input • No state government level
Built environment	Urban form and density Infrastructure	• Increasing demand for infrastructure • Failing Infrastructure • Ageing assets	• New airport • Remaining service life of infrastructure	• Infrastructure keeping pace with increasing population • Housing choice preferences and expectations
Technological/ scientific	Ongoing technological development New technologies	• Advancement in telecommunications • Renewable energy production • Transport technology	• Investment in commercial R&D	• Floating technology (residential housing on water) • Energy transportation technologies—efficiency, alternatives

numbers, will also alter the region's vulnerability and adaptive capacity to natural hazards and climate change. In addition to biophysical and socio-economic drivers, stakeholders also identified drivers classified under the political and governance category. Although these drivers only accounted for 15% of all drivers identified across all levels of uncertainty, they were considered by the participants to be among the top three most influential drivers in two out of the three workshops. The perceived importance of political and governance drivers among workshop participants highlights the need to review the current decision-making processes to enable the successful design and implementation of adaptation options. The identification and subsequent enhanced understanding about how these drivers of change will influence coastal areas comprise an important first step for the development of future scenarios for the region (Moss et al. 2010; Schoemaker 1991). These scenarios will be developed in collaboration with stakeholders and be used as a *test bed* for both current and future climate change adaptation policies. Ultimately, they will provide insights for strategic management of coastal areas within the region, particularly focused toward longer-term time frames, which enable the incorporation of spatial and temporal uncertainties inherent to climate change.

In summary, the multimethod approach used in these case studies enabled the exploration of the key determinants of adaptive capacity as well as the key drivers of change for the region. This approach allowed the identification of adaptive capacity determinants for the region, which related to (1) infrastructure provision (e.g., coastal protection); (2) emergency response capacity (e.g., to major events such as storm surges and floods); and (3) the changing socioeconomic characteristics of the region (e.g., an ageing population). This approach also revealed the critical biophysical, socioeconomic, and political drivers of change that are likely to influence the choice of adaptation options in SEQ.

CHALLENGES AND OPPORTUNITIES FOR CLIMATE ADAPTATION WITHIN AUSTRALIAN COASTAL REGIONS

Adapting to climate change is one of the key challenges identified for coastal regions. Yet, it is a process that occurs in concert with a diverse range of other issues that affect the sustainability of these regions such as the over-extraction of natural resources, pollution, and social inequality. In particular, these regions already experience high rates of population growth, coastal development, and frequent extreme weather events that increase the potential for exposure to multiple stressors (Department of Climate Change and Energy Efficiency 2010; Hennessy et al. 2007). The range of potential stressors and consequent impacts identified for coastal regions demonstrates that adapting to climate change requires an understanding of the likely impacts of climate change and the range of stakeholders that may be affected. In particular, it will be important to be aware of the diversity of stakeholder perceptions in relation to climate change and to ensure the development of adaptation strategies that reflect a diverse range of needs, aspirations, and capacities within a region.

Acknowledging that stakeholder needs vary across spatial and temporal scales and changing socioecological contexts highlights the value of understanding current contexts as well as using scenarios to prepare for those of the future. The participatory and collaborative approaches presented in this chapter contribute to a richer understanding of coastal stakeholders as nonuniform and dynamic entities. By developing shared understandings of systems, the approach also engages stakeholders in social learning processes and provides a foundation for developing the adaptive capacity of participants and their respective sectors.

This work is part of broader ongoing research that is examining the SEQ region's vulnerability to climate change with the intent to develop practical and cost-effective adaptation strategies to assist decision-makers in state and local government, industry, and the community. Local government, in particular, will play an increasingly important role in preparing and assisting their communities to adapt to future climates. Their collaboration in this research, including the involvement of their key personnel in these workshops, should position them well to meet these increasingly important challenges.

POSTSCRIPT

Several journal articles have been published since the drafting of this chapter and relate to the findings of the research. See, in particular, Bussey et al. (2012), Richards et al. (2012), Roiko et al. (2012), Thomsen et al. (2012), Keys et al. (2013), McAllister et al. (2013), and Serrao-Neumann et al. (2013).

ACKNOWLEDGMENTS

This research is part of the SEQ Climate Adaptation Research Initiative, a partnership between the Queensland and Australian governments, the CSIRO Climate Adaptation National Research Flagship, Griffith University, University of the Sunshine Coast, and University of Queensland. The initiative aims to provide research knowledge to enable the region to adapt and prepare for the impacts of climate change.

REFERENCES

Australian Bureau of Statistics. (2001). *Population Projections by SLA (ASGC 2001), 2002–2022*, Canberra, Australia: Commonwealth Department of Health and Ageing.

Australian Bureau of Statistics. (2004). *Year Book Australia*, Canberra, Australia: Commonwealth of Australia.

Barnett, J. (2002). "Environmental change and human security in Pacific island countries," *Development Bulletin* 58, School of Anthropology, Geography and Environmental Studies, Melbourne, Victoria, Australia: University of Melbourne, pp. 28–32.

Burton, D., Mallon, K., Laurie, E., and Bowra, L. (2009). *Scoping Climate Change Risk for MBRC: A Climate Risk Report*, Moreton Bay, Queensland, Australia: Moreton Bay Regional Council.

Bussey, M., Carter, R.W., Keys, N., Carter, J., Mangoyana, R., Matthews, J., Nash, D. et al. (2012). "Framing adaptive capacity through a history–futures lens: Lessons from the South East Queensland Climate Adaptation Research Initiative," *Futures*, 44: 385–97.

Castelletti, A. and Soncini-Sessa, R. (2007). "Bayesian networks and participatory modelling in water resource management," *Environmental Modelling & Software*, 22 (8): 1075–88.

Catenacci, M. and Giupponi, C. (2010). *Potentials and Limits of Bayesian Networks to deal with Uncertainty in the Assessment of Climate Change Adaptation Policies*, 2010-007. Italy: FEEM Note di Lavoro, Fondazione Eni Enrico Mattei. http://www.feem.it/userfiles/attach/2010231030364NDL2010-007.pdf (accessed 14 November 2011).

Department of Climate Change and Energy Efficiency. (2010). *Climate Change Risks to Australia's Coast: A First Pass National Assessment*, Canberra, Australia: Commonwealth of Australia.

Department of Infrastructure and Planning. (2009). *South East Queensland Regional Plan 2009–2031*, Brisbane, Queensland, Australia: Department of Infrastructure and Planning.

Godet, M. (1994). *From Anticipation to Action: A Handbook of Strategic Prospecting*, United Nations: UNESCO Publishing.

Gold Coast City Council. (2009). *Climate Change Strategy 2009–2014: Setting Direction, Enabling Action*, Queensland, Australia: Gold Coast City Council.

Hennessy, K., Fitzharris, B., Bates, B.C., Harvey, N., Howden, S.M., Hughes, L., Salinger, J., and Warrick, R. (2007). "Australia and New Zealand. Climate change 2007: Impacts, adaptation and vulnerability," in M.L. Parry, O.F. Canziani, J.P. Palutikof, P.J. van der Linden, and C.E. Hanson (eds.), *Contribution of Working Group II to the Fourth Assessment Report of the Intergovernmental Panel on Climate Change*, Cambridge: Cambridge University Press. http://www.ipcc.ch/ (accessed August 2013).

Inayatullah, S. (2009). "Questioning scenarios," *Journal of Futures Studies*, 13(3): 75–80.

Keys, N., Bussey, M., Thomsen, D.C., Lynam, T., and Smith, T.F. (2013). "Building adaptive capacity in South East Queensland, Australia," *Regional Environmental Change*. DOI:10.1007/s10113-012-0394-2.

Lazarow, N.S., Smith, T.F., and Clarke, B. (2008). "Coasts," in D. Lindenmayer, S. Dovers, M.H. Olson, and S. Morton (eds.), *Ten Commitments: Reshaping the Lucky Country's Environment,* Canberra, Australia: CSIRO Publishing.

Low Choy, D.C. (2006). "Coastal NRM challenges: Meeting regional challenges through local government planning processes," in N. Lazarow, R. Souter, R. Fearon, and S. Dovers (eds.), *Coastal Management in Australia: Key Institutional and Governance Issues for Coastal Natural Resource Management and Planning.* Queensland, Australia: Cooperative Research Centre (CRC) for Coastal Zone, Estuary and Waterway Management, supported by The Australian National University and the National Sea Change Taskforce.

McAllister, R.R.J., Smith, T.F., Lovelock, C.E., Low Choy, D., Ash, A.J., and McDonald, J. (2013). "Adapting to climate change in South-East Queensland, Australia," *Regional Environmental Change*. DOI:10.1007/s10113-013-0505-8.

Millet, S.M. (2009). "Should probabilities be used with scenarios?," *Journal of Futures Studies*, 13(4): 61–8.

Minnery, J. (2001). "Inter-organisational approaches to regional growth management: A case study in South East Queensland," *The Town Planning Review*, 72: 25–44.

Mosadeghi, R., Tomlinson, R., Mirfenderesk, H., and Warnken, J. (2009). "Coastal management issues in Queensland and application of the multi-criteria decision making techniques," *Journal of Coastal Research*, SI56: 1252–6.

Moss, R.H., Edmonds, J.A., Hibbard, K.A., Manning, M.R., Rose, S.K., van Vuuren, D.P., Carter, T.R. et al. (2010). "The next generation of scenarios for climate change research and assessment," *Nature*, 463: 747–56.

Norman, B. (2009). "Principles for an intergovernmental agreement for coastal planning and climate chnage in Australia," *Habitat International*, 33(3): 293–9.

O'Connor, K.B., Stimson, R.J., and Taylor, S.P. (1998). "Convergence and divergence in the Australian space economy," *Australian Geographical Studies*, 36: 205–22.

Parry, M.L., Canziani, O.F., Palutikof, J.P., van der Linden, P.J., and Hanson, C.E. (eds.); IPCC (Intergovernmental Panel on Climate Change). (2007). *Climate Change 2007: Impacts, Adaptation and Vulnerability. Contribution of Working Group II to the Fourth Assessment Report of the Intergovernmental Panel on Climate Change*, Cambridge: Cambridge University Press. http://www.ipcc.ch/ (accessed August 2013).

Queensland Water Commission. (2010). *South East Queensland Water Strategy*, Brisbane, Queensland, Australia: The State of Queensland, Queensland Water Commission.

Richards, R., Sanó, M., Roiko, A., Carter, R.W., Bussey, M., Matthews, J., and Smith, T.F. (2012). "Bayesian belief modeling of climate change impacts for informing regional adaptation options," *Environmental Modelling and Software*, 44: 113–121.

Roiko, A., Mangoyana, R.B., McFallan, S., Carter, R.W., Oliver, J., and Smith, T.F. (2012). "Socio-economic trends and climate change adaptation: the case of South East Queensland," *Australasian Journal of Environmental Management*, 19(1): 35–50.

Schoemaker, P.J.H. (1991). "When and how to use scenario planning," *Journal of Forecasting*, 10: 549–64.

Schriefer, A. and Sales, M. (2006). "Creating strategic advantage with dynamic scenarios," *Strategy & Leadership*, 34(3): 31–42.

Serrao-Neumann, S., Crick, F., Harman, B., Sano, M., Sahin, O., van Staden, R., Schuch, G., Baum, S., and Low Choy, D. (2013). "Improving cross-sectoral climate change adaptation for coastal settlements: Insights from South East Queensland, Australia," *Regional Environmental Change*. DOI:10.1007/s10113-013-0442-6.

Smit, B. and Wandel, J. (2006). "Adaptation, adaptive capacity and vulnerability," *Global Environmental Change: Human and Policy Dimensions*, 16(3): 282–92.

Smith, T.F., Lynam, T., Preston, B.L., Matthews, J., Carter, R.W.B., Thomsen, D.C., Carter, J. et al. (2010). "Towards enhancing adaptive capacity for climate change response in South East Queensland," *The Australasian Journal of Disaster and Trauma Studies,* Palmerston North, New Zealand: Massey University. http://trauma.massey.ac.nz/ (accessed November 2011).

Smith, T.F., Myers, S., Thomsen, D.C., and Rosier, J. (2011). "Integrated coastal zone management and planning," in W. Gullet, C. Schofield, and J. Vince (eds.), *Marine Resources Management*, pp.109–21. Australia: LexisNexis Butterworths.

Smith, T.F., Preston, B., Brooke, C., Gorddard, R., Abbs, D., McInnes, K., Withycombe, G., Morrison, C., Beveridge, B., and Measham, T.G. (2009). "Managing coastal vulnerability: New solutions for local government," in E. Moksness, E. Dahl, and J. Støttrup (eds.), *Integrated Coastal Zone Management*. Oxford: Wiley-Blackwell.

Smith, T.F. and Thomsen, D.C. (2008). "Understanding vulnerabilities in transitional coastal communities," in L. Wallendorf, L. Ewing, C. Jones, and B. Jaffe (eds.), *Proceedings of Solutions to Coastal Disasters 2008 Conference*, pp.980–9, Turtle Bay, Oahu, Hawaii, April 13–16, 2008. American Society of Civil Engineers.

Sunshine Coast Regional Council. (2010). *Sunshine Coast Climate Change and Peak Oil Strategy 2010–2020: Our Place, Our Future*, Queensland, Australia: Sunshine Coast Regional Council.

Thomsen, D.C., Smith, T.F., and Keys, N. (2012). "Adaptation or manipulation? Unpacking climate change response strategies," *Ecology and Society*, 17(3): 20. http://www.ecologyandsociety.org/vol17/iss3/art20/ (accessed July 2013).

Tobey, J., Rubinoff, P., Robadue, J., Jr., Ricci, G., Volk, R., Furlow, J., and Anderson, G. (2010). "Practicing coastal adaptation to climate change: lessons from integrated coastal management," *Coastal Management*, 38: 317–35.

Tompkins, E.L. and Adger, W.N. (2004). "Does adaptive management of natural resources enhance resilience to climate change?," *Ecology and Society*, 9(2): 10. http://www.ecologyandsociety.org/vol9/iss2/art10/ (accessed March 2012).

Vincent, K. (2007). "Uncertainty in adaptive capacity and the importance of scale," *Global Environmental Change: Human and Policy Dimensions*, 17(1): 12–24.

Walker, P. (1988). "Problem solving with INFLUENCE," unpublished manuscript, Canberra, Australia: CSIRO Division of Wildlife and Ecology.

Willekens, M., Van Poucke, L., and Maes, F. (2010). *Exploratory Scenario Approach*, IMCORE (Innovative Management for Europe's Changing Coastal Resource) Project. Developing socio-economic scenarios. http://imcore.files.wordpress.com/2009/10/establishing-future-visions-becn-exploratory-scenario-approach.pdf (accessed November 2011).

Chapter 13

From coping to resilience

The role of managed retreat in highly developed coastal regions of New Zealand

Andy Reisinger, Judy Lawrence, Georgina Hart, and Ralph Chapman

Abstract: Sea-level rise is an inevitable consequence of a warming world and will continue long after global average temperatures may have been stabilized. Nonetheless, significant uncertainty remains about the rate and magnitude of this rise over the coming decades to centuries, in particular, the contribution from polar ice sheets. Most responses to sea-level rise consist of coping mechanisms, such as raising minimum floor levels applied to existing developments, or preventative measures, such as coastal hazard lines applied to greenfield developments. However, the effectiveness of such responses, if they are employed in a static way and not supported by additional policies that recognize the dynamic nature of coastal hazards, is expected to diminish as sea level continues to increase in the long term (beyond 2100) and may exceed process model-based projections even within the twenty-first century. Here, we aim to advance the discussion of managed retreat as an additional tool to promote the resilience of highly developed coastal regions and to increase the flexibility of local response options. We provide an overview of policies to support managed retreat that link with different socioeconomic contexts, community preferences, and timescales for implementation. We explore the potential implications of these alternative approaches for two case study sites in New Zealand and highlight the technical and institutional elements that would support the implementation of managed retreat in practice. We conclude that, given the risk from and uncertainties about sea-level rise, as well as the long time frames to implement managed retreat, further active development of policy tools and the information base required for managed retreat would contribute to the resilience of coastal communities.

INTRODUCTION

Sea-level rise is an inevitable consequence of a warming world and will continue for centuries even if global average temperatures are stabilized (Pachauri et al. 2007). Nonetheless, planning for rising sea levels is difficult because no practical upper bound for sea-level rise over the twenty-first century can be given at present, owing to the still limited understanding of the rate at which polar ice sheets will shrink in a warming climate (cf. Chapter 2). Decisions on how to manage climate-related coastal hazards, therefore, have to remain responsive not only to changing societal and environmental pressures but also to new information about future risks (Allison et al. 2009; Nicholls and Cazenave 2010; Pachauri et al. 2007; Richardson et al. 2009).

The long-term nature and uncertainty of sea-level rise projections have profound implications for the at least 600 million people globally living in coastal regions currently within 10 m above sea level (McGranahan et al. 2007). Sea-level rise at the upper end of or even exceeding current process model-based projections would place these rapidly increasing coastal populations at risk from erosion and episodic inundation. Impacts would be exacerbated with increases in storm surges and fluvial flooding near river mouths, salination of water supplies, rising water tables, and loss of ecosystems in the *coastal squeeze* between rising sea levels and immobile infrastructure (Adger et al. 2007; Nicholls and Cazenave 2010; Nicholls et al. 2007).

In New Zealand, about 65% of the population and major infrastructure are located within 5 km of the coast (Statistics New Zealand 2010). Greater affluence, increased value of coastal land, and permissive planning approaches with a focus on private property rights have supported ongoing coastal urban development. The Resource Management Act (1991) and the Local Government Act (2002) have provisions for coastal management, but they have not been effective in curbing either intensive or extensive ribbon development along the coast due to competing near-term economic and lifestyle interests (Freeman and Cheyne 2008; Healy 2003; Reisinger et al. 2011).

Even though *protect*, *accommodate*, and *retreat* are generally regarded as the principal options for coastal hazard management (Carter et al. 1994; Ministry for the Environment 2009), most responses employed to date in New Zealand are predicated on a static perspective, designed to *hold the line* by protecting against or accommodating *current* coastal hazards and/or based on *current* knowledge about future risks. As sea level will continue to rise in the long-term and could exceed process model-based projections even within the twenty-first century, such static responses could become increasingly ineffective or incur rising economic, social, and environmental costs.

This chapter focuses on identifying the conditions under which managed retreat could become a realistic response option to reduce vulnerability to sea-level rise, based on case studies from two coastal settlements in the Auckland region of the North Island of New Zealand. It outlines the potential challenges for the two case study locations and their special features and discusses the coastal management approaches employed to date and their potential limitations in dealing with long-term sea-level rise and the potential for more rapid rises. Elements that could allow managed retreat to become part of the toolbox of response options are discussed, along with the key factors that could support or hinder its implementation. It concludes with practical recommendations for steps to improve the long-term resilience of at-risk locations.

FUTURE CHALLENGES UNDER CLIMATE CHANGE

The most recent scientific assessment of climate change by the Intergovernmental Panel on Climate Change (IPCC) in 2007 provided a range of model-based projections of sea-level rise to the end of the twenty-first century, with increases ranging from 18 to 59 cm by the 2090s relative to 1990 (Meehl et al. 2007). However, the IPCC noted that dynamic processes related to ice flow could increase the vulnerability of the polar ice sheets to warming and increased sea-level rise and that such processes have been observed but are not included in current models and could result in sea-level rise substantially larger than IPCC model-based projections. Even an increase of 1.6 to 2 m by 2100 (relative to 1990) cannot be ruled out entirely (see, e.g., Allison et al. 2009; Nicholls and Cazenave 2010; Nicholls et al. 2011a).

The scientific literature is not yet able to assign any likelihood to such large sea-level rises and differing views about their plausibility remain (see, e.g., Lowe and Gregory 2010; Rahmstorf 2010). However, the inability to rule out large sea-level rise even by 2100 is a critical consideration in risk-based coastal management, as the long-term resilience of coastal settlements depends on their ability to respond flexibly and adapt their response strategies over time as sea level continues to rise for centuries into the future (Nicholls et al. 2011b).

We explore the implications of and response options to long-term and potentially accelerated sea-level rise in the Auckland region. Auckland is New Zealand's largest city and has more than 1.3 million people, about one-third of the country's population. Substantial parts of Auckland's coastline have been intensively developed. Population is projected to grow to about two million people by 2035, driving intensification of existing urban settlements and further greenfield development (Auckland Regional Council 2010).

Two contrasting case study locations in the Auckland region were chosen. Mission Bay and Kohimarama (Auckland City) are close to downtown Auckland, with high-value established residential settlements and iconic, high use recreational beaches. By contrast, Kawakawa Bay (Manukau City) is a small settlement on the rural outskirts of the Auckland region, with

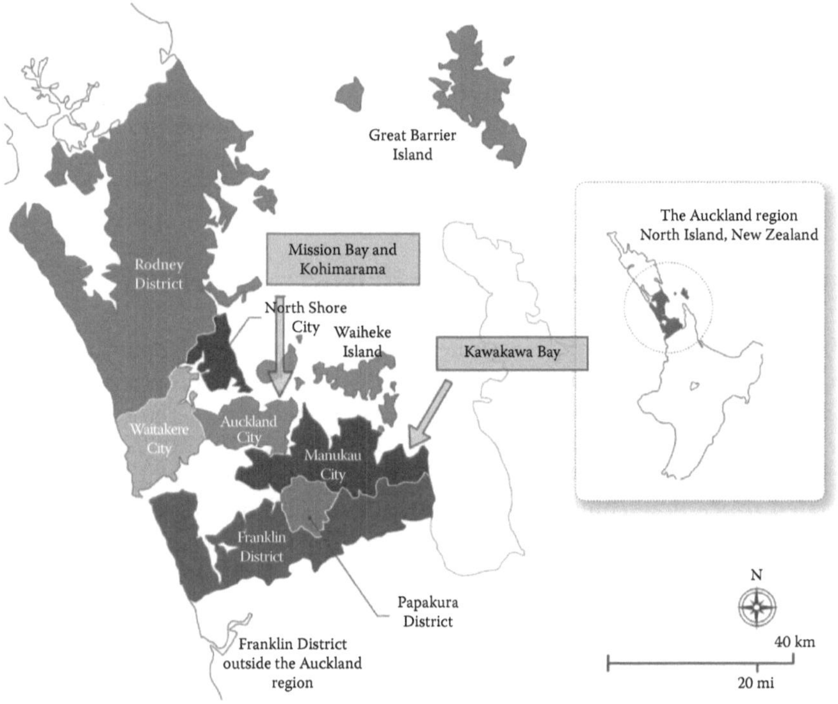

Figure 13.1 Map of the case study locations within the Auckland region and of Auckland region within the North Island of New Zealand. (Data from Auckland Regional Council.)

lower socioeconomic status and comparatively low recreational utilization, but with high amenity value to its community. The location of the two case studies is shown in Figure 13.1.

Both areas are low lying with a historical record of land-based flooding and coastal erosion. This prompted hard and soft coastal protection works at both locations along with requirements for minimum freeboard and flood risk area zoning. As a result of these measures, present-day risk from coastal inundation or erosion to private properties, coastal roads, and reserves at both locations is considered very low (Auckland City Council 2003, 2006; Manukau City Council 2007).

The challenges from potentially large future sea-level rise for the case study sites are illustrated in Figure 13.2a and b, which show the areas subject to storm-tide inundation with a 1% annual exceedance probability (AEP) for sea-level rises of 0.5, 1.0, 1.5, and 2.0 m (D. Ramsay, personal communication, based on methodology described in Ramsay et al. 2008a, 2008b). Flood hazard zones with 1% probability are identified in the Auckland Regional Policy Statement as zones where further

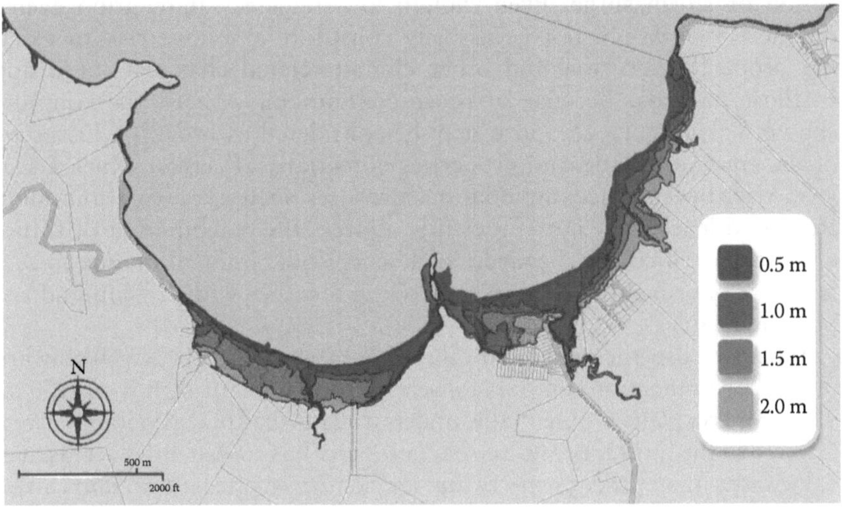

(a)

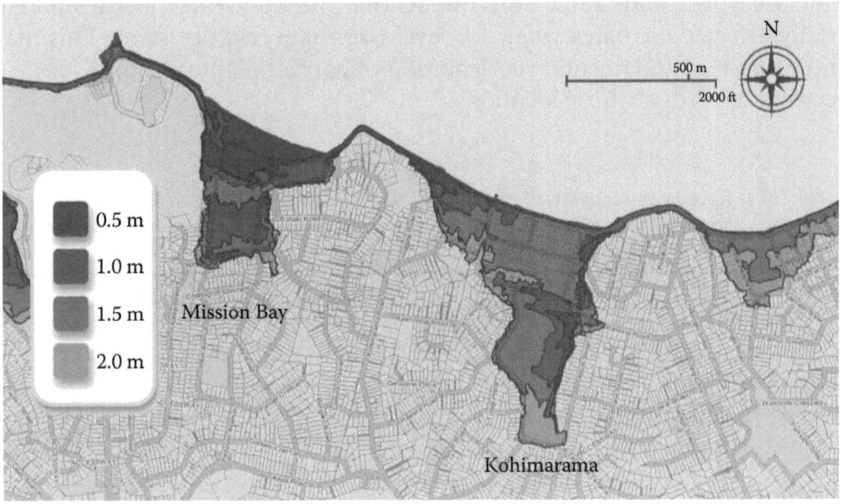

(b)

Figure 13.2 (a) Map of Kawakawa Bay, showing areas subject to storm-tide inundation with 1% annual probability (lower bound estimates), for sea-level rise of 0.5, 1.0, 1.5, and 2.0 m relative to present day levels. (b) Storm-tide inundation contours as in (a) for Mission Bay and Kohimarama.

development should be discouraged unless flood hazards can be fully mitigated without adversely affecting neighboring properties (Auckland Regional Council 1999).

The model results shown in Figure 13.2a and b represent static storm-tide inundation levels and do not account for erosion or changes in wave action or storm surge. Even though low-frequency inundation events on their own would not necessarily constitute a serious risk to existing properties, erosion and other climate-related changes would add to those hazards. Because of those confounding effects, the contours shown in Figure 13.2a and b may be considered as broadly illustrative of the amount of land and properties potentially affected by future sea-level rise, because accommodation measures such as raising minimum freeboard levels may not sufficiently address the combination of future climate-related coastal hazards (such as episodic inundation due to sea-level rise, storm surge, and river flooding together with erosion and rising water tables).

Table 13.1 lists the total rated value of properties (in 2010 NZD) seaward of 1% AEP inundation contours for sea-level rise of 0.5, 1.0, 1.5, and 2.0 m. The property values potentially underestimate the total economic assets affected in the future, because coastal property has consistently increased in value faster than other property and the consumer price index, and can be expected to continue to do so given projected population growth (Freeman and Cheyne 2008). Table 13.1 reveals that the value of properties potentially affected escalates when sea level rises in excess of 0.5 m. This may necessitate a fundamental reorientation of current planning approaches to coastal hazards at those locations.

Table 13.1 Rated capital value of property (2010 NZD 000s) potentially affected by sea-level rise of 0.5–2 m relative to average present-day levels, based on 1% AEP inundation contours

	Relative local sea-level rise			
	0.5 m	*1 m*	*1.5 m*	*2.0 m*
Kawakawa Bay				
1% AEP lower estimate	38,284	63,149	92,544	97,484
1% AEP upper estimate	50,374	78,649	96,704	112,978
Kohimarama				
1% AEP lower estimate	59,630	126,850	250,380	284,500
1% AEP upper estimate	84,870	188,240	269,880	293,770
Mission Bay				
1% AEP lower estimate	82,235	156,510	190,750	205,230
1% AEP upper estimate	107,960	175,545	201,850	207,035

Source: Auckland Regional Council.

Note: The upper and lower estimates represent upper and lower estimates of 1% AEP local relative sea levels, due to localized storm-tide changes and differences in baseline sea level along the coastline.

CURRENT APPROACHES TO MANAGING SEA-LEVEL RISE

Local government in New Zealand has the duty and responsibility to control land uses to avoid or mitigate coastal hazards, but many local authorities are yet to begin actively managing for the range of potential sea-level rise (Lawrence and Allan 2009).

Responses to coastal erosion and inundation in New Zealand in the past focused largely on hard protection works, but the negative impacts of hard protection on coastal ecosystems and sediment budgets have seen an increasing preference for soft protection over the past decade (Bell et al. 2001; Ministry for the Environment 2008a). As more information about potential sea-level rise has been generated, including through guidance from central government (Ministry for the Environment 2004, 2008a, 2009), councils have begun a new generation of regional and district plans that include provisions to avoid and mitigate the effects of future natural hazards to coastal property. These plans mostly aim to restrict development within coastal hazard zones or seaward of setback lines, or to accommodate increasing hazards through minimum ground and freeboard levels, floodable basements, and removable structures.

Combined with existing hard and soft protection measures, those responses are generally designed to *hold the line* or accommodate sea-level rise, mostly within the comparatively narrow range projected by models as assessed by IPCC (2007), and for time horizons of 50 or 100 years that the courts have deemed sufficient based on current knowledge (Ministry for the Environment 2008a). Due to the absence or cost of detailed modeling, councils are often forced to make limiting assumptions about changes in other drivers of coastal hazards such as potential changes in erosion rates, storm frequencies, and wave climate associated with climate change.

Both Auckland and Manukau City Councils, through their district plans, aim to limit risks from existing flooding and coastal hazards (based on 1% AEPs) through minimum building site and freeboard requirements. Hard protection measures exist at both case study sites in the form of seawalls, and additional beach nourishment is used in part to compensate for damage to and limited success of the seawalls and to retain beach amenities. As sea levels continue to rise, these options will face practical limits and growing economic disadvantages in the form of restricted local sand supply, costs of incremental reinforcement, rising water tables, increased need for stormwater and wastewater pumping, and growing exposure to catastrophic failure of hard protection measures.

Neither Auckland nor Manukau City district plans currently have explicit provisions for sea-level rise or other climate change effects applicable to the case study sites, although existing freeboard provisions include significant buffer above current sea levels. Both councils commissioned modeling of exposure to storm-tide inundation under sea-level rise scenarios (Ramsay et al. 2008a, 2008b); however, the implications of potential sea-level rise in excess of 1 m by 2100 were not in the studies' briefs. Neither study modeled or estimated changes in erosion risks nor storm surge associated with these sea-level changes, due to fundamental data and cost limitations. A separate technical report commissioned by Auckland Regional Council on erosion risks did consider the effects of climate change, but only in the form of sea-level rise of 0.5 m by 2100 (Auckland Regional Council 2009).

Given the amalgamation of the eight councils in the Auckland region into one large unified council at the end of 2010 (see Local Government [Auckland Council] Act 2009), neither Auckland nor Manukau City Councils changed their current district plans in response to this information. Auckland City Council noted that the future Auckland Council would consider the full range of response options available to communities (Auckland City Council 2010; Craig 2009). By contrast, Manukau City Council recommended increases in minimum freeboard provisions, which are reduced for buildings with an expected lifetime of less than 50 years (Hassan et al. 2008). Specific policy recommendations currently await the completion of the Unitary Plan for Auckland, but climate change has been identified as a risk in the (strategic, nonbinding) Auckland Plan.

The permanence of coastal settlements far beyond the lifetime of individual buildings, and the rapid increase in property values in areas potentially affected by future sea-level rise, raises significant questions about the sustainability of current approaches to managing coastal hazards. When viewed against a potential sea-level rise in excess of 1 m, or even more beyond 2100, a continued focus on static *hold the line* and constrained accommodation policies in the two case study locations is problematic. Any plausible static coastal setback line has the potential to be exceeded at some point as sea level continues to rise at highly uncertain rates and will, therefore, only postpone but not eliminate the need for protection or retreat policies for areas formerly regarded as *safe*.

Sea-level rise of 1 m or more would fundamentally change the nature of the case study locations and current management approaches. Even if existing properties could be retained through upscaled hard protection measures and they could be afforded by the communities, this would imply the complete eventual loss of the beach amenity (and its environmental) values that represent major assets for both those communities (see Figure 13.3) and the wider Auckland community.

At the same time, interim protection works and policies support the impression of property owners and developers that coastal hazards are being dealt with effectively and thus support the establishment and

Figure 13.3 Illustration of the environmental and social value of the public beach in front of high-value residential homes and main access road at Mission Bay, Auckland City, New Zealand, on a fine summer day (left) and during a major storm surge (right) on January 23, 2011 (estimated storm surge AEP 1.1%; R. Bell, personal communication). (Courtesy of G. Hart [left] and B. Eitelberg [right].)

further intensification of high-value coastal properties behind hazard zones. In turn, this strongly increases the pressure on councils to continue mitigating increasing coastal hazards through enhanced protection, locking in a development pattern that becomes less and less resilient. The resulting *coastal squeeze* raises the question of whether a more dynamic and flexible adaptive approach could be implemented that would allow those communities to better maintain the social and environmental values of their coastal environment, despite uncertainties about the rate of future sea-level rise and concurrent development pressures on the coast.

TOWARD ADAPTIVE MANAGEMENT: OPTIONS FOR MANAGED RETREAT

General considerations and options

Managed retreat generally implies a long-term, strategic decision to allow the shoreline to migrate inland in response to sea-level rise and attendant erosion, and proactive management of the removal of affected assets, rather than protecting the existing shoreline. This is intended to limit economic costs associated with ongoing and increasing protection, to reduce the risk of protection failures during storm events, to preserve important ecological habitats, and to maintain recreational spaces and visual amenity of the coast. Managed retreat could, therefore, enable an adaptive management approach that can reduce the otherwise inevitable long-term squeeze of the coastal environment as sea levels rise (Abel et al. 2011; Carter et al. 1994;

Department of Climate Change 2009; Feenstra et al. 1998; Neal et al. 2005; Titus and Neuman 2009; Titus et al. 2009).

Retreat, however, necessarily involves a cost and, if the retreat is to occur proactively before developed land is lost to the sea, it would require attenuation or extinction of existing use rights held by private property owners (Neal et al. 2005). Although this is anticipated in New Zealand's Resource Management Act (Berry and Vella 2010), the reality is that community support for managed retreat policies that affect existing use rights will depend on who bears those costs, how the loss of assets and existing use rights are managed, and who benefits from natural shoreline retreat. Therefore, the question of whether a community views retreat as an adaptation option or as a failure of adaptation to sea-level rise depends critically on details of the implementation of any retreat scheme, whether alternative land is available, and the underpinning value systems of communities and individuals within communities (Adger et al. 2009; Alexander et al. 2012).

The New Zealand Coastal Policy Statement (NZCPS) provides overarching national objectives and policies that direct and guide coastal management. The first coastal policy statement required coastal management to "preserve the natural character of the coastal environment from inappropriate use and development" (NZCPS 1994: 4). This has been interpreted mostly as a preference for soft, rather than hard protection measures, but not to limit protection per se, let alone a basis for retreat from increasing coastal hazards. A board of inquiry to review a revised coastal policy statement recommended more explicit emphasis on managed retreat as a response option and clearer national-level guidance on how to consider and manage competing interests in the coastal zone (Kenderdine et al. 2010). The revised coastal policy statement (NZCPS 2010) now identifies structural protection measures only as an option of last resort and highlights the need for transition mechanisms that enable a shift from current static and short-term coastal management responses to managing evolving risks from climate change, including by requiring a hazard risk assessment over at least 100 years.

The two locations selected for this case study offer an opportunity to develop and apply a typology of managed retreat policy *packages* that illustrate the various costs and negative consequences of retreat as well as benefits across the community and thus could respond to different preferences expressed by communities.

Table 13.2 qualitatively characterizes differing approaches suggested in the literature (Abel et al. 2011; Attwater et al. 2008; Few et al. 2007; Kapiti Coast District Council 2010; Kenderdine et al. 2010; Kwadijk et al. 2010; National Oceanic and Atmospheric Administration 2010; Neal et al. 2005; Turbott 2006). We highlight the distribution of the costs of retreat, the stages and timeframe over which retreat would be achieved, the certainty with which retreat would take place, the degree to which negative consequences for the

Table 13.2 Qualitative options for managed retreat, including their distributional effects, barriers, enabling factors, time frame, and limitations

Primary measure	*Implementation detail*	*Distributional effect: Who bears costs?*	*Key barriers and enabling factors*	*Time frame and certainty of retreat*	*Limitations*
Prohibit/limit protection	Council decision in-principle not to upgrade or maintain, or to remove, existing protection Prohibition of privately funded or maintained protection works	Dependent on implementation details and combination with other measures	*Barrier:* Community resistance and re-litigation as consequences start to bite *Enabler:* National-level guidance that ongoing protection is the exception, with decision-making criteria	Years to many decades Depends on timing and nature of sea-level rise and combination of other retreat policies	Does not achieve retreat in itself but allows the physical process to occur and thus triggers the need to move or abandon properties
Information on inundation and erosion risk	Preparation and publication of risk maps for range of future sea levels by councils Notification of future risk on Land/Project Information Memoranda (LIMs and PIMs) for range of future sea levels	*Councils (community)* for preparation of risk maps *Owners of at-risk properties* if properties lose value; but no clear evidence that risk disclosure reduces property values in near term even for present-day risks	*Barriers*: Legal challenges to technical details; modeling erosion difficult; high costs for councils with long coastline *Enabler:* Legislation to require preparation and disclosure of maps, including sea levels; central agency to fund or do risk assessments	First-pass modeling could be completed within a decade Many decades for complete abandonment of properties	Does not result in retreat per se but a precursor for retreat Resulting retreat not likely to be consistent across properties unless additional planning controls put in place
User pays for protection and insurance, based on actual risk	Removal of government-guaranteed insurance (through Earthquake Commission) for catastrophic coastal erosion and inundation	*Owners of at-risk properties* whose costs of protection increase, or who can no longer afford protection	*Barrier:* Long history of government-guaranteed insurance for flood and other natural hazards *Enabler:* Unwillingness of community to pay for protection of a few	Dependent on physical characteristics of coast as well as attitudes of property owners	User-pays approach removes community values from coastal management No recognition of ability to pay and its social implications

(*Continued*)

Table 13.2 (Continued) Qualitative options for managed retreat, including their distributional effects, barriers, enabling factors, time frame, and limitations

Primary measure	*Implementation detail*	*Distributional effect: Who bears costs?*	*Key barriers and enabling factors*	*Time frame and certainty of retreat*	*Limitations*
	Funding of protection works only through targeted rates levied on at-risk properties		properties; central government decision to change government-guaranteed insurance		Could still result in loss of beach amenity if property owners decide to meet protection costs
Change betterment limits to Earthquake Commission Act	Enable Earthquake Commission to make proactive payouts for flood proofing, relocation, and possible land purchase, rather than replacement of like with like	*Property owners* no longer able to reinstate their properties *Central government* liability could increase in short term	*Barrier:* Priority not given to long-term cost reduction *Enabler:* Legislation change to Earthquake Commission Act	Legislative change could be given effect quickly Would reduce exposure to risk over decades as damages occur	Reactive approach as only effective after buildings have been damaged; planning controls needed to shift from damage-driven to proactive approach
Regulations limiting further development or reinstatement	Define hazard zones where further development or reinstatement after storm damages is prohibited Progressively extinguish existing use rights for properties at risk (subject to additional criteria and trigger points such as erosion events)	*Property owners* whose development potential is limited *Property owners* who lose existing use rights either following damage or because trigger point has been reached	*Barrier:* Legal challenges to hazard zones; interpretation of hazard zones as static and thus defining a *safe* space for landward development *Enabler:* Central government guidance on hazard zone definition; funding for risk assessments; collaboration between regional and district councils	Hazard zones limit further development immediately Time-bound hazard zones (e.g., *area at risk by 2100*) need to be redrawn when sea-level rise projections are revised or planning horizons change	Hazard lines can encourage development behind hazard line Disproportionate negative effects on property owners with lower socioeconomic status

Relocation, removal, dismantling	Require relocation of buildings within property boundaries Require removal of buildings from coastal properties and transport to other blocks of land Dismantle buildings	*Property owners* face costs for removal and loss of equity in existing property, unless supported by other funds	*Barrier:* Removal from beach-front blocks of land implies loss of status *Enabler:* Clear trigger points for relocation/removal/dismantling-time frames and council plans; community buy-in	Near-term up to many decades depending on rate of inundation and erosion	Lack of alternative blocks of land to receive removed buildings in highly developed suburbs Depending on the proximity of the at-risk property to the coast, on-site relocation may only be effective in the short term
Rezoning	Change land use through re-zoning to public spaces and facilities, together with extinction of existing use rights following suitable triggers (such as end of lifetime of current owner)	*Property owners* whose ability to sell or extend their property is curtailed and limited to their own occupation or rental	*Barrier:* Could result in depressed prices for residential properties and change character of suburbs *Enabler:* Locations with mixed uses now may have greater potential for changing the mix; increase in public amenity values	Zoning changes could be signaled and implemented over time frames ranging from immediate to many decades	Additional measures would be needed to achieve complete retreat or transition to soft-buffer land uses, including mechanism for removal of buildings
Council purchase	Councils purchase at-risk properties (on voluntary or compulsory basis) and either dismantle or rent/lease back	*Community* bears primary cost if funded from rates or general taxes	*Barrier:* Difficulty of defining market value for at-risk properties; high costs for some locations; forced purchase difficult to enforce	Flexible; could state the timing of acquisition well into the future, or make it dependent on	May have highly disparate effects depending on social or cultural significance of properties to owners

(*Continued*)

Table 13.2 (Continued) Qualitative options for managed retreat, including their distributional effects, barriers, enabling factors, time frame, and limitations

Primary measure	*Implementation detail*	*Distributional effect: Who bears costs?*	*Key barriers and enabling factors*	*Time frame and certainty of retreat*	*Limitations*
	properties for fixed terms	*Property owners* who saw properties as intergenerational investment bear secondary cost *Community and property owners* if shared approach	*Enabler:* Wealthy community with strong community values; central government fund; legislation that facilitates forced purchases under certain conditions; tax incentives for property sales; community fund	trigger points like sea-level rise or dune movement/ erosion Can be proactive rather than reactive (i.e., don't need to wait for erosion to occur)	Risks encouraging further beach-front development since it implies certainty of future value Ostensible wealth transfer from community to select individuals if funded by community
Community-managed funds	Compulsory or voluntary contributions to funds to • Support property owners with future relocation or removal • Support councils in purchase of properties	*At-risk property* owners if contributions are voluntary; non contributing owners would need to be excluded from benefitting from the fund *Community*, if property owners not at risk contribute to fund	*Barriers:* Equitable dispersal of fund difficult; uncertainty of investment markets; conflicts with goals to limit rates increases *Enabler:* Coordinated funding mechanisms across councils or at regional level and/or managed by central government	Funds could seek contributions immediately, and/ or revise level of contributions over time Timing of fund pay out is flexible, but purpose and trigger points need to be decided well in advance	Willingness for voluntary contributions may be limited if the timing of fund drawdown is greater than average duration of property ownership

Note: It is assumed that in principle, management of coastal resources and hazards remains the responsibility of local government. Note that options are complementary; no single option will suffice on its own.

most vulnerable parts of society could be addressed, and the degree to which strong central government direction and/or support would be needed to achieve community-level outcomes.

Commonalities and differences between the case study locations

Interviews with council officers, coastal management experts, and representatives of community groups were conducted to compare and contrast the potential influence of socioeconomic conditions on the feasibility of various retreat policies highlighted in Table 13.2.

Some measures were identified as almost universally necessary, though not sufficient, to enable managed retreat regardless of the socioeconomic characteristics of the location. These measures include the decision to gradually remove, or at least not maintain or upgrade, hard and eventually soft protection measures and full disclosure of resulting future risks to properties for a wide range of potential sea levels. Linking such risk assessments to the extent of sea-level rise, rather than a specific planning time horizon, would reduce the need to revise risk assessments and response measures whenever new projections of future sea-level rise become available. However, councils and property owners would still need to monitor the magnitude and rate of observed and projected sea-level rise to turn such risk assessments into planning tools and to ensure that appropriate responses are implemented at agreed trigger points (see Table 13.2).

Respondents also agreed that any retreat strategy would need to engage affected communities very early to discuss the fundamental choices that communities will have to make and the consequences of those choices. Such an engagement would have to consider the full community-wide economic, social, environmental, and cultural implications of continued protection. However, current levels of skepticism about climate change and resulting long-term sea-level rise, together with varying risk preferences, could make a constructive public debate on such issues difficult and could result in forced and unplanned responses in the future. Respondents stressed that community engagement could only occur once a certain minimum understanding of the fundamental issues had been achieved, including a cost–benefit analysis (see Box 13.1) and risk assessment by the Auckland City Council.

Differences between the case study locations arise from their socioeconomic structure, level of development, and background geography. Lower income levels and socioeconomic status could make retreat a more likely long-term outcome for Kawakawa Bay, and lower density of development could allow limited removal of at-risk properties as long as spatial planning measures retain the necessary room to move. The lower value of properties

BOX 13.1 COST–BENEFIT ANALYSIS OF MANAGED RETREAT: ISSUES AND IMPLICATIONS

Adoption of a managed retreat strategy signals that in the interests of the wider community, property owners' land-use rights are limited in time. The net costs and benefits of such an approach consist of mostly monetary and direct costs to private property owners from loss of land, and mostly indirect and intangible benefits to the community from retaining a natural coastline, including amenity, environmental, and cultural values. In New Zealand, councils can (but are not legally required to) redistribute the costs associated with extinction of existing use rights by partly or fully compensating property owners for their loss (Berry and Vella 2010).

Proactive retreat implies that costs may arise earlier, and potentially much earlier, than the community benefits, which raises the question of appropriate discount rates to be applied in cost–benefit analyses. Because councils represent the ongoing community, they can in principle be expected to take a longer-term view, which would imply a lower discount rate than private or commercial coastal property owners may have. A lower discount rate would result in a greater preparedness to incur up-front investment cost for the sake of protecting coastal amenity and environmental values over a longer time frame.

Cost–benefit assessments need to value the benefits of a naturally evolving coastline and weigh them against the costs of losing the use of developed land. Alternative techniques for valuing intangible and nonmarket benefits are well established in the academic literature (Jansson et al. 1994; Vatn 2005) but can lead to widely differing results and require specialist engagement. This presents a significant barrier to smaller councils and a risk that intangible benefits are not considered at the same level as the direct monetary costs of retreat.

potentially affected by sea-level rise could make council, or wider community-funded, property purchases more feasible. One interviewee suggested, however, that the affected individuals in this more remote settlement would have relatively little influence, which might enable a council to extinguish existing use rights with only limited financial assistance.

By contrast, the high-density development in Mission Bay and Kohimarama makes neither on-site relocation nor removal of properties to other locations within the same suburb feasible, unless a major redevelopment cycle for the entire suburb (including shifting the main coastal road) was undertaken at some point in future. Due to the high value of properties at risk, and the generally high socioeconomic status of residents, council purchase of affected properties was seen as not feasible, nor was it regarded as politically acceptable for the wider community to recompense wealthy individuals for the loss of their private assets. At the same time, the greater political influence of residents in these suburbs would make it more difficult for the council to withhold protection and to extinguish existing use rights without financial assistance.

Community-managed strategic funds could address barriers arising from compensation issues in a location such as Mission Bay and Kohimarama. However, there are diverging views about obtaining community support, especially because those paying into the fund initially may not witness its eventual benefit many decades in the future. Concurrent pressure on the council to limit rates (i.e., local government tax) increases could make it politically difficult to create mandatory contributions to new and additional long-term funds, unless such funds were required by legislation to enable long-term structural changes in response to climate change.

There is precedent and experience in the Auckland region of managed retreat funded by a council for wider community benefit. Waitakere City Council, now part of the new Auckland Council, undertook the *Twin Streams* project to restore an ecological corridor and reduce flooding and pollution, which required purchase and removal of houses from an area affected by repeated river flooding (Trotman 2008). The project, including the purchase of houses, was funded from a NZD45 million capital grant from Infrastructure Auckland, a regional public fund for significant infrastructure services of community benefit. The success of the *Twin Streams* project was due largely to its collaborative nature and wider community benefits, but also the availability of regional funding to purchase 78 properties at market value.

National-level challenges, barriers, and enabling factors

A common challenge to managed retreat is that it requires strategic and consistent local authority decisions based on community support and a commitment to their implementation over a number of decades. Short-term changes in retreat policies could lead to higher costs and loss of option value (e.g., the option to retain recreational spaces) and provide less ability to balance costs and benefits between affected individuals and the wider community and future generations. Given the typically short planning cycles of local government in New Zealand,* central

* The longest statutory time horizon is 10 years for long-term plans under the Local Government Act and all plans under the Resource Management Act. Many asset management plans have longer time horizons.

government legislation, guidance, and support appear essential to avoid repeated litigation of council plans by individuals and local interest groups. Small councils in New Zealand struggle to undertake even basic inundation assessments, let alone more comprehensive risk* and cost–benefit assessments. This limits their ability to even initiate a strategic conversation with their communities about how to manage long-term development in the coastal zone. In such situations, additional funding and technical support would be necessary and could build on existing funding mechanisms.

Coastal hazard zones are increasingly used by councils to control land use in areas subject to coastal hazards. However, the use of coastal hazard zones to manage the risk of sea-level rise is a double-edged sword. Hazard notification is essential to signal the exposure of high-value assets in at-risk areas. If hazard zones are part of an overall policy approach that recognizes the *dynamic* nature of coastal hazards, and thus envisages the zones and associated land-use restrictions shifting over time, they can signal that human uses of coastal land may be time-limited, depending on the rate of sea-level rise. However, relying on hazard zones as the main risk management tool within an overall static approach to risk could encourage intensification in areas that are not yet affected by coastal hazards but may be in future. This would load an increasing burden on future generations as sea levels continue to rise, as structural change in those densely developed areas becomes more and more difficult.

The current arrangements for provision of disaster insurance through the Earthquake Commission Act (1993)† in New Zealand create some barriers to the consideration of changing risk. Its policy of replacing *like-with-like* following extreme storm events removes opportunities for managed retreat from hazards through removal or relocation and reduces incentives for councils to avoid and mitigate natural hazards before they occur. The limited time horizons employed in plans under the Resource Management Act and the Local Government Act further support a focus on near-term priorities consistent with interests of individuals rather than the long-term interests of the community (Ministry for the Environment 2008b). Different time frames and cultures under these legislative mandates, as well

* We use the term *risk* as indicating the product of both probability and consequence of an event (IPCC 2007). A full risk assessment has to consider not just options to protect against well-defined events but how to deal with a broad range of more or less likely, and more or less severe, extremes whose statistical frequency of occurrence and resulting damages will change over time.

† Private insurance for property in New Zealand enables property owners to draw on the government-guaranteed Earthquake Commission fund for storm or flood damage to residential land up to a dollar limit beyond which the insurer pays for damage incurred. Those with private property insurance are levied for this purpose.

as under the Building Act (2004), result in disjointed planning approaches and can have the effect of perpetuating risk to natural hazards, including sea-level rise (Lawrence and Allan 2009), and thus can result in maladaptation (Barnett and O'Neill 2010).

If regulations established under the Earthquake Commission Act were changed to include a limited range of *betterment* options, rather than reinstatement in prescribed circumstances, this would enable exposure to be reduced over time. However, because assistance under the act is only triggered after damage has occurred and only applies to damage from storm-related flooding and erosion and not to gradual changes, it would need to be complemented with planning approaches that proactively seek to reduce exposure. Given the strong push from reinsurance companies to take climate change seriously (e.g., Lloyds 2010), a proactive approach will be necessary to avoid insurance companies refusing to cover at-risk properties and those who cannot afford to pay ending up with stranded assets.

Existing use rights are seen by some as another barrier. They can be extinguished through regional plans prepared by regional councils, which must be given effect to by district councils under the Resource Management Act (RMA) (Berry and Vella 2010). To date, Canterbury is the only region to do so, where buildings damaged by coastal hazards are only allowed to be reconstructed if the land area has not eroded below a certain size; as a result, development will gradually be forced back from the coast as erosion occurs (Environment Canterbury 2005). A number of district councils have requirements for new buildings to be relocatable or removable in response to increasing coastal hazards, but these requirements tend to be controversial and have yet to be tested in practice and at the scale that could be required under future sea-level rise (Kapiti Coast District Council 2010; Kenderdine et al. 2010).

The merging of the Auckland councils into a single unitary council, including its development of a single spatial plan, provides an opportunity for consistency across the whole region through region-wide strategic assessments and the development as well as implementation of coherent methods and rules. This could include mechanisms for the extinction of existing use rights in specific cases and thus encourage more strategic risk management decisions.

Although our case study focused on private properties, we note that significant public assets are also potentially affected by long-term sea-level rise, in particular the major access road to and through Mission Bay and Kohimarama. Decisions to either maintain this road or shift it inland would require major strategic planning decisions and investment by Auckland Council and would have significant implications for the management of private assets located behind this road.

CONCLUSION

Dealing with the risks posed by long-term sea-level rise is only one facet of the challenges that climate change poses for local government in New Zealand. Near-term development interests, the need to ensure the long-term sustainability of communities and their often limited resources, and remaining skepticism about climate change can conspire against an effective response. Sequential risk management approaches based on bottom-up vulnerability assessments can help reduce, but do not eliminate such barriers, and the relative roles of community-based consultation processes and more prescriptive central government guidance and legislation continue to be debated (Lawrence et al. 2013; Reisinger et al. 2011).

Century scale but highly uncertain rates of sea-level rise presents particular challenges for long-term coastal management and proactive councils are beginning to engage with the economic, legal, and institutional issues this raises. Ongoing sea-level rise means that a strategy for managed retreat will become necessary in many locations sooner or later. Moreover, councils as the trustees of the community's long-term interests are obliged to adopt a long planning horizon. There can, however, be strong pressures against even considering managed retreat as a response option where individual coastal property owners may have short planning horizons and a strong preference for ongoing community-funded protection, or where governments at different levels may have competing objectives and priorities for coastal development (Abel et al. 2011).

Clearly, a decision to retreat from increasing coastal hazards cannot be taken lightly, as it could entail significant costs, disrupt communities, and severely affect individuals. The distributional issues to be faced by councils in considering options for managed retreat are real. They require councils to balance the rights of individuals and the collective interests of the wider community and to weigh direct monetary costs against less tangible social, environmental, and cultural values associated with a natural coastal environment.

Based on the views and information canvassed from participants in the case studies reported here, and analysis of the options, requirements, and barriers for managed retreat, we suggest that the following three measures are of highest priority to increase the long-term resilience of New Zealand communities to sea-level rise.

- Conduct across New Zealand a first-pass assessment of the risks and consequences of a broad range of potential sea-level rises up to at least 1.5 m above current levels and investigate in-principle options for and consequences of protection or accommodation up to those levels. The results of this first-pass assessment should be publicly available, including through detailed maps.

- Initiate conversations between local government and their communities about the long-term future of the coastal environment under various sea-level rise scenarios and about the implications of choosing pathways of continued protection/accommodation or gradual retreat as and when sea-level rise migrates the shoreline inland.
- Start a national-level conversation involving district and regional councils and central government to consider opportunities for aligning legislative mandates for the management of natural hazards that will increase over time due to climate change. This should include the Resource Management Act and its NZCPS, the Local Government Act, the Building Act, and the Earthquake Commission Act, but could also consider funding mechanisms to support long-term structural adjustment to changing natural hazards.

It will require far-sighted local government authorities to reach early community agreement on which community and private assets should be protected or *retreated*, what restrictions to place on investment in assets that cannot be protected or where further investment would jeopardize the resilience of the community over time, the stages of retreat, and the allocation of financial support.

Even where high values of private properties seem to make managed retreat unlikely, we argue that it would be foolhardy to dismiss retreat as a response option. Even if ongoing protection of those assets appears economically feasible due to growing wealth of communities, following such a course of action would make those locations more, rather than less, vulnerable to climate change while the social, environmental, and cultural values of public beaches and associated coastal spaces gradually become squeezed out of these communities.

ACKNOWLEDGMENTS

This research was funded by the New Zealand Foundation for Research, Science and Technology under contract VICX0805 "Community Vulnerability and Resilience." Inundation modeling was provided by Doug Ramsay and colleagues at the National Institute of Water and Atmospheric Research (NIWA) and GIS analysis was performed by Scott Schimmel (Victoria University of Wellington). The authors are grateful to Auckland Regional Council and Auckland and Manukau City Councils for their ongoing input and support, including GIS data for inundation modeling and to the planners and coastal management experts who shared their views and experiences through interviews conducted for this study. Additional comments from colleagues Rob Bell, Ann Magee, and Sylvia Allan as well as two anonymous reviewers helped improve drafts of this manuscript.

REFERENCES

Abel, N., Gorddard, R., Harman, B., Leitch, A., Langridge, J., Ryan, A., and Heyenga, S. (2011). "Sea level rise, coastal development and planned retreat: analytical framework, governance principles and an Australian case study," *Environmental Science & Policy*, 14(3): 279–88.

Adger, W.N., Agrawala, S., Mirza, M.M.Q., Conde, C., O'Brien, K.L., Pulhin, J., Pulwarty, R., Smit, B., and Takahashi, K. (2007). "Assessment of adaptation practices, options, constraints and capacity," in M.L. Parry, O.F. Canziani, J.P. Palutikof, C.E. Hanson, and P.J. van der Linden (eds.), *Climate Change 2007: Impacts, Adaptation and Vulnerability. Contribution of Working Group II to the Fourth Assessment Report of the Intergovernmental Panel on Climate Change*, Cambridge: Cambridge University Press. http://www.ipcc.ch/ (accessed August 2013).

Adger, W.N., Dessai, S., Goulden, M., Hulme, M., Lorenzoni, I., Nelson, D.R., Naess, L.O., Wolf, J., and Wreford, A. (2009). "Are there social limits to adaptation to climate change?," *Climatic Change*, 93(3): 335–54.

Alexander, K.S., Ryan, A., and Measham, T.G. (2012). "Managed retreat of coastal communities: Understanding responses to projected sea level rise," *Journal of Environmental Planning and Management*, 55(4): 409–33.

Allison, I., Bindoff, N.L., Bindschadler, R.A., Cox, P.M., de Noblet, N., England, M.H., Francis, J.E. et al. (2009). *The Copenhagen Diagnosis, 2009: Updating the World on the Latest Climate Science*, Sydney, New South Wales, Australia: The University of New South Wales Climate Change Research Centre.

Attwater, C., Carley, J., and Witte, E. (2008). "Establishing trigger points for adaptive response to climate change," Paper presented to the *IPWEA (Institute of Public Works Engineering Australia) National Conference on Climate Change Response: Responding to Sea-Level Rise*, Coffs Harbour, New South Wales, Australia, August 3–5.

Auckland City Council. (2003). *Kohimarama Beach Replenishment Project*, Report for the Auckland City Council prepared by Beca Carter Hollings & Ferner Ltd., Auckland, New Zealand: Auckland City Council.

Auckland City Council. (2006). *Kohimarama Beach Replenishment Project*, Report for the Auckland City Council prepared by Beca Carter Hollings & Ferner Ltd., Auckland, New Zealand: Auckland City Council.

Auckland City Council. (2010). *City Development Committee Meeting Minutes, 4 February 2010, Item 10: The Effects of Sea-Level Rise in Coastal Environments.* http://www.aucklandcity.govt.nz/ (accessed November 2011).

Auckland Regional Council. (1999). *Auckland Regional Policy Statement*, Auckland, New Zealand: Auckland Regional Council. http://www.arc.govt.nz/ (accessed November 2011).

Auckland Regional Council. (2009). *Regional Assessment of Areas Susceptible to Coastal Erosion, Technical Report No. 009, February 2009*, Auckland, New Zealand: Auckland Regional Council.

Auckland Regional Council. (2010). *State of the Auckland Region: Our Region, Our Future*, Auckland, New Zealand: Auckland Regional Council. http://www.arc.govt.nz/ (accessed November 2011).

Barnett, J. and O'Neill, S. (2010). "Maladaptation," *Global Environmental Change*, 202: 211–13.

Bell, R.G., Hume, T.M., and Hicks, D.M. (2001). *Planning for Climate Change Effects on Coastal Margins*, Wellington, New Zealand: Ministry for the Environment. http://www.mfe.govt.nz/publications/ (accessed November 2011).

Berry, S. and Vella, J. (2010). "Planning Controls and Property Rights—Striking the Balance," Paper presented to the Resource Management Law Association, Roadshow 2010. http://www.rmla.org.nz/ (accessed November 2011).

Building Act. (2004). Wellington, New Zealand: New Zealand House of Representatives. http://www.legislation.govt.nz/ (accessed December 2011).

Carter, T.R., Parry, M.L., Harasawa, H., and Nishioka, S. (1994). *IPCC Technical Guidelines for Assessing Climate Change Impacts and Adaptations*, London: University College London and Tsukuba, Japan: National Institute for Environmental Studies.

Craig, C. (2009). *The Effects of Sea-Level Rise in Coastal Environments*, report presented to Auckland City Development Committee, Auckland, New Zealand: Auckland City Council.

Department of Climate Change. (2009). *Climate Change Risks to Australia's Coast: A First-Pass National Assessment*, Canberra, Australia: Government of Australia.

Earthquake Commission Act. (1993). Wellington, New Zealand: New Zealand House of Representatives. http://www.legislation.govt.nz/ (accessed December 2011).

Environment Canterbury. (2005). *Regional Coastal Environment Plan for the Canterbury Region*, Christchurch, New Zealand: Environment Canterbury. http://ecan.govt.nz/publications/ (accessed November 2011).

Feenstra, J., Burton, I., Smith, J., and Tol, R.S.J. (eds.) (1998). *Handbook on Methods for Climate Change Impact Assessment and Adaptation Strategies*, Amsterdam, the Netherlands: UNEP and Vrije Universiteit Amsterdam.

Few, R., Brown, K., and Tompkins E.L. (2007). "Public participation and climate change adaptation: Avoiding the illusion of inclusion," *Climate Policy*, 7(1): 46–59.

Freeman, C. and Cheyne, C. (2008). "Coasts for Sale: Gentrification in New Zealand," *Planning Theory and Practice*, 9(1): 33–56.

Hassan, M., Qian, Z., and Ramsay, D. (2008). "Policy to manage impacts of climate change in Manukau City," Paper presented to the *IPWEA (Institute of Public Works Engineering Australia) National Conference on Climate Change Response: Responding to Sea-Level Rise*, Coffs Harbour, New South Wales, Australia, August 3–5.

Healy, T. (2003). "Urgent need to protect New Zealand's coastal landscapes," in P. Kench and T. Hume (eds.), *Proceedings of the Coasts and Ports Australasian Conference 2003*, Auckland, New Zealand, September 9–12.

Jansson, A., Hammer, M., Folke, C., and Costanza, R. (eds.) (1994). *Investing in Natural Capital. The Ecological Economics Approach to Sustainability*, Washington, DC: Island Press.

Kapiti Coast District Council. (2010). *Natural Hazards and Managed Retreat: Discussion Document to Support the District Plan Review 2010/2011*, Paraparaumu, New Zealand: Kapiti Coast District Council. http://www.kapiticoast.govt.nz/ (accessed November 2011).

Kenderdine, S., Edmonds, K., Woollaston, P., and Gage, R. (2010). *Proposed New Zealand Coastal Policy Statement. Board of Inquiry Report and*

Recommendations, Volumes 1 and 2. Wellington, New Zealand: Department of Conservation. http://www.doc.govt.nz/ (accessed November 2011).

Kwadijk, J.C.J., Haasnoot, M., Mulder, J.P.M., Hoogvliet, M.M.C., Jeuken, A.B.M., van der Krogt, R.A.A., van Oostrom, N.G.C. et al. (2010). "Using adaptation tipping points to prepare for climate change and sea-level rise: A case study in the Netherlands," *Wiley Interdisciplinary Reviews: Climate Change*, 1(5): 729–40.

Lawrence, J. and Allan, S. (2009). *A Strategic Framework and Practical Options for Integrating Flood Risk Management to Reduce Flood Risk and the Effects of Climate Change*, Wellington, New Zealand: Ministry for the Environment.

Lawrence, J., Sullivan, F., Lash, A., Ide, G., Cameron, C., and McGlinchey, L. (2013). "Adapting to changing climate risk by local government in New Zealand: Institutional practice barriers and enablers," *Local Environment*, 1–23. doi: 10.1080/13549839.2013.

Lloyds. (2010). *Coastal Communities and Climate Change—Maintaining Future Insurability*, London: Lloyds Reinsurance.

Local Government Act. (2002). Wellington, New Zealand: New Zealand House of Representatives. http://www.legislation.govt.nz/ (accessed November 2011).

Local Government (Auckland Council) Act. (2009). Wellington, New Zealand: New Zealand House of Representatives. http://www.legislation.govt.nz/ (accessed November 2011).

Lowe, J.A. and Gregory, J.M. (2010). "A sea of uncertainty," *Nature Reports Climate Change*, 1004: 42–3.

Manukau City Council. (2007). *Kawakawa Bay Shoreline Protection and Enhancement: Assessment of Environmental Effects*, Report prepared by Tonkin & Taylor Ltd. for the Manukau City Council, Manukau, New Zealand: Manukau City Council.

McGranahan, G., Balk, D., and Anderson, B. (2007). "The rising tide: Assessing the risks of climate change and human settlements in low elevation coastal zones," *Environment & Urbanization*, 19: 17–37.

Meehl, G.A., Stocker, T.F., Collins, W.D., Friedlingstein, P., Gaye, A.T., Gregory, J.M., Kitoh, A. et al. (2007). "Global climate projections," in S. Solomon, D. Qin, M. Manning, Z. Chen, M. Marquis, K.B. Averyt, M. Tignor, and H.L. Miller (eds.), *Climate Change 2007: The Physical Science Basis. Contribution of Working Group I to the Fourth Assessment Report of the Intergovernmental Panel on Climate Change*, Cambridge; New York: Cambridge University Press. http://www.ipcc.ch/ (accessed August 2013).

Ministry for the Environment. (2004). *Coastal Hazards and Climate Change: A Guidance Manual for Local Government in New Zealand*, A report prepared by D. Ramsay and R. Bell (NIWA) for the Ministry for the Environment, Wellington, New Zealand: Ministry for the Environment.

Ministry for the Environment. (2008a). *Coastal Hazards and Climate Change: A Guidance Manual for Local Government in New Zealand (2nd edition)*, A report revised by D. Ramsay and R. Bell (NIWA) for the Ministry for the Environment, Wellington, New Zealand: Ministry for the Environment.

Ministry for the Environment. (2008b). *Meeting the Challenges of Future Flooding in New Zealand*, Wellington, New Zealand: Ministry for the Environment.

Ministry for the Environment. (2009). *Preparing for Coastal Change: A Guide for Local Government in New Zealand*, Wellington, New Zealand: Ministry for the Environment.

National Oceanic and Atmospheric Administration. (2010). *Managed Retreat Strategies*, Australia: NOAA. http://coastalmanagement.noaa.gov/ (accessed September 2010).

Neal, W.J., Bush, D.M., and Pilkey, O.H. (2005). "Managed retreat," in M. Schwartz (ed.), *Encyclopaedia of Coastal Science*, the Netherlands: Springer.

New Zealand Coastal Policy Statement (NZCPS). (2010). Wellington, New Zealand: Department of Conservation. http://doc.org.nz/ (accessed November 2011).

Nicholls, R.J. and Cazenave, A. (2010). "Sea-level rise and its impact on coastal zones," *Science*, 328(5985): 1517–20.

Nicholls, R.J., Hanson, S.E., Lowe, J.A., Warrick, R.A., Lu, X., Long, A.J., and Carter, T.R. (2011b). *Constructing Sea-Level Scenarios for Impact and Adaptation Assessment of Coastal Areas: A Guidance Document*, Supporting material, Intergovernmental Panel on Climate Change Task Group on Data and Scenario Support for Impact and Climate Analysis (TGICA), the United Kingdom; Germany; the United States: IPCC Data Distribution Centre.

Nicholls, R.J., Marinova, N., Lowe, J.A., Brown, S., Vellinga, P., de Gusmão, D., Hinkel, J., and Tol, R.S.J. (2011a). "Sea-level rise and its possible impacts given a 'beyond 4°C world' in the twenty-first century," *Philosophical Transactions of the Royal Society A: Mathematical, Physical and Engineering Sciences*, 369(1934): 161–81.

Nicholls, R.J., Wong, P.P., Burkett, V.R., Codignotto, J.O., Hay, J.E., McLean, R.F., Ragoonaden, S., and Woodroffe, C.D. (2007). "Coastal systems and low-lying areas," in M.L. Parry, O.F. Canziani, J.P. Palutikof, C.E. Hanson, and P.J. van der Linden (eds.), *Climate Change 2007: Impacts, Adaptation and Vulnerability. Contribution of Working Group II to the Fourth Assessment Report of the Intergovernmental Panel on Climate Change*, Cambridge: Cambridge University Press. http://www.ipcc.ch/ (accessed August 2013).

NZCPS 1994. *New Zealand Coastal Policy Statement 1994*, Wellington, New Zealand: Department of Conservation.

NZCPS 2010. *New Zealand Coastal Policy Statement 2010*, Wellington, New Zealand: Department of Conservation.

Pachauri, R.K. and Reisinger, A. (eds.); Intergovernmental Panel on Climate Change (IPCC). (2007). *Climate Change 2007: Synthesis Report. Contribution of Working Groups I, II and III to the Fourth Assessment Report of the Intergovernmental Panel on Climate Change*, Geneva, Switzerland: IPCC. http://www.ipcc.ch/ (accessed August 2013).

Rahmstorf, S. (2010). "A new view on sea-level rise," *Nature Reports Climate Change*, 1004: 44–5.

Ramsay, D.L., Altenberger, A., Bell, R., Oldman, J., and Stephens, S.A. (2008a). *Review of Rainfall Intensity Curves and Sea Levels in Manukau City. Part 2: Sea Levels*, Client report by NIWA, Manukau, New Zealand: Manakau City Council.

Ramsay, D.L., Stephens, S.A., Oldman, J., and Altenberger, A. (2008b). *The Influence of Climate Change on Extreme Sea Levels around Auckland City*, Client report by NIWA. Auckland, New Zealand: Auckland City Council.

Reisinger, A., Wratt, D., Allan, S., and Larsen. H. (2011). "The role of local government in adapting to climate change: Lessons from New Zealand," in J.D. Ford and L. Berrang-Ford (eds.), *Climate Change Adaptation in Developed Nations*, the Netherlands: Springer.

Resource Management Act. (1991). Wellington, New Zealand: New Zealand House of Representatives. http://www.legislation.govt.nz/ (accessed November 2011).

Richardson, K., Steffen, W., Schellnhuber, H.J., Alcamo, J., Barker, T., Kammen, D.M., Leemans, R., Liverman, D., Munasinghe, M., Osman-Elasha, B., Stern, N., and Wæver, O. (2009). *Climate Change—Global Risks, Challenges & Decisions. Synthesis Report*, Copenhagen, Denmark: University of Copenhagen.

Statistics New Zealand. (2010). *Are New Zealanders Living Closer to the Coast? Internal Migration.* http://www.stats.govt.nz/ (accessed February 2010).

Titus, J.G., Anderson, K.E., Cahoon, D.R., Gesch, D.B., Gill, S.K., Gutierrez, B.T., Thieler, E.R., and Williams, S.J. (eds.), *Coastal Sensitivity to Sea-Level Rise: Focus on the Mid-Atlantic Region*, Washington, DC: U.S. Environmental Protection Agency.

Titus, J.G. and Craghan, M. (2009). "Shore protection and retreat," in J.G. Titus, K.E. Anderson, D.R. Cahoon, D.B. Gesch, S.K. Gill, B.T. Gutierrez, E.R. Thieler, and S.J. Williams (eds.), *Coastal Sensitivity to Sea-Level Rise: focus on the Mid-Atlantic Region*, Washington, DC: U.S. Environmental Protection Agency.

Titus, J.G. and Neuman, J.E. (2009). "Implications for decisions" in J.G. Titus, K.E. Anderson, D.R. Cahoon, D.B. Gesch, S.K. Gill, B.T. Gutierrez, E.R. Thieler, and S.J. Williams (eds.), *Coastal Sensitivity to Sea-level Rise: Focus on the Mid-Atlantic Region*, Washington, DC: U.S. Environmental Protection Agency.

Trotman, R. (2008). *Project Twin Streams to 2007: Waitakere City Council's Story*, Report for the Waitakere City Council, Auckland, New Zealand: Waitakere City Council. http://www.waitakere.govt.nz/ (accessed November 2011).

Turbott, C. (2006). *Managed Retreat from Coastal Hazards: Options for Implementation*, Technical Report 2006/48, Hamilton, New Zealand: Environment Waikato.

Vatn, A. (2005). *Institutions and the Environment*, Cheltenham: Edward Elgar.

Part V

Climate change and the coastal zone

Small islands

Chapter 14

A tale of two atoll nations

A comparison of risk, resilience, and adaptive response of Kiribati and the Maldives

Carmen Elrick-Barr, Bruce C. Glavovic, and Robert Kay

Abstract: It is widely recognized that small, low-lying island nations are highly vulnerable to the impacts of climate change. This chapter provides a comparative meta-analysis of climate change adaptation in two atoll nations: Kiribati in the Pacific and the Maldives in the Indian Ocean. The analysis focuses on how biophysical constraints coupled with political, cultural, and socioeconomic conditions shape adaptation pathways. Kiribati and the Maldives present, at first glance, a very similar set of issues and challenges. The two countries, however, are remarkably different in terms of their biophysical setting; political history; and cultural, social, and economic context and, consequently, their respective capacities to adapt to climate change. Despite these inherent differences, there appear to be common pathways or trajectories for adaptation. Nevertheless, important questions remain: Are current adaptation actions leading to more resilient or restrictive futures? What is the preferred adaptation pathway? How effective will the chosen pathway be for reducing vulnerability and building adaptive capacity to long-term climate change? Significant conceptual and practical challenges must be addressed in order to effectively answer these fundamental questions.

INTRODUCTION

Small island nations are critically vulnerable to the impacts of climate change (IPCC 2007). The atoll nations of Kiribati in the Pacific and the Maldives in the Indian Ocean are two such examples. They are constellations of dispersed low-lying islands that have limited access to freshwater (no rivers or streams), poor quality soil (in respect of agricultural productivity), limited natural resources, and livelihood options. Atoll communities are dependent on maintaining healthy coastal–marine ecosystems that are imperilled by climate change impacts. Retreat options within national borders are limited because these are mid-oceanic nations. Notwithstanding these commonalities, there are a number of inherent differences in biophysical, historical, political, cultural, and socioeconomic characteristics. It is proposed that these differences shape the selection of adaptation strategies, which in turn will influence future vulnerability to the impacts of climate change. Although this hypothesis may appear straightforward, to date there has been limited investigation of the drivers influencing the practical selection of adaptation options for low-lying atoll nations. We aim to contribute to this knowledge gap through a meta-analysis of the potential adaptation trajectories of these two island nations.

Island nations, particularly coral atolls, are renowned for their natural beauty and the vital and highly valued ecosystem services they provide. Coral ecosystems are hot spots for marine biodiversity, provide essential habitat for significant fisheries, protect coastlines from erosion and storm surges, are the basis for major tourist and recreational activities, provide food, and sustain the livelihoods of island communities. The beauty and tranquillity of atolls can mask, at least to outside observers, the dynamic and complex relationships between island livelihoods and the coastal–marine ecosystems upon which these communities depend. It is the interplay of these social-ecological features that shapes prevailing and prospective vulnerability, resilience, and adaptive capacity of atoll nations.

WHAT MAKES THIS PLACE SPECIAL?

Distinctive biophysical features

Kiribati consists of 33 atolls with a total land area of approximately 820 km^2 (Thomas 2003) (Figure 14.1). The Maldives has 1,190 islets grouped into 26 atolls (Global Environment Fund 2010) (Figure 14.2). Both nations are spread over vast ocean areas. The Maldivian islands are distributed north to south over 90,000 km^2, while the islands of Kiribati are spread over a massive 3.55 million km^2 of ocean. Biophysical conditions vary both within and between the island nations. Selected data comparing each nation are shown in Table 14.1.

These island nations are low lying, with over 80% of land in the Maldives less than 1 m above mean sea level (Ministry of Home Affairs, Housing and Environment 1999) and the islands of Kiribati approximately 3–4 m above mean sea level. The rate of sea-level rise, in both Kiribati and the Maldives, is consistent with the global average; however, there is significant seasonal and interannual variability (Church et al. 2001; Woodroffe and McLean 1992) (Figure 14.3). Coral atolls are by nature prone to change. They are created and shaped by cycles of deposition and erosion. The health of offshore coral systems is, therefore, of critical importance in ensuring a steady supply of sediment to the shoreline.

The Maldives is regularly exposed to windstorms, heavy rainfall, drought, sea swells, and storm surges. The southern atolls (including Malé, the capital)

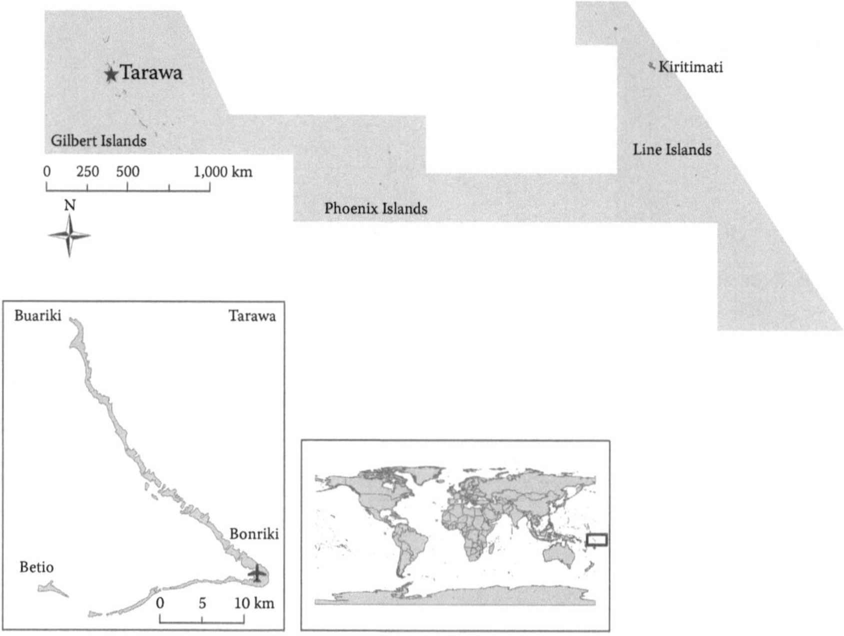

Figure 14.1 Kiribati location map.

are less exposed to cyclonic winds and storm surges than the northern atolls due to their proximity to the equator (UNDP 2006). The islands of Kiribati are not exposed to cyclonic activity; rather, climate is dominated by fluctuations in rainfall and other variables (including sea level) associated with the El Nino Southern Oscillation phenomenon. Rainfall is extremely variable, both spatially and temporally. Mean annual rainfall ranges from less than 1,000 mm in eastern Kiribati, to over 3,000 mm in western Kiribati (Porteous and Thompson 1996). A consequence of variable rainfall is that Kiribati is prone to drought (Thomson et al. 2008). In the absence of rivers and lakes, regular and reliable rainfall is essential to maintain supplies of fresh water.

Distinctive socioeconomic features

The Kiribati people (*I-Kiribati*) are a sea-faring people of Micronesian descent with a close affiliation to the ocean. The first European contact in Kiribati was in the form of whaling, slave trading, and the arrival of merchant vessels in the sixteenth century (History of Nations 2004). Kiribati was a British colony from 1916 to 1979 and today western-style governance arrangements are overlaid upon traditional property rights and governance structures. The effectiveness of these arrangements is dependant upon relationships established with traditional leaders through

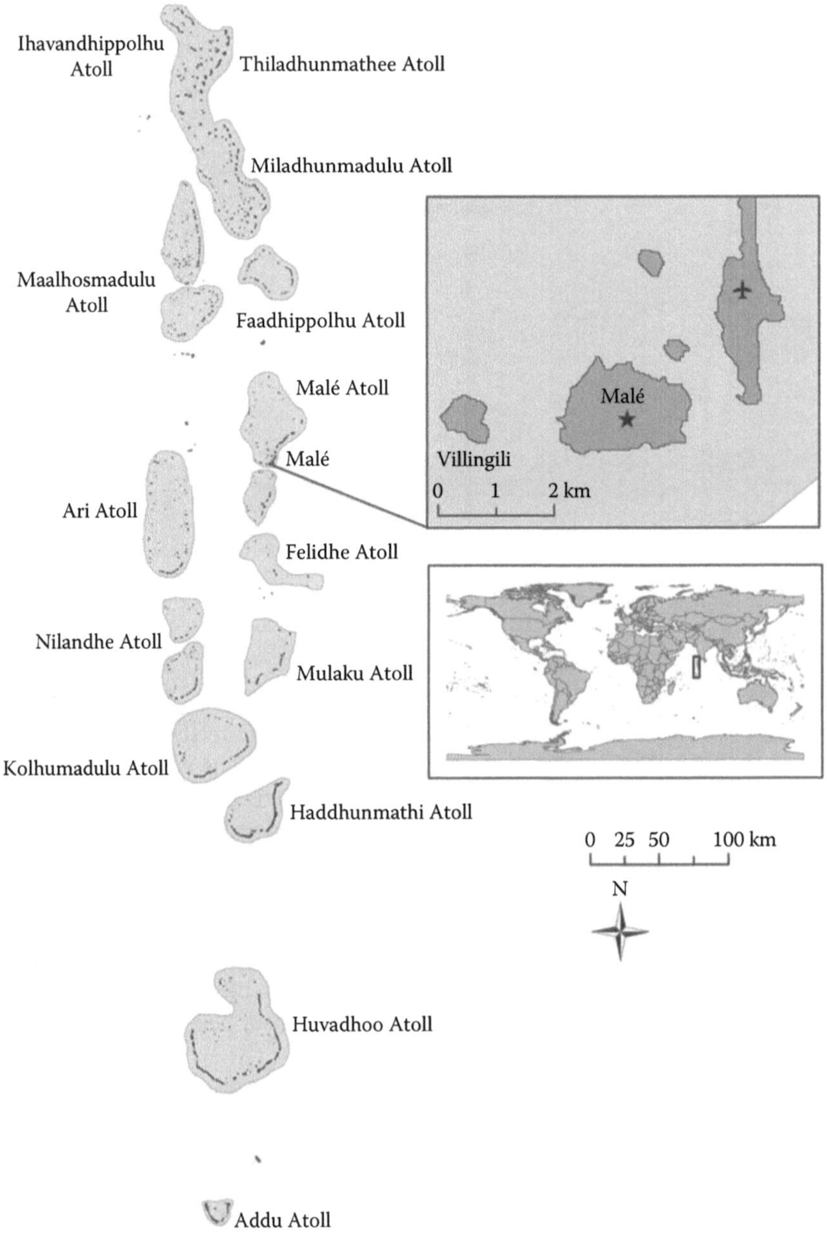

Figure 14.2 Maldives location map.

Table 14.1 Selected biophysical data, comparative assessment

	Kiribati	*Maldives*
Ratio of land to sea	0.02%	0.3%
Land area	726 km²	300 km²
Sea area	3.55 million km²	115,300 km²
Climate	Tropical	Tropical monsoon
Mean annual rainfall	2100 mm	2,124 mm
Average land height above mean sea level	4 m	1 m
Land movement (capital atoll)	Static (0.01 mm per year uplift)	Static

Source: Anderson, R.C., *Coral Reefs*, 17, 339–41,1998; AusAID, *Sea level and Climate: Their Present State*, Kiribati Pacific Country Report, June 2006, Australian Agency for International Development (AusAID), Canberra, Australia, 2006; Government of Kiribati, *First National Communication to the United Nations Framework Convention on Climate Change*, http://unfccc.int/, 1999; KMS, *Current and Future Climate of Kiribati*, Pacific Climate Change Science Program, International Climate Change Adaptation Initiative, Pacific Climate Change Science Program Partners 2011, http://www.cawcr.gov.au, 2011; Stevenson, S., "Land use planning in the Maldives: Creating sustainable and safe island communities," http://www.fao.org, n.d.; Thompson, C. et al., *High Intensity Rainfall and Drought*, A report prepared for the Kiribati Adaptation Programme Phase II: Information for Climate Risk Management, National Institute for Water and Atmospheric Research Ltd, 2008; United Nations Statistics Division (UNSTATS), "Environment statistics country snapshot: Kiribati," http://unstats.un.org, 2010; UNSTATS, "Environment statistics country snapshot: Maldives," http://unstats.un.org, 2010b.

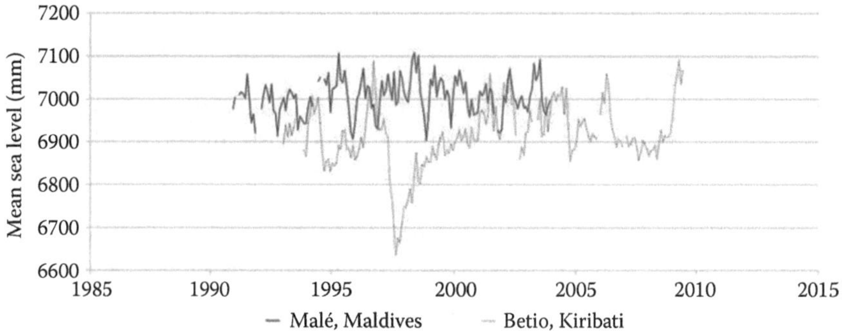

Figure 14.3 Trends in mean sea level: Malé, Maldives, and South Tarawa (Betio), Kiribati.

the *unimane* or *unimane* associations, particularly in the outer islands (University of the South Pacific 1985). *Unimane* associations are groups of traditional leaders created to consider and build consensus on local issues. Traditionally spiritual people, today 54% of the population is Roman Catholic, 39% Protestant, and 2% Baha'I (Lahmeyer 2004). Missionaries played a strong role in shaping religious beliefs, especially with the arrival of the Evangelical movement in 1857 (Kuruppu 2009); however, traditional beliefs (*Tabuenea*[*]) remain interwoven with the now established Christian religions. Adaptation to climate change is influenced by the role the church plays in contemporary society, with impacts that are both positive (e.g., through the delivery of free education and promotion of women's empowerment) and negative (e.g., expectations of financial contributions that can impoverish, cause family tensions, and restrict investment in adaptation) (Kuruppu 2009).

Religious beliefs and practices are markedly different in the Maldives. Islam was adopted in the Maldives in 1153 and today almost the entire population is Sunni Muslim (Lahmeyer 2004). Notwithstanding the geographic dispersal of islands, Maldivians are bound together by a common religion and language (Dhivehi); however, there is significant cultural heterogeneity and livelihood diversity as well as social, economic, and political inequality in the Maldives (Fulu 2007). The Maldives was a British protectorate from 1887 to 1965 but was governed as an independent Islamic sultanate from 1153 to 1968. With the abolition of the monarchy in 1968, Ibrahim Nasir assumed the presidency for 10 years, leaving government structures largely intact. His presidency ended after political and economic turmoil. In 1978, Maumoon Gayoom became president. His 30-year dictatorial reign ushered in a period of economic growth. Political engagement was, however, virtually impossible for all but a small elite, stifling the development of civil society. There was growing political opposition to the Gayoom regime resulting in violent protests in 2004 and 2005. This led to political reforms and the democratic election of Mohamed Nasheed in 2008. But the democratically elected government faced significant challenges in addressing the Gayoom legacy, with persistent collusion between the economic and political elite, and barriers to building a strong democratic culture and civil society. A constitutional crisis and political protests took place in 2010, and civil unrest flared up again in early 2012, underscoring the challenges being faced in the transition to democracy.[†]

Economic development in Kiribati and the Maldives aligns with the migration, remittances, aid, and bureaucracy (or government employment) model of economic development, which contrasts with economies based on commerce, industrial production, service industries, and trade. Both nations rely heavily on imports, external aid, and foreign remittances because of limited natural endowments (Thomas 2003). Economic prosperity is, however, significantly different between the two nations (see Figure 14.4 and Table 14.2). The Maldives was to be removed from the United Nations list of *least developed countries* before the tsunami struck on December 26, 2004, setting the country back by approximately two decades of investment and

[*] Belief in spirits and sorcery.

[†] For a more detailed description of the historical and cultural development of these nations, see Maloney (1980), Ridgell (1995), Romero-Frias (1999), and Thomas (2003).

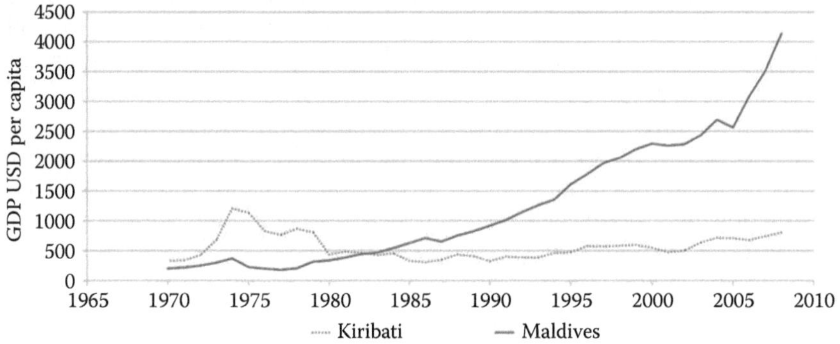

Figure 14.4 Historic gross domestic product (GDP) *per capita*, Kiribati and the Maldives.

Table 14.2 Selected socioeconomic data, comparative analysis

	Kiribati	*Maldives*
Population (2008)	96,558	309,575
Percentage of population in capital city	44%	33%
Population density (country)	123/km²	1,046/km²
Population density (capital)	2,558/km²	37,035/km²
Total GDP (USD) (billions)	0.09	1.48
GDP per capita (USD)	USD1,414	USD4,135
Percentage of GDP official development assistance	20%	4%
Percentage of GDP remittances	7%	0.25%
Dominant religion	Roman Catholic (55%) Protestant (36%)	Islam
Independence	July 12, 1979	July 26, 1965
Constitution	July 12, 1979	August 7, 2008
Primary sector (capital)	Government employment, copra, and fishing	Tourism, commercial fishing, port-related activities and shipping, government employment
Primary sector (outer islands)	Subsistence livelihoods	Tourism, subsistence fishing, and farming
Human development index	0.515 (1998)	0.77 (2007)

Source: Bureau of Public Affairs, *Republic of Maldives: Background Note*, http://www.state.gov, 2011; Kiribati National Statistics Office, *Census of Population and Housing 2005: Summary of Main Indicators*, http://www.spc.int, 2006; UNDP, "International human development indicators," http://hdrstats.undp.org, 2010; UNIDO, "Maldives fact sheet," http://www.unido.org, 2010a; UNIDO, "Kiribati fact sheet," http://www.unido.org, 2010b; WHO, *Maldives and Family Planning: An Overview*, The Department of Family and Community Health, World Health Organization, Regional Office for South-East Asia, New Delhi, India, n.d.

economic development. In contrast to Kiribati, the Maldives has a substantial tourism industry that operates through government leases of micro-atolls (Scheyvens 2011; Tourism Act of the Maldives 1999). Directly and indirectly, tourism contributes approximately 74% of gross domestic product (GDP) (World Bank 2005).

The capital atolls of Kiribati (Tarawa) and the Maldives (Malé) are, in many respects, different to their respective outer islands (Figure 14.5). The distinction may be termed *the center–periphery divide*, with population

Malé (Elrick-Barr 2011); Tarawa (Kay 2007)

Housing: Malé (Glavovic 2006); Tarawa (Kay 2009)

Main road Malé (Glavovic 2005); Tarawa (Elrick-Barr 2007)

Figure 14.5 Malé (capital of Maldives) and Tarawa (capital of Kiribati).

density in the capitals significantly higher than the outer islands. Changes in lifestyle, brought about by high rates of population growth, population distribution imbalances, and a move toward western-oriented materialism, are placing increasing demands on the environment and natural resources, particularly in the capital atolls of both island nations (Thomas 2003). In urban Tarawa, there are a high number of illegal dwellings and a strong push for government to increase housing availability (Pacific Region Infrastructure Facility 2011). Uncontrolled urbanization has been linked to environmental degradation as well as civil unrest, unemployment, crime, poverty, inadequate housing, and pressures on a range of public services from education to health (Connell 2011). Some commentators argue that such socioenvironmental problems pose a grave and immediate threat that needs to be tackled in parallel with efforts to address climate change impacts (Storey and Hunter 2010). The picture in the center (the capital atoll) is markedly different to that of the periphery (outer islands) in both cases. The inhabitants of the Kiribati and the Maldivian outer islands rely on a combination of subsistence fisheries and agriculture, with the Maldivian outer islands also receiving remittances from and employment in the tourism industry. The remoteness and inaccessibility of these outer islands poses a significant challenge for basic service delivery and imposes massive diseconomies of scale.

LOOKING FORWARD, WHAT MAKES THIS PLACE AT PARTICULAR RISK FROM FUTURE CLIMATE CHANGE AND OTHER POTENTIAL THREATS?

The risk and vulnerability landscape

Climate change poses an extreme risk to both Kiribati and the Maldives (Elrick and Kay 2009; Government of Kiribati 1999, 2010a; Ministry of Environment, Land and Agricultural Development 2007; Ministry of Home Affairs, Housing and Environment 2001). The coral atoll ecosystems of Kiribati and the Maldives are distinctive by virtue of their ecological significance, natural beauty, and as the foundation upon which communities depend for their livelihoods. These ecosystems and their dependent communities, and indeed the very survival of these nations, are, however, extremely vulnerable to climate change impacts (IPCC 2007).

With about three-quarters of the Maldives less than 1 m above mean sea level, even a marginal rise in sea level poses a grave threat. Low-lying small islands are also vulnerable to changing storm and wave climates. Coral ecosystems are sensitive to ocean acidification and temperature increases, and fragile rain-fed ecosystems are vulnerable to climate change-forced shifts in mean-annual rainfall and drought events. Existing physical vulnerabilities will be exacerbated by future climate changes, particularly as urbanization increases. To date, urbanization has resulted in a transition from traditional forms of housing and land tenure to developments that are location specific and capital intensive, particularly on the capital atolls.

The location of Kiribati and the Maldives close to the equator and outside the zone of direct tropical cyclone occurrence means that they are not currently exposed to significant storm surges. Despite this, storm events that generate only minor surges coinciding with high tides can inundate significant areas of land, simply due to the low elevation (Simpson et al. 2009). This can result in both inundation and saltwater intrusion in the freshwater lens (Elrick and Kay 2009). This physical vulnerability has been acknowledged not only through the practices of the local communities (where makeshift seawalls are increasingly used as a mechanism to protect private property), but also at the national level through various government initiatives.

In Kiribati, the role of environmental management in reducing physical exposure to current and future climate variability is recognized. However, the broader impacts of climate change and their effects on social vulnerability, including security, sovereignty, and economy, have received less attention to date. The 2004 Indian Ocean Tsunami brought the physical and social vulnerability of the Maldives into sharp relief. The tsunami had a devastating impact on coastal infrastructure, those whose livelihoods are dependent on fisheries and subsistence farming, and the overall economy, which is heavily dependent on tourism for foreign exchange. Tsunami damage was estimated to be greater than USD 470 million—equivalent to 62% of GDP. This triggered a concerted national and international response including the recent relocation of the population of an entire atoll (IFRCRCS 2010) and appears to have redefined perceptions about vulnerability and the significance of climate change.

On average, Maldivians have higher GDP per capita than those living in Kiribati and hence a degree of economic resilience. The new era of democratic politics in the Maldives opens up opportunities for citizens to have more say in future social choices. However, years of dictatorial rule have disempowered civil society and this, coupled with the center–periphery divide, means overreliance on the state. Building resilience and adaptive capacity will, therefore, be dependent on building social capital through more effective community engagement in planning for and deciding on climate change adaptation options. In Kiribati, traditional cultural practices remain intertwined with western governance arrangements. This coupled with challenges to development such as overpopulation, unclear land rights, and enforcement of multilayered and sectoral requirements of the planning system act to exacerbate vulnerability and increase barriers to adaptation (Ministry of Environment, Land and Agricultural Development 2007). Therefore, actions to increase resilience and adaptive capacity need to be aligned with and build upon cultural values and traditions if these interventions are to be effective (Kuruppu 2009; Kuruppu and Liverman 2011; MacKenzie 2004).

The outer islands of Kiribati have been managing and responding to environmental changes for centuries through traditional coping mechanisms. Although these areas receive limited financial assistance to support management activities, the islands deal with events such as storms and/or prolonged drought through surplus production, food storage, and preservation; planting resilient crops; and intercommunity cooperation (Barnett and Campbell 2010). These activities are critical in supporting the communities to adapt to changing environmental conditions. However, the extent to which these traditional approaches will prevail in the face of projected climate changes is poorly understood (Hay and Onario 2006). There is an entrenched cultural tradition to *safeguard* specialist traditional knowledge. For example, knowledge on navigation, weather forecasting, and building practices is closely guarded and rarely shared beyond the family group (MacKenzie 2004). This is an important constraint given the role that such knowledge can play in adaptation.

Some of the outer islands of the Maldives are experiencing population and development pressures similar to those experienced in the capital of Kiribati. Urbanization is a key driver of environmental change, as residents and the government seek to reclaim land and implement coastal defense measures to contain erosion. In contrast to Kiribati, the relative wealth of Maldives, and access to alternative sources of external donor funding, has enabled it to fund significant engineering interventions including large-scale coastal reclamations and protective works, particularly in the vicinity of Malé.

In summary, there is pronounced risk to climate change-induced impacts due to the physical vulnerability of these atoll nations. Moreover, this risk is superimposed on unsustainable practices that are already having severe impacts on the integrity of atoll ecosystems and the livelihoods of dependent communities. This physical vulnerability is compounded by social vulnerability that is shaped by geography, political history, cultural beliefs and practices, and socioeconomic circumstances. The wide dispersal of islands hampers provision of basic services and causes severe eonomic diseconomies, compounded by limited economic and livelihood opportunities, reliance on imports, and the concentration of critical infrastructure in a few physically vulnerable localities, notably on the capital atolls.

Adaptation pathways

The exposure and sensitivity of Kiribati and the Maldives to variations in climate is not new. The islands have been undertaking planned and autonomous adaptation for centuries, as evidenced by the survival of populations on what may be considered inhospitable environments exposed to droughts and other extreme events (Barnett and Campbell 2010). However, the magnitude of projected climate changes, including mean sea-level rise, presents a level of change outside historic ranges of natural variability. As a result, addressing the impacts of climate change has been a priority issue for Kiribati and the Maldives for more than a decade, commencing with ratification of the United Nations Framework Convention on Climate Change (UNFCCC) in 1992 (Maldives) and 1995 (Kiribati), followed by preparation of an Initial National Communication (INC), a requirement of UNFCCC signatories. The INC provides a framework for planned adaptation by assessing projected changes in climate and identifying key vulnerabilities (Table 14.3). The planned adaptation journey of each island nation, as aligned to UNFCCC reporting requirements and large donor-funded adaptation projects, is reviewed here (Figure 14.6).

The Maldives INC noted that hard defenses "may seem the only realistic option along well-developed coasts" to reduce vulnerability to storm surge, coastal erosion, and tsunamis, "where vital infrastructure and human settlement are at immediate risk" (Ministry of Home Affairs, Housing and Environment 2001: 80). Further, a *safer island strategy* was proposed as a mechanism to safeguard islanders from climatic and nonclimatic (tsunami) hazards by moving communities from high-risk areas to specially engineered *safe* islands (Bertaud 2002; Ministry of Home Affairs, Housing and Environment 2001). The larger islands would be designed to sustain communities through emergencies and disasters via provision of vertically elevated areas and stocks of water, food, provisions, and basic services in an emergency (e.g., telecommunications, health, and transport infrastructure). In contrast, the Kiribati INC was less prescriptive on proposed adaptation options and focused on activities that would enhance environmental management more broadly.

Table 14.3 Adaptation priorities, as noted in the INC

Kiribati adaptation priorities	*Maldivian adaptation priorities*
Integration of environmental considerations into economic planning	Land loss and beach erosion
Enact an environmental framework law and effectively implement	Infrastructure damage
Enhance awareness of climate change	Damage to coral reefs
Water management	Impacts on the economy—in particular the tourism sector
Development of a coastal management plan	Food security
Agricultural management	Water resources

Source: Government of Kiribati, *First National Communication to the United Nations Framework Convention on Climate Change*, http://unfccc.int/, 1999; MHAHE, *First National Communication of the Republic of Maldives to the United Nations Framework Convention on Climate Change*, Government of the Republic of Maldives, Malé, Maldives, 2001.

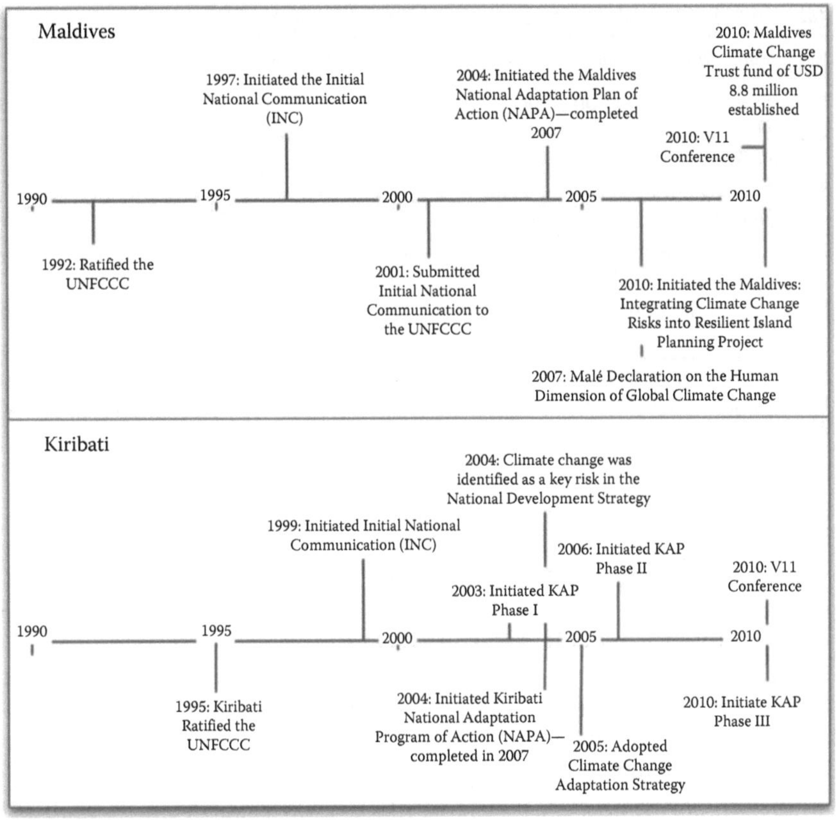

Figure 14.6 Climate change adaptation timeline, Maldives and Kiribati.

Planned adaptation continued in Kiribati and the Maldives guided through preparation of a National Adaptation Programme of Action (NAPA) under the UNFCCC (Table 14.4). NAPAs identify priority activities to respond to the *urgent and immediate adaptation needs*. Urgent and immediate needs are defined as those for which further delay would increase vulnerability and/or costs at a later stage (UNFCCC 2010).

The Maldivian NAPA presents a sustained preference for structural engineering solutions and a commitment to the safer island strategy (Ministry of Environment, Energy and Water 2007), which has since been criticized for not fully considering all natural hazards (Riyaz and Park 2009) and being overreliant on a *hard adaptation pathway*. The hard adaptation pathway is expensive, capital intensive, generates major social and environmental impacts, and is inflexible in the face of unanticipated changes (Sovacool 2011). Despite such criticisms, it has been proposed that alternative and/or

Table 14.4 Top five adaptation priorities, as noted in the NAPA

Kiribati adaptation priorities	*Maldivian adaptation priorities*
Water resource adaptation project	Build capacity for coastal protection, coastal zone management, and flood control
Coastal zone management and resilience enhancement for adaptation	Consolidate population and development
Strengthening environmental, climate change information and monitoring	Introduce new technologies to increase local food production
Project management institutional strengthening for NAPA	Acquire support for the speedy and efficient implementation of safer island strategy
Upgrading of meteorological services	Develop coastal protection for airports and development focus islands

Source: MEEW, *National Adaptation Program of Action*, Government of the Republic of Maldives, Malé, Maldives, 2007; MELAD, *National Adaptation Program of Action*, Government of Kiribati, Tarawa, Kiribati, 2007.

complementary strategies for dealing with long-term climate change could be incorporated within the strategy.

In April 2010, the Maldives initiated the project Integrating Climate Change Risks into Resilient Island Planning in the Maldives. The USD 9.3 million project originally titled Integrate Climate Change Risks into the Maldives Safer Island Development Programme reflects a transition to a more strategic approach to adaptation, driven by what has been termed an *evolution of government thinking* on the safer islands strategy (Global Environment Fund 2010). This evolution recognizes the merits of a *soft adaptation pathway*, which aims to strengthen the resilience of inhabited islands more generally through a range of adaptation measures that draw on natural *infrastructure* (e.g., dune rehabilitation), small-scale, low-impact technologies and practices, and seeks to empower local communities and build on local assets, and provides flexibility for decentralized decision-making to respond to change (Global Environment Fund 2010; Sovacool 2011). This pathway stands in contrast to the *hard adaptation pathway*, which relies mainly on the physical protection of a few *safe islands* based on artificial protective structures and infrastructure that generate significant environmental impacts, are complex and capital intensive, use processes and technologies that are imported, and are inflexible to surprise (Sovacool 2011). The evolution in adaptation thinking perhaps reflects a desire to take a holistic view of adaptation that considers a range of resilience building and vulnerability reduction measures, focussed on policy mainstreaming at the national level—as well as continuing to recognize the merits of structural engineered approaches under appropriate circumstances.

In Kiribati, the NAPA commenced in conjunction with the Kiribati Adaptation Project (KAP) (Figure 14.6). KAP was one of the first adaptation-focussed projects implemented by a multilateral development bank and targeted a number of climate change vulnerability issues, including water availability and coastal erosion. The need to transition from sole reliance on hard defenses to a suite of measures that build resilience and adaptive capacity was recognized and adopted in the KAP program. Although regulations to monitor implementation of hard defenses (such as seawalls) exist, these are often bypassed. Further, personnel responsible for granting approval have limited capacity to access a range of fit-for-purpose structural options, and there is little consideration of nonstructural options. The transition from a focus on hard structural adaptive measures to resilience and capacity building measures is similar to that of the Maldives, where a large externally funded climate change adaptation project proposed the transition from hard defenses to a broader-based multidimensional approach.

Climate change impacts could, however, render each nation uninhabitable because it is infeasible (due to cost, practicability, or other constraints) to build coastal engineering defenses around enough atolls or through surpassing of critical thresholds of protective structures (Ramesh 2008). Each nation will then have to resort to long-term retreat options through international migration. Consideration has been given to this option at the political level* (Government of Kiribati 2010b). Large-scale migration and complete atoll protection are not, however, within the scope of donor-supported adaptation assessment projects so they are essentially independent initiatives. This poses conceptual and practical challenges to harmonize such high-level publicly driven initiatives (which, perhaps correctly, reduce the long-term options for each country to either *build large seawalls around atolls or move to other countries*) and the technical adaptation analyses supported by external donors, which seek to address multiple adaptation options with perhaps shorter time frames (which also include significant components to address short-term development imperatives). These tensions of time-dependant prioritization of adaptation actions are ever present and are proving extremely difficult to conceptualize and develop practical adaptation actions that achieve the co-benefits of enhancing development priorities in the short term and reducing long-term climate change vulnerability (Persson et al. 2009). Addressing these difficulties will be dependent on building collaborative partnerships between government, international collaborators, and key actors in the private sector and civil society so that robust social choices can be made.

* Both countries are members of the "Vulnerable 11" (V11), a group of countries that have identified themselves as critically vulnerable to the impacts of climate change. The V11 has signed a declaration calling for enhanced international efforts to support vulnerable countries through the long-term challenges that climate change may pose (Government of Kiribati 2010a).

WHAT CAN BE LEARNT FROM PAST EXPERIENCE?

This overview of adaptation pathways, priorities, and projects provides a starting point to reflect on drivers behind the selection of particular adaptation strategies. Although it is appreciated that significant gaps remain in this meta-analysis, reflecting, among other things, the emergent status of vulnerability assessments and adaptation planning, some initial lessons can be drawn.

The adaptation trajectories of Kiribati and the Maldives can be viewed along a time frame continuum, commencing with inter-island migration and diversity in agricultural practice in response to historic *nonclimate change* environmental stresses. Subsequently, in conjunction with a movement toward western-orientated development, there has been a progression toward alternate adaptation practices to manage biophysical changes—primarily through the application of structural coastal defenses: the hard adaptation pathway. Interestingly, the application of structural defense differs markedly between and within both countries based mainly on density of infrastructure, land-use arrangements, and access to human and financial resources. In the outer islands of Kiribati, customary land tenure arrangements are predominant, enabling communities to relocate in response to coastal change. Structural defense systems are not widely applied, due in part to limited access to the resources needed for construction. Conversely, in Tarawa, there is a complex overlay of traditional and western-style land tenure systems, high population density, and high levels of development. Thus, structural options are sought to protect private property, government buildings, and fixed infrastructure such as sealed roads and utilities. In the Maldives, Malé is entirely protected by a donor-funded seawall. Although inhabited Maldivian outer islands have varying degrees of coastal protection, some that have previously been impacted by extreme events (such as the 2004 Tsunami) are entirely protected (ringed), but designed to a lower protection standard than Malé and Hulhule island (the airport atoll) (Kench et al. 2003). The suitability of the hard adaptation pathway for achieving sustainable and resilient development is increasingly questioned, often by the donor community or government rather than the local populace for whom alternate options for protecting private property are severely limited (indeed, seawalls can often be a stimulus for short-term economic development through the ability to reclaim land). Consequently, there has been a transition toward soft adaptation pathways that build local resilience, adaptive capacity, and prepare for a changing but uncertain future. Mainstreaming climate change into decision-making, enhancing partnerships, research, and capacity building are some of the resilience-building measures proposed and being adopted in Kiribati and the Maldives.

The transition from traditional coping mechanisms to structural defense and subsequently to resilience-building activities appears to follow a similar trajectory within and between both atoll nations, despite their different socioeconomic and biophysical characteristics. The primary difference is the stage along the adaptation trajectory that each atoll nation has reached, which appears to be driven by, among other factors, (1) exposure to extreme hazard events; (2) the transition from subsistence to western-oriented development, and tourism development in particular; (3) the influence of external financial resources and requirements for its expenditure; and (4) the nature of the interactions between government, the private sector, and communities across these atoll nations that shape how difficult social choices are made.

Important questions remain: Are current adaptation actions leading to more resilient or restrictive futures? What is the preferred adaptation pathway? How effective will the chosen pathway be for reducing vulnerability and building adaptive capacity to long-term climate change?

MOVING FORWARD

Are current adaptation actions leading to more resilient or restrictive futures? Adaptation funding appears to have been based on an underlying assumption that inputs (e.g., technical impact assessments, information gathering and sharing, and capacity building) will directly inform outputs (e.g., effective adaptive decisions). Although technical assessments are undertaken, linking the outcomes of these assessments to practical adaptation actions or even the key messages delivered by senior government officials is extremely difficult—especially given the magnitude of the adaptation challenge and the conflicts between short-term needs, development imperatives, and the extremely limited suite of long-term adaptation choices. Although the underlying drivers for this disconnect require further analysis, it would appear that they could be driven by political imperatives to have bold strategies to address the impacts of climate change (such "as *safer-island*)," and in so doing demonstrate *leadership* in addressing the issue, and/or an inadequate conceptual understanding of adaptation research, especially in what constitutes *effective* long-term adaptation in atoll contexts (Hedger et al. 2008).

What is the preferred adaptation pathway? The choice between hard and soft adaptation pathways is far from simple. Sovacool (2011) highlights three key considerations. First, these pathways present different risks and raise vexing questions about the choice of infrastructure (small-scale, low-impact, low-cost local technologies versus large-scale, capital-intensive works) and who should choose and lead the preferred adaptation trajectory? Second, are these pathways mutually exclusive or complementary? There is a range of potential incompatibilities between the two pathways, for example: the potential clash between imported complex technologies and local cultural beliefs and practices; the requisite institutional frameworks may not be able to coexist; the logistical, human, and financial resources for the hard path may limit options for the soft path; and the hard path may not be physically compatible with retaining the natural and social infrastructure that is key to the soft adaptation pathway. There may be complementary possibilities such as engineered protective works around highly populated and intensively developed locales, such as the capital atolls, while simultaneously relying on intact natural processes to attenuate the impacts of hazards in more remote and rural islands. Adaptation interventions could be sequenced to experiment with options; adapt and adjust to changing circumstances, for example, trial soft options at a small scale; and supplement this with hard options if appropriate. In other words, the preferred adaptation trajectory might be one that maximizes the synergistic benefits of using both hard and soft options in appropriate locales at appropriate scales and times. Third, these two pathways generate different opportunities, costs, and benefits for different stakeholders. The choice of pathway or some mix of the two, therefore, needs to be made taking into account the differential impacts on diverse stakeholders and vulnerable groups in particular.

How effective will the chosen pathway be for reducing vulnerability and building adaptive capacity to long-term climate change? If projected climate change impacts are experienced, as is becoming increasingly clear given ongoing greenhouse gas emission rates, low-lying islands are likely to be under severe stress and the viability of both soft and hard adaptation pathways is questionable. Under such dire circumstances, the whole population may have to migrate to other, less vulnerable countries. In the short term, there is a range of adaptation options that can be (and are being) implemented, which will increase the resilience to the impacts of climate change. What remains unclear is how Kiribati and the Maldives will transition from resilience-building adaptation options in the short term to an uncertain longer-term future and if there is a more *effective* approach to transition than is currently being undertaken.

It is clear that short-term and urgent development and environmental priorities in each country need to be mainstreamed into climate change adaptation strategies, and *vice versa*. Traditionally, adaptation options in the coastal zone have included retreat, accommodate, or protect. In atoll contexts, where the accommodate option is untenable, the question is raised as to whether the current frameworks for coastal adaptation that frame impact assessment and adaptation planning are appropriate. It is our view that atoll-specific frameworks, both conceptual and practical, are urgently required. Such frameworks need to address, in an integrated manner, immediate priorities and longer-term climate change risks. Vexing questions are raised by the challenges facing atoll nations in this era of global change. Answers to such questions need to be explored and creative solutions developed to reduce vulnerability and build resilience and adaptive capacity. The outcomes have profound consequences for the sovereignty, security, and sustainable development of atoll nations and regions that include small island states (Locke 2009; Ware 2005; Weir and Virani 2010).

CONCLUSION

Kiribati and the Maldives present, at first glance, a very similar set of issues and challenges with respect to climate change adaptation. They are both atoll nations comprising small islands that are extremely low lying and dispersed across vast ocean areas. They are both extremely vulnerable to climate change. But they are remarkably different in terms of their biophysical, historical, political, and socioeconomic characteristics and, consequently, their respective capacities to adapt to climate change. Despite these inherent differences, there appear to be synergies in the choice of adaptation pathways in the short, medium, and longer terms. The question remains, how effective are the actions being taken and will the trajectory of adaptation lead to more resilient or restrictive futures? This comparative analysis demonstrates that studies of this type, especially if undertaken through a shared analytical framework, would provide a valuable tool in exploring commonalities and differences between apparently similar nations. Such information may help to improve the design of adaptive responses both at a national level and within the context of international support.

ACKNOWLEDGMENTS

We thank Luke Dalton for assistance with background data collection and the creation of maps. We also thank Mick Kelly for his important editorial and review comments.

REFERENCES

Anderson, R.C. (1998). "Submarine topography of Maldivian atolls suggests a sea level of 130 metres below present at the past glacial maximum," *Coral Reefs*, 17: 339–41.

Asian Development Bank. (2009). *Maldives Fact Sheet*. http://www.adb.org (accessed August 2010).

AusAID. (2006). *Sea Level and Climate: Their Present State*, Kiribati Pacific Country Report, June 2006. Canberra, Australia: Australian Agency for International Development (AusAID).

Barnett, J. and Campbell, J. (2010). *Climate Change and Small Island States: Power, Knowledge and the South Pacific*, London: Earthscan.
Bertaud, A. (2002). "A rare case of land scarcity: The issue of urban land in the Maldives." http://www.alain-bertaud.com (accessed October 2010).
Bureau of Public Affairs. (2011). *Republic of Maldives: Background Note.* http://www.state.gov (accessed October 2011).
Caribsave. (2010). *The CARIBSAVE Partnership: Protecting and Enhancing the Livelihoods, Environments and Economies of the Caribbean Basin.* http://caribsave.org/ (accessed January 2011).
Church, J.A., Gregory, J.M., Huybrechts, P., Kuhn, M., Lambeck, K., Nhuan, M.T., Qin, D., and Woodworth, P.L. (2001). "Changes in sea level," in J.T. Houghton, Y. Ding, D.J. Griggs, M. Noguer, P.J. Van der Linden, X. Dai, K. Maskell, and C.A. Johnson (eds.), *Climate Change 2001: The Scientific Basis: Contribution of Working Group I to the Third Assessment Report of the Intergovernmental Panel on Climate Change*, Cambridge; New York: Cambridge University Press. http://www.ipcc.ch (accessed August 2013).
Connell, J. (2011). "Elephants in the Pacific? Pacific urbanisation and discontents," *Asia Pacific Viewpoint*, 52(2): 121–35.
Elrick, C. and Kay, R. (2009). *Mainstreaming of an Integrated Climate Change Adaptation Based Risk Diagnosis and Response Process into Government of Kiribati: Final Report*, Report prepared for the KAP Project, Phase II, Perth, Australian: Coastal Zone Management Pty Ltd.
Fulu, E. (2007). "Gender, vulnerability, and the experts: Responding to the Maldives tsunami," *Development and Change*, 38(5): 843–64.
Global Environment Fund. (2010). "Integrating climate change risks into resilient island planning in the Maldives," A project document request for CEO Endorsement/Approval from The Least Developed Countries Fund For Climate Change (LDCF), Global Environment Fund. http://www.thegef.org/gef/project_detail?projID=3847.
Government of Kiribati. (1999). *First National Communication to the United Nations Framework Convention on Climate Change.* http://unfccc.int/ (accessed October 2010).
Government of Kiribati. (2010a). "Tarawa climate change conference issues ambo Declaration," *Climate Change in Kiribati: The Climate Change Portal of the Office of the President of Kiribati*, Tarawa, Kiribati: Office of the President. http://www.climate.gov.ki (accessed January 2010).
Government of Kiribati. (2010b). "Kiribati climate change strategies," *Climate Change in Kiribati: The Climate Change Portal of the Office of the President of Kiribati*, Tarawa, Kiribati: Office of the President. http://www.climate.gov.ki (accessed January 2010).
Graphic Maps. (2010). *World Atlas: Countries of the World*, Graphic Maps. http://www.worldatlas.com (accessed October 2010).
Hay, J.E. and Onorio, K. (2006). *Kiribati Country Environmental Analysis: Incorporating Environmental Considerations into Economic and Development Planning Processes*, Philippines: Asian Development Bank.
Hedger, M., Mitchell, T., Leavy, J., Greeley, M., and Downie, A. (2008). *Desk Review: Evaluation of Adaptation to Climate Change from a Development*

Perspective, Institute of Development Studies (IDS). http://www.esdevaluation.org (accessed February 2009).

History of Nations. (2004). *History of Kiribati*. http://www.historyofnations.net/ (accessed January 2011).

IFRCRCS. (2010). *Disaster Management Tsunami Option in Maldives: Dhuvaafaru the Rebirth of a Community*. http://www.ifrc.org (accessed January 2011).

IPCC. (2007). "Summary for policymakers," in *Climate Change 2007: The Physical Science Basis. Contribution of Working Group I to the Fourth Assessment Report*, S. Solomon, D. Qin, M. Manning, Z. Chen, M. Marquis, K.B. Averyt, M. Tignor, and H.L. Miller (eds.), Cambridge; New York: Cambridge University Press. http://www.ipcc.ch/ (accessed August 2013).

Kench, P., Owen, S., Resture, A., Ford, M., Trevor, D., Fowler, S., Langrine, J. et al. (2010). *Improving Understanding of Local-Scale Vulnerability in Atoll Island Countries: Developing Capacity to Improve In-Country Approaches and Research*, Asia-Pacific Network for Global Change Research. https://www.apn-gcr.org (accessed April 2012).

Kench, P., Parnell, K., and Brander, R. (2003). "A process based assessment of engineered structures on reef islands of the Maldives," Paper presented at the *Coasts and Ports Australasian Conference 2003*. http://eprints.jcu.edu.au (accessed April 2012).

Kiribati Meteorology Service (KMS). (2011). *Current and Future Climate of Kiribati*, Pacific Climate Change Science Program, International Climate Change Adaptation Initiative, Pacific Climate Change Science Program Partners 2011. http://www.cawcr.gov.au (accessed April 2012).

Kiribati National Statistics Office. (2006). *Census of Population and Housing 2005: Summary of Main Indicators*. http://www.spc.int (accessed October 2010).

Kuruppu, N. (2009). "Adapting water resources to climate change in Kiribati: The importance of cultural values and meanings," *Environmental Science & Policy*, 12: 799–809.

Kuruppu, N. and Liverman, D. (2011). "Mental preparation for climate adaptation: The role of cognition and culture in enhancing adaptive capacity of water management in Kiribati," *Global Environmental Change*, 21: 657–69.

Lahmeyer, J. (2004). *Kiribati General Country Data*. http://www.populstat.info (accessed January 2011).

Locke, J.T. (2009). "Climate change-induced migration in the Pacific region: Sudden crisis and long-term developments," *The Geographic Journal*, 175(3): 171–80.

MacKenzie, U.N. (2004). *Social Assessment Final Report*, Report prepared for the Kiribati Adaptation Project.

Maldives Culture. (n.d.). "Maldives culture." www.maldivesculture.com (accessed January 2011).

Maloney, C. (1980). *People of the Maldive Islands*, New Delhi, India: Orient Longman.

Ministry of Environment, Energy and Water (MEEW). (2007). *National Adaptation Program of Action*, Malé, Maldives: Government of the Republic of Maldives.

Ministry of Environment, Land and Agricultural Development (MELAD). (2007). *National Adaptation Program of Action*, Tarawa, Kiribati: Government of Kiribati.

Ministry of Home Affairs, Housing and Environment (MHAHE). (1999). *Second National Environmental Action Plan*, Malé, Maldives: Government of the Republic of Maldives.

Ministry of Home Affairs, Housing and Environment (MHAHE). (2001). *First National Communication of the Republic of Maldives to the United Nations Framework Convention on Climate Change*, Malé, Maldives: Government of the Republic of Maldives.

Pacific Region Infrastructure Facility. (2011). *Kiribati Sustainable Towns Programme: Phase 2: 2010–2013*. http://www.theprif.org (accessed December 2011).

Persson, A., Klien, R., Siebert, C.K., Atteridge, A., Muller, B., Hoffmaister, J., Lazarus, M., and Takama, T. (2009). *Adaptation Finance under a Copenhagen Agreed Outcome*, Research report, Stockholm, Sweden: Stockholm Environment Institute.

Porteous, A.S. and Thompson, C.S. (1996). *The Climate and Weather of the Rawaki and Northern Line Islands of Eastern Kiribati*, Wellington, New Zealand: NIWA Science and Technology Series No. 42.

Ramesh, R. (2008). "Paradise almost lost: Maldives seek to buy a new homeland," *The Guardian*, November 10. http://www.guardian.co.uk (accessed January 2011).

Ridgell, R. (1995). *Pacific Nations and Territories: The Islands of Micronesia, Melanesia, and Polynesia*, Honolulu, HI: Bess Press.

Riyaz, M. and Park, K. (2009). "'Safer island concept' developed after the 2004 Indian Ocean tsunami: A case study of the Maldives," *Journal of Earthquake and Tsunami*, 4(2): 135–43.

Romero-Frias, X. (1999). *The Maldive Islanders: A Study of the Popular Culture of an Ancient Ocean Kingdom*, Barcelona, Spain: Nova Ethnographia Indica.

Scheyvens, R. (2011). "The challenge of sustainable tourism development in the Maldives: Understanding the social and political dimensions of sustainability," *Asia Pacific Viewpoint*, 52(2): 148–64.

Simpson, M.C., Scott, D., New, M., Sim, R., Smith, D., Harrison, M., Eakin, C.M. et al. (2009). *An Overview of Modelling Climate Change Impacts in the Caribbean Region with Contribution from the Pacific Islands*, Barbados, West Indies: United Nations Development Programme (UNDP).

Sovacool, B.K. (2011). "Hard and soft paths for climate change adaptation," *Climate Policy*, 11: 1177–83.

Stevenson, S. (n.d.). "Land use planning in the Maldives: Creating sustainable and safe island communities." http://www.fao.org_(accessed October 2010).

Storey, S. and Hunter, S. (2010). "Kiribati: An environmental 'perfect storm'," *Australian Geographer*, 41(2): 167–81.

Thomas, F.R. (2003). "Kiribati: Some aspects of human ecology, 40 years later," *Atoll Research Bulletin No 501*, Washington, DC: National Museum of Natural History, Smithsonian Institution.

Thompson, C., Mullan, B., Burgess, S., and Ramsay R. (2008). *High Intensity Rainfall and Drought*, A report prepared for the Kiribati Adaptation Programme Phase II: Information for Climate Risk Management, National Institute for Water and Atmospheric Research Ltd.

Tourism Act of the Maldives. (1999). http://www.tourism.gov.mv/downloads/Tourism_Act(Law2-99).pdf (accessed January 2011).

UNDP. (2006). *Developing a Disaster Risk Profile for Maldives: Volume 1*, A report prepared by RMSI, India, for the United Nations Development Program. Malé, Maldives: UNDP.

UNDP. (2010). "International human development indicators." http://hdrstats.undp.org (accessed October 2010).

UNFCCC. (2010). *National Adaptation Programmes of Action.* http://unfccc.int (accessed January 2011).

UNIDO. (2010a). "Maldives fact sheet." http://www.unido.org (accessed October 10, 2010).

UNIDO. (2010b). "Kiribati fact sheet." http://www.unido.org (accessed October 2010).

United Nations Statistics Division (UNSTATS). (2010a). "Environment statistics country snapshot: Kiribati." http://unstats.un.org (accessed October 2010).

United Nations Statistics Division (UNSTATS). (2010b). "Environment statistics country snapshot: Maldives." http://unstats.un.org (accessed October 2010).

University of the South Pacific. (1985). "Kiribati: A changing atoll culture" cited in, Resture, J. (2008). *Origins and Culture.* http://www.janeresture.com (accessed January 2011).

Ware, H. (2005). "Migration and conflict in the Pacific," *Journal of Peace Research*, 42(4): 435–54.

Webb, A.P. and Kench, P.S. (2010). "The dynamic response of reef islands to sea-level rise: Evidence from multi-decadal analysis of island change in the Central Pacific," *Global and Planetary Change*, 72: 234–46.

Wier, T. and Virani, Z. (2010). *Three Linked Risks for Development in the Pacific Islands: Climate Change, Natural Disasters and Conflict*, Pacific Centre for Environment and Sustainable Development Occasional Paper No. 2010/3.

World Bank. (2010). "Maldives economic update: Economic policy and poverty team South Asia region," April 2010. http://siteresources.worldbank.org (accessed October 2011).

World Health Organisation. (n.d.). *Maldives and Family Planning: An Overview*, The Department of Family and Community Health, World Health Organization, Regional Office for South-East Asia, New Delhi, India.

Woodroffe, C.C. and McLean, R.F. (1992). *Vulnerability to Accelerated Sea-Level Rise: A Preliminary Study*, Canberra, Australia: Department of Arts, Sports, Environment and Territories, Government of Australia.

Chapter 15

Planning for coastal change in Caribbean small islands

Gillian Cambers and Sharon Roberts-Hodge

Abstract: Viewed from economic, social, and environmental perspectives, the Caribbean islands are among the world's most vulnerable small island regions. Dependent to a large extent on the tourism industry, with high population densities and numerous environmental and development issues, they face significant challenges, which are likely to be exacerbated with climate change. In particular, highly valued coastal areas are facing significant threats from erosion and pollution as well as development-related challenges. This chapter discusses how planning for coastal change has evolved in the Caribbean islands, and in Anguilla in particular, over the past 20 years. The development and implementation of new coastal planning guidelines, which to some extent take into account future climate change, are discussed. These guidelines focus on changing present practices such as by building a *safe* distance from the beach and using wood and piled structures in beachfront buildings. Future opportunities for mainstreaming climate change into the government agenda include incorporating wise practices into the planning process; using visual images of local phenomena, such as beach erosion, to illustrate long-term, *abstract* phenomena such as climate change and variability; involving the private sector, especially the insurance industry in development planning; maximizing windows of opportunity such as those that open up after extreme events; and emphasizing the economic impacts of poor planning practices.

IDENTITY OF PLACE IN THE CARIBBEAN ISLANDS

Caribbean islands are among the most beautiful and most accessible tropical regions in the world. With their relatively easy access from major European, North and South American travel markets, they have increasingly become a destination of choice for many visitors over the past four decades. As a result, tourism is now a major driver of the economy in most Caribbean islands. For example, over the 20-year period 1988–2007, the contribution of travel and tourism to the gross domestic product in the insular Caribbean increased by more than three times, from USD 11.36 billion to USD 38.3 billion (World Travel and Tourism Council 2010).

The natural resource base of the Caribbean islands is critical for the region's tourism development. The industry is primarily located close to the coast and is heavily dependent on the tropical climate and the presence of sandy beaches and scenic coastal areas with clean, clear seas free from pollution and abundant in marine life. Besides tourism infrastructure, most of the islands' major cities, towns, and villages, as well as the transport network, are located near the coast. Despite the economic importance of the natural resource base, anthropogenic degradation of coastal and marine resources has been identified as a serious problem in the Caribbean over recent decades, for example, Cartagena Convention for the Protection and Development of the Marine Environment in the Wider Caribbean Region adopted in 1983, entered into force in 1986 (Richards and Bohnsack 1990; UNEP 2010). Activities such as beach and dune sand mining, removal of mangroves, destruction of seagrass beds and coral reefs, and overfishing are still serious issues in almost every Caribbean island.

The insular Caribbean represents the ring of islands enclosing the Caribbean Sea and related bodies of water and includes Trinidad and Tobago, the Lesser Antilles, The Bahamas, Virgin Islands, Greater Antilles, as well as those islands east of Central America and north of South America. The islands are immensely varied, ranging from low sandy cays to high volcanic islands and with land areas of less than 1 km^2 to more than 100,000 km^2. Offshore bathymetry ranges from deep ocean trenches to barrier reefs.

Anguilla, the Caribbean island featured in this chapter, is a British overseas territory and is one of the most northerly of the Leeward Islands. It lies east of Puerto Rico and the Virgin Islands and is located directly north of St. Martin (Figure 15.1). It consists of the main island, Anguilla, approximately 26 km long and 5 km wide, together with a number of much smaller islands and cays with no permanent population. The capital is The Valley. The total land area of the territory is 91 km^2 with a population of approximately 15,962 (Anguilla Statistics Department 2010).

Anguilla was chosen as the focus of this case study because of its history of coping with coastal changes dating back many years and especially since Hurricane Luis in 1995. Important lessons have been learnt in Anguilla that have relevance to other small islands in the Caribbean and worldwide. Furthermore, the authors have had direct experience with the design and implementation of the planning measures discussed in this chapter.

Due to its poor soil and low rainfall, agriculture in Anguilla has always been a challenge. Tourism has been the mainstay of the economy and is characterized by developments often managed in partnership with multinational companies. This has led to the territory's economy growing rapidly during the late 1980s and early 1990s. During the period 1993–1997,

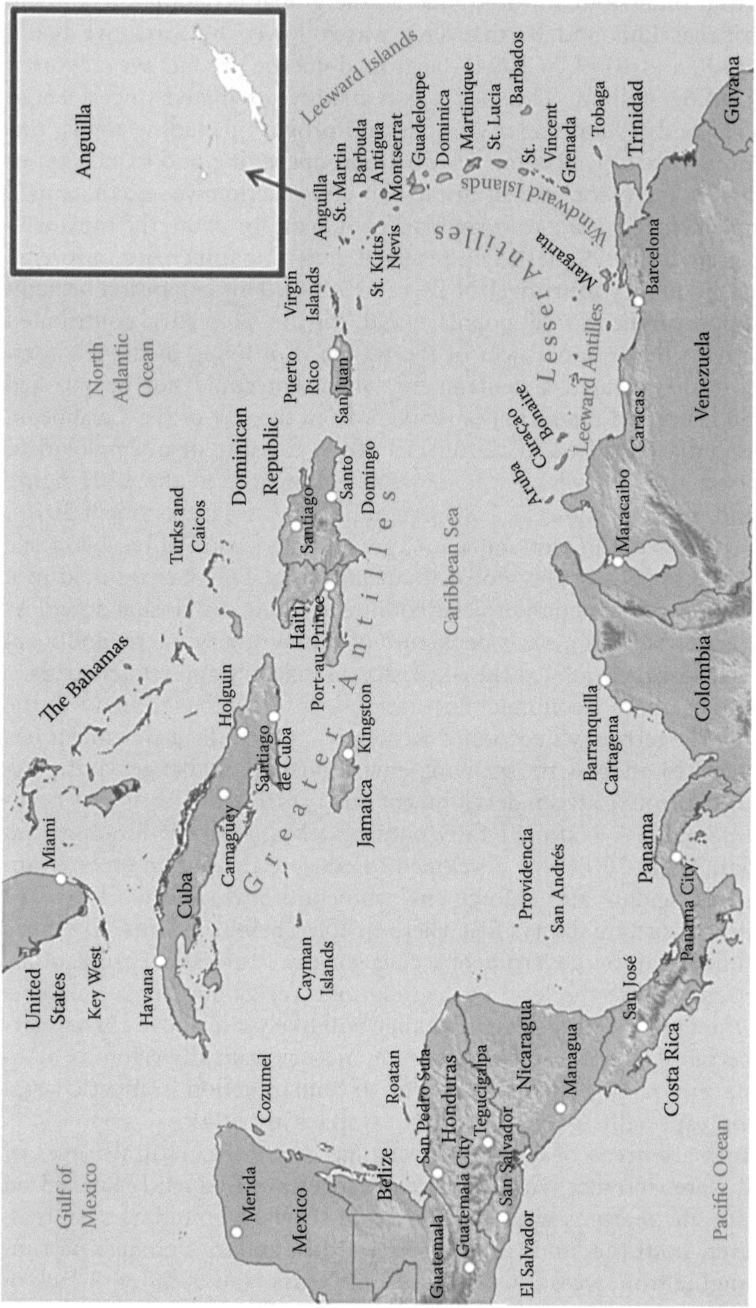

Figure 15.1 Map of the Caribbean Sea and its islands (island of Anguilla inset). (Adapted from K. Musser, Map of the Caribbean Sea and its islands. http://commons.wikimedia.org/wiki/File:Caribbean_general_map.png, 2011; and OpenStreetMap contributors, Online community of Mapping. https://www.openstreetmap.org/, 2010.)

however, the pace of economic growth slowed due to the Gulf Crisis, competition from other tourism destinations, the global economic slowdown, and Hurricanes Luis and Bertha. This was followed by a relative boom between 2000 and 2009. In 2004, the gross domestic product was reported as XCD 394.67 million. This boom was mainly attributed to residential development and several major construction projects including roads, private vacation villa-type accommodation, and upgrading and expansion of existing hotels and associated utilities. This construction was accompanied by the implementation of a tourism marketing plan. In 2006, the territory's economy grew by 10.9%, largely a result of a continual increase in foreign investments in luxury tourism (UN-ECLAC 2006). This expansion brought new job opportunities to the populace and, for the most part, contributed significantly to the enhancement of the standard of living in the territory. However, it also created a demand for labor that could not be satisfied locally and generated migration of workers from the rest of the Caribbean, as well as India (UN-ECLAC 2006). In 2008, the rate of unemployment was reported as 1.5%, which was low when compared to the 2001 figure of 6.7% and the 2002 figure of 7.8% (Anguilla Statistics Department 2010). From 2010 onward, the economy has experienced a period of recession due mainly to the impact of the global financial crisis. This has resulted in a number of large construction projects coming to a halt and closing down. As a result, there has been a mass departure of immigrant workers, and local citizens have lost their jobs in the construction industry and related areas.

Throughout these economic fluctuations, tourism has remained the mainstay of the territory's economy. Recognizing this, the government has seen the need to address the growing environmental challenges and pressures resulting from tourism development and preserve the territory's natural resource base. A National Environmental Management Strategy and Action Plan 2005–2009 was developed to educate the public on environmental issues, update and enforce environmental laws, and work toward environmental sustainability. Elsewhere in the Caribbean islands, efforts are also underway by governments, civil society, regional organizations, and others to reduce the level of degradation. Yet as human populations increase, the rate of anthropogenic change will likely rise above current levels. Furthermore, it is difficult to separate, in a scientifically rigorous manner, the natural changes from those due to human action in most coastal ecosystems, especially where monitoring systems are weak.

Shorelines are areas of continuous change where the natural forces of wind and water interact with the land. Beaches are frequently labeled on maps as specific features with landward and seaward boundaries. In reality, however, both the landward and seaward boundaries change on timescales ranging from weeks through seasons, years, and decades. A lack of understanding about this fundamental dynamism lies at the center of most beach development issues in the Caribbean (UNEP 2010).

This chapter focuses particularly on beach ecosystems, which are immensely varied in the insular Caribbean and range from black sand beaches in Montserrat to the coralline beaches of Anguilla and include the silica sand beaches of the east coast of Barbados and mud and silt beaches of the Guyana coast.

CLIMATE CHANGE RISKS AND CARIBBEAN COASTAL AREAS

The Caribbean UK Overseas Territories (Anguilla, British Virgin Islands, Cayman Islands, Montserrat, and Turks and Caicos Islands), together with their independent neighbors, are vulnerable to the adverse effects of climate change* as well as climate variability and severe weather events (Box 15.1). The territories are all small islands with narrow economic bases and limited natural resource bases and human capacity (UK Department for International Development 2007). The UK Overseas Territories, because of the nature of their relationship to the United Kingdom, are not obligated to submit National Communications to the United Nations Framework Convention on Climate Change or prepare National Adaptation Programmes of Action to Climate Change. Nor have they directly benefitted from a series of regional programs funded by the Global Environment Facility since 1998 and implemented in the independent Caribbean Community (CARICOM) countries to assist them in planning for adaptation to climate change. This situation has changed somewhat since 2007 with the development and implementation of an ongoing project to integrate climate change adaptation into the national development planning in UK Overseas Territories. This is being implemented in collaboration with the CARICOM Climate Change Centre.

As discussed earlier in this chapter, for several decades, the Caribbean islands have been experiencing significant environmental degradation, which is closely linked to development pressures. For example, in Barbados, beachfront land has attracted developers, particularly on the west and south coast over the past 50 years. Coastal swamps, which acted as sediment filters, have been cleared. Stabilizing beach vegetation has been removed affecting the natural movement of sand. New tourism developments have resulted in the increased production of sewage and wastewater; until recently, most of this was discharged into the sea with little to no treatment increasing the level of nutrients in nearshore waters. Changing farming practices, particularly in the sugarcane industry, have resulted in

* The definition of climate change used by the United Nations Framework Convention on Climate Change is used throughout this chapter, namely "a change of climate which is attributed directly or indirectly to human activity that alters the composition of the global atmosphere and which is in addition to natural climate variability observed over comparable time periods."

BOX 15.1 KEY CLIMATE CHANGE RISK ISSUES FACING UK OVERSEAS TERRITORIES

Freshwater availability due to reduced precipitation, increased evaporation, and saline intrusion from sea-level rise:

- Likely to impact water resources, tourism, and agriculture sectors.
- May constrain economic activity and cause health concerns.

Degradation of marine and coastal ecosystems due to sea-level rise and changes in ocean temperature and pH:

- Likely to impact fisheries and tourism sectors.
- Tourism and fisheries sectors account for a high proportion of the gross domestic product and are highly dependent on healthy coastal and marine ecosystems.
- Already coral bleaching is becoming more frequent and severe.

Increased coastal inundation as a result of sea-level rise:

- Likely to impact tourism and agriculture.
- Loss of beaches will impact tourism activities located in the coastal area and sea turtle conservation.

Adapted from UK Department for International Development 2007.

increased sediment loads and more pesticides and fertilizers reaching the marine environment. Because corals thrive in clean, clear, shallow waters that are virtually free of sediment, nutrients, and toxins, all these factors are stressing Barbados' coral reefs and in turn causing increased beach erosion (Braithwaite et al. 2008).

It has been shown that, globally, approximately 70% of the world's sandy beaches are eroding (Bird 1985, 1987). This statistic appears to be repeated in the Caribbean. Beach monitoring data covering five to 15 years of regular beach profile monitoring, conducted at three-month intervals, at 113 beaches (200 profile sites) on eight islands, including Anguilla, show an average erosion trend of 0.5 m yr^{-1} over the period 1985–2000, with elevated rates in those islands impacted by a higher number of hurricanes (Cambers 2009). In Anguilla, the average rate of erosion over the period 1991–2000 was 1.26 m yr^{-1}. During this period, the territory was impacted by three major hurricanes: Luis in 1995, Georges in 1998, and Lenny in 1999.

Beach erosion is not a regular occurrence, sometimes several years may elapse with only seasonal changes, followed by significant erosion during a particular storm event. The erosion in the insular Caribbean has been attributed to both anthropogenic factors, for example, sand mining, reef degradation, and poorly planned coastal development and sea defenses, and natural causes, for example, winter swells and hurricanes. Sea-level rise is another causative factor of beach erosion and the inundation of low lying lands (Leatherman and Beller-Sims 1997). However, it is difficult to distinguish the quantitative impact of sea-level rise from the other causes of beach erosion. Between 1985 and 2000, tropical storms and hurricanes in the insular Caribbean appeared to be the dominant factor influencing the erosion, with beaches often failing to return to their pre-hurricane levels (Cambers 2009). Dune retreat and disappearance has also been widely documented in the Caribbean islands (Cambers 1998, 2005). Cliff retreat and changes in rocky shores are less well documented. However, there appears to be an increase in the exposure of beachrock ledges associated with widespread beach erosion (Cambers 1998).

Climate change adds another dimension of risk factors for beach ecosystems and coastal areas, in particular, projected sea-level rise, ocean acidification, increasing sea surface temperatures, saltwater intrusion, and stronger hurricanes. Present projections based on Fourth Assessment Report of the Intergovernmental Panel on Climate Change (IPCC 2007) are that it is likely that future tropical cyclones (typhoons and hurricanes) will become more intense, with larger peak wind speeds and more heavy precipitation associated with ongoing increases in sea surface temperatures. There is less confidence in projections of a global decrease in numbers of tropical cyclones. Table 15.1 summarizes the projected global climate change by 2099 based on the Fourth Assessment Report of the IPCC (2007).

Analysis of global tide-gauge data shows that the rate of sea-level rise increased in the late nineteenth and early twentieth century, with a global average rate of about 1.7 mm yr^{-1} over the twentieth century (Church et al. 2010). Since 1993, the average rate of rise as measured by satellite altimeters is 3.2 mm yr^{-1}. Tide gauge data indicate a slightly smaller rate of rise (3.0 mm yr^{-1}) over the same period (Pacific Climate Change Science Program 2010). In many Caribbean islands, it is anticipated that sea-level rise is contributing to increased beach erosion, although the data are insufficient to allow quantification of the contribution of sea-level rise to overall erosion rates (Cambers 2009). The Bruun Rule (Bruun 1962) is used to project future shoreline position in response to sea-level rise, is based on the concept of a sandy beach and nearshore environment, and has limitations for coral reef coastlines. There is ongoing debate about models of shoreline change in coral reef settings, including linkages between reef productivity and beach erosion. For example, Webb and Kench (2010) have shown that

Table 15.1 Summary of projected global climate changes by 2099

Parameter	*Projected change*
Temperature	Increase of between 1.1°C and 6.4°C
Sea-level rise	Increase of between 0.18 and 0.59 m[a]
Ocean acidification	Decrease in pH of 0.14–0.35 units (resulting in increased acidity)
Snow and ice extent	Decrease in areal extent of ice and snow
Extremes: heat waves and heavy precipitation	More extreme events
Tropical cyclones	Stronger tropical cyclones
Precipitation	Changes vary regionally, some areas getting drier, some wetter

Source: IPCC, "Summary for policymakers," in S. Solomon et al. (eds.), *Climate Change 2007: The Physical Science Basis. Contribution of Working Group I to the Fourth Assessment Report of the Intergovernmental Panel on Climate Change,* Cambridge University Press, Cambridge; New York, http://www.ipcc.ch/, 2007.

[a] Sea-level rise projected change does not include the full effect of changes in ice sheet flow because a basis in the published literature is lacking.

some coral atoll islands in the Pacific have shown little change in overall area over a 19- to 61-year period despite rising sea level.

Coral reefs are an important part of many Caribbean beach systems. Increased ocean acidification and coral bleaching events have the potential to seriously impact the health and sustainability of tropical reef ecosystems in the medium term (10–30 years), as well as livelihoods dependent on coastal fisheries.

Based on the measured beach changes discussed by Cambers (2009), stronger hurricanes may be the most significant risk to Caribbean beaches in the immediate future. Research indicates that there is a cyclical pattern in Atlantic hurricanes with decades of frequent and intense hurricane activity followed by decades with less frequent and less intense activity. For instance, major hurricanes (category 3 and higher) were more frequent in the 1950s and 1960s as compared to the period 1971–1994 (Emanuel 2005). The period 1995 to the present has seen an increase in the frequency and intensity of hurricanes, with 2005 recorded as the most active year in history, with 23 named storms, 13 of which were hurricanes and seven were major hurricanes. Researchers disagree as to whether the increased intensity and frequency of hurricanes over the past 30 years is part of this cyclical pattern or a result of climate change, or both (Emanuel 2005). The Fourth Assessment Report of the IPCC projects that future hurricanes will become more intense (IPCC 2007). Table 15.2 shows the number of hurricanes passing within 110 km of Anguilla since 1950 and clearly demonstrates the high level of interdecadal variability. It should be noted that satellite information has only been available since the 1970s.

Table 15.2 Hurricanes passing within 110 km of Anguilla (18° 20′ North and 63° 08′ West)

Decade	*Year*	*Hurricane category*[a]	*Hurricane name*
1950–1959	1950	3	Dog
	1954	1	Alice
1960–1969	1960	4	Donna
	1966	1	Faith
1970–1979	No hurricanes recorded		
1980–1989	No hurricanes recorded		
1990–1999	1995	4	Luis
	1996	1	Faith
	1998	2	Georges
	1999	1	Jose
	1999	4	Lenny
2000–2009	2000	1	Debbie
	2008	2	Omar

Source: Caribbean Hurricane Network, http://stormcarib.com/, 2009.

[a] Category 1: wind speed 118–152 km/h; category 2: wind speed 153–176 km/h; category 3: wind speed 177–208 km/h; category 4: wind speed 209–248 km/h; category 5: wind speed 249+ km/h.

In Anguilla, beach erosion is recognized as a serious challenge by the Department of Physical Planning. The effects of beach erosion are particularly evident after high wave energy events such as a tropical storm, hurricane, or high winter swell, when coastal buildings may be impacted by the waves and in some cases foundations may be undermined (Figure 15.2). As noted earlier (Cambers 1997b, 2009), beach profiling has shown that in some cases beach levels do not return to pre-hurricane conditions.

LEARNING FROM PAST COASTAL PLANNING EXPERIENCES

From September 4–6, 1995, Anguilla was impacted by Hurricane Luis. Following surveys of the hurricane's impact, it quickly became apparent that most of the severe damage was in the coastal area where the port facilities, hotels, tourism villas, condominiums, restaurants, and bars were located (Cambers 1996). As a result, the government resolved to put in place measures to protect the territory's coastline and economy, as well as personal livelihoods, from such events in the future. Coastal development guidelines were among the measures proposed.

Coastal development setbacks were already an intrinsic part of the planning guidelines in Anguilla (Cambers 1997a). A coastal development setback may be defined as a prescribed distance to a coastal feature, such as

Figure 15.2 Beach bar at Shoal Bay East in danger of being undermined by the waves in 2010; in 2002, the building was 12 m from the active beach zone. (Courtesy of S. Roberts-Hodge.)

the line of permanent vegetation, within which all or certain types of development are prohibited. Such setbacks provide buffer zones so the beach has the space to erode or accrete and maintain its protective function during high wave events, without the need for expensive sea defense structures. They also allow for improved vistas and access along the beach and provide privacy for coastal property owners and beach users alike.

With the help of overseas development assistance, new guidelines for coastal development setbacks were designed specifically for Anguilla in 1996 (Cambers 1996); these were adapted for more general use in 1997 (Cambers 1997a). These guidelines sought to ensure that new development would be situated a *safe* distance from the active beach zone. Most of the Caribbean islands already had coastal development setback guidelines included in their planning laws, these ranged from 15 m from high water mark in the British Virgin Islands to 30 m from high water mark in Barbados (Cambers 1997a).

The 1996 and 1997 guidelines were based on the expectation that the frequency and intensity of hurricanes would likely increase in the Caribbean (as a result of climate variability) and took into account an expected sea-level rise of 3 mm yr^{-1} by 2100 (one of the lower estimates in the 1990s). The methodology, detailed in Cambers (1997a), calculated a setback distance for each beach based on historical changes at that beach, projected changes likely to result from a major hurricane, coastline recession resulting from sea-level rise in the next 30 years, and specific geographical and planning factors at a particular beach, for example, presence or absence of a healthy barrier reef. The period of 30 years was selected because it represented the average economic life of a building or structure.

Using this methodology, coastal development setback distances were calculated for each beach in several eastern Caribbean islands, including Anguilla, Antigua and Barbuda, St. Kitts and Nevis, and St. Lucia, using the permanent vegetation line as the point of reference (Cambers et al. 2008; UNESCO 2003). However, the preparation of the planning guidelines was only a first step, as implementation of the guidelines has been an evolving process over more than 10 years.

In Anguilla, like many other coastal tourism destinations, developers come to the island with a vision of building their resorts and vacation homes as close to the white sand beach and the sound of the surf as possible. In an attempt to attract large developments and economic gain to the island, planning guidelines and standards are sometimes compromised in their passage through the decision-making process, often the result of political expediency. Further down the line, when the project has been built and operating, a storm surge or hurricane event may occur and cause severe beach erosion. This can be a significant challenge for both the developer (assuming the same corporation that developed a resort becomes its operator) and the regulating government body. If the erosion is very severe, the development may have to shut down temporally to make repairs. Grave economic losses occur when this happens as millions of dollars are required to mitigate the damage. In small island territories such as Anguilla, there is a rippling effect throughout the community because workers are out of jobs for months on end and many households are subsequently affected.

Experiences such as these on Anguilla have been a catalyst for initiating change as far as coastal setback guidelines and standards are concerned, especially since Hurricane Luis in 1995. Since 1996, the Land Development Control Committee (LDCC), the statutory body responsible for granting planning approval, has developed and applied the following process for determining the minimum acceptable setback distances for structures as measured from the line of permanent vegetation.

First, a visit to the proposed site is conducted. Each application for coastal development is considered on its own merit using the following four factors: (1) the setback distances recommended in Cambers (1996); (2) the setback distance proposed by the particular developer; (3) the size of the parcel of land; and (4) the setback distance of any other existing buildings close to the proposed site. Once the LDCC is satisfied that a structure built adhering to the derived setback distances would not compromise the free movement of sand on the beach or interfere with pedestrian movement on the beach or rocky shore, a setback distance for the development can be determined and the development, after fulfilling all other planning requirements, may be approved.

The LDCC has in some cases permitted reduced coastal setback distances for wooden structures. However, such structures must be built on piles, be less than 370 m^2 in size and for use as bars or restaurants. Because these buildings are not used as dwellings, the potential risk to human life is minimized and the piles allow the water to penetrate underneath the structure without undermining it during a high wave event and/or storm surge. It is recognized, however, that such structures may ultimately be undermined and lost if erosion continues.

The implementation of this methodology has brought positive results. Structures set back a *safe* distance from the active beach zone endure. Such developments are living proof that a key to sustainable development of beachfront areas is building structures that are adequately set back away from the line of permanent vegetation and active area of the beach. However, other hotels, for example, at Rendezvous Bay where sand dunes have been removed and lowered, have repeatedly experienced damage from weather systems. The process described above is, however, dependent on there being limited political interference and that the developer constructs the building in the manner approved by the LDCC.

In the past few years, erosion on some beaches has accelerated dramatically resulting in the beach in front of certain establishments virtually disappearing overnight and causing decks and balconies to hang precariously above the water. This is of concern, because valuable beach areas are quickly disappearing. For this reason, the Department of Physical Planning has recognized that a policy needs to be drafted to reflect the modified setback distances that have been customized and applied since 1996. It is necessary that this policy also addresses coastal *hot spots* (areas that are experiencing increased erosion) and applies setback distances that have been used and have subsequently become accepted in the policy mainstream.

LOOKING TO THE FUTURE

For many residents of small islands, climate change and climate variability are relatively new terms and difficult to understand. During a 2002 opinion survey about environment and development issues in three small islands (Palau in the Pacific, Seychelles in the Indian Ocean, and St. Kitts and Nevis in the Caribbean), climate change was not raised as an issue by any of the survey respondents (UNESCO 2002). Although scientists have been studying the phenomenon known as climate change for decades, it has only really entered the public domain during the first decade of the twenty-first century and particularly since the Fourth Assessment Report of the IPCC was published in 2007.

One practical recommendation is to enhance society's level of understanding about climate change. This will involve working with governments, nongovernmental organizations, the private sector, professional groups, church groups, and many others. In Anguilla, elected representatives, many government officials, and the public at large show an interest in climate change and are keen to gain more knowledge about the phenomenon, but it will take more evidence to totally convince them that climate change is real and can greatly impact their lives. Providing practical, locally relevant examples of climate change and climate variability, such as changes in hurricane patterns and increased beach erosion, as have been presented in this chapter, is one way to make the topic relevant; especially for small tropical islands, such as Anguilla, most of the likely impacts of climate change are likely to be adverse, so it is important to present practical ways of coping with the anticipated impacts, such as by building new coastal infrastructure further inland from the line of permanent vegetation. Enlisting the help of those whose establishments have been damaged as a result of coastal erosion to advocate for improved coastal planning measures is one proposal.

Climate change is an issue that will be relevant for the rest of our lives, so incorporating climate change into the school curriculum (into science, social science, geography, and other areas) is an important part of the process of adapting to climate change. Already some countries have integrated aspects of climate change into their school curricula, but much more needs to be done. International organizations, such as the United Nations Educational, Scientific and Cultural Organization (UNESCO) are providing support for such initiatives through leadership in the Decade of Education for Sustainable Development (2005–2014). Another program, Sandwatch, which focuses on helping children and youth monitor their beaches and take practical measures to address beach-related issues and enhance the coastal environment, has integrated climate change into several beach monitoring protocols (Cambers and Diamond 2010).

In the Caribbean region, changes in hurricane frequency and intensity are likely to be the most immediate and obvious effect of climate variability and climate change. Some regional and international organizations are integrating disaster risk management and climate change adaptation into one agenda. For example, the Caribbean Development Bank in 2009 published revised disaster risk management guidelines that included adaptation to climate change. Many of the measures being put in place for disaster risk management such as vulnerability reduction are also a part of climate change adaptation. Thus, combining the two agendas, where appropriate, provides ways of maximizing limited human resources and existing institutional frameworks. In the Pacific, some island nations are also integrating the two agendas, for example, the Kingdom of Tonga recently published a Joint National Action Plan on Climate Change Adaptation and Disaster Risk Management (Ministry of Environment and Climate Change 2010).

Climate change is adding to the list of sustainable development issues that need immediate and ongoing action in the Caribbean region. Integrating climate change into existing programs and activities is one way forward for the region as a whole and for individual territories such as Anguilla.

OPPORTUNITIES FOR MAINSTREAMING CLIMATE CHANGE INTO THE GOVERNMENT PLANNING AGENDA

From a scientific perspective, the evidence for climate change as a result of human activities is overwhelming; with global warming *very likely* due to the observed increase in anthropogenic greenhouse gas emissions (IPCC 2007). In the IPCC report, *very likely* has a 90%–99% likelihood of occurrence. The question must be asked, however, how does this translate to the situation on the ground for a small Caribbean territory such as Anguilla? One issue relates to distinguishing between climate change (due to forcing by greenhouse gases resulting from anthropogenic activities) and climate variability (a result of natural variation in climate). This issue was discussed above in relation to changes in hurricane frequency and intensity in the Caribbean basin since 1995. Possibly, from a planning perspective in a small Caribbean island, the important consideration is that the observational record shows that hurricanes have become more frequent and more intense since 1995 and that this is likely to continue for the next two decades.

Key opportunities for mainstreaming climate change and climate variability in the government planning agenda are described in the subsequent paragraphs.

Opportunity 1: Changing actions and practices

Beach erosion has been a serious issue in many Caribbean islands for several decades and climate change is only one of the causative factors. Climate change does, however, provide governments and planners with added justification to change the way we do things. Changing attitudes is difficult and seemingly impossible to achieve; changing actions and practices is more feasible. In Anguilla, there is a need to change practices now such as building further back from the line of permanent vegetation and using wood and piled structures as much as possible.

Opportunity 2: Using visual examples of environmental change

Observations in Anguilla show visual examples of severe beach erosion. Structures that were placed too close to the vegetation line are in danger of being destroyed or have had portions of the structure already damaged due to storm surges and winter swells (locally known as ground seas). At Shoal Bay East on the north coast of Anguilla, the average rate of erosion between 1991 and 2000 was 0.8 m yr^{-1} (UNESCO 2003). In recent years, several coastal bars and restaurants have been directly affected by beach erosion. Such examples provide visual evidence for all to see and can help people relate to somewhat abstract phenomena like climate change and variability when they visit their favorite bar or restaurant and see the beautiful wide beach in front has virtually disappeared overnight and been replaced by sea or bare rocks.

Opportunity 3: Involving the private sector

In recent years, climate change has begun to occupy a prominent place in the development agenda, and much external development assistance is being directed toward climate change adaptation. Adapting to climate change is a part of the sustainable development agenda and as such it is necessary to involve all the government and nongovernmental players, not just planning and environmental agencies. The private sector also has an important role to play in sustainable development. For example, the insurance industry plays a key role in development and their due diligence practices could be usefully turned to the advantage of sound coastal planning, especially in the face of added stressors such as climate change.

Opportunity 4: Maximizing windows of opportunity

Climate change is a long-term gradual phenomenon taking place unseen and slowly over decades and centuries. The political agenda usually spans four to five years, and many people live their lives on a day-to-day basis. The challenge lies in how to match the different agendas and time frames. For the majority of local Anguillans, the concept of planning ahead is foreign. However, opportunities do exist in the immediate time frames: for example, after a major hurricane directly impacts an island, there is usually a period of six to nine months, a window of opportunity, when people are likely to really appreciate that building close to the beach is an unwise practice. For example, in Anguilla, after the damage wrought by Hurricane Luis in 1995, the Government resolved to put in place measures to protect their coastlines. Had the measures already been designed, they would likely have been approved by the Legislative Council. A window of opportunity was missed.

Opportunity 5: Emphasizing the economic impact of poor planning practices

Individuals, governments, and corporations react to the loss of economic earnings. For instance, in the island of Nevis, part of the Federation of St. Kitts and Nevis, the flagship hotel closed for two years after damage incurred during Hurricane Omar in 2008 (eTurboNews.com 2010). This resulted in serious unemployment and loss of earnings for individuals and government. This hotel was built in 1989 near the coast in a low-lying wetland and had incurred serious damage during several previous hurricanes. There is a need to publicize the economic impacts of poor planning practices and how they can be exacerbated by climate change and climate variability.

REFERENCES

Anguilla Statistics Department. (2010). *Preliminary Country Poverty Assessment Findings*, Unpublished report, Government of Anguilla, British West Indies.

Bird, E.C.F. (1985). *Coastline Changes: A Global Review*, New York: Wiley.

Bird, E.C.F. (1987). "The modern prevalence of beach erosion," *Marine Pollution Bulletin*, 18(4): 151–7.

Braithwaite, A., Oxenford, H., and Roach, R. (2008). *Barbados: A Coral Paradise*, Barbados: Miller Publishing.

Bruun, P. (1962). "Sea level rise as a cause of shore erosion," *Journal of Waterways and Harbours Division*, 88(1–3): 117–30.

Cambers, G. (1996). *The Impact of Hurricane Luis on the Coastal and Marine Resources of Anguilla: Coastal Development Setback Guidelines*, Unpublished report prepared for British Development Division in the Caribbean.

Cambers, G. (1997a). "Planning for coastline change. Guidelines for construction setbacks in the eastern Caribbean islands," *Coastal Region and Small Island Info (4)*, Paris, France: UNESCO. http://www.unesco.org/ (accessed January 2012).

Cambers, G. (1997b). "Beach changes in the Eastern Caribbean islands: Hurricane impacts and implications for climate change," *Journal of Coastal Research Special Issue (Island States at Risk: Global Climate Change, Development and Population)*, 24, 29–38.

Cambers, G. (1998). *Coping with Beach Erosion,* Paris, France: UNESCO Coastal Management Sourcebooks (1) UNESCO Publishing. http://www.unesco.org/ (accessed January 2012).

Cambers, G. (2005). "Caribbean islands coastal ecology and geomorphology," in M.L. Schwartz (ed.), *Encyclopaedia of Coastal Science*, New York: Springer Publishing.

Cambers, G. (2009). "Caribbean beach changes and climate change adaptation," *Aquatic Ecosystem Health & Management*, 12(2): 168–76.

Cambers, G. and Diamond, P. (2010). *Sandwatch: Adapting to Climate Change and Educating for Sustainable Development*, Paris, France: UNESCO.

Cambers, G., Richards, L., and Roberts-Hodge, S. (2008). "Conserving beaches and planning for coastal change in the Caribbean islands," *Tiempo*, 66(1): 18–24. http://www.tiempocyberclimate.org/ (accessed August 2013).

Caribbean Development Bank. (2009). *Disaster Risk Management Strategy and Operational Guidelines*. http://www.caribank.org (accessed October 2010).

Caribbean Hurricane Network. (2009). http://stormcarib.com/ (accessed November 2010).

Church, J.A., Woodworth, P.L., Aarup, T., and Wilson, W.S. (eds.) (2010). *Understanding Sea-Level Rise and Variability*, Chichester: Wiley-Blackwell Publishing.

Emanuel, K. (2005). "Increasing destructiveness of tropical cyclones over the past 30 years," *Nature*, 436: 686–8.

eTurboNews.com. (2010). "Four seasons resort Nevis announces reopening after Hurricane Omar." http://www.eturbonews.com/ (accessed April 2011).

IPCC. (2007). "Summary for policymakers," in S. Solomon, D. Qin, M. Manning, Z. Chen, M. Marquis, K.B. Averyt, M. Tignor, and H.L. Miller (eds.), *Climate Change 2007: The Physical Science Basis. Contribution of Working Group I to the Fourth Assessment Report of the Intergovernmental Panel on Climate Change*, Cambridge; New York: Cambridge University Press. http://www.ipcc.ch/ (accessed August 2013).

Leatherman, S.P. and Beller-Sims, N. (1997). "Sea level rise and small island states: An overview," *Journal of Coastal Research Special Issue (Island States at Risk, Global Climate Change and Population)*, 24: 1–16.

Ministry of Environment and Climate Change and National Emergency Management Office Tonga. (2010). *Joint National Action Plan on Climate Change Adaptation and Disaster Risk Management 2010–2015*, Suva, Fiji: SOPAC. http://www.sprep.org/ (accessed January 2012).

Musser, K. (2011). Map of the Caribbean Sea and its islands. http://commons.wikimedia.org/wiki/File:Caribbean_general_map.png. OpenStreetMap contributors. (2010). Online community of Mapping. https://www.openstreetmap.org/.

Pacific Climate Change Science Program. (2010). *Climate Variability and Change in the Pacific islands and East Timor*, Canberra, Australia: Australian Government. http://www.cawcr.gov.au/projects/PCCSP/ (accessed January 2011).

Richards, W.J. and Bohnsack, J.A. (1990). "The Caribbean sea: A large marine ecosystem in crisis," in K. Sherman, L.M. Alexander, and B.D. Gold (eds.), *Large Marine Ecosystems: Patterns, Processes, and Yields*, Washington, DC: American Association for the Advancement of Science.

UNESCO. (2002). "Small islands voice," Opinion surveys. http://www.unesco.org/csi/smis/siv/sivindex.htm (accessed April 2011).

UNESCO. (2003). *Wise Practices for Coping with Beach Erosion*. Anguilla; Paris, France: UNESCO Environment and Development in Coastal Regions and Small Islands. http://www.unesco.org/csi/publica.htm (accessed January 2012).

United Kingdom Department of International Development. (2007). *Enhancing Capacity for Adaptation to Climate Change in the UK Overseas Territories*, Project memorandum. http://www.ukotcf.org/pdf/climateChangeMemo260307.pdf (accessed January 2011).

United Nations Economic Commission for Latin America and the Caribbean (UN-ECLAC). (2006). *Caribbean Knowledge Management Centre*. http://ckm-portal.eclacpos.org/dev/ai (accessed January 2012).
United Nations Environment Programme (UNEP). (2010). *Global Environment Outlook: Latin America and the Caribbean GEO LAC 3*, Panama City, Panama: UNEP.
Webb, A.P. and Kench, P.S. (2010). "The dynamic response of reef islands to sea level rise: Evidence from multi-decadal analysis in the Central Pacific," *Global and Planetary Change*, 72(3): 234–46.
World Travel and Tourism Council. (2010). Economic Data Search Tool. http://www.wttc.org/focus/research-for-action/economic-data-search-tool/ (accessed October 2010).

Part VI

Climate change and the coastal zone

South America

Chapter 16

A risk-based and participatory approach to assessing climate vulnerability and improving governance in coastal Uruguay

Gustavo J. Nagy, Mónica Gómez-Erache, and Robert Kay

Abstract: The complex geophysical environment of the Rio de la Plata basin in Uruguay is under stress from existing pressures such as changing hydro-climatic and wind regimes, sea-level rise, extreme events, growing population, and associated increases in development. This chapter presents the emerging lessons learned from ongoing work undertaken on coastal climate change vulnerability and adaptation in Uruguay (Climate Change Adaptation Project) and implemented through an integrated coastal zone management framework (EcoPlata Program), which provides a strategy on coastal governance. The case study describes a four-stage methodology, across a range of spatial scales, including a risk-based approach to assessing impact; geophysical modeling; a stakeholder-driven Vulnerability Reduction Assessment; and multicriteria approaches to adaptation within a participatory bottom-up and top-down process. Effective coastal adaptation in Uruguay requires that technical knowledge be merged with lessons learnt through an adaptive management cycle to meet both short-term decision objectives and long-term adaptation goals.

INTRODUCTION

The Uruguayan coast

The Uruguayan coast is 670 km in length with 450 km lying within the Rio de la Plata estuary and the remaining 220 km on the Atlantic Ocean (Figure 16.1). The coast contains a variety of environments with many unique characteristics, including sandy beaches with alternating rock outcrops and small inlets and lagoons. This diverse landscape, together with the biodiversity it sustains, provides goods and services critical to the national economy and sustains sectors such as fishing, tourism, harbor development, and agriculture (Gómez-Erache and Martino 2008; Robayna 2009).

Uruguay is an *upper middle income country* (World Bank 2010) and a full democracy (Economist Intelligence Unit 2010; Legatum Institute 2010).

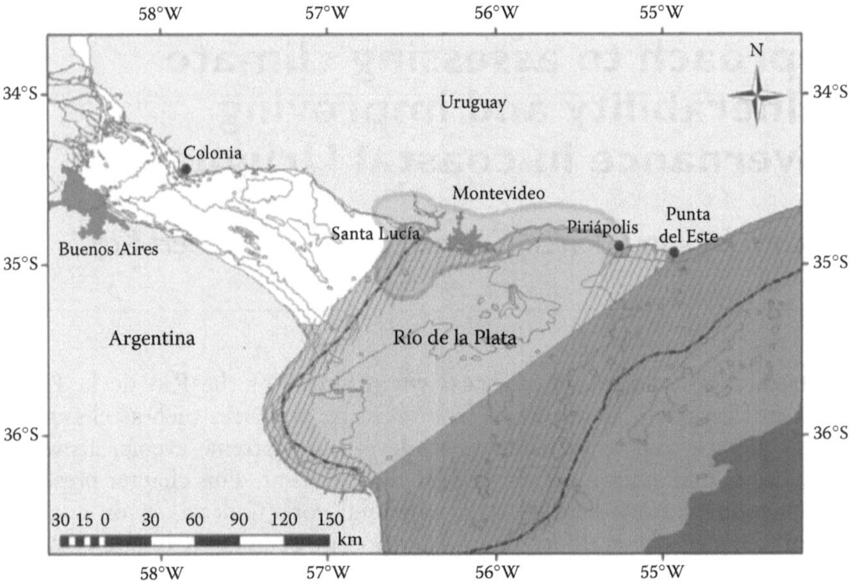

Figure 16.1 The Rio de la Plata estuary and frontal zone pilot site. (Adapted from Brazeiro, A. et al., *Aquatic Priority Areas for the Conservation and Management of the Ecological Integrity of the Río de la Plata and its Maritime Front*, Reporte Científico del Proyecto Freplata 1-2.2, PNUD/GEF RLA/99/G31, www.freplata.org, 2004.) (Note: The pilot site includes the estuarine waters front [inner black line and lined light gray] and the coastal zone [mostly sandy beaches] from Santa Lucia to Piriápolis–Punta del Este.)

Most socioeconomic and human development indices (gross domestic product [GDP], poverty, unemployment, investments, and debt), except for education (OECD 2010b), show continuous and robust improvement at both national and coastal scales following the country's severe economic crisis in 2002 (OECD 2010a; UNDP 2011).

An estimated 70% of Uruguayans live in coastal areas, with an estimated 34% of the coast being urbanized, including the capital city of Montevideo. Over two-thirds of economic activities and income generated in Uruguay are directly or indirectly related to the coast. Also, globally relevant biodiversity sites are located in coastal areas. Consequently, coastal processes coexist alongside a variety of activities that compete for space and resources, the subsequent result being coastal degradation (Gómez-Erache et al. 2010).

Impacts of climate change

There are growing concerns in Uruguay regarding social and ecological issues related to coastal natural resource degradation due to increasing pressures resulting from poor development planning. In addition, since the early 1980s, the impacts from climate and environmental changes have become pronounced (Box 16.1; see also Nagy et al. 2005, 2007, 2008b, 2008c; Unidad de Cambio Climatico 2005).

Predicted climate changes for the area suggest that coastal areas in Uruguay face the prospect of rising sea levels, changes in the intensity of storm surges, and changes in wind and hydroclimatic patterns. Additionally, these changes will be overlaid on areas already stressed by existing climatic pressures. Magrin et al. (2007) and Dasgupta et al. (2007) have identified Uruguay as one of the most exposed Latin American countries to coastal climate change. A recent Regional Economic Climate Change Study (ECLAC 2010) identified

BOX 16.1 ENSO EVENTS: SEA LEVEL, RIVER FLOW, WINDS, AND COASTAL SEDIMENTARY BALANCE

Sea level along the Uruguayan coast fluctuates ±0.1–0.2 m (superimposed on sea-level rise) due to ENSO-related anomalies in precipitation, river flow and winds. Wind-induced flooding reaches 2, 3, and 4 m on yearly, decadal, and historical timescale basis, respectively. Wind regime changes were reported during El Niño, which was confirmed for ENSO-related wind anomalies from 1950 to 2004. An increase in E–SE–SSE–N and decrease in SW winds (called *Pamperos*) during El Niño, and increase in SW winds during La Niña have also been identified. These anomalies correlated well to observed fluctuations in the interannual sedimentary budget along the sandy beaches of Montevideo over the past five decades. Coastal erosion was observed during La Niña because SW winds generate short-period waves reverting the direction of the littoral drift, which is not compensated during El Niño-related wind anomaly (stabilization or sedimentation).

Source: Adapted from Nagy, G.J. et al., "Desarrollo de la Capacidad de Evaluación de la Vulnerabilidad Costera al Cambio Climático: Zona Oeste de Montevideo como Caso de Estudio," in V. Barros, A. Menéndez, and G.J. Nagy (eds.), *El Cambio Climático en el Río de la Plata, Project Assessments of Impacts and Adaptation to Climate Change (AIACC)*, CIMA-CONYCET-UBA, Buenos Aires, Argentina, 2005; Nagy, G.J. et al., *Revista Medio Ambiente y Urbanización, Cambio climático: Vulnerabilidad y Adaptación en ciudades de América Latina*, 67, 77–93, 2007; Bidegain, M. et al., "Tendencias climáticas, hidrológicas y oceanográficas en el Río de la Plata y costa Uruguaya," in V. Barros, A. Menéndez, and G.J. Nagy (eds.), *El Cambio Climático en el Río de la Plata, vol. 14*, Centro de Investigaciones del Mar y la Atmósfera, Universidad de Buenos Aires, Buenos Aires, Argentina, 2005; Bidegain, M. et al., *Escenarios de cambio climático y*

(Continued)

del nivel medio del mar e impactos de los mismos en áreas costeras y evaluación de la vulnerabilidad geológica costera, Reporte de la Facultad de Ciencias al Proyecto GEF-Implementación Medidas de Adaptación al Cambio Climático en Areas Costeras del Uruguay (ACCC), Unidad de Cambio Climático (UCC), MVOTMA, Uruguay, 2009, http://www.cambioclimatico.gub.uy/; Gutiérrez, O., *Sedimentary Dynamics in the Uruguayan Coast: evolution and trends of urban beaches within global change framework*, Masters thesis, Faculty of Sciences, University of the Republic (UdelaR), Montevideo, Uruguay, 2011.

Uruguay as the most exposed Latin American country in terms of percentage of exposed population to a sea-level rise of 1 m, namely, 30% of the country's population. The estimated cost of climate change on coastal resources for sea-level rise and wind-induced flooding of +0.3, 0.5, and 1 m represents 2%, 4%, and 12% of 2008 GDP, respectively (ECLAC 2010; Sención 2009).

These factors have resulted in a concerted response by management authorities to address both current management challenges and to look forward to addressing the future challenges of climate change. The first of the two key initiatives in this regard, the EcoPlata Integrated Coastal Zone Management Program (EcoPlata), is focused on coastal zone governance; and the second initiative, a Global Environmental Facility funded project, Implementing Pilot Sites of Adaptation Measures to Climate Change in Coastal Areas of Uruguay (ACCC), is focused on adaptation of coastal resources to climate change. EcoPlata and ACCC build on previous initiatives on capacity building for climate change such as Comisión Nacional sobre el Cambio Global (CNCG 1997) country study; Assessments of Impacts and Adaptations to Climate Change (Nagy et al. 2008a, 2008b; UCC 2005); and Program of General Measures for Mitigation and Adaptation to Climate Change in Uruguay (UCC 2004).

THE ECOPLATA PROGRAM AND THE ACCC PROJECT

Initiative overview

The EcoPlata Program (Gómez-Erache et al. 2010) and ACCC Project are seeking to address the complexities of current coastal management issues in the region. Additionally, both initiatives aim to develop strategies to effectively manage future climate change impacts by promoting a participatory and adaptable management model, developed over many years by EcoPlata. This model is based on technical and scientific research and capacity building of institutions and local stakeholders, so that knowledge can be integrated into the design and application of policies and collective action.

The EcoPlata and ACCC initiatives are involved in developing activities to assess and reduce vulnerability to climate change impacts and strengthen institutional arrangements, governance, community resilience, and adaptive capacity. There is a close coordination between programs that are both working to analyze challenges and opportunities to integrate protected areas management, sustainability, risk reduction, and adaptive capacity. Further details of EcoPlata and ACCC are shown in Box 16.2.

BOX 16.2 DESCRIPTION OF ECOPLATA AND ACCC GOALS AND ACTIVITIES

The EcoPlata Program represents a long-term initiative aimed at strengthening institutional capacity, the scientific community, managers, and public in general, in all issues relative to ICZM strategy (see www.ecoplata.org). EcoPlata has promoted and participated in the identification, measurement, and registration of useful indicators for monitoring coastal management. The program evolution phases and goals are synthesized as follows:

- *1991–1996.* Identification of sectors, threats, and opportunities: Scientific capacity building, multidisciplinary perspective, and interinstitutional approach
- *1997–2001.* Political, socioeconomic, and environmental context: Improvement of ICZM; support for research, planning, and policy-making; and public awareness
- *2002–2005.* Evaluation and analysis: ICZM consolidation, training, sustainability, and pilot areas
- *2006–2009.* Implementation: Connecting knowledge to action, environmental conservation, ICZM institutionalization
- *2010–2015.* Consolidation, replication, and scaling up: Governance, public participation, sustainability, and coordination of public institutions that have jurisdiction over the coastal zone

The ACCC Project (long title: Implementing Pilot Climate Change Adaptation Measures in Coastal Areas of Uruguay) aims to strengthen the country's adaptive capacity and contribute to the resilience of its coastal ecosystems in the face of climate change. The project is working at three levels. The first is a nationwide scale and incorporates climate change considerations in the regulation and processes of land planning. The second is concerned with implementation

(Continued)

of specific measures at municipal levels (i.e., departments). These measures can subsequently be included in the land planning processes in order to protect vulnerable coastal ecosystems and biodiversity. The third level of project work aims to promote the launching and replication of successful pilot project measures implemented at municipal levels and the widest possible community practices for adaptation on the estuarine frontal system (Figure 16.1) and Laguna de Rocha (Rocha estuarine lagoon system, Figure 16.2) pilot sites. Adaptation and climate risk management experiences will be replicated by spreading knowledge and employing evaluation and monitoring systems that will help track the effectiveness of adaptation initiatives over time.

Source: Adapted from EcoPlata, "Hacia una estrategia nacional para la gestión Integrada de la Zona Costera 2010–15," lineamientos para la discussion, EcoPlata, Febrero 2010; Gómez-Erache, M. et al., *The Sustainability of Integrated Management in the Coastal Zone of Uruguay: Connecting Knowledge to Action*, EcoPlata, Montevideo, Uruguay, 2010, http://www.ecoplata.org/documentos/; ACCC, "Implementing Pilot Climate Change Adaptation Measures in Coastal Areas of Uruguay," Project document prepared for UNDP/GEF Trust Fund, 2008, http://www.adaptationlearning.net/.

Importantly, the aim of developing an integrated approach to addressing future climate change impacts is to build on the current coastal management governance arrangements developed through EcoPlata. This aims to ensure that potential adaptive responses analyzed through the ACCC Project can be implemented effectively at the local scale and through national coastal governance initiatives to address the critical problem of coastal climate change and risk management in Uruguay that build on the work of EcoPlata. This has been conceptualized through the development of a series of specific *pilot site* analyses outlined in the following section.

Pilot site analysis

In developing the project concept, two pilot sites were selected for analysis: (1) the Rio de la Plata estuarine frontal zone or salinity front where fresh- and saltwaters mix (see Figure 16.1) and (2) Rocha coastal lagoon (see Figure 16.2). These sites were chosen because they (1) represent unique and diverse demonstration sites that are both nationally significant in their own right and can also provide information on how future adaptive actions can be *scaled up*; (2) are rich in biodiversity and resources; and (3) provide an example of a highly populated (former site) and low populated area (latter site).

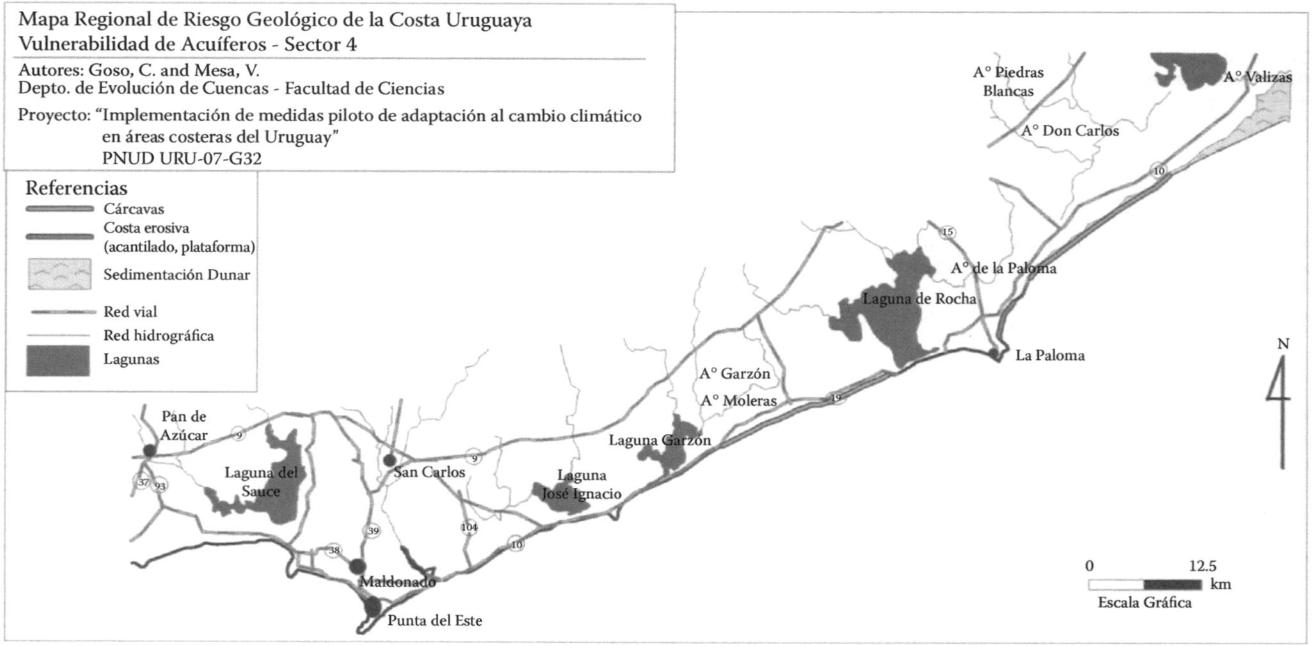

Figure 16.2 Atlantic estuarine lagoons system (eastern coast). (Adapted from Bidegain, M. et al., *Escenarios de cambio climático y del nivel medio del mar e impactos de los mismos en áreas costeras y evaluación de la vulnerabilidad geológica costera*, Reporte de la Facultad de Ciencias al Proyecto GEF-Implementación Medidas de Adaptación al Cambio Climático en Areas Costeras del Uruguay [ACCC], Unidad de Cambio Climático [UCC], MVOTMA, Uruguay, 2009, http://www.cambioclimatico.gub.uy/; SNAP, *Proyecto de ingreso del área Laguna de Rocha al Sistema Nacional de Áreas Protegidas [SNAP]*, DINAMA, Montevideo, Uruguay, Febrero 12, 2010.) (Note: Erosive beaches [dark gray] and gullies [gray] are shown to the east and west of Laguna de Rocha, respectively. Sea-level data are taken at La Paloma, a few kilometers to the east of Laguna de Rocha.)

Table 16.1 Current and future climatic and sea-level scenarios for the Uruguayan coast

Variable/period *Trend/variability*	*Long-term trend*	*1961–2003*	*2000–2008*	*2025*	*2050*	*2085*
Sea-level pressure	$\downarrow$ 3 HPa	Small change. $\downarrow$ 0.5 HPa > eastward and $\uparrow$ variability (Δ)		$\uparrow$ according to ECHAM-5. $\downarrow$ according to HADCM-3.		
Wind	< Wind intensity since 1961 especially since 2000			Likely increase in ESE-SSE.		
ESE-SSE	> ESE-SSE (became prevailing by 1998) and E and N frequency increases during El Niño (Gutierrez 2011).					
SW	≤ SW, as well as < N. SW frequency increases during La Niña (Gutierrez 2011)					
River flow (Δ%) Freshwater inflow to the Rio de la Plata (Q_F)	≥ 30 $\uparrow$ Δ $\uparrow$ Floods	≥ 20	Q_F (Q_P + Q_U):−9	Uncertain. Projected precipitation and temperature fields have different balances. Most literature agrees on uncertainty and a change by −5% to +20% since 1990. Likely increase of $Q_U > Q_P$.		
Paraná River (Q_P) Uruguay River (Q_U)	≥30% $\uparrow$ Δ $\uparrow$ Floods $\downarrow$ Hydro-climatic homogeneity of the basin by ca. 2002		Q_P: +1 Q_U: −35. Trend reverted by 2003–2004.			
Water/sea level (m)				Expected SLR +30–60 cm plus/minus river flow and/or wind effects. Likely change in seasonal regime and storm surges.		
Colonia	≥0.20 m		Trend reverted (Mv) by 2004 due to $\downarrow Q_U$.			
Montevideo (Mv)	≥0.11 m	+0.08 m				
Punta del Este	≥0.11 m, short time series (34 yrs), discontinuous, with high rate 3.3 mm/yr.					
La Paloma, Rocha	≥0.17 m, short time series (53 yrs), 3.3 mm/yr.					

Source: Bidegain, M. et al., *Escenarios de cambio climático y del nivel medio del mar e impactos de los mismos en áreas costeras y evaluación de la vulnerabilidad geológica costera*, Reporte de la Facultad de Ciencias al Proyecto GEF-Implementación Medidas de Adaptación al Cambio Climático en Areas Costeras del Uruguay (ACCC), Unidad de Cambio Climático (UCC), MVOTMA, Uruguay, 2009, http://www.cambioclimatico.gub.uy/.

The climate change drivers used in the pilot site impact analysis are shown in Table 16.1. Potential changes to the functioning of the Rio de la Plata estuarine frontal zone (the zone of freshwater–saline transition) will have both significant socioeconomic and environmental impacts. The majority of Uruguay's population and economic assets are located on the estuary–front coast (i.e., the metropolitan area of Montevideo, see Figure 16.1).

In addition, the aquatic salinity frontal zones are sensitive ecological areas, important for fisheries and spawning habitats (Brazeiro and Defeo 2006; Defeo et al. 2009), despite their current overexploitation (INFOPESCA 2001). These frontal zones are key habitats for biodiversity and functional integrity of the Rio de la Plata estuary. The location and productivity of artisanal fisheries are strongly related to the location of the frontal zones. Where these zones occur is linked to seasonal wind patterns and seasonal and interannual fluctuations of the Rio de la Plata tributaries discharges (Nagy et al. 2008b, 2008c; Box 16.1). There is evidence that climate change will continue to affect biodiversity, with consequences for distribution and alterations in reproduction times (Defeo et al. 2004).

The eastern Atlantic coastal lagoons system, within which the Laguna de Rocha is situated, is a biodiversity region of global relevance (Ramsar site wetlands). It is characterized by a natural connectivity between lagoons, their basins, and the sea (sand barrier), a Coastal Protected Area in Laguna de Rocha (Figure 16.2), and by exposed small rural and fishing communities. Importantly, the local municipality holds great concerns regarding potential climate change impacts in the area. These factors in combination make this an ideal site to develop participatory planning processes, implement adaptation measures, and help increase community resilience to better cope with climate change impacts.

TAKING ACTION TOWARD ADAPTATION—A FOUR-STEP APPROACH

In order to assess the potential risks of climate change on the pilot sites, four *cascading* supporting streams of activity were undertaken, given as follows:

1. Vulnerability mapping to consider key system drivers
2. Baseline Vulnerability Reduction Assessment (VRA)
3. Development of a customized risk management conceptual model (MESA)
4. Multicriteria approaches for selecting adaptation options

These four activities combined created a unique approach for addressing some of the key climate change issues facing the pilot site and identifying adaptation options. Applied for the first time in Latin America, one of the key aspects of this approach, later acknowledged as a success factor by project stakeholders, was the embedding of climate change into key institutions involved through the application of participatory approaches previously applied within the EcoPlata ICZM Programme.

The ACCC Project has developed together with EcoPlata Program and Territorial Climate Change Project (CCT) all the four supporting streams of activity in the metropolitan area of Montevideo (three municipalities). Also, it provided advice to Rocha's Local Government (IDR) and the National System of Protected Areas (SNAP) regarding the incorporation of climate issues and risks within a management plan for the Laguna de Rocha. Finally, the project is advising the authorities in charge of Environmental Impact Assessment (EIA) on the incorporation of climate risks into national-level EIA procedures for coastal areas. Three of the activity streams are outlined in greater detail subsequently.

Vulnerability reduction assessment

The ACCC VRA undertaken followed the process required of UNDP-implemented adaptation projects (Droesch et al. 2008). However, the project team customized the process to provide baseline information on stakeholders' perception of vulnerability and adaptive capacity across the country's coastal departments, and additional information useful for the project. The evaluation was impact-oriented in order to understand the stakeholders' perception of vulnerability. Relative change in reference to a baseline was assessed. Data were also collected on perceptions from the recent past by asking stakeholders to think about their perception of climate change 5–10 years ago (Seijo 2010).

The VRA surveys were undertaken during six workshops where stakeholders (community and municipal managers, elected officials, nongovernmental organizations, and local representatives) were informed by ACCC experts about climate issues and risks. Then, they were requested to assess the main coastal climate threats and impacts, as well as the barriers, needs, and opportunities to adapt.

Survey results showed that stakeholder perceptions and identification of climate vulnerability, risks, and impacts were similar to expert judgment for the Laguna de Rocha site (Table 16.2). Consulted experts were from academia (University of the Republic), national government (Directorate of the Environment), and nongovernmental organizations. The process undertaken combined meetings, workshops, and interviews where experts were requested to qualify impacts and risks on a 1–5 scale.

Importantly, the VRA process was a useful tool for engaging stakeholders in the project and also in engaging them on the issue of climate change impacts overall.

Table 16.2 Drivers and climate variables in Laguna de Rocha site based on an expert judgment survey (1–5 scale)

Relative importance of drivers of change and climatic and oceanographic variables (1–5)[a]	*Subsystems*						
	L-W	*B*	*SBa*	*MZ*	*SBe*	*CC*	*X*
Precipitation	4–5	5	4	3–4	2–3	3	3.7
Temperature	4	3–4	1–2	3	2–3	4–5	3.3
Land-use (basin and coast)	4	5	4–5	2–3	4–5	4–5	4.0
Wind	3–4	2–3	4–5	3–4	3–4	1–2	3.3
Tide	2	1	2–3	2	2–3	1	1.8
River flow (entering the Rio de la Plata)	2–3	1	2–3	3–4	2	2	2.2
Sea-level rise	3–4	1–2	5	2–3	3	2–3	2.9
Average (X)	3.1	2.8	2.7	2.9	2.9	2.8	

Source: Nagy, G.J., *Modelo Evolutivo Sistémico de Adaptación (MESA): Bases generales, Sitio piloto Frente Salino y franja costera, Sitio piloto Laguna de Rocha y Adyacencias*, Reporte de proyecto GEF, Implementación Medidas de Adaptación al Cambio Climático en Areas Costeras del Uruguay (PACCC), MVOTMA, Unidad de Cambio Climático (UCC), Montevideo, Uruguay, 2010b, http://www.cambioclimatico.gub.uy.

Note: Although experts prioritize long-term warming, land-use, and sea-level changes, the VRA shows that stakeholders prioritize short-term extreme events, inundation, and erosion.

Subsystems: B, basin; CC, connectivity corridors; L-W, lagoon-wetlands; MZ, marine zone; SBa, sand barrier; SBe, sandy beach.

[a] Conservation of biodiversity of global relevance and habitats is the overarching goal of both ACCC and SNAP, whereas it is very important for local government.

Risk management model

The ACCC Project is developing a conceptual risk-based systemic evolution adaptation model (in Spanish, Modelo Evolutivo Sistémico de Adaptación, or MESA) for the pilot sites of the estuarine frontal zone and adjacent coast and Rocha coastal lagoon (Figures 16.1 and 16.2). The approach to the MESA is based on experience from Australia in developing geomorphic models of coastal response to climate change (Kay 2009; Kay et al.; Chapter 11). This approach seeks to integrate knowledge of historic coastal system sensitivity to coastal drivers with consensus-led evaluation of potential future impacts. The approach seeks to combine expert judgment, stakeholder's experience, and analysis of potential future change to key variables that will likely drive coastal change (Table 16.1; Box 16.2) and that are themselves subject to stakeholder analysis to develop subjective scores of the relative importance of drivers (Table 16.2).

The MESA approach was implemented in the Laguna de Rocha pilot site to adapt the management of a protected coastal area to cope with the risks of climate change in a globally relevant biodiversity site (CCT 2010; IDR 2010; Nagy 2010b; SNAP 2010). Two phases were used: (1) technical with National System of Protected Areas (SNAP) and the municipality's experts;

and (2) participatory with stakeholders, including through the VRA. The successive activities were to define/identify (1) partners and common objectives; (2) climatic risks; (3) stakeholders and champions; (4) impact and success criteria; (5) management horizons; (6) risk management approach and thresholds for acceptable/unacceptable risks at management horizons; and (7) selection of measures by stakeholders (Table 16.3). The thresholds and risk management horizons were still in the early stages of development while this chapter was being completed (December 2011).

Table 16.3 Selected adaptation actions/measures and implementation status by late 2011

Adaptation action/measure	*Priority*	*Level of application(s)*	*Type of measure(s)*	*Status*
Soft works to maintain and stabilize the living coastline	Short term	Departmental	Biophysical	Selected
Incorporation of vulnerability and adaptation to climate change in the management plan of protected area Laguna de Rocha	Short term	Local: Laguna de Rocha Pilot site	Management	Early
Management protocol for the open–close system of the bar of the Laguna de Rocha	Short term	Local: Laguna de Rocha Pilot site	Management/ biophysical	Early
List of globally relevant migratory birds threatened sites and list of invasive species and vulnerable coastlines	Short term	Local and departmental	Information/ monitoring	Selected
Zoning, assessment, and mapping of habitats and ecosystems vulnerable to climate issues	Short term	National and departmental	Information/ monitoring	Early
Monitoring of beach profiles along the coast: erosion, retreat, and sediment balance	Short term	All scales	Monitoring	Advanced
Preparation of a guide to incorporate climate issues in the Environmental Impact Assessment-Strategic Environmental Assessment (EIA-SEA)	Short term	National	Management	Early
Preparation of a guide to incorporate climate issues in local land-use plans	Medium term	Departmental	Management	Early

(Continued)

Table 16.3 (Continued) Selected adaptation actions/measures and implementation status by late 2011

Adaptation action/measure	*Priority*	*Level of application(s)*	*Type of measure(s)*	*Status*
Identify biological corridors in the Eastern coastal lagoons	Medium term	Departmental and national	Management/ monitoring	Selected
Incorporation of climate issues to the closed fishing areas according to the Eco-System Approach to Fisheries, Food and Agriculture Organization of the United Nations (FAO)	Medium term	National	Information/ management/ monitoring	Selected
Incorporation of climate issues and risks to the plans of comanagement of protected fishing areas defined by the La Dirección Nacional de Recursos Acuáticos (DINARA)	Medium term	Local: Frontal zone pilot site	Information/ management/ monitoring	Selected
Awareness and dissemination to decision makers and elected representatives	Medium term	All scales	Outreach	Early
Economic impact assessment for the observed and expected climate issues	Medium term	All scales	Information	Early
Adequacy of codes/standards of construction considering climate issues	Medium term	All scales	Management	Selected
Train local journalists	Long term	All scales	Outreach	Early
Training of National Emergency System managers and elaboration of guidelines for the management of climate risks	Long term	Departmental to national	Outreach/ management	Early
Environmental Education on climatic risks	Long term	All scales	Outreach	Early
Support for implementation of management plans for natural pasture (incorporating climate issues) in Laguna de Rocha for feeding and nesting sites for birds	Long term	Local	Information/ management	Selected
Community Emergency Plans for climatic extreme events	Long term	Local	Management/ monitoring	Selected
Preparation of a guide to mainstream climate risk management in the coastal tourism sector	Long term	National and departmental	Outreach/ management	Early

Note: Climate issues include climate change, variability, and extremes.

Multicriteria approach to adaptation option selection

The final step was the prioritization of adaptation actions. The process applied was based on nine criteria drawn from international adaptation projects that have used multicriteria analysis for the selection of adaptation options, for example, the Kiribati Adaptation Project, Phase II (Elrick et al. 2010). The criteria were weighted according to the time frames defined under the ACCC Project, as detailed in Table 16.4. Three lists were compiled with recommended measures to be implemented in the short term (nine months), in the medium term (18 months), and with implementation measures that addressed most project objectives. Key aspects of the methodology applied (Brizikova et al. 2008; Nagy 2010a; Nagy et al. 2011; NOAA 2010) are as follows:

Table 16.4 Adaptation measures decision-making matrix

Proposed measure prioritization criteria	*Example of criteria (from 37) to assess and rank the measure*	*Used scales*
1. Social acceptability	Is it likely that community oppose/support?	1–5
2. Political acceptability	Is there a champion?	1–2/1–5
3. Technical pertinency/applicability	Does it solve a problem or a symptom?	1–5
4. Implementation capacity (resources, administrative)	Does the implementation need external support?	1–2/1–5
5. Legal/normative/administrative viability	Is there an adequate norm to regulate the implementation?	1–2
6. Economic/financial analysis and viability	Is the initial cost of implementation proportional to the impacts of the problem and the benefits?	1–5
7. Environmental pertinency/viability	Is the measure compatible with ecological sustainability?	1–5
8. Urgency, risk, and uncertainty	Is the risk likely in the short-, middle-, long-term?	1–5
9. No-regret and multiple benefits	Does it contribute to nonclimatic issues?	1–2/1–5

Source: Nagy, G.J., *Borrador de Criterios de Evaluación, Selección y Priorización de Medidas de Adaptación*, Reporte de proyecto GEF Implementación Medidas de Adaptación al Cambio Climático en Areas Costeras del Uruguay (PACCC), MVOTMA, Unidad de Cambio Climático (UCC), Montevideo, Uruguay, 2010a, http://www.cambioclimatico.gub.uy; Brizikova, L. et al., *Canadian Communities' Guidebook for Adaptation to Climate Change: Including an Approach to Generate Mitigation Co-Benefits in the Context of Sustainable Development*, Environment Canada and University of British Columbia, Vancouver, British Columbia, Canada, 2008; National Oceanic and Atmospheric Administration (NOAA), *Adapting to Climate Change: A Planning Guide for State Coastal Managers*, National Oceanic and Atmospheric Administration, Silver Spring, MD, 2010.

Note: Stakeholders rank each criterion using 1–5 or yes/no scale. Thirty-seven criteria have been proposed.

- Multiple sources of input—experts, government staff, and key community-level stakeholders
- Transparency of criteria (and wide-ranging criteria)
- Consideration of implementation barriers

The measures selected were then categorized into the following:

- *Information*: Knowledge and scientific information
- *Outreach*: Training, outreach, and awareness
- *Management*: Policies, programs, and plans
- *Monitoring*: Systems monitoring, modeling, and early warning based on knowledge and information
- *Biophysical*: Biophysical intervention, such as soft engineering (anti-wind fences, dune vegetation); ecosystem restoration; and physical management of bars, beaches, and dunes.

As of December 2011, the ACCC Project has used the above processes to select a suite of prioritized adaptation measures. These measures, together with their implementation status, are shown in Table 16.3. Importantly, the scale of implementation was also identified during this process. Most short-term options are in the early stages of implementation. These actions included planning soft adaptation measures and mapping and assessment activities. These outcomes indicate the emergence of a strong foundation on which to base longer-term actions. Long-term actions were focused on education and outreach activities, such as developing a guide to mainstream climate risk management in the coastal tourism sector.

EMERGING LESSONS LEARNED

Analysis of the combined experience of EcoPlata and ACCC, particularly the emerging results from the adaptation action selection and implementation process, has revealed some significant lessons learned.

First, coastal adaptation efforts need to build on, and support, existing frameworks for ICZM efforts to strengthen coastal zone management. For example, there is a need to implement a comprehensive suite of coastal management regulations and eliminate the overlapping jurisdictions of different management agencies. Second, the enhanced coordination in assessment of extreme event-related impacts was the main driver of increased awareness. For instance, the National System to Respond to Climate Change and Variability (SNRCC 2010) was created only after the severe drought during 2008–2009. However, it focuses on current climate variability.

Third, providing a strong scientific basis and understanding around coastal processes and climate change has proved to be very effective in moving the adaptation agenda forward in a country. It has also helped to incorporate climate change risks into national and regional legislation and into policies. The mainstreaming of climate change issues has been a significant success of the engagement process undertaken for the VRA and provides a key lesson learned about early engagement with stakeholders. Further information on these lessons is provided in Gómez-Erache et al. (2010) and Nagy (2010a).

Lastly, there are broad lessons emerging with respect to how the use of risk assessments and associated technical analysis is working collaboratively with efforts to build on ICZM initiatives. Using the typology of five climate change risk assessment *generations* proposed by Jones and Preston (2010), Uruguayan experience suggests that early work in identification and awareness raising of the problem during the mid- to late 1990s (i.e., the first and second risk assessment generations) has evolved through the second and third generations of risk analysis and risk evaluation until 2010. It is planned that by the end of the ACCC Project, the fourth generation of risk management that seeks to mainstream risk management into effective adaptation policy and practice will have been completed. This should provide a platform to support the fifth and final generation of adaptation action implementation and monitoring that seeks to evaluate the benefits of well-defined and well-justified adaptation actions.

MOVING FORWARD WITH ADAPTATION PLANNING IN URUGUAY

The experience in Uruguay in developing coastal adaptation options has highlighted the direction for future technical work and potential *key success factors* in enhancing implementation of future adaptation activities. Some suggested future activities follow:

- Further research on physical processes and thresholds; determine when and why a sector and/or resource will be irreversibly impacted.
- Complete an update of climate and nonclimate data, including following up baselines and stakeholder perceptions, as both are needed for adaptive management.
- Expand the number of local-scale vulnerability assessments and adaptation plans because they are close to local stakeholders.
- Merge top-down and bottom-up approaches to integrate stakeholder perceptions and knowledge about the risks from climate change to help inform future adaptation planning.

- Develop and prioritize a step-by-step planning and implementation process of affordable, easy, cost-effective, and accepted measures, in order to facilitate further actions that are more difficult.
- Plan and implement adaptation actions aimed at attaining multiple goals for stakeholders, including impact assessments, capacity building, mapping, monitoring, and participative decision-making.
- Identify local *champions* (individuals or organizations) who are able to promote the need for climate change adaptation activities and to form effective bridges between local communities, academia, and national agencies. Based on our experience, this is a key factor that has emerged from the participatory approach adopted.
- Assess both the market and nonmarket assets and economic impacts of climate risks on coastal resources and livelihoods in order to achieve a balance between economic development and maintaining ecological values. This would ideally be completed through an informed, participative process.

Successful implementation of the process outlined above in the pilot sites also sets the stage for extending the pilot analysis to a national adaptation response. Further improvement of coordination between programs, academia, agencies, and local governments is still needed. The choice of adaptation measures followed an international best practice approach, being both an informed top-down and a participatory bottom-up process. The decisions made and adaptation measures selected were agreed on the basis of ACCC implementation time horizons, overall acceptability and viability, cost benefit, and effectiveness. This is no guarantee that the identified measures will all be achieved, but a very positive sign is that several have already commenced implementation.

POSTSCRIPT

Since this chapter was written in 2011, Uruguay's socioeconomic context has continued its overall growth at a slower pace. Also, the country's status has been updated from *upper middle income country* to *high income country* (World Bank 2013). The process of implementation of the proposed adaptation measures (see Table 16.4) has progressed steadily. Forty percent of these measures are advanced or already implemented by July 2013 (e.g., the management plan of Laguna de Rocha and the management protocol of the bar [Rodriguez-Gallego et al. 2012], the list of globally relevant migratory birds and the list of invasive species [Brugnoli et al. 2012], the mapping of vulnerable ecosystems or the soft works). Thirty-eight percent of the measures have improved, and 22% remain unchanged (e.g., the economic climate impact assessment).

The most important advances since early 2011 have been

1. The completion of the VRA allowed elicitation of stakeholders' perception of climate threats, as well as the barriers, supports, and opportunities for adaptation (Nagy et al. 2014),
2. The improvement of biophysical adaptation interventions (soft measures) such as dune vegetation at Rocha and their replication within two new municipalities,
3. The local and national authorities will incorporate the bar protocol within the Lagoon Management Plan (Nagy et al. 2014), and
4. New information and monitoring measures to forecast and monitor the influence of El Niño/Southern Oscillation (ENSO) variability on river flow, salinity, and beach water quality status are being implemented with the collaboration of the Municipal Government of Montevideo, Uruguay.

REFERENCES

ACCC. (2008). "Implementing pilot sites of adaptation measures to climate change in the Uruguayan coastal areas," Project document prepared for UNDP/GEF Trust Fund. http://www.adaptationlearning.net/ (accessed January 2012).

Bidegain, M., Caffera, R.M., Pshennikov, V., Lagomarsino, J.J., Nagy, G.J., and Forbes, E.A. (2005). "Tendencias climáticas, hidrológicas y oceanográficas en el Río de la Plata y costa Uruguaya," in V. Barros, A. Menéndez, and G.J. Nagy (eds.), *El Cambio Climático en el Río de la Plata, vol. 14*, Buenos Aires, Argentina: Centro de Investigaciones del Mar y la Atmósfera, Universidad de Buenos Aires.

Bidegain, M., de los Santos, B., de los Santos, T., de los Santos, M., Goso, C., Pshennikov, V., and Severov, D.N. (2009). *Escenarios de cambio climático y del nivel medio del mar e impactos de los mismos en áreas costeras y evaluación de la vulnerabilidad geológica costera*, Reporte de la Facultad de Ciencias al Proyecto GEF-Implementación Medidas de Adaptación al Cambio Climático en Areas Costeras del Uruguay (ACCC), Unidad de Cambio Climático (UCC), MVOTMA, Uruguay. http://www.cambioclimatico.gub.uy/ (accessed January 2012).

Brazeiro, A., Acha, E.M., Mianzán, H.W., Gómez-Erache, M., and Fernández, V. (2004). *Aquatic Priority Areas for the Conservation and Management of the Ecological Integrity of the Río de la Plata and its Maritime Front*, Reporte Científico del Proyecto Freplata 1-2.2, PNUD/GEF RLA/99/G31. www.freplata.org.

Brazeiro, A. and Defeo, O. (2006). "Bases ecológicas y metodológicas para el estudio de un Sistema Nacional de Áreas Marinas Protegidas en Uruguay," in R. Menafra, L. Rodríguez-Gallego, F. Scarabino, and D. Conde (eds.), *Bases para la conservación y el manejo de la costa uruguaya*, Montevideo, Uruguay: Vida Silvestre Uruguay.

Brizikova, L., Neale, T., and Burton, I. (2008). *Canadian Communities' Guidebook for Adaptation to Climate Change: Including an Approach to Generate Mitigation Co-Benefits in the Context of Sustainable Development*, Vancouver, British Columbia, Canada: Environment Canada and University of British Columbia.

Brugnoli, E., Guerrero, J., Carvajales, A., Lanfranconi, A., and Muniz, P. (2012). *Especies Invasoras y Cambio Climático: Identificación y Estudios de caso sobre medidas de Adaptación al Cambio Climático en Áreas Costeras del Uruguay*, Working paper, GEF Project "Implementing pilot adaptation measures in coastal areas of Uruguay," UCC-DINAMA-UNDP, Oceanografía y Ecología Marina, IECA, Facultad de Ciencias, UdelaR, Montevideo, Uruguay.

CCT (Proyecto ART Cambio Climático Territorial). (2010). *Síntesis del Taller Metropolitano de Adaptación al Cambio Climático para el sector Costas*, documento de Proyecto ART Cambio Climático Territorial.

Comisión Nacional sobre el Cambio Global (CNCG). (1997). *Assessment of Climate Change Impacts in Uruguay: Uruguay Climate Change Country Study Final Report*, Montevideo, Uruguay: CNCG.

Dasgupta, S., Laplante, B., Meisner, C., Wheeler, D., and Yan, J. (2007). *The Impact of Sea Level Rise on Developing Countries: A Comparative Analysis*, Policy Research Working Paper 4136, Washington, DC: World Bank. http://econ.worldbank.org/ (accessed January 2012).

Defeo, O., de Álava, A., Gómez, J., Lozoya, J.P., Martínez, G., Riestra, G., Amestoy, F., Martínez, G., Horta, S., Cantón, V., and Batallés, M. (2004). "Hacia una implementación de áreas marinas protegidas como herramientas para el manejo y conservación de la fauna marina costera en Uruguay," *Jornada de Comunicación Científica del PDT*, 1: 81–7.

Defeo, O., Horta, S., Carranza, A., Lercari, D., de Alava, A., Gómez, J., Martínez, G., Lozoya, J.P., and Celentano, E. (2009). *Hacia Un Manejo Ecosistémico De Pesquerías: Areas Marinas Protegidas En Uruguay*, Montevideo, Uruguay: Facultad de Ciencias-DINARA.

Droesch, A.C., Gaseb., N., Kurukulasuriya, P., Mershon, A., Moussa, K.M., Rankine, D., and Santos, A. (2008). *A Guide to the Vulnerability Reduction Assessment*, UNDP Working Paper, UNDP Community-Based Adaptation Programme. http://www.undp-adaptation.org/ (accessed January 2012).

ECLAC. (2010). *Economics of Climate Change in Latin America and the Caribbean: Summary 2010*, Santiago, Chile: Comisión Económica para América Latina y el Caribe.

Economist Intelligence Unit (EIU) (2010). *Democracy Index 2010*. http://graphics.eiu.com/PDF/Democracy_Index_2010_web.pdf (accessed January 2012).

EcoPlata. (2010). "Hacia una estrategia nacional para la gestión Integrada de la Zona Costera 2010–15," lineamientos para la discussion, EcoPlata, Febrero 2010.

Elrick, C., Kay, R., and Travers, A. (2010). *Integrating Climate Change and Coastal Zone Management, CZM White Paper No.1, A Capacity Driven Approach in the Republic of Kiribati*. http://www.coastalmanagement.com/resources.html (accessed January 2012).

Gómez-Erache, M.G., Conde, D., and Villarmarzo, R. (2010) *The Sustainability of Integrated Management in the Coastal Zone of Uruguay: Connecting Knowledge to Action*, Montevideo, Uruguay: EcoPlata. http://www.ecoplata.org/documentos/ (accessed January 2012).

Gómez-Erache, M.G. and Martino, D. (2008). "Zona Costera," in *GEO Uruguay 2008*, Montevideo, Uruguay: UNEP DINAMA, CLAES, EcoPlata. http://www.ecoplata.org/documentos/ (accessed January 2012).

Gutiérrez, O. (2011). Sedimentary dynamics in the Uruguayan coast: Evolution and trends of urban beaches within global change framework, Masters thesis, Montevideo, Uruguay: Faculty of Sciences, University of the Republic (UdelaR).

INFOPESCA. (2001). *Informe Proyecto Gestión Marítima- Componente Pesquero*, Diagnóstico de los recursos pesqueros en Uruguay Informe presentado al Banco Mundial, Washington, DC.

Intendencia Departamental de Rocha (IDR). (2010). *Informe Ambiental Estratégico.* Plan de ordenamiento Territorial "Lagunas Costeras." Sector comprendido entre Laguna de Rocha y Laguna Garzón. Intendencia Departamental de Rocha, Julio, 2010.

Jones, R.N. and Preston, B.L. (2010). *Adaptation and Risk Management*, Climate Change Working Paper No. 15, Melbourne, Victoria, Australia: Centre for Strategic Economic Studies Victoria University.

Kay, R.C. (2009). *Implementing Pilot Climate Change Adaptation Measures in Coastal Areas of Uruguay Reviewing Project Progress to Date and Advice on Priorities for the Annual Operational Work Plan 2010*, A report (PROJECT URU/07/G32), Perth, Australia: Coastal Zone Management Pty Ltd.

Legatum Institute. (2010). *The Legatum Prosperity Index 2010.* http://www.prosperity.com/ (accessed January 2012).

Magrin, G., García, C.G., Choque, D.C., Giménez, J.C., Moreno, A.R., Nagy, G.J., Nobre, C., and Villamizar, A. (2007). "Latin America" in M.L. Parry, O.F. Canziani, J.P. Palutikof, P.J. van der Linden, and C.E. Hanson (eds.), *Climate Change 2007: Impacts, Adaptation and Vulnerability. Contribution of Working Group II to the Fourth Assessment Report of the Intergovernmental Panel on Climate Change*, Cambridge; New York: Cambridge University Press. http://www.ipcc.ch/ (accessed August 2013).

Nagy, G.J. (2010a). *Borrador de Criterios de Evaluación, Selección y Priorización de Medidas de Adaptación*, reporte de proyecto GEF Implementación Medidas de Adaptación al Cambio Climático en Areas Costeras del Uruguay (PACCC), Montevideo, Uruguay: MVOTMA, Unidad de Cambio Climático (UCC). http://www.cambioclimatico.gub.uy (accessed January 2012).

Nagy, G.J. (2010b). *Modelo Evolutivo Sistémico de Adaptación (MESA): Bases generales, Sitio piloto Frente Salino y franja costera, Sitio piloto Laguna de Rocha y Adyacencias*, Reporte de proyecto GEF, Implementación Medidas de Adaptación al Cambio Climático en Areas Costeras del Uruguay (PACCC), Montevideo, Uruguay: MVOTMA, Unidad de Cambio Climático (UCC). http://www.cambioclimatico.gub.uy (accessed January 2012).

Nagy, G.J., Bidegain, M., Caffera, R.M., Blixen, F., Ferrari, G., Lagomarsino, J.J., Norbis, W., López, C.H., Ponce, A., Presentado, M.C., Pshennikov, V., Sans, K., and Sención, G. (2008a). "Climate and water quality in the estuarine and coastal fisheries of the Rio de la Plata," in N. Leary, J. Adejuwon, V. Barros, I. Burton, J. Kulkarni, and R. Lasco (eds.), *Climate Change and Adaptation*, London: Earthscan.

Nagy, G.J., Bidegain, M., Caffera, R.M., Norbis, W., Ponce, A., Pshennikov, V., and Severov, D.N. (2008b). "Fishing strategies for managing climate variability and change in the Estuarine Front of the Rio de la Plata," in N. Leary, J. Adejuwon, V. Barros, I. Burton, J. Kulkarni, and R. Lasco (eds.), *Climate Change and Adaptation*, London: Earthscan.

Nagy, G.J., Gómez-Erache, M.G., and Fernández, V. (2007). "El Aumento del Nivel del Mar en la costa uruguaya del Río de la Plata: Tendencias, vulnerabilidades y medidas para la adaptación," *Revista Medio Ambiente y Urbanización, Cambio climático: Vulnerabilidad y Adaptación en ciudades de América Latina*, 67: 77–93.

Nagy, G.J., Gómez-Erache, M.G., and Seijo, L. (2011). *Prioritization of Adaptation Measures in the Uruguayan Coastal Zone*, Project report for GEF funded project: Implementing pilot sites of adaptation measures to climate change in coastal areas of Uruguay (PACCC), Montevideo, Uruguay: MVOTMA, Unidad de Cambio Climático (UCC). http://www.cambioclimatico.gub.uy (accessed January 2012).

Nagy, G.J., Ponce, A., Pshennikov, V., Silva, R., Forbes, E.A., and Kokot, R. (2005). "Desarrollo de la Capacidad de Evaluación de la Vulnerabilidad Costera al Cambio Climático: Zona Oeste de Montevideo como Caso de Estudio," in V. Barros, A. Menéndez, and G.J. Nagy (eds.), *El Cambio Climático en el Río de la Plata, Project Assessments of Impacts and Adaptation to Climate Change (AIACC)*, Buenos Aires, Argentina: CIMA-CONYCET-UBA.

Nagy, G.J., Seijo, L., Bidegain, M., and Verocai, J.E. (2014). "'Stakeholders' climate perception and adaptation in coastal Uruguay," *International Journal of Climate Change Strategies and Management*, 6(1): 63–84.

Nagy, G.J., Severov, D.N., Pshennikov, V.A., De los Santos, M., Lagomarsino, J.J., Sans, K., and Morozov, E.G. (2008c). "Rio de la Plata estuarine system: Relationship between river flow and frontal variability," *Advances in Space Research*, 41: 1876–81.

National Oceanic and Atmospheric Administration (NOAA). (2010). *Adapting to Climate Change: A Planning Guide for State Coastal Managers*, Silver Spring, MD: National Oceanic and Atmospheric Administration.

OECD. (2010a). *Latin American Economic Outlook 2010*, Development Centre of the Organisation for Economic Co-Operation and Development. Washington, DC: OECD Publishing.

OECD. (2010b). *Strong Performers and Successful Reformers in Education*. http://www.PISA.oecd.org (accessed January 2012).

Robayna, A. (2009). "Presión antrópica en la costa uruguaya: Análisis sobre indicadores sobre turismo y transporte," *Informe EcoPlata*, Junio.

Rodriguez-Gallego, L., Nin, M., Suarez, C., and Conde, D. (2012). *Consultoría Técnica Para Apoyar el Proceso de Elaboración del Plan de Manejo del Paisaje Protegido Laguna de Rocha. Propuesta de Plan de Manejo*, Working paper, Sistema Nacional de Áreas Protegidas (SNAP), Futuro Sustentable S.A., Diciembre 2012.

Seijo, L. (2010). *Evaluación de la Reducción de la Vulnerabilidad (VRA)*, Reporte de proyecto GEF Implementando Medidas de Adaptación al Cambio Climático en Aéreas Costeras del Uruguay (PACCC), Montevideo, Uruguay: MVOTMA, Unidad de Cambio Climático (UCC). http://www.cambioclimatico.gub.uy (accessed January 2012).

Sención, G. (2009). *Economic Assessment of the Impacts of Climate Change in Coastal Resources in Uruguay*, Uruguay's report to ECLAC-RECC 2009.

SNAP. (2010). *Proyecto de ingreso del áreas Laguna de Rocha al Sistema Nacional de Áreas Protegidas (SNAP)*, Montevideo, Uruguay: DINAMA, Febrero 12, 2010.

SNRCC. (2010). *Plan Nacional de Respuesta al cambio Climático: Diagnóstico y lineamientos estratégicos. Sistema nacional de Respuesta al Cambio Climático y la Variabilidad*, Montevideo, Uruguay: MVOTMA.

Unidad de Cambio Climatico (UCC). (2004). *General Program of Measures for Mitigation and Adaptation to Climate Change in Uruguay*, Montevideo, Uruguay: MVOTMA/DINAMA, UCC. http://www.cambioclimatico.gub.uy (accessed January 2012).

Unidad de Cambio Climatico (UCC). (2005). *Uruguay's Second National Communication to the Parties in the UNFCCC*, Montevideo, Uruguay: Unidad de cambio climatico, DINAMA, MVOTMA.

United Nations Development Program (UNDP). (2011). "Human development statistical index." http://hdr.undp.org/en/statistics/hdi/ (accessed January 2012).

World Bank. (2010). "World development indicators database 2009." http://data.worldbank.org/indicator (accessed January 2012).

World Bank. (2013). World development indicators database 2012." http://data.worldbank.org/indicator (accessed August 2013).

Chapter 17

The promise of coastal management in Brazil in times of global climate change

Marcus Polette, Dieter Muehe, Mario L.G. Soares, and Bruce C. Glavovic

Abstract: In recent decades, Brazil's coast has been subject to rapid urbanization, industrialization, port expansion and development, tourism development, offshore oil and gas exploitation, and activities ranging from fishing to shrimp farming and coastal conservation. There is escalating conflict between many of these activities and inequitable and environmentally destructive practices are commonplace, especially near urban centers. Climate change is expected to intensify these conflicts, and sustainable coastal development will become even more elusive unless coastal communities build adaptive capacity and resilience. This chapter explores the promise, challenges, and opportunities that Brazil's coastal management provisions present for mainstreaming climate change and building resilience and sustainability. Brazil's coastal management regime offers considerable potential for integrating climate change adaptation into local community planning and decision-making. Challenges and opportunities for translating legal rhetoric into practical reality are identified. It is recommended that a new program be established to pilot climate change adaptation projects in each of the 17 coastal states of Brazil with a focus on (1) raising public awareness about the coast and climate change through active social learning processes; (2) creating meaningful opportunities for public participation in coastal management efforts; and (3) integrating and mainstreaming coastal management, disaster risk reduction, and climate change adaptation efforts.

INTRODUCTION

The Brazilian coast is a domain of incredible environmental, cultural, and economic importance that is subject to unsustainable practices that will be exacerbated by climate change. This chapter explores the promise, challenges, and opportunities that Brazil's coastal management provisions present for mainstreaming climate change and building resilient and sustainable coastal communities.

Climate change poses a significant threat to coastal biodiversity, especially for climate-sensitive ecosystems like coral reefs, mangroves, and estuaries, and the ecosystem goods and services that underpin coast-dependent livelihoods (Copertino et al. 2010; Dominguez 2004; Muehe 2006, 2010; Neves and Muehe 2008a, 2008b; Nicolodi and Petermann 2010). According to Muehe (2010), climate change impacts are likely to be differentially experienced along the coast, with coastal erosion exacerbated by projected sea-level rise affecting the entire coast to varying degrees; extreme events becoming increasingly problematic, especially along the southern coast; and low-lying densely populated urban centers, like Recife in the state of Pernambuco, being extremely vulnerable to coastal erosion and inundation (Nicolodi and Petermann 2010). Based on Intergovernmental Panel on Climate Change emission scenarios, Margulis et al. (2010) estimate that material losses along the Brazilian coast due to climate change could be in the order of USD 75–115 billion. Preliminary estimates suggest that adaptive measures to reduce such losses could cost less than USD 2 billion if implemented before 2050: clearly demonstrating the imperative to proactively chart adaptation pathways.

Little attention has, however, been focused on barriers and opportunities for adapting to climate change at the coast. Most work to date has focused on Brazil's contribution to climate change mitigation and to a lesser extent on projected impacts and vulnerability in terrestrial settings (e.g., Copertino et al. 2010; Salazar et al. 2007). There is emerging research and action at the local level, for example, a multidisciplinary coastal zone network has been established in the National Institute of Sciences and Technology for Climate Change and studies have been initiated on aspects of vulnerability to climate change and adaptation at the coast (see, e.g., Copertino et al. 2010), and some coastal cities are exploring actions to address concerns about climate change (Barbi and Ferreira 2013; Costa et al. 2010; Ferreira et al. 2011). Much remains to be done. This chapter aims to contribute to this endeavor by exploring how adaptation can be mainstreamed into coastal management efforts. The chapter starts by describing Brazil's coastal setting and the issues facing coastal communities in times of global climate change. The next section explores climate risks and the vulnerability of the coast and its communities to climate change impacts and the implications for building adaptive capacity, resilience, and sustainability. Attention is then focused on the institutional framework for coastal management in Brazil and how historic efforts have shaped coastal sustainability; and what the prospects, challenges, and opportunities are in the face of climate change. Finally, recommendations are made for realizing the promise of coastal management by fostering adaptive capacity, resilience, and sustainability.

THE BRAZIL COAST: SETTINGS, ISSUES, AND PROSPECTS

This section describes the geographic, demographic, social, economic, cultural, and institutional context of the Brazilian coast, with a focus on what makes this coast distinctive, the issues facing coastal communities, and their prospects for the future.

The geographic setting

The Brazilian coast extends for over 8,500 km and encompasses more than 500,000 km^2 of varying climatic and biophysical conditions that give rise to diverse coastal ecosystems and habitats—from mud coasts and mangroves in the north to sandy coasts in the south, as well as coral reefs, wetlands, estuaries, coastal lagoons, beaches, dunes, headlands, and rocky shores. Taking into account geological, geomorphological, and climatological aspects, and building on previous studies on the physiography of the Brazilian coast, Muehe (2010) portrays geomorphological aspects relevant to climate change in Figure 17.1.

Demographic, economic, social, cultural, and institutional setting

The coastal zone is divided into 400 municipalities in 17 coastal states but makes up only about 5% (257,148 km^2) of the land area of the country. More than 50 million people, over 25% of the population, live in this narrow strip along the seashore, concentrated in urban centers, notably Belém, Fortaleza, Recife, Salvador, and Rio de Janeiro. Up to 40% of the population could be considered coastal if one takes into account urban centers, such as São Paulo, within 50 km of the seashore (Polette and Seabra 2011). Importantly, the coastal population is growing rapidly, increasing by over 70% between 1980 and 2010 (Figure 17.2).

The average population density in Brazil is 22.3 inhabitants/km^2, but in coastal municipalities the average population density is 5.5 times greater (123 inhabitants/km^2), a figure that changes dramatically in popular tourist destinations during the summer season. Population size and density varies markedly around the coast with the most densely populated localities being Pernambuco (803 inhabitants/km^2) and Rio de Janeiro (656.5 inhabitants/km^2). Large areas of the coast, especially in the north, are sparsely populated. Coastal municipalities have about 17.4 million households, of which 9.2% are used only in summertime. About 36% of coastal municipalities have up to 20,000 inhabitants; 31% have 20,000–50,000 inhabitants; 13% have 50,000–100,000 inhabitants; 17% have 100,000–500,000 inhabitants; 2% have 500,000 to one million inhabitants; and 1.5% have

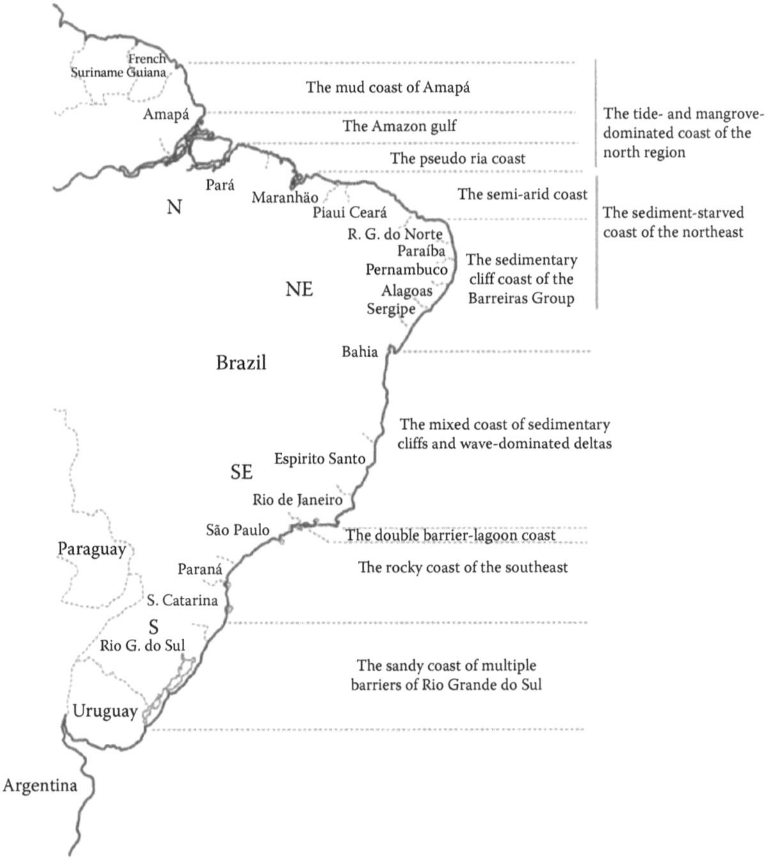

Figure 17.1 Geomorphological features of the Brazilian coast. (Data from Muehe, D., *Pan-American Journal of Aquatic Sciences*, 5(2), 173–83, 2010, http://www.panamjas.org/artigos.php?id_publi=183.)

over one million inhabitants. About 34% of the coastal population live in the six largest coastal cities with more than one million inhabitants each (see Table 17.1).

The significance of the coast to Brazilian society, economy, and culture is evident in the many ways in which the coast shapes people's identity and livelihoods. Economic activities at the coast account for about 73% of the national gross domestic product (Asmus and Kitzmann 2004). The diversity and juxtaposition of coastal activities and cultures is striking. Contrast the iconic playgrounds of Rio's beaches with the abject poverty of those living in slums (favelas); subsistence fishing versus port and industrial development; and oil and gas exploitation versus marine protected areas. Heavily urbanized and industrialized centers contrast starkly with

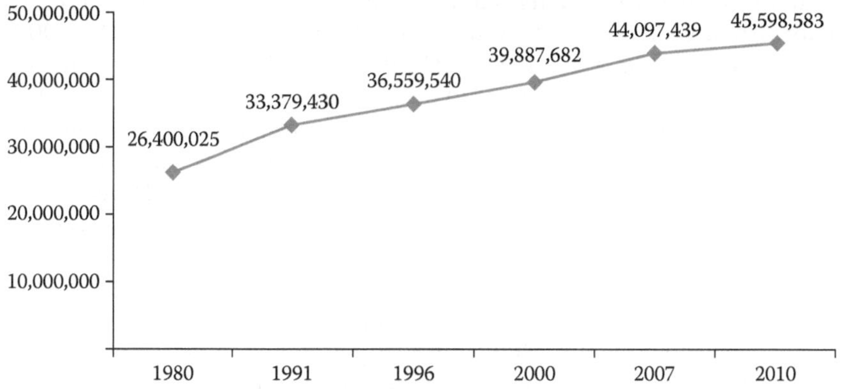

Figure 17.2 Population growth in the Brazilian coast 1980–2010. (Data from IBGE, "Censo Demográfico Brasil 2010," 2011, www.ibge.br, compiled by Polette.)

Table 17.1 Brazilian coastal regions, number of municipalities, and population

Population/ region	*<20,000*	*20,000–50,000*	*50,000–100,000*	*100,000–500,000*	*500,000–1,000,000*	*>1,000,000*	*Total*
North	17	21	5	6	–	1	50
Northeast	82	64	27	22	5	4	204
Southeast	12	20	12	29	3	1	77
South	33	19	7	9	1	–	69
Total	144	124	51	66	9	6	400

Source: IBGE, "Censo Demográfico Brasil 2010," 2011, www.ibge.br, compiled by Polette.

the vast expanses of coastal wilderness. Coastal communities are characterized by diverse cultural influences, from native Brazilians to those of African origin and the predominant Portuguese influence that is textured by many other European cultures including Italian, Spanish, German, and Polish among others. Brazil is also home to the largest population of people of Japanese origin outside of Japan, concentrated mainly in the states of São Paulo and Paraná. The Brazilian coast is thus a place of geographic, economic, social, cultural, and institutional diversity and contrasts.

Coastal issues in the context of climate change

Coastal ecosystems are among the most exploited of Brazil's biomes. Over the last five to six decades, many coastal communities have experienced intensified and conflicting usage of coastal resources and rapid urbanization and industrialization. Severe environmental and social impacts are widespread due to the transformation, degradation, overexploitation, and pollution of coastal ecosystems and resources, especially in the vicinity of major urban centers (e.g., Ferreira et al. 2009; Jablonski and Filet 2008; Marroni and Asmus 2005; Reis 2002). For example, inadequate provision of sewage infrastructure (up to 80% of Brazil's urban population lacked access to a sewage system at the turn of the century) causes serious pollution of coastal waters especially in localities that experience a dramatic influx of tourists over the summer (Jablonski and Filet 2008). Development practices frequently ignore the important roles played by coastal ecosystems in meeting the needs of coastal communities. For example, many developments, including housing and tourism developments, are located in or adversely impact sensitive coastal systems, for example, dunes, wetlands, and mangroves, and diminish their capacity to provide vital coastal ecosystem goods and services. Box 17.1 presents a case study about the vulnerability of mangrove forests in Rio de Janeiro to a changing climate.

BOX 17.1 VULNERABILITY OF MANGROVES IN RIO DE JANEIRO TO CLIMATE CHANGE

Mangroves provide a range of ecosystem goods and services that benefit coastal communities, yet they are sensitive to coastal development pressures that are likely to be made worse by climate change. In order to evaluate the likely responses of mangrove forests in the Metropolitan Region of Rio de Janeiro (RJMA) to climate change, especially given mean sea-level rise projections, Soares et al. (2011) studied the remaining mangroves in the RJMA and the role played by coastal management and urban development in shaping the vulnerability of these forests.

Mangroves are expected to contract toward the coastal plain, as already observed in some mangrove forests in the region (Soares 2009; Soares et al. 2005). However, the survival of these forests under projected climate change depends to a large extent on patterns of future urban development. Based on scenarios of rising mean sea level, the response of mangroves to these changes and the dynamics of urban development in the RJMA, the mangroves were classified as having low, medium, and high vulnerability.

Mangroves having low vulnerability are typically associated with a nonurbanized coastal plain or with very low urbanization, making it possible for accommodation and/or retraction in the face of rising mean sea level. Forests classified as highly vulnerable were those located in regions with no land available for accommodation and/or retraction, such as those near the mountains or on plains that are highly urbanized or with a physical barrier to their

(Continued)

retreat toward the mainland (e.g., roads and urban infrastructure). Those with medium vulnerability included forests in areas prone to urbanization but where there is still area for retraction on the border of the coastal plain even when urbanization has occurred in the inner parts of the plain.

Soares et al. (2011) identified and evaluated 28 segments of mangrove forests in the RJMA. There are only three remaining mangrove forests that can be classified as having low vulnerability to projected sea-level rise because much of this region is rapidly urbanizing. Many forests were identified as being highly vulnerable to sea-level rise, including many of the remaining mangroves in Guanabara Bay and associated coastal lagoons. These forests are highly vulnerable for two main reasons: (1) the land adjacent to the mangroves is characterized by high rates of urbanization and (2) the mangroves are bordered by land with steep relief, which acts as a natural barrier to migration.

This study reveals the need to integrate long-term coastal planning and management with urban planning provisions to ensure that vital coastal ecosystem goods and services, such as those provided by mangrove forests, are sustained for the benefit of both current and future generations in the face of climate change. It also underscores the need to mainstream climate change adaptation into all urban planning and coastal management efforts and to keep open adaptation pathways for coastal ecosystems and their dependent human communities.

Urban development has often taken place in low-lying areas that are prone to coastal erosion, flooding, and storm events, and sea-level rise will make matters worse (Brazil is ranked seventh in the world for total land area in low-elevation coastal areas, see McGranahan et al. 2007). A plethora of other impacts result from coastal development, industrial processes, and extractive activities. From the 1990s, activities associated with oil and gas expanded rapidly: from geophysical research to the construction of oil rigs and pipelines, increased traffic, and provision of related infrastructure at the coast, a trend that has been reinforced since the Petroleum Act 1997. With the discovery of Pre-Salt (the western hemisphere's largest oil field discovery in recent decades that is located in the Santos Basin, 250 km off the coast of Rio de Janeiro), massive investment and related initiatives were kick-started to realize the promise of this major new economic opportunity described as the Pre-Salt Cycle (Seabra et al. 2011). Notwithstanding significant economic and associated benefits, the social and environmental impacts of these activities have been profound, especially in localities where rapid, large-scale oil and gas-related development has occurred. In some cases, small coastal communities and their municipal authorities have been overwhelmed and transformed by industrial-scale oil and gas development, with attendant environmental, social, and economic impacts that are not evenly distributed due to the sudden influx of people and associated demand for housing, transportation, and other basic needs; haphazard and uncontrolled physical development; inadequate provision of basic infrastructure and services (e.g., sewage systems); environmental impacts that negatively affect preexisting activities such as fishing; and intensified social conflict (Marroni 2011; Marroni and Asmus 2013; Seabra et al. 2011).

Significant impacts have also resulted from the rapid expansion of coastal tourism developments in recent decades, many of which have focused on short-term financial gain without due consideration of environmental and social impacts (e.g., Dias et al. 2013; Ferreira et al. 2009), transforming the strip along the seashore of popular tourist destinations (see Figure 17.3).

Inland activities, such as agriculture, have also caused severe impacts along some stretches of the coast, typically by adding sediments and/or polluting waterways that negatively affect the healthy functioning of estuaries with serious consequences for coast-dependent livelihoods such as fisheries.

Economic interests are invariably prioritized over social and environmental concerns, and the resultant unsustainable development is evident at the coast, especially around urban centers, posing a significant challenge for coastal communities. For example, the government's Growth Acceleration

Figure 17.3 Aerial view of Balneário Camboriú municipality in south Brazil where its coastal fringe is completely urbanized. (Courtesy of Polette.)

Plan, which seeks to maximize trade and economic growth and modernize public infrastructure, coupled with pursuit of government-supported private megaprojects through public–private partnerships, has resulted in what some describe as a *legal land grab* of coastal areas that are home to native Brazilians, quilombolas (descendants of escaped Africo-Brazilian slaves), and peasants (Pedlowski 2013). Despite legal provisions aimed at securing the rights of coastal landowners and squatters, thousands of families are being evicted without compensation. Pedlowski (2013) describes the impacts of the forced removal of small farmers to facilitate the construction of the Açu Superport Industrial Complex in Rio de Janeiro state. In a related study, Ditty and Rezende (2014) demonstrate that this megaproject is neither just nor socially sustainable when taking into account the rights and interests of coastal communities, including artisanal fishers, impacted by this project. What is the prevailing legislative and institutional framework to manage these coastal issues?

Institutional and legal framework for coastal management and climate change adaptation in Brazil

Regulation of Brazilian coastal activities dates back to colonial times. Coastal management in the modern era is, however, usually associated with the government focus on exploiting coastal and marine resources that emerged in the 1970s and, in particular, the introduction of the Brazilian Constitution in 1988 and a subsequent series of laws, decrees, and regulations that advocate, among other things, democracy, decentralization, participation, sustainable development, and integrated management of the coastal and marine environment (Jablonski and Filet 2008; Marroni and Asmus 2013; Muñoz 2001; Wever et al. 2012). Prior to 1988, decades of centralized government decision-making had focused on economic growth and all but ignored environmental sustainability and equity considerations. The post-1988 institutional structures and legislative provisions arguably provide a robust framework for coastal management and for mainstreaming climate change adaptation into planning and decision-making in coastal communities (see Figure 17.4). The challenge, however, is to translate the well-intended legislative rhetoric into practical reality. The institutional framework for coastal management, and its promise, challenges, and opportunities for mainstreaming climate change adaptation, is described in more detail subsequently.

In summary, the coast is legally defined as a national heritage that includes the seashore and the area subject to land–sea–air interactions extending 50 km inland and up to the 12 nautical mile territorial sea limit. Coastal management is carried out under the auspices of the legally enforced Brazilian National Plan of Coastal Management (PNGC) supported by state and municipal plans, and ecological-economic zoning for some portions of

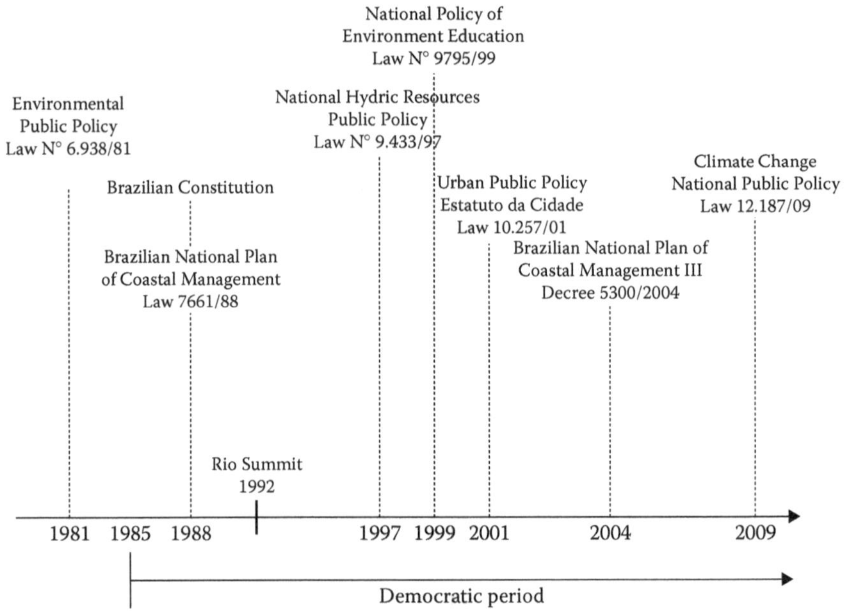

Figure 17.4 Timeline of some key legislative provisions shaping contemporary coastal management and climate change efforts in Brazil. (Drawn by Polette.)

the coast, as well as various other legal, planning, and conservation measures to sustainably manage coastal ecosystems and resources. The aim of the PNGC can be defined broadly as seeking to manage coastal activities in an integrated, decentralized, and participatory manner to protect, conserve, and, where necessary, rehabilitate coastal ecosystems and resources. Coastal management at the federal level, through provisions of the PNGC, is the responsibility of the Ministry of the Environment, with coastal states and municipalities responsible for establishing their own coastal zone management plans in keeping with the requirements and guidelines set out in federal provisions. Despite more than 25 years of developing, refining, and implementing this coastal management regime, unsustainable practices are commonplace and conflicting interests are intensifying in many localities, namely, subsistence versus commercial fisheries; traditional communities versus resort developers; local communities versus proponents of industrial projects; and oil and gas versus other coastal activities. Climate change is expected to intensify these conflicts and compound unsustainable practices but a concerted effort to address climate change is very recent compared to coastal management provisions.

Brazil introduced climate change legislation, the National Policy on Climate Change (NPCC), with the adoption of Law 12.187/09 (Brasil 2009). Attention has been focused on mitigation and, in particular, reducing greenhouse gas emissions through changes in land use, such as biofuels and afforestation, as well as more efficient use of natural resources. Little attention has, however, been focused on climate change adaptation and the complex interactions between climate change, coastal livelihoods, vulnerability, resilience, and sustainability, and key considerations such as water, energy, food security, and disaster risk (cf. Castello 2011; Copertino et al. 2010; Ferreira et al. 2011; Nicolodi and Petermann 2010; Pereira et al. 2013; Pittock 2011). There is potential to refine and better integrate the provisions of the NPCC (as well as various Presidential decrees, multi-stakeholder forums such as the Brazilian Forum on Climate Change, the Inter-Ministerial Committee on Climate Change coordinated by the Office of the President and the inter-departmental Executive Group on Climate Change coordinated by the Ministry of the Environment) with the long-standing institutional structures and provisions that guide coastal management efforts. There is potential to ensure more effective implementation of existing laws, policies, and related provisions such as the NPCC and coastal provisions, for example, PNGC and the National Coastal Zone Management Plan (Law 7661/88) (Brasil 1988), as well as urban development provisions among others. These provisions establish principles, objectives, guidelines, and instruments, and also define roles and responsibilities for government at all levels as well as the private sector and civil society, that are foundational for building adaptive capacity, resilience, and sustainability at the coast. It is therefore timely to explore how to bridge gaps, eliminate contradictions, and realize synergies between existing legislative and institutional provisions that govern sustainability and climate change at the coast. A starting point is to understand the nature of climate risks and vulnerabilities at the coast.

CLIMATE RISKS, VULNERABILITY, AND ADAPTING TO CLIMATE CHANGE AT THE COAST

This section describes climate risks and the vulnerability of coastal communities, and adaptation endeavors underway at the local level.

Climate risks and vulnerability to climate change

Climate change is expected to have significant impacts on Brazil's coastal biodiversity and the range of ecosystem goods and services that underpin coast-dependent livelihoods and coastal development more generally. For example, climate change impacts on the Discovery Coast and Abrolhos shelf, off Bahia and Espírito Santo states in northeastern Brazil, are expected to include a decline in nearshore fisheries; an increase in sediments delivered to coral reefs; increasing vulnerability to coastal erosion;

negative interactions between local climate, fire, and forest fragmentation; negative impacts on estuaries; declining water yield; and a decline in areas suitable for some crops, such as coffee (Pereira et al. 2013).

Large regions of the coast have low population levels and little development, and consequently, climate risk is concentrated mainly in urban centers, especially low-lying, densely populated areas characterized by social vulnerability and complex critical infrastructure such as the metropolitan regions of Belém, the capital cities of the northeast, Rio de Janeiro and Santos (Muehe 2010; Nicolodi and Petermann 2010). Low-lying coastal plains, especially near river mouths, are susceptible to flooding and coastal erosion (see Figure 17.5). Urban centers in such locations are particularly at risk in the face of rising sea level, an increase in the frequency and intensity of coastal storms, and secondary impacts such as groundwater contamination and the spread of vector-borne diseases, especially those areas already prone to drainage problems, coastal erosion, and severe storms (Muehe 2010). For example, the metropolitan area of Recife is one of the most vulnerable regions to sea-level rise due to its physical location and exposure to flooding and coastal erosion. Costa et al. (2010) estimate that with a 0.5 m rise in sea level, at least 39.32 km^2 of the municipal area would be a potential flood zone, and with a 1 m rise in sea level this would increase to 53.69 km^2. They estimate that 81% of the urban development situated less than 30 m from the shoreline and 5 m above ground level would be severely impacted by sea-level rise, and 45.7% of this coast is highly vulnerable.

Figure 17.5 Itajaí municipality in Santa Catarina, south Brazil, has had seven major floods in the last 30 years. In 2008, almost 85% of the city was flooded. (Drawn by Polette.)

Those whose livelihoods are dependent on climate-sensitive ecosystems like coral reefs, mangroves, and estuaries, for example, artisanal fishers, are especially vulnerable to projected climate change (Faraco et al. 2010; Leão et al. 2010). More broadly speaking, it is anticipated that marginalized and poor coastal communities are likely to bear the brunt of the negative impacts of climate change (see, e.g., studies in São Paulo state by Ferreira et al. 2011; Martins and Ferreira undated; Mendonça 2010; Souza 2003, 2009; Vieira et al. 2010) because they are more likely to live in at-risk localities and have low coping capacity (see, e.g., Borelli 2008; Martins and Ferreira undated; Mello et al. 2010). Securing sustainable coastal livelihoods presents complex governance challenges because climate change impinges on coastal ecosystem goods and services, livelihood opportunities, public safety, and human security and rights. In a study of the climate vulnerabilities of fishing communities in the Patos Lagoon in southern Brazil, Kalikoski et al. (2010) found that fishing communities that have high levels of self-organization are better able to reduce vulnerability to adverse climatic conditions but that few communities have developed the capacity to cope with climate-driven changes in the abundance and availability of fish. Erosion of traditional resource use systems and practices together with limited external institutional support for small-scale fishers and declining fish stocks in recent decades have progressively increased the vulnerability of fishing communities in this locality and this is further compounded by the uncertainty of future impacts.

Climate risks and vulnerability are significantly influenced by socioeconomic and environmental stressors. A study by Martins and Ferreira (undated) of the vulnerability and adaptive capacity of coastal communities along the northern coast of São Paulo state found that four decades of intensive urbanization and related industrial and tourism development have caused severe social and environmental problems and increased vulnerability to climate change impacts. Prevailing institutions fail to address the root causes and drivers of vulnerability and this presents a significant governance challenge for the future. The capacity of small municipalities in particular to address the underlying drivers of climate risk and adapt to climate change is extremely weak (Ferreira et al. 2011; Martins and Ferreira 2010, undated).

Adapting to climate change at the local level

Municipal authorities play a pivotal role in building adaptive capacity at the local level. There is, however, little evidence of climate change adaptation being explicitly and systematically addressed by local government in Brazil. Barbi and Ferreira (2013) studied climate change policies and practices in Brazilian cities and found that local climate change provisions are isolated initiatives within the national setting, with few cities actively addressing climate change mitigation and adaptation in an integrated and meaningful way. Where such efforts are underway, it usually involves actors from different segments of society, active participation in transnational cooperation networks on climate change, and multisectoral implementation. More typically, climate risks are being addressed through sectoral activities, for example, through measures to address flood and/or coastal erosion risks, landslides, and other perils that are likely to be exacerbated by climate change (Barbi and Ferreira undated). Efforts by local government to build adaptive capacity, in São Paulo for example, have been limited to information and awareness campaigns, emergency relief and evacuation, and in some cases resettlement of vulnerable families to safer areas in anticipation of future impacts (Martins and Ferreira undated). De Oliveira (2009) found that climate change initiatives in São Paulo had synergistic interdependencies with nonclimate change-related goals, policies, and programs, as well as economic and stakeholder interconnections, for example, greenhouse gas reduction initiatives are linked to pollution control measures. But there is almost complete disregard for the adaptation imperative. Natural hazard risks likely to be exacerbated by climate change, such as flooding, could be reduced with proactive adaptation measures. The lack of adaptation planning exposes vulnerable communities to progressively worse impacts over time.

Municipal authorities have at their disposal a number of statutory tools that enable them to avoid development in high-risk locations, maintain building standards, secure public health and safety, and build adaptive capacity, including financial provisions (e.g., collecting taxes and license fees enables municipalities to dedicate resources to reducing climate risk but it is commonly held that a shortage of funds in local government is a serious limitation); engineering and public infrastructure provisions; local community and urban development provisions (e.g., land-use planning, zoning, and real estate registration); building standards; public health and hygiene provisions (e.g., water supply, sanitation, solid waste, public health services, and pollution control); social services (e.g., measures relating to affordable housing, schooling, and vulnerable groups such as youth and elderly); civil defense and emergency management (e.g., measures to assist with disaster preparedness and emergency response); and public administration and human resources. To date, these tools and provisions have not been effectively used to reduce vulnerability and adapt to projected climate change (Martins and Ferreira undated).

According to Ferreira et al. (2011), recently enacted climate change provisions at the federal level offer promise but prevailing institutional structures and practices at the city level are inadequate for adapting to climate change. Moreover, confronting unsustainable coastal development raises fundamental ethical dilemmas that challenge the prevailing *status quo*, posing significant challenges for coastal communities and leaders across all sectors of Brazilian society. Climate change is expected to exacerbate vulnerability to the impacts of unsustainable patterns of coastal development and path dependencies driven by historic and prevailing practices that restrict adaptive capacity or are even maladaptive. Demographic and coastal development trends have deepened the vulnerability of social groups who live in high-risk locations and are susceptible to a range of coastal hazards that will be compounded by climate change. Prevailing municipal governments are not able to address the underlying causes of vulnerability and lack the capacity to stem unsustainable coastal practices. There is an urgent need to build adaptive capacity at the municipal level, of local government and coastal communities, and vulnerable groups in particular. How might this be done? Clearly, this is more than a technical matter, notwithstanding important information needs and research gaps, and includes significant ethical, political, social, economic, and institutional considerations. What is the promise of coastal management provisions and practices for mainstreaming climate change adaptation into coastal community planning and decision-making?

COASTAL MANAGEMENT IN BRAZIL: A FRAMEWORK FOR BUILDING ADAPTIVE CAPACITY, RESILIENCE, AND SUSTAINABILITY AT THE COAST

This section describes coastal management provisions in Brazil and their evolution over time. It then outlines the key challenges and opportunities revealed by more than 25 years of effort to implement this coastal management regime. Finally, it highlights considerations that could help or hinder efforts to mainstream climate change adaptation into planning and decision-making at the coast.

Coastal management provisions in Brazil

Focused attention on exploiting the natural resources of the coastal and marine realms was initiated from the early 1970s and led to an initial policy for managing the coast in 1974. Only with the promulgation of the Constitution of the Federated Republic of Brazil were environmental considerations, including environmental rights, addressed with legislative backing and the coastal zone established as a realm of national patrimony (Jablonski and Filet 2008; Marroni and Asmus 2013; Muñoz 2001). Significant legislative and institutional provisions have been introduced since that time, establishing a robust coastal management regime. Table 17.2 summarizes key milestones in the evolution of contemporary coastal management in Brazil.

The National Policy for Sea Resources (PNRM) established in 1980 provides the foundation for implementing coastal management through a

Table 17.2 Evolution of coastal management in Brazil

Year	*Institutional provisions*
Early 1970s	Environmental review of state planning undertaken.
1973	Special Secretary of Environment of the Presidency: Initiation of environmental perspective in state planning.
1974	Inter-Ministerial Commission for Sea Resources (CIRM): Focused attention on use of coastal and marine resources through issuance of a national policy for the Brazil coast (Decree no. 74 577).
1980	National Policy for the Sea Resources (PNRM): Implemented through pluri-annual plans and programs outlined by the CIRM.
1981	The National Policy of Environments (PNMA): Aims to preserve, enhance, and restore environmental quality.
1982	Subcommission of Coastal Management established in CIRM.
1982–1985	Sectoral Plan for Sea Resources (PSRM I): Structured research, analysis, and prospecting of sea resources in interest of Brazilian society and economy.
1986–1989	PSRM II: Defines objectives for addressing socioeconomic challenges at the coast. Consider scientific and technical capacity of organizations and human resources involved in projects.
1987	National Program of Coastal Management (GERCO) formulated by CIRM: Defines zoning methodology and institutional model for coastal zone management.
1988	The Federal Constitution of Brazil defines constitutional provisions and the institutional framework for decentralization, participatory, and integrated governance in Brazil, including provisions for environmental management, and declares the coastal zone a national patrimony (Article 225, x 4). The National Plan of Coastal Management (PNGC) was promulgated (Law 7.661/88) with political and judicial support of CIRM and the National Environmental Council (CONAMA). The coastal zone is defined as *the geographic space of interaction of air, sea, and land* including its renewable and nonrenewable resources and encompassing a marine and a terrestrial strip. It requires licensing for any activity that may affect the natural characteristics of the coastal zone through an obligatory Environmental Impact Report. It provides the framework for integrated, decentralized, and participatory coastal management and defines policy tools (e.g., Federal Coastal Zone Plan of Action, State and Municipal Coastal Management Plans, Information and Environmental Monitoring System, Environmental Quality Report, and Ecological-Economic Coastal Zoning). Regulates the licensing of activities that potentially impact the coastal environment, and provides regulations for environmental impact reports.
1990	Resolution CIRM 001/90 approves PNGC I, defining the methodological basis, institutional model and tools for GERCO. Environment Secretary of the Presidency established (Law 8.028).
1990–1993	PSRM III: Validation of II PSRM. Study implications of confirming United Nations Convention about Sea Rights (UNCSR), establishing as a main target the investigation and rational exploration of resources in the Exclusive Economic Zone.

(Continued)

Table 17.2 (Continued) Evolution of coastal management in Brazil

Year	*Institutional provisions*
1992	Special Secretary of Environment (SEMAM) in Environment Ministry.
1992	Coastal Extractive Reserves (RESEX) officially established: Aims to protect nature through controlled resource use; protects the livelihoods of traditional populations; and integrates traditional ecosystem users into national development.
1994–1998	PSRM IV: Adequacy of PSRM III. Implement Program for the Study of Sustainable Potential of Live Resources Capture of the Exclusive Economic Zone—REVIZEE.
1997	Resolution 005 of CIRM establishes PNGC II which creates the Group of Integration of the Coastal Management and the Subgroup of Integration of the State Program. CONAMA (National Environmental Council) Resolution 237: Establishes the standards of competence for licensing, according to the range of the impact of each enterprise and also, in some cases, to its location.
1999–2003	PSRM V: Update of IV PSRM.
2001	Project ORLA (a tool of the PNGC): Initiated to promote decentralized, integrated, and participatory shoreline management and prioritize partnership building and conflict resolution between competing activities at the local level.
2002	New Brazilian Civil Code: Defines seas and rivers as *public property* for common use. Decree 4340: Establishes the National System of Conservation Units. CONAMA Resolution 303: Defines *areas of permanent preservation* including lagoons, mangroves, dunes, beaches, and breeding areas of wild fauna, and in *restingas*. CONAMA Resolution 312: Defines restrictions for shrimp farming. National System of Conservation Units (SNUC): Provides the legal framework for different categories of protected areas. Special Secretariat for Aquaculture and Fisheries (SEAP): Aims to provide resource management assistance and socioeconomic support to fishing communities; and build capacity to plan and manage fisheries, including artisanal fisheries.
2003	Coastal Agency is created and aims to promote integrated coastal management.
2004	Decree 5300 regulates the National Coastal Management Plan and other instruments for coastal zone management (including Federal Coastal Zone Plan of Action, State and Municipal Coastal Management Plans, Information System, Environmental Monitoring System, Environmental Quality Report, and Ecologic-Economic Coastal Zoning). Refines the definition of the coastal zone to include the *territorial sea* up to 12 nautical miles and a terrestrial strip of municipalities along the shoreline. The terrestrial strip also includes nonlittoral municipalities contiguous to large cities; estuarine counties; and those that have infrastructure or activities of significance to the coastal zone, including those up to 50 km from the seashore. It defines the *coastline* as a fringe of the coastal zone covering its marine portion up to 10 m depth and landwards varying from 50 m in urban areas to 200 m in nonurban ones, from the tidal line or

(*Continued*)

Table 17.2 (Continued) Evolution of coastal management in Brazil

Year	*Institutional provisions*
	from the inland boundary of existing coastal ecosystems. The Federal Ministry of the Environment remains the coordinating body for the PNGC with planning and implementation responsibilities decentralized to the state and local level.
2004–2008	PSRM VI: PNRM update, with focus on sustainable development, preparation of human resources, and incentive to research.
2008–2011	PSRM VII: Recognizes the role of oceans in global climate change. Focus on the need for cooperation between Government, academia, civil society, and private sector to facilitate sustainable use of sea resources.
2012–2015	PSRM VIII: Introduces a novel approach for integrated and participatory management of sea resources, involving several ministries, science and private sector. Highlights importance of data availability for society. Identifies conservation of sea resources as a priority. Stimulates development of human resources and the international cooperation. Particular focus on the natural resources of the coastal zone.

Source: Jablonski, S. and Filet, M., *Ocean & Coastal Management*, 51, 536–43, 2008; Marroni, E.V. and Asmus, M.L., *Ocean & Coastal Management*, 76, 30–7, 2013; Muñoz, J.M.B., *Coastal Management*, 29, 137–56, 2001; Wever, L. et al., *Ocean & Coastal Management*, 66, 63–72, 2012.

pluriannual planning process that provides executing authorities with operational guidelines for a period of four years. The plans are developed by the Executive Secretary of the Inter-Ministerial Commission for Sea Resources (CIRM) and regular updates enable ongoing improvements. In practice, the centralized execution of the PNRM is complemented by provisions that seek to foster active participation of several Ministries, coastal states, municipalities, civil society, the private sector, and science community to enable effective resolution of coastal concerns. The PNGC planning process is a hierarchical approach to coastal plan-making with lower tier plans needing to conform to the guiding principles and prescriptions of higher level plans. The Municipal Plan of Coastal Management (PMGC) addresses local issues by giving effect to the intentions of the relevant State Plan of Coastal Management and overarching PNGC. The PMGC needs to be informed by and well connected to other local level plans dealing with other matters that have a bearing on local coastal planning and decision-making. The PNGC is thus implemented through a hierarchical set of state and municipal coastal management plans and complementary management tools, including an Information System for the Coastal Management Plan; Coastal Zone Environmental Monitoring System; Environmental Quality Report for Coastal Zone; Coastal Ecological-Economic Zoning; Coastal Zone Macro-diagnostic; and Project Orla. Project Orla seeks to guide coastal development in coastal municipalities, reduce losses due to coastal erosion, and is a key mechanism for enabling coastal stakeholders, including federal, state,

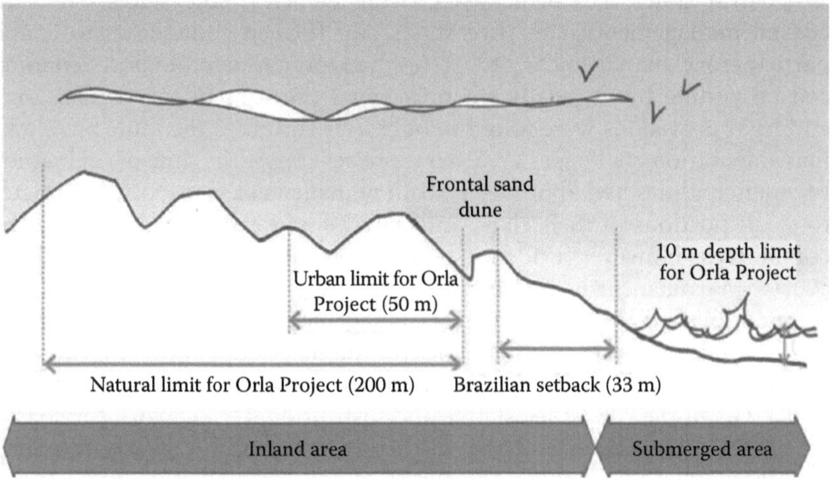

Figure 17.6 Boundaries of Project Orla in Brazil. (After Muehe, D., *Erosão e Progradação do Litoral Brasileiro,* Ministério do Meio Ambiente, Brasília, Brazil, 476pp., 2006.)

and local government, as well as partners in civil society, the private sector, and scientific community, to communicate and coordinate their activities. Project Orla encompasses an offshore limit to the 10 m isobath and a setback line of 50–200 m landward from the beach (extending beyond the 33 m coastal setback in most areas that is managed by the federal authority Superintendência do Patrimônio da União) (see Figure 17.6).

Coastal management efforts in Brazil have evolved considerably because initial provisions were made in the 1970s and with more recent PNGC and Sectoral Plan for Sea Resources (PSRM) provisions that have been amended over time. There has been a marked shift from the initial focus on mapping the biophysical characteristics of the coast to inform technical decision-making to a focus on better integrating coastal management and securing meaningful participation of key coastal stakeholders in planning and decision-making, chiefly through Decree 5.300/04 in 2004 which focused attention on the local level through Project Orla and, more recently, through PSRM VIII, the Sectoral Plan for Sea Resources for 2012–2015 (see Table 17.2).

To what extent have these efforts to legislate and institutionalize coastal management in Brazil been successful?

Effective translation of the good intentions outlined in various coastal management provisions into practical reality in local communities has been elusive (see Asmus et al. 2006; Diegues 1999; Jablonski and Filet 2008; Marroni and Asmus 2013; Muñoz 2001; Souza 2009; Wever et al. 2012). In short, effective implementation remains weak. Jablonski and Filet (2008) found that after 20 years of implementation effort, only eight of

17 coastal states had dedicated coastal agencies, nine states had state coastal management plans, five states had formed committees to enable participatory management, 10 states had prepared ecological-economic coastal zoning but typically for only small portions of the coastal zone, and legal provisions were found in only four states. At the municipal level, implementation challenges are even more severe. At the time of their study, 56 municipalities had approved coastal management plans, often restricted to small portions of their coast and only 10 had established management committees. However, evidence of engagement by local communities in coastal management is growing. In some states, conservation bodies have been established to oversee coastal management for large coastal regions. For example, an area encompassing five municipalities about 140 km long and 10 km wide is under such a regime on the north coast of the state of Bahia. Given the robust legislative and institutional framework for coastal management in Brazil, and the vastness of the coast, one would expect that there would be ample opportunity to resolve conflicting interests and progress resilience and sustainability. Yet media reports, the activities of the Public Prosecutor's Office and reviews, and independent studies demonstrate that the gap between good intentions and reality on the ground is pervasive and persistent.

What are the main challenges and opportunities for realizing the promise of coastal management provisions in Brazil?

Challenges and opportunities for implementing coastal management provisions in Brazil

We reviewed published research on Brazil's coastal management experience (Asmus et al. 2006; Belchior 2008; Diegues 1999; Glaser and Da Silva Oliveira 2004; Glaser and Krause 2005; Jablonski and Filet 2008; Julian 2003; Macedo et al. 2013; Marroni and Asmus 2013; Muñoz 2001; Polette and Viera 2006; Polette et al. 2006; Reis 2002; Scherer et al. 2011; Souza 2009; Wever et al. 2012) and identified the following challenges and associated opportunities for realizing the promise of coastal management:

- Economic interests outweigh concerns about coastal sustainability
- Government-led, top-down versus a bottom-up approach to coastal management
- Fragmented legislative provisions and difficulties in enforcement and compliance
- Institutional coordination
- Capacity at the local level
- Meaningful participation in coastal management
- Information, communication, and public awareness

Economic interests outweigh concerns about coastal sustainability

Notwithstanding constitutional and legislative provisions to secure environmental rights and promote sustainable development, in practice, economic interests override concerns about public safety, equity, and environmental sustainability—reflected among other things in government prioritization of economic growth and fast-track development projects at the coast. Furthermore, provisions to secure environmental sustainability are seldom well integrated with provisions to foster social sustainability, for example, justice, poverty, and equity. Coastal conservation efforts, for example, are typically disconnected from the interests of coast-dependent resource users, who are often poor, such as those dependent on mangroves for their livelihoods, triggering poor compliance with coastal management provisions and unsustainable practices (e.g., Glaser 2003; Glaser and Krause 2005). Efforts to initiate comanagement arrangements are being explored but powerful local economic interests tend to predominate and local coast-dependent communities tend to be marginalized in strategic decision-making. Poor coast-dependent communities typically experience marginalization by dominant development interests on the one hand, and, on the other hand, top-down conservation initiatives, such as establishing marine protected areas, however well intended, adversely impact local resource users and deepen poverty and inequity in already marginalized communities (e.g., Diegues 1999, 2008). These conflicting interests between different sectors and within and between federal, state, and local spheres of government, and between different coastal stakeholders, that become pronounced at the local level, are likely to be compounded with climate change. As such conflicts become more intense, there is an opportunity to mobilize affected parties in pursuit of outcomes that foster adaptive capacity, resilience, and sustainability. Concerted efforts to apply the provisions already entrenched in coastal management legislation, policies, and plans offer potential to reconcile these contending interests.

Government-led, top-down versus a bottom-up approach to coastal management

Coastal management efforts have been dominated by government-led, top-down legislative prescriptions and regulations to control coastal activities. Global coastal management best practice underscores the need to enable coastal stakeholders to work together to manage divergent interests in an integrative, devolved, and participatory manner. There is clear evidence in among other things Project Orla and PSRM VIII of the intention to shift the focus from reliance on government-driven top-down practices toward a more integrated governance approach based on meaningful participation by coastal stakeholders. There is opportunity to better integrate top-down and bottom-up processes of coastal planning and decision-making in Brazil. However, overcoming the dominant hierarchical nature of coastal management remains fraught in practice.

Fragmented legislative provisions and difficulties in enforcement and compliance

Realizing the promise of devolved, integrated, and participatory coastal management is challenging in practice, especially given the compartmentalized and sector-based nature of legislation and the difficulties encountered in enforcement and compliance. In some cases, overlapping and/or unclear mandates hamper effective implementation. Legislative intentions and practical realities and priorities in local communities are often mismatched, for example, legal obligations at federal and state levels to protect environmental rights and interests are often second to economic interests and development pressures at the local level. Key legislative matters needing attention to realize the promise of coastal management provisions include the following: (1) reduce the overlap between laws, policies, plans, and provisions relevant to coastal planning and decision-making; (2) improve the integration of public policies relevant to the coastal zone; (3) ensure consistency in interpretations of legal provisions and mechanisms, with a key role to be played by the Public Prosecution Ministry; (4) promulgate local legislation where such provisions will enable local-level actions to improve adaptive capacity, resilience, and sustainability; (5) adopt measures and practices to ensure more effective enforcement of and compliance with legal provisions; (6) resolve conflicting interests arising from the exercise of private property rights in the coastal zone which is a national patrimony that also includes a range of *common* and *diffuse rights*; (7) improve the efficiency of the judicial system; and (8) provide clarity about the roles and responsibilities of public institutions and other coastal stakeholders. Many consider the prevailing legislative regime for coastal management in Brazil to be robust. Much, however, remains to be done to better align legislation relevant for managing the coast with legal provisions in other sectors that have an influence on efforts to build coastal communities that are resilient and sustainable in the face of climate change. In particular, there is a compelling need, first, to foster better integration of sectoral and institutional structures and practices and, second, to facilitate collaboration at the local level so that key stakeholders can work together to reconcile divergent interests.

Institutional coordination

Integration within and between different sectoral agencies (horizontal integration) and spheres of government (vertical integration) remains poorly developed in Brazil, despite the provisions in legislation. Responsibilities for matters relating to coastal management, urban development, nature conservation, climate change, and so on are spread across spheres and sectors of government and need to be better coordinated and integrated. The structured PSRM review process provides an opportunity to improve horizontal and vertical integration of efforts to build adaptive capacity, resilience, and sustainability.

Capacity at the local level

In a study of coastal management in the Amazon, Szlafsztein (2012) found that there was poor progress in implementing coastal management provisions because there was weak support from local communities and society at large; weak institutional coordination; and inadequate technical, human, and financial resources especially at the local level. Differential effectiveness of implementation efforts can be attributed at least in part to capability differences that are quite marked between wealthier southern states compared to less well-resourced northern states. Constraints in funding, human, administrative, and technical resources, including access to information, are, however, commonplace at the local level. Officials in local government are expected to take on new responsibilities in coastal management without necessarily having the background or competencies required, and key decision-making powers often reside at higher levels of government. At the local level in particular, economic interests prevail over concerns about coastal sustainability. In many localities, there is scant regard for the law, poor accountability, corruption, elite capture, and generally low levels of compliance and enforcement of existing provisions. According to Gerhardinger et al. (2011), the main flaws of efforts to decentralize marine conservation and coastal management in Brazil stem from lack of leadership and inadequate institutional and financial capabilities. In a study of coastal municipal planning in Rio Grande in southern Brazil, Pereira (2012), for example, found that there were persistently low budgets available at the local level to develop and implement plans for environmental concerns, including the coast, with budgets as low as 0.5% of the total annual budget for coastal cities. Among other things, coastal management capabilities at the local level need to be developed with enabling support from higher levels of government, including securing a more prominent place for coastal management on the public agenda; enabling more effective coordination by the Ministry of the Environment; and strengthening the institutionalization of coastal management at all levels of government, especially at the local level.

Meaningful participation in coastal management

Marroni and Asmus (2013) suggest that the gap between legislative intention and local reality arises in part because local communities have not perceived that they can play an active and engaged role in local coastal management because they see this as the domain of government. Much remains to be done to facilitate more active involvement in decisions to foster sustainable coastal development and to mainstream climate change adaptation into day-to-day decision-making at the local level. In general, dialogue, communication, and cooperation within and between coastal governance actors from government, civil society, the private sector, and the scientific community remains weak and in need of improvement. Nonetheless, there is evidence of active participation by stakeholders in some coastal management efforts, and a variety of networks and forums are being established to bring together government authorities, local coastal resource users, nongovernmental organizations, and scientists.

Active participation of local communities is being fostered through comanagement arrangements in marine extractive reserves (RESEXMar) along the coast (Da Silva 2004; Glaser and Krause 2005; Wever et al. 2012), engaging at least 60,000 small-scale fishermen by the late 2000s (Moura et al. 2009). The RESEX approach seeks to involve traditional resource users in development initiatives that secure the right of local communities to define the rules for managing local resource use that reconciles cultural, environmental, and economic interests on a sustainable basis, in partnership with government, nongovernmental organizations, and academia (Moura et al. 2009).

Although efforts are being made to foster more meaningful public participation in coastal management, effective involvement remains embryonic and inadequate. Local and regional authorities tend to have benefitted from decentralization initiatives but local communities and coast-dependent resource users typically remain marginalized by at best passive consultation efforts. Overcoming mistrust and resolving conflicting interests remains challenging in many coastal regions. Efforts are being made through PSRM VIII and Project Orla to encourage more active involvement of stakeholders in local coastal management efforts. The emergence of community-based initiatives is encouraging given the opportunities such forums create for dialogue, communication, and cooperation between key stakeholders and coastal governance actors more generally.

Information, communication, and public awareness

Coastal management efforts in Brazil, and the PNGC in particular, have stimulated sound technical analysis through mapping, geo-referencing, and meta-data management and the development of an information system, ecological-economic zoning, and State of Environment Reporting in relation to the coast. However, information gaps and inadequate exchange and coordination of information within and between government, civil society, the private sector, and academia are commonplace. To compound matters, effective coordination within scientific disciplines remains a challenge.

Persistent difficulties center on the lack of qualified human resources within the three levels of government to understand coastal ecosystem dynamics, as well as the management system itself, and the sharing of this information with coastal stakeholders and raising public awareness and understanding of coastal issues in times of global climate change. More specific challenges include the following: (1) descriptive surveys focused on ecological, economic, and social components typically fail to advance understanding of underlying dynamic processes and their interactions and interdependencies; (2) gaps in fundamental scientific understanding about some aspects of the coastal zone; (3) inadequate understanding of the value of coastal ecosystems goods and services and how to incorporate such understanding into coastal decision-making; (4) insufficient understanding of the barriers and opportunities to advance devolved, integrated, and participatory coastal management in the face of climate change; (5) the need to ensure that scientific information and understanding addresses priority management concerns and can be effectively used to inform coastal planning and decision-making processes; (6) poor access to information and low levels of public awareness about local coastal issues and how to make better coastal management decisions to build adaptive capacity, resilience, and sustainability; and (7) the need to build capacity to monitor changing coastal circumstances and use this information more effectively in coastal planning and day-to-day decision-making. Given the foregoing, what will help and what will hinder efforts to mainstream climate change adaptation into coastal management efforts?

What will help and what will hinder climate change adaptation at the coast in Brazil?

Key determinants of adaptive capacity have been suggested by a range of scholars (see, e.g., Adger 2003; Engle and Lemos 2010; Folke et al. 2002; IPCC, 2007; Pelling and High 2005; Smit and Wandel 2006; Yohe and Tol 2002), and these can be summarized as different forms of capital—the presence of which will influence how feasible it is to mainstream climate change adaptation and build resilience.

Political capital: It includes governance practices, legitimate leadership, devolved decision-making, meaningful participation, capable decision-making, and management capacity, among other things, that foster inclusive, transparent, and accountable public decision-making that empowers coastal communities. Societies and communities with robust political capital are more likely to create social and governance institutions that sustainably manage coastal resources, foster equity, secure human rights, reduce risk, and adapt to global climate change.

Social capital: It includes norms, reciprocal relationships, social ties, and networks that build trust and facilitate cooperation and community development. Social capital is shaped by among other things underlying societal values, the nature of state–civil society relations, local coping networks, social mobilization, and the density and effectiveness of institutional relationships.

Human capital: It includes the stock of knowledge, competencies, and attributes such as education, health, entrepreneurship, and skills that enable people to contribute to economic and social life. Knowledge is multifaceted and encompasses scientific, technical, political, local, traditional, and tacit knowledge(s). Local traditions, customs, and social memory about resource use are also relevant because they enable people to respond to environmental feedback and build adaptive capacity. Also relevant are capabilities to acquire relevant data, analyze, transfer, and exchange data and technology, the effectiveness of communication networks, freedom of expression, and the capacity to innovate. Human capital can thus include both the *technical* and *cultural* dimensions of knowledge and experience that shape public perceptions and awareness of and attitudes toward coastal risks and sustainability, and recognizes that learning takes place in a social context.

Economic capital: It includes opportunities for private gain and accumulation of financial wealth to create jobs and build a vibrant economy that sustains coastal livelihoods. Prioritization of short-term profit and economic growth at the expense of ecological sustainability, public safety, and social equity, however, escalates disaster risk and reduces livelihood options and community resilience and sustainability. Attention needs to be focused on alternative economic models that take into account the value of natural capital, coastal risks, and the complex interconnected nature of social-ecological systems, as well as income and wealth distribution, economic marginalization, accessibility and availability of financial instruments (e.g., insurance and credit), and financial and fiscal incentives for risk management.

Physical capital: It includes investment in factors of production such as equipment and machinery as well as critical infrastructure at the coast (including energy supply and management, telecommunications, water supply, food supply, public health, protective works, financial services), buildings and roads that are vital for economic productivity and public safety.

Natural capital: It includes diverse, healthy, and productive coastal ecosystems that are essential for providing vital goods and services, including potentially life-saving *natural defenses* (e.g., intact dunes, mangrove forests) against climate change-driven extreme events and/or slow onset change.

In practice, these dimensions of adaptive capacity are not easily separated, and these *capitals* are mutually reinforcing, with the governance and institutional dimensions being especially important. In short, the more democratic, devolved, participatory, and integrated coastal governance structures and processes are, the stronger the anticipated adaptive capacity and the better are prospects for building resilience and sustainability.

These considerations have become the underlying tenets of the coastal management framework in Brazil; the challenge is to translate legislative intention into practical reality by strengthening this suite of capitals.

There is evidence that some Brazilian communities have developed the institutional capacity to manage complex environmental problems and may be well placed to adapt to climate change. For example, Pereira et al. (2009) analyzed river basin management from 1999 to 2008 in the São João River catchment on the southeast coast of Brazil. They found that environmental management practices had improved significantly once severe water pollution problems mobilized local leaders to address the problem. A number of factors were central to building effective institutions that could pursue an adaptive management approach: (1) enabling national and state water laws; (2) stakeholder engagement (including municipal government, non-governmental organizations, and private companies) and a common vision; (3) a collective identity and local ownership; and (4) an independent financing mechanism. Investing in activities that reduced vulnerability and improved livelihoods garnered community support and enabled participants to see *quick wins* that built confidence. Communicating in *plain language* and demonstrating the effectiveness of *no- and low-regrets* options, notwithstanding uncertainty about climate change impacts, was considered key to mainstreaming adaptation efforts. In sum, better understanding needs to be developed about anticipated climate change scenarios and locally specific impacts and risks, and more integrated risk reduction and adaptation measures need to be implemented (Nicolodi and Petermann 2010).

Brazil's coastal management regime, and the PSRM process in particular, provides a robust institutional framework for nurturing the political, social, human, economic, physical and natural capital necessary for adapting to climate change and building resilient, sustainable coastal communities. The challenge remains the translation of legislative rhetoric into practical reality.

PRACTICAL RECOMMENDATIONS FOR ADAPTING TO CLIMATE CHANGE AT THE COAST IN BRAZIL

Based on the foregoing analysis, three practical recommendations are made for building adaptive capacity, resilience, and sustainability in coastal communities in Brazil. We recommend that a new program be established to pilot a climate change adaptation project in each of the 17 coastal states of Brazil, with a focus on the following:

First, *raise public awareness about the coast and climate change through active social learning processes*. Vulnerability to climate change impacts is not only a function of exposure but also the susceptibility of individuals, groups, and communities at the coast to sudden shock events and their capacity to proactively adapt to slow onset change. Building a shared understanding of climate risks, vulnerabilities, and barriers and opportunities to charting adaptation pathways is therefore essential. Serrao-Neuman et al. (2013) provide three key insights about the nature of research processes that foster social learning to enable adaptation to climate change: (1) include key stakeholders in the research process so that all parties benefit from the coproduction of knowledge and to shaping the research process in light of changing circumstances and scientific and community priorities. (2) Different types of knowledge (cf. publishable scientific knowledge, local and meta-knowledge) play distinct but potentially reinforcing roles in the learning and adaptation process. (3) The timing and duration of collaborative research endeavors has a significant bearing on the extent to which findings inform policy and practice and thus can help or hinder the adaptation process. These insights are reinforced by practical experience. A study of integrated water-management institutions in the Paraíba do Sul River basin in southeast Brazil by Kumler and Lemos (2008) showed that social learning was key to facilitating institutional reform and to ensuring the effectiveness and sustainability of new provisions. Key enabling factors included a shared understanding of the institutional mission, possibilities, and problems; trust; and the ability to work together. Effective governance through social learning was demonstrated by the institution's capacity to adapt in the face of severe drought, underscoring the foundational importance of robust political, social, and human capital.

Second, *create opportunities for meaningful public participation in coastal management efforts* to mobilize community capabilities to address the coastal issues they currently face and proactively plan adaptation pathways. Authentic participation strengthens social and political capital and stimulates governance innovations that enable adaptation, resilience, and sustainability. Active participation in the Pintadas Adaptation project, inland of the northeastern coast, demonstrates the importance of integrating poverty reduction and adaptation measures to secure local buy-in as well as support for mainstreaming adaptation into regional and national policies, and stimulating local and regional development (Simões et al. 2010). Meaningful opportunities to participate and tangible benefits are key to catalyzing and sustaining engagement in such projects and to avoid overreliance on government or external organizations. Testing, replicating, and mainstreaming participatory pilot projects is important given the multidimensional nature and inherent uncertainties of climate change and the need to tailor adaptation to locality-specific vulnerabilities and climate variability.

Third, *integrate and mainstream coastal management, disaster risk reduction, and climate change adaptation efforts*. Tompkins et al. (2008) found that four factors are critical to reducing disaster risk and building long-term adaptive capacity in climate vulnerable localities like Ceará state in northeastern

Brazil: (1) governance practices need to be flexible, learning-based, and responsive; (2) governance actors need to be committed, reform-minded, and politically active; (3) disaster risk reduction, as well as coastal management and climate change adaptation measures, needs to be mainstreamed into other political, social, and economic processes; and (4) a long-term commitment to reducing disaster risk, and building adaptive capacity, resilience, and sustainability, needs to be inculcated. They found that disaster responses in Ceará had failed to address the root causes and drivers of vulnerability and consequently those already marginalized and exposed to coastal hazards are at risk of long-term climate-driven changes. They conclude that a two-tiered approach—based on structural reforms that promote good governance as well as effective disaster risk reduction measures—is key to creating an institutional environment that builds long-term adaptive capacity to climate change.

CONCLUSION

Climate change has the potential to cause severe environmental, social, and economic impacts for coastal communities in Brazil because it directly affects their safety, livelihoods, and future prospects. Climate change is expected to exacerbate environmental risks and amplify extreme weather events and coastal hazards. Climate change impacts will not only affect exposed low-income groups, but those with independent access to assets and less dependence on coastal resources are less likely to be adversely affected.

Brazil's coastal management regime offers considerable potential for mainstreaming climate change adaptation into the planning and decision-making undertaken by coastal communities. Realizing this promise requires better understanding of the challenges and opportunities for translating legal rhetoric into practical reality. Notwithstanding the slow-onset nature of climate change, there is an urgent need to build institutional capacity to better understand and address climate change impacts at the coast and chart adaptive pathways.

We identified the following main challenges for realizing the promise of coastal management in Brazil: (1) Economic interests outweigh concerns about coastal sustainability; (2) a government-led, top-down coastal management approach prevails and needs to be better integrated with bottom-up coastal management approaches; (3) legislative provisions are fragmented and enforcement and compliance are difficult to achieve in practice; (4) despite legislative provisions, much remains to be done to improve institutional coordination; (5) capacity to undertake devolved, integrated, and participatory coastal management is extremely limited at the local level; (6) concerted effort is needed to ensure more meaningful public participation in coastal management processes; and (7) information about the coast and future challenges and opportunities for adapting to climate change needs to be developed and shared to raise public awareness and mobilize adaptive action. These challenges are far from simple because deeply held values about the primacy of economic interests over other societal concerns, including public safety and environmental and social sustainability, need to be confronted. Systemic governance reforms are required to enable new ways of working that stimulate more effective partnerships between coastal stakeholders to adapt to climate change.

Coastal communities need to secure and strengthen their political, social, human, economic, physical, and natural capital in order to build adaptive capacity and resilience. Three interdependent practical recommendations are made to this end based on the establishment of a new program that pilots climate change adaptation projects in each of the 17 coastal states of Brazil, with a focus on (1) raising public awareness about the coast and climate change through active social learning processes; (2) creating meaningful opportunities for public participation in coastal management efforts; and (3) integrating and mainstreaming coastal management, disaster risk reduction, and climate change adaptation efforts.

ACKNOWLEDGMENT

We thank Pedro Fidelman, CRN Research Fellow, Sustainability Research Centre, University of the Sunshine Coast, Queensland, Australia, for reviewing a draft of this chapter.

REFERENCES

Adger, W.N. (2003). "Social capital, collective action, and adaptation to climate change," *Economic Geography*, 79(4): 387–404.

Asmus, M.L. and Kitzmann D. (2004). "Gestão costeira no Brasil: Estado atual e perspectivas," Programa de Apoyo a la Gestión Integrada en la Zona Costera Uruguaya, EcoPlata, 63p.

Asmus, M.L., Kitzmann, D., Laydner, C., and Tagliani, P.R.A. (2006). "Gestão costeira no Brasil: instrumentos, fragilidades e potencialidades," *Gestão Costeira Integrada*, 5: 52–7.

Barbi, F. and Ferreira, L.C. (2013). "Climate change in Brazilian cities: Policy strategies and responses to global warming," *International Journal of Environmental Science and Development*, 4(1): 49–51.

Barbi, F. and Ferreira, L.C. (undated). "Risks and political responses to climate change in Brazilian coastal cities," Unpublished, www.anppas.org.br/encontro6/anais/ARQUIVOS/GT11-1276-1036-20120629100324.pdf (accessed January 2014).

Belchior, C.C. (2008). "Gestao Costeira Integrada e Estudo de Caso do Projeto ECOMANAGE na Regiao Estuarina de Santos-Sao Vicente, SP, Brazil PROCAM," Universidade de São Paulo, São Paulo, Brazil, 108p.

Borelli, E. (2008). "Gerenciamento costeiro e qualidade de vida no Litoral de São Paulo," *Cadernos IPPUR* 22(1): 79–97.

Brasil. (1988). "Plano Nacional de Gerenciamento Costeiro," Lei 7661/88.

Brasil. (2009). "Política Nacional de Mudanças Climáticas," Institui a Política Nacional sobre Mudança do Clima—PNMC e dá outras providências.

Castello, M.G. (2011). "Brazilian policies on climate change: The missing link to cities," *Cities*, 28: 498–504.

Copertino, M.S., Garcia, A.M., Muelbert, J.H., and Garcia C.A.E. (2010). "Introduction to the special issue on climate change and Brazilian coastal zone," *Pan-American Journal of Aquatic Sciences*, 5(2): 1–8.

Costa, M.B.S.F., Mallmann, D.L.B., Pontes, P.M., and Araujo, M. (2010). "Vulnerability and impacts related to the rising sea level in the Metropolitan Center of Recife, Northeast Brazil," *Pan-American Journal of Aquatic Sciences*, 5(2): 341–9.

Da Silva, P.P. (2004). "From common property to co-management: Lessons from Brazil's first maritime extractive reserve," *Marine Policy*, 28: 419–28.

De Oliveira, J.A.P. (2009). "The implementation of climate change related policies at the subnational level: An analysis of three countries," *Habitat International*, 33: 253–9.

Dias, J.A., Cearreta, A., Isla, F.I., and de Mahiques, M.M. (2013). "Anthropogenic impacts on Iberoamerican coastal areas: Historical processes, present challenges, and consequences for coastal zone management," *Ocean & Coastal Management*, 77: 80–8.

Diegues, A.C. (1999). "Human population and coastal wetlands: Conservation and management in Brazil," *Ocean & Coastal Management*, 42(2–4): 187–210.

Diegues, A.C. (2008). "Marine protected areas and artisanal fisheries in Brazil," International Collective in Support of Fishworkers. www.aquaticcommons.org/1565/1/Samudra_mon2.pdf (accessed January 2014).

Ditty, J.M. and Rezende, C.E. (2014). "Unjust and unsustainable: A case study of the Açu port industrial complex," *Marine Policy*, 45: 82–8.

Dominguez, J.M.L. (2004). "The coastal zone of Brazil: An overview," *Journal of Coastal Research*, SI 39: 16–20.

Engle, N.L. and Lemos, M.C. (2010). "Unpacking governance: Building adaptive capacity to climate change of river basins in Brazil," *Global Environmental Change*, 20: 4–13.

Faraco, L.F.D., Andriguetto-Filho, J.M., and Lana, P.C. (2010). "A methodology for assessing the vulnerability of mangroves and fisherfolk to climate change," *Pan-American Journal of Aquatic Sciences*, 5(2): 205–23.

Ferreira, J.C., Silva, L., and Polette, M. (2009). "The coastal artificialization process. Impacts and challenges for the sustainable management of the coastal cities of Santa Catarina (Brazil)," *Journal of Coastal Research*, SI. 56: 1209–13.

Ferreira, L.C., Andrade, T.H.N., Martins, R.D.A., Barbi, F., Ferreira, L.C., Mello, L.F., Urbinatti, A.M., and Souza, F.O. (2011). "Governing climate change in Brazilian oastal cities: Risks and strategies," *Journal of US-China Public Administration*, 8(1): 51–65.

Folke, C., Carpenter, S., Elmqvist, T., Gunderson, L., Holling, C.S., and Walker, B. (2002). "Resilience and sustainable development: Building adaptive capacity in a world of transformations," *Ambio*, 31(5): 437–40.

Gerhardinger, L., Godoy, E., Jones, P., Sales, G., and Ferreira, B. (2011). "Marine protected dramas: The flaws of the Brazilian national system of marine protected areas," *Environmental Management*, 47(4): 630–43.

Glaser, M. (2003). "Interrelations between mangrove ecosystem, local economy and social sustainability in Caeté Estuary, North Brazil," *Wetlands Ecology and Management*, 11(4): 265–72.

Glaser, M. and Da Silva Oliveira, R. (2004). "Prospects for the co-management of mangrove ecosystems on the North Brazilian coast: Whose rights, whose duties and whose priorities?" *Natural Resources Forum*, 28(3): 224–33.

Glaser, M. and Krause, G. (2005). "Integriertes Küstenmanagement im föderalen Brasilien: Institutionelle, sektorale und legale Strukturen und die Grenzen der partizipativen Planung," in: Glaeser, B. (ed.), *Küste—Ökologie—Mensch: Integriertes Küstenmanagement als Instrument nachhaltiger Entwicklung*. München, Germany: Oekom Verlag, pp. 37–54.

IBGE. (2011). "Censo Demográfico Brasil 2010." www.ibge.br (accessed January 2014).

IPCC. (2007). "*Climate Change 2007: Impacts, Adaptation and Vulnerability. Contribution of Working Group II to the Fourth Assessment Report of the Intergovernmental Panel on Climate Change*," in M.L. Parry, O.F. Canziani, J.P. Palutikof, P.J. van der Linden, and C.E. Hanson (eds.), Cambridge, UK: Cambridge University Press. http://ipcc.ch/ (accessed January 2014).

Jablonski, S. and Filet, M. (2008). "Coastal management in Brazil—A political riddle," *Ocean & Coastal Management*, 51: 536–43.

Julian, C. (2003). "Prospects for co-management in Indonesia's marine protected areas," *Marine Policy*, 27(5): 389–95.

Kalikoski, D.C., Neto, P.Q., and Almudi, T. (2010). "Building adaptive capacity to climate variability: The case of artisanal fisheries in the estuary of the Patos Lagoon, Brazil," *Marine Policy*, 34: 742–51.

Kumler, L.M. and Lemos, M.C. (2008). "Managing waters of the Paraíba do Sul river basin, Brazil: A case study in institutional change and social learning," *Ecology and Society* 13(2): 22. http://www.ecologyandsociety.org/vol13/iss2/art22/ (accessed January 2014).

Leão, Z.M.A.N., Kikuchi, R.K.P., Oliveira, M.D.M., and Vasconcellos, V. (2010). "Status of Eastern Brazilian coral reefs in time of climate changes," *Pan-American Journal of Aquatic Sciences*, 5(2): 224–35.

Macedo, H.S., Vivacqua, M., Rodrigues, H.C.L., and Gerhardinger, L.C. (2013). "Governing wide coastal-marine protected areas: A governance analysis of the Baleia Franca Environmental Protection Area in South Brazil," *Marine Policy*, 41: 118–25.

Margulis, S., Dubeux, C.B., and Marcovitch, J. (eds.) (2010). "*Economia da Mudança do Clima no Brasil: Custos e Oportunidades*," São Paulo, Brazil: IBEP Gráfica.

Marroni, E.V. (2011). "O Pré-Sal e a Soberania Marítima do Brasil," in: *Uma Nova Geopolítica para o Atlântico Sul. Anais do IV Seminário Nacional de Ciência Política. Teoria e Metodologia em Debate*, pp. 2179–7153. Programa de Pós-Graduação em Ciência Política da Universidade Federal do Rio Grande do Sul (UFRGS).

Marroni, E.V. and Asmus, M.L. (2005). "Gerenciamento Costeiro: Uma proposta para o fortalecimento comunitário na gestão ambiental," União Sul-Americana de Estudos da Biodiversidade, Pelotas, Brazil, 149p.

Marroni, E.V. and Asmus, M.L. (2013). "Historical antecedents and local governance in the process of public policies for coastal zone of Brazil," *Ocean & Coastal Management*, 76: 30–7.

Martins, R.D.A. and Ferreira, L.C. (2010). "Enabling climate change adaptation in urban areas: A local governance approach," *INTERthesis*, 7(1): 241–75.

Martins, R.D.A and Ferreira, L.C. (undated) "Coastal cities and climate change: Urbanisation, vulnerability and adaptive capacity on the northern coast of the São Paulo State, Brazil," Unpublished research article. http://www.edocs.fu-berlin.de/docs/servlets/MCRFileNodeServlet/FUDOCS_derivate_000000001640/Martins_Ferreira.pdf;jsessionid=CD761A91193D6EB93921A8A74068FF08?hosts (accessed January 2014).

McGranahan, G., Balk, D., and Anderson, B. (2007). "The rising tide: Assessing the risks of climate change and human settlements in low elevation coastal zones," *Environment & Urbanization*, 19(1): 17–37.

Mello, A.Y.I., D'Antona, A.O., Alves, H.P.F., and Carmo, R.L. (2010). "Análise da Vulnerabilidade Socioambiental nas Áreas Urbanas do Litoral Norte de São Paulo," *Proceedings of the V Encontro ANPPAS*, Florianópolis, Brazil.

Mendonça, M. (2010). "A vulnerabilidade da urbanização do Centro Sul do Brasil frente à variabilidade climática," *Mercator*, 9(1): 135–51.

Moura, R.L.D., Minte-Vera, C.V., Curado, I.B., Francini Filho, R.B., Rodrigues, H. D.C.L., Dutra, G.F., Alves, D.C., and Souto, F.J.B. (2009). "Challenges and prospects of fisheries co-management under a marine extractive reserve framework in northeastern Brazil," *Coastal Management*, 37(6): 617–32.

Muehe, D. (2006). *Erosão e Progradação do Litoral Brasileiro*, Brasília, Brazil: Ministério do Meio Ambiente (MMA), 476pp.

Muehe, D. (2010). "Brazilian coastal vulnerability to climate change," *Pan-American Journal of Aquatic Sciences*, 5(2): 173–83. http://www.panamjas.org/artigos.php?id_publi=183 (accessed January 2014).

Muñoz, J.M.B. (2001). "The Brazilian National Plan for Coastal Management (PNGC)," *Coastal Management*, 29: 137–56.

Neves, C.F. and Muehe, D. (2008a) "Mudança do clima e zonas costeiras brasileiras," *Parcerias Estratégicas*, 27: 217–95.

Neves, C.F. and Muehe, D. (2008b) "Vulnerabilidade, impactos e adaptação a mudanças do clima: A zona costeira," *Parcerias Estratégicas*, 27: 205–17.

Nicolodi, J.L. and Petermann, R.M. (2010). "Potential vulnerability of the Brazilian coastal zone in its environmental, social, and technological aspects," *Pan-American Journal of Aquatic Sciences*, 5(2): 184–204.

Pedlowski, M.A. (2013). "When the State Becomes the Land Grabber: Violence and Dispossession in the Name of 'Development' in Brazil," *Journal of Latin American Geography*, 12(3): 91–111.

Pelling, M. and High, C. (2005). "Understanding adaptation: What can social capital offer assessments of adaptive capacity?" *Global Environmental Change*, 15: 308–19.

Pereira, B. (2012). "Os Planos Ambientais de Gestão Integrada no estabelecimento da Governança Ambiental Costeira: Instrumentos e Práticas no âmbito da

Gestão Ambiental Municipal," Dissertação de Mestrado, Universidade Federal Do Rio Grande (FURG), Brazil.

Pereira, L.F.M., Barreto, S., and Pittock, J. (2009). "Participatory river basin management in the São João River, Brazil: A basis for climate change adaptation?," *Climate and Development*, 1(3): 261–68.

Pereira, R., Donatti, C.I., Nijbroek, R., Pidgeon, E., and Hannah, L. (2013). "Climate change vulnerability assessment of the Discovery Coast and Abrolhos Shelf, Brazil," Conservation International, 79p. http://static.weadapt.org/knowledge-base/files/1230/51dadbe0339d7climate-change-vulnerability-assessment-report-discovery-coast-and-abrolhos-shelf.pdf (accessed January 2014).

Pittock, J. (2011). "National climate change policies and sustainable water management: Conflicts and synergies," *Ecology and Society* 16(2): 25. http://www.ecologyandsociety.org/vol16/iss2/art25/ (accessed January 2014).

Polette, M., Rebouças, M.G.N., Filardi, L.A.C., and Vieira, F.P. (2006). "Rumo à Gestão Integada e Participativa de Zonas Costeiras no Brasil: Percep2ões da Comunidade Científica e do Terceiro Setor," *Gestão Costeira Integrada*, 5: 43–8.

Polette, M. and Seabra, A. (2011). "Governança Costeira No Brasil: Os Desafios Do Ciclo Do Pré-Sal," 19p.

Polette, M. and Vieira, P.F. (2006). *Avaliação do processo de gerenciamento costeiro no Brasil: Bases para discussão*, Tese de Pós doutoramento, UFSC, Brazil, 286p.

Reis, E.G. (2002). "Gestão Costeira Integrada," Programa Train Sea Coast Brasil.

Salazar, L.F., Nobre, C.A., and Oyama, M.D. (2007). "Climate change consequences on the biome distribution in tropical South America," *Geophysical Research Letters*, 34: L09708.

Scherer, M.E.G., Asmus, M.L., Filet, M., Sanches, M., and Poleti, A.E. (2011). "El manejo costero en Brasil: análisis de la situación y propuestas para uma posible mejora," in Farinós, J. (ed.), *La gestión integrada de zonas costeras. Algo más que uma ordenación del litoral revisada?*, vol. 9. Valência, Spain: IIDL/Publicacions de la Universitat de València, pp. 161–73.

Seabra, A.A., Freitas, G.P., Polette, M., and Casilla, T.A.D.V. (2011). "A Promissora Província Petrolífera do Pré-Sal," *Revista Direito GV*, 7(1): 57–74.

Serrao-Neuman, S., Di Giulio, G.M., Ferreira, L.C., and Choy, D.L. (2013). "Climate change adaptation: Is there a role for intervention research?" *Futures*, 53: 86–97.

Simões, A.F., Kligerman, D.C., La Rovere, E.L., Maroun, R.R., Barata, M., and Obermaier, M. (2010). "Enhancing adaptive capacity to climate change: The case of smallholder farmers in the Brazilian semi-arid region," *Environmental Science & Policy*, 13: 801–8.

Smit, B. and Wandel, J. (2006). "Adaptation, adaptive capacity and vulnerability," *Global Environmental Change*, 16: 282–92.

Soares, M.L.G. (2009). "A conceptual model for the response of mangrove forests to sea level rise," *Journal of Coastal Research*, SI 56: 267–71.

Soares, M.L.G., Almeida, P.M.M., Cavalcanti, V.F., Estrada, G.C.D., and Santos, D.M.C. (2011). "Vulnerabilidade dos manguezais da Região Metropolitana do Rio de Janeiro face às mudanças climáticas," in *Megacidades, vulnerabilidades e mudanças climáticas: Região Metropolitana do Rio de Janeiro*, CST/INPE e NEPO/UNICAMP (eds.), Rio de Janeiro, Brazil.

Soares, M.L.G., Tognella-De-Rosa, M.M.P., Oliveira, V.F., Chaves, F.O., Silva Jr., C.M.G., Portugal, A.M.M., Estrada, G.C.D., Barbosa, B., and Almeida, P.M.M. (2005). *Environmental changes in South America in the last 10k years: Atlantic and Pacific controls and biogeophysical effects: Ecological impacts of climatic change and variability: Coastal environments—Mangroves and salt flats*. Report to the Inter-American Institute on Global Change (IAI). 62p.

Souza, C.R.G. (2003). "The coastal erosion risk zoning and the São Paulo Plan for Coastal Management," *Journal of Coastal Research*, SI35: 530–47.

Souza, C.R.G. (2009). "A Erosão Costeira e os Desafios da Gestão Costeira no Brasil," *Gestão Costeira Integrada*, 9(1): 17–37.

Szlafsztein, C.F. (2012). "The Brazilian Amazon coastal zone management: Implementation and development obstacles," *Journal of Coastal Conservation*, 16: 335–43.

Tompkins, E.L., Lemos, M.C., and Boyd, E. (2008). "A less disastrous disaster: Managing response to climate-driven hazards in the Cayman Islands and NE Brazil," *Global Environmental Change*, 18: 736–45.

Vieira, B.C., Fernandes, N.F., and Filho, O.A. (2010). "Shallow landslide prediction in the Serra do Mar, São Paulo, Brazil," *Natural Hazards and Earth System Sciences*, 10: 1829–37.

Wever, L., Glaser, M., Gorris, P., and Ferrol-Schulte, D. (2012). "Decentralization and participation in integrated coastal management: Policy lessons from Brazil and Indonesia," *Ocean & Coastal Management*, 66: 63–72.

Yohe, G. and Tol, R.S.J. (2002). "Indicators for social and economic coping capacity—Moving toward a working definition of adaptive capacity," *Global Environmental Change: Human and Policy Dimensions*, 12(1): 25–40.

Part VII

Climate change and the coastal zone

Europe

Chapter 18

Toward adaptive management in coastal zones

Experience from the eastern coastline of England

R. Kerry Turner and Tiziana Luisetti

Abstract: Over recent centuries, the eastern coastal zone of England has been subject to dynamic change as the result of the combined pressure of, *inter alia*, land-use change, urbanization and industrialization, and episodic natural storm events. There is now a growing weight of evidence to suggest that climate change-related impacts, such as sea-level rise and more frequent and extreme storm surges, pose significant additional threats to human communities, economic and cultural assets, and ecosystems and habitats. Historical adaptive strategies, dominated by engineered structures of various types, will need to be augmented and/or changed if both people and nature are to build resilience to future stress and shock. This chapter considers past coastal management principles and practices and the emergence of new coastal policy practice in England, drawing lessons for future, more adaptive management strategies such as managed realignment. Although economic analysis such as cost–benefit analysis can help to more clearly identify future trade-offs, successful adaptation will require governance reforms and a new approach to matters of stakeholder engagement, compensation and social justice and equity perceptions and realities.

INTRODUCTION

Over the past few centuries, European coastal lowlands and estuaries have been dominated by land reclamation to make space for economic activity and its supporting infrastructure. Consequently, a suite of economic and other assets are at risk because of future coastal change. In the United Kingdom, these include around GBP 150 billion worth of property value at risk from flooding and GBP 10 billion worth of property value at risk from erosion (Parliamentary Office of Science and Technology 2009). Flood and erosion protection has been implemented principally through the construction of hard seawalls and related structures, as well as the drainage of wetlands. Loss of coastal wetlands has been exacerbated because of the *coastal squeeze* phenomenon as intertidal habitats are constrained between the sea and the seawall. The precise configuration of wetland pressures and consequent damage varies around Europe. Generalizing, in the European North Sea coastal zone, industrial development combined with agricultural intensification has historically been responsible for the majority of wetland loss, around 60% of total wetland area (United Nations Environment Programme 2004).

Increased cliff erosion and more extensive, as well as more frequent, flooding will speed up the degradation of existing defenses (often hard seawalls) putting at risk, privately owned houses, business and cultural, and other ecosystem-based benefits on the coastline of the United Kingdom (O'Riordan et al. 2008). The availability and adequacy of compensation measures and funds are highly contested matters and proposals and practice to ameliorate the growing lack of trust in coastal communities have been slow to emerge. As a response to difficulties posed by coastal process change and human welfare, more adaptive coastal governance is currently evolving, albeit in a somewhat reactive and piecemeal fashion. Currently, the response process is inevitably fragmented and there has been a lack of plan and policy integration among public, private, and voluntary organizations concerned with coastal adaptation (Nicholson-Cole and O'Riordan 2009).

In this chapter, we first summarize the distinctive features of the southern North Sea coastal zone in England and assess the environmental change drivers and pressures that are serving to increase risk from flooding and erosion. Past coastal management principles and practice are then critically reviewed in order to draw lessons for the new policy and governance regime based on a better appreciation of sustainability goals that is emerging.

THE NORTH SEA COASTAL ZONE: SOCIAL-ECOLOGICAL GOVERNANCE FEATURES OF THE BRITISH COASTS

European context

Many estuaries in northern Europe share a similar history of reclamation and biogeochemistry (Andrews et al. 2006). As part of the European Union, North Sea estuaries share the same governance regime and have to conform to European Union directives. The Water Framework Directive (European Commission 2000) requires that European Union member states introduce quality objectives for all water bodies (covering the whole catchment area), including coastal waters. The Habitats Directive (European Commission 1992b) is another relevant directive for coastal areas, and especially wetland areas. This specifically establishes Marine Special Areas of Conservation and requires compensation for displaced habitats, including those lost through natural or seminatural causes such as sea-level rise or coastal erosion. The Habitats and the Birds Directives (European Commission 1992a, 1992b) together aim to create a network of designated areas (known as Natura 2000) to protect habitats and species of community-wide importance.

British coastal policy structure

In England and Wales, two national policy initiatives in coastal management, the programme Making Space for Water (MSFW) (Department for Environment, Food and Rural Affairs 2005) and the Flood and Water Management Act (Parliament of the United Kingdom of Great Britain and Northern Ireland 2010), are currently being implemented. The latter sets out the government's proposals to make flood and coastal erosion risk management in England and Wales more adaptable, in line with the objectives of the MSFW programme, which addresses future climate change risks (Parliamentary Office of Science and Technology 2009).

The MSFW programme was launched by the British Department for Environment, Food and Rural Affairs in 2005. It represents a policy shift from hard defenses and *holding the line* policies to risk management. Risk management implies the recognition that risk can be reduced but not eliminated and, hence, this involves more sustainable forms of flood protection such as managed realignment. At the local level, Shoreline Management Plans have been the official mechanism for guiding decision-making about coastal protection for subsections of the coastline. Shoreline Management Plans were first implemented in the mid-1990s, to provide a focused assessment of the risks associated with coastal processes and the technical feasibility of policies that work with natural processes. The second generation of Shoreline Management Plans are currently in production, a number having been completed, and will cover the entire 6000 km coast of England and Wales (Environment Agency 2010; Parliamentary Office of Science and Technology 2009).

British coastal governance structure

The governance of coastal areas for England and Wales operates at three levels: political (central government departments), executive (statutory but nondepartmental agencies, such as the Environment Agency), and civic organizations and coastal communities (other organizations concerned with the coastal management) (Figure 18.1). The first level includes the Department for Environment, Food and Rural Affairs, which provides the overall policy as well as the supervision of the nondepartmental agencies; the Department for Communities and Local Government, which produces planning policy and local government finance; and the Treasury, which provides general funding to the Department for Environment, Food and

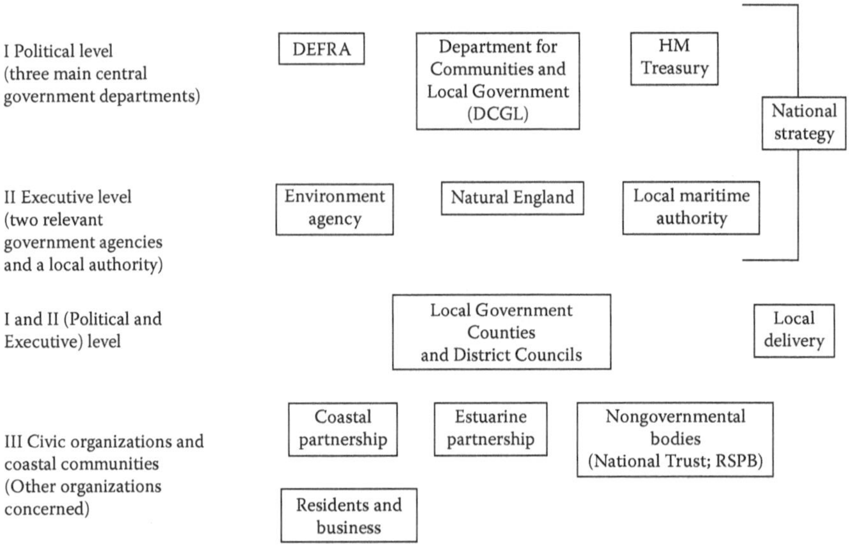

Figure 18.1 Levels of governance and actors involved in the coastal management of England and Wales. (Adapted from O'Riordan, T. et al., *Journal of the Academy of Social Sciences*, 3, 145–57, 2008.)

Rural Affairs, and the nondepartmental agencies and local authorities. Local government county and district councils operate both at the political and at the executive level as they are concerned with local planning, economic development, and the provision of social well-being. At the second level, we find the two relevant governmental agencies, the Environment Agency and Natural England and the Local Maritime Authority. The Environment Agency takes care of strategic coastal management. Natural England is concerned with the wildlife and habitat conservation of marine and coastal sites. In governance terms, the problem is that these agencies have different core objectives and statutory duties, but both have an interest in plans to manage the shoreline. Both recognize the importance of the economic valuation of coastal ecosystem services. The Local Maritime Authority is a political body with elected membership that has power over coastal planning and protection and economic management. At the third level, there are local residents and communities and nongovernmental bodies such as the National Trust and the Royal Society for the Protection of Birds that own coastal land managed as natural reserves (O'Riordan et al. 2008).

In practice, the neatly tiered governance structure operates in a much less ordered way. An intricate web interconnects the three levels and both coordination and organizational problems often arise. These difficulties lead, in turn, to local people questioning policy and financial decision-making in the coastal zone and to take their own actions in certain circumstances, for example, by constructing their own ad hoc coastal defense works.

THE NORTH SEA COASTAL ZONE: ENVIRONMENTAL THREATS AND SOCIOECONOMIC PRESSURES

The so-called drivers–pressures–state changes–impacts–response (DPSIR) framework has proved useful in the investigation of relevant historic and current causes, both socioeconomic and natural, and consequences of coastal change. The DPSIR framework was first developed by the OECD (1993) and has been successively refined and adapted to the context of coastal zone management by, among others, Turner et al. (1998). The DPSIR is a useful device for clarifying the role that socioeconomic drivers play in inducing pressures on the environment (over varying timescales and across a range of spatial scales) that result in state changes (often ecosystems degradation or loss) and consequent impacts on the welfare of people and communities locally, regionally, and, sometimes, globally. Efforts to modify the impacts through policy responses produce feedback effects within the drivers/pressures systems (Turner et al. 1998).

Coastal zone management is hindered by, among other factors, the scale mismatch problem, which has intensified as the process of globalization has itself accelerated. The socioeconomic drivers of environmental change in coastal zones, such as global trade and related finance flows, the growth in world population and the migration of people to the more hospitable coastal zones, are increasingly regional and global in scale and the local population may have little leverage over them. Within this demographic shift is the extra consequence that the poorest people and poorest communities facing the increased risks are the least able to cope (Millennium Ecosystem Assessment 2005; Nissanke and Thorbecke 2010; Turner and Fisher 2008). Coastal zone issues are often conditioned by an historical legacy, for example, the buildup of contaminants in estuarine and coastal sediments from past industrial/urban development and chronic eutrophication from intensive agriculture and/or inadequate sewage treatment facilities. That legacy impacts negatively on ecosystem services provision and this can be subsequently difficult and costly to ameliorate, for example, by improving soil quality/productivity, cleaning aquifers or modifying coastal defense structures, requiring at least catchment-scale action (Cave et al. 2003).

As far as climate change is concerned, of particular importance for coastal areas is the effect of rising global temperature on sea level, amplifying the risk of coastal erosion and flooding. Sea-level rise and vertical land movements associated with the melting of ice sheets (glacial isostatic adjustment) together are a major concern for some coastal regions in the United Kingdom: England, Wales, and Shetlands are subsiding and Scotland is lifting (Department for Environment, Food and Rural Affairs 2009). The latest climate projections report puts absolute sea-level rise (not including land movement) over the period 1990–2095 in the range 12–76 cm on the coastline of the United Kingdom (Department for Environment, Food and Rural Affairs 2009a). Risks associated with storm surges are also a serious concern. When surges occur at or near a high tide, they can cause major flooding, weaken cliffs, and increase the risk of erosion and collapse. Such an event was experienced on the southern coastline of the North Sea in 1953 when an extreme sea surge caused considerable loss of life and damage to properties in England and the Netherlands (Royal Commission on Environmental Pollution 2010).

A DPSIR framework for the coastal zone of the North Sea is presented in Figure 18.2.

LESSONS LEARNT: BRITISH COASTAL AREAS CASE STUDIES

Despite the scientific uncertainty over the precise consequences and the impacts of climate change, future projections of sea-level rise are being taken into account, at least on a precautionary basis, and adaptation action is being considered in the context of the British coastline. There are several ongoing experimental projects. The Thames Estuary 2100 project, based on a flexible approach to planning and climate risks, is an attempt to move from reactive flood defense to proactive flood risk (Environment Agency 2009). Other experiments relate to managed realignment of the coastline, the breaching of existing hard defenses letting the water flood the land behind to recreate coastal habitats, such as salt marshes (Royal Commission on Environmental Pollution 2010). Salt marsh areas provide a range of ecosystem services and welfare benefits, which can be enhanced by the recreation of formerly reclaimed land. Benefits provided include carbon storage benefits; fisheries' productivity via nursery areas; recreation and amenity benefits, such as walking, bird watching, and other recreational activities; and existence value benefits linked to biodiversity maintenance. Some of these usages are, however, competitive and can lead to conflicts among the stakeholders involved. From the management standpoint, trade-offs need to be made between greater resource efficiency, social justice, equity, and compensation objectives (Turner et al. 2007), especially because managed realignment involves sacrifice of some previously reclaimed coastal land, usually agricultural land, in order to reduce the threats of coastal erosion and flooding along the coast.

Traditionally, in the North Sea coastal areas, reclaimed land for agriculture, infrastructure, and urban centers has been protected by means of hard sea defenses with the promise of hold the line protection. Currently,

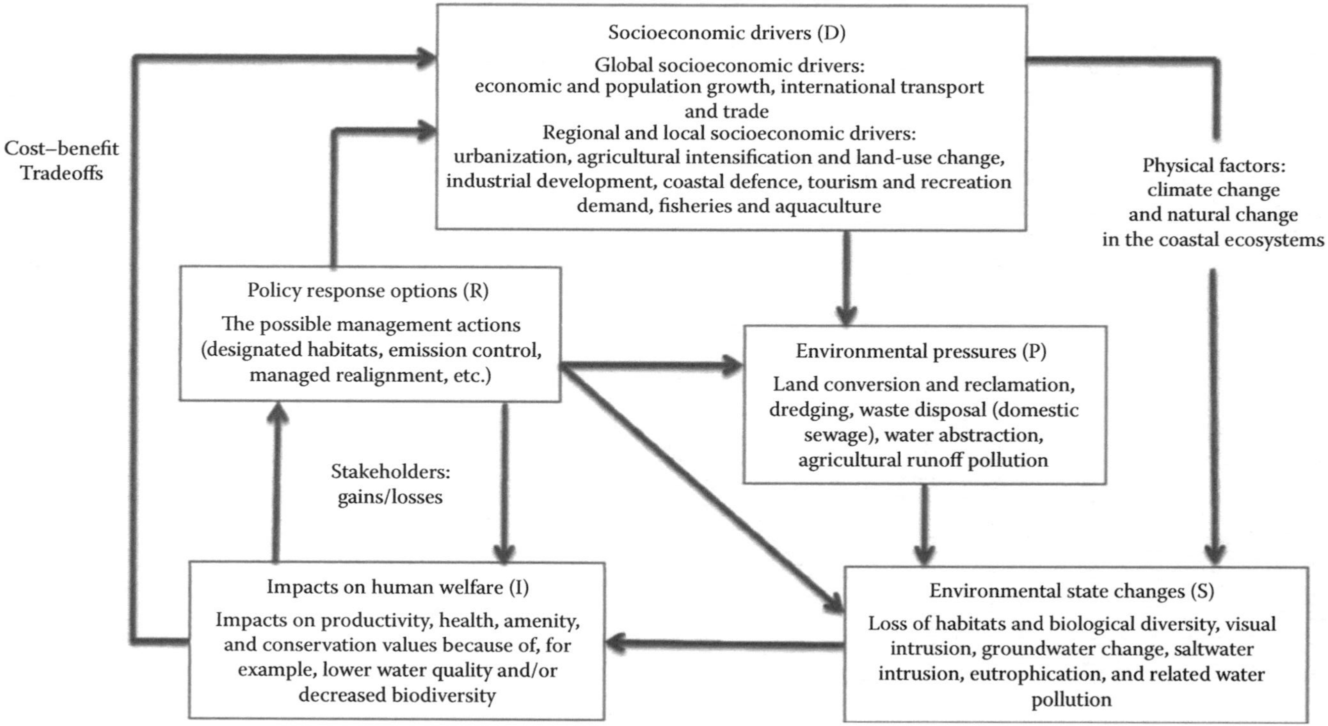

Figure 18.2 Drivers–pressures–state changes–impacts–response (DPSIR) for the North Sea coastal zone. (Adapted from Turner, R.K. et al., *The Geographical Journal*, 164, 269–81, 1998; Turner, R.K. et al., *The Geographical Journal*, 169, 99–116, 2003.)

the English coastline is protected by over 2000 km of flood defenses (Turner et al. 2007). But this same strategy has, through coastal squeeze, led to the loss of coastal wetlands. Wetlands across the United Kingdom and Europe have been lost or are under threat despite the existence of various international agreements (such as the Ramsar Convention) and national conservation policies. This situation has multiple causes, not just coastal squeeze, including the public nature of many wetland products and services, user externalities imposed on other stakeholders, and policy intervention failures due to a lack of consistency among policies being enacted across different sectors of the economy. These causes are also all related to *information failures* (e.g., a lack of appreciation of the full range of ecosystem services provided by healthy functioning wetlands), which in turn can be linked to the complexity and *invisibility* of spatial relationships between ground and surface waters and wetland vegetation (Turner et al. 2000).

In the United Kingdom, there is a complex relationship between coastal flood defense policy, stakeholder perceptions of flood risk, and ecosystem change. In the latter decades of the twentieth century, flood defense investments served to reduce the fear of flooding and, in the farming community, this perception encouraged land drainage and conversion schemes, with negative consequences for biodiversity. With the growing recognition of climate change and its threats and the need for greater food security, perceptions have again begun to alter with calls for more extensive sea defense and fluvial flooding investments (including tidal barriers). It has becoming increasingly apparent, however, that if current climate change predictions prove to be correct, engineered hard defenses as the only form of coastal protection are unlikely to be sustainable in the long run. Mixed approaches, in which only high-value areas continue to be protected and the rest of the coast line is left free to adapt to change more naturally, seem to be a better solution (Turner et al. 2007). Concern over national food security may, however, once again encourage more agricultural activity with negative spill-over effects on wetlands and water resources, including groundwater, as irrigated cropping regimes are further developed and calls for more flood protection investments could increase.

The Broads coastal wetlands (see Box 18.1) are located in the counties of Norfolk and Suffolk and form a complex wetland area comprised of shallow lakes in which fresh and saltwater are mixed. The Broads, together with the more southerly located Essex salt marshes, have been affected by generic environmental change drivers and pressures such as land-use change and agricultural development. But both areas have also faced specific drivers and pressures, such as pollution of water bodies and groundwater and increasing fluvial and saline flooding risk in the Broads and disappearing intertidal habitats because of coastal squeeze and, possibly, increasing storm surge flooding in the case of the salt marshes in Essex (Figure 18.3).

BOX 18.1 THE BROADS CASE STUDY

The Norfolk and Suffolk Broads are located on the east coast of England. They are shallow reed-fringed lakes, formed by peat excavation, that in mediaeval times were parts of an extensive water communication network. The surrounding area contains many prized biodiversity conservation sites (Broads Authority 2004). The rivers and connected broads are intensively used for recreational boating, involving around 100 boatyard operators, bringing with it the problem of congestion and noise pollution at various locations in the systems (Brouwer et al. 2002). The level of boating activity may at times cause conflict between stakeholders, especially between the interests of navigation, agricultural production, and biodiversity conservation and recreation/amenity. The range of stakeholders includes farmers and landowners, boating interests (hire boat/boat building and recreational boating groups), conservation agencies and local nongovernmental agencies, the environmental regulation agency (wide pressures including waste disposal and flood protection), local residents, holiday makers, service industry (hotels, pubs, restaurants), tourism agencies, and water companies. Climate change threats and food security concerns are likely to further complicate the situation. An ongoing flood alleviation program is seeking to maintain current protection levels, but it will not be sufficient to combat the more extreme climate change impact predictions.

The Broads Authority is the official agency that manages the Broads. At times, it has to operate by making pragmatic trade-offs, subject to European Union Directives and national legislation constraints. That was the case, for example, in Hickling Broad. This water body has a legal navigation channel but, over the years, boating has become possible over a large part of the lake. More recently, due to better management, the water quality has been improved and aquatic plant growth has accelerated, making some sections of the water body virtually inaccessible to navigation. As part of its sustainable development commitments, the Broads Authority must have regard for livelihoods and local enterprise, and it also has a statutory duty to maintain navigation access. The increasingly dense bed of aquatic plant (a rare species of stonework covered by the Habitat Directive) began to obstruct boating. Nonpowered boats and electric craft were particularly affected and the risk was a return to more diesel-powered craft, which, by generating pollution, would go against the Broads Authority's environmentally friendly boating initiative. A protracted process of dialogue with a range of interest groups and other government agencies eventually led to a compromise partial plant cutting program (Turner et al. 2003).

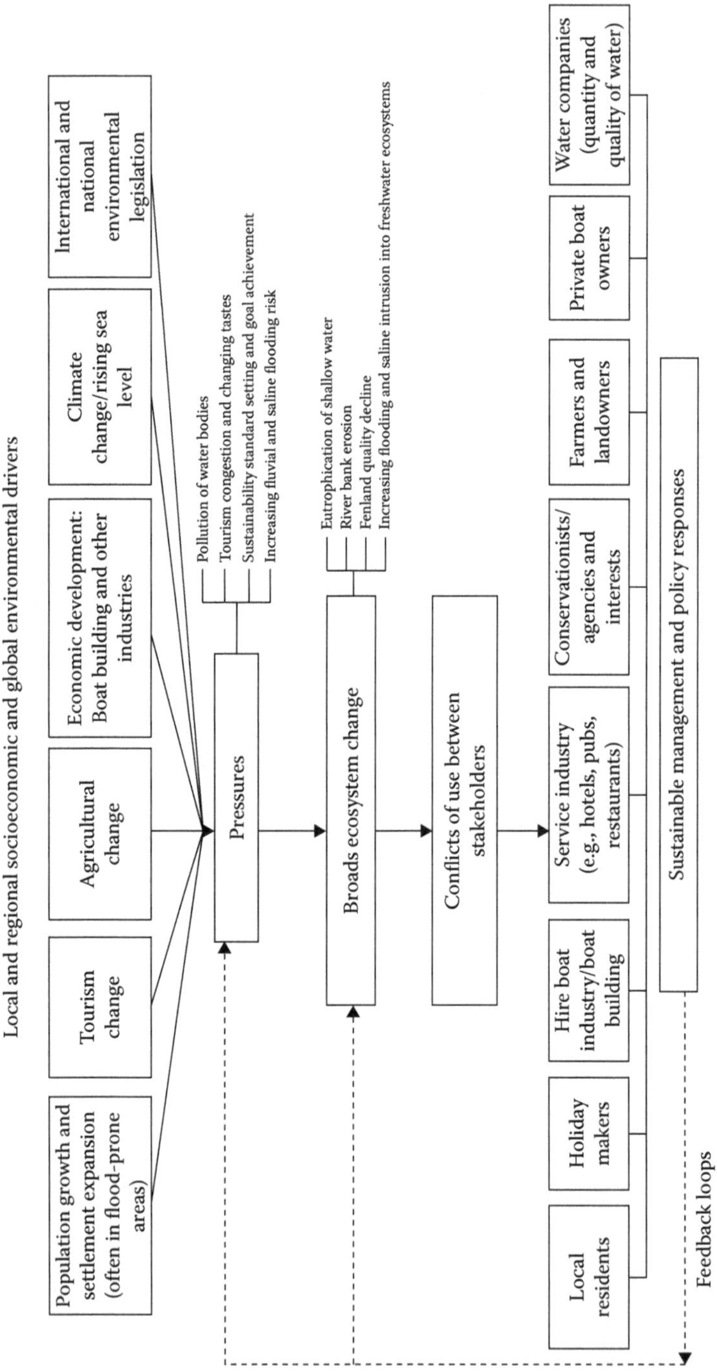

Figure 18.3 Direct and indirect pressures and related use conflicts in the Broads. (Data from Turner, R.K. et al., *Regional Environmental Change*, 4, 86–99, 2004.)

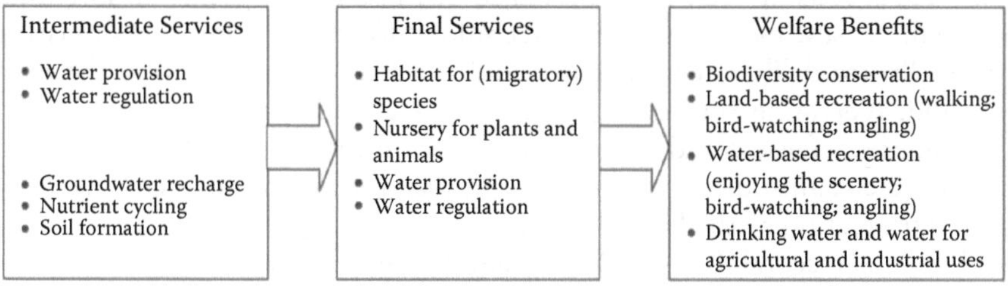

Figure 18.4 Classification of ecosystem services provided by the Broads.

Figure 18.3 sets out the full range of direct and indirect causes of wetland pressure and related use conflicts that have arisen in Broadland. Figure 18.4 brings together the at-risk ecosystem services and benefits provided by the Broads, based on the ecosystem services approach proposed by Fisher et al. (2009).

The long-term future provision of the ecosystem services provided by the Broads will be conditioned above all else by future flood defense policy and practice as the twin threats of saline inundation and fluvial flooding increase. If, in the future, climate change-related sea-level rise predictions prove correct, current hard defenses will not be able to provide an ongoing uniform level of protection across Broadland. Efforts are underway to selectively maintain/improve flood defenses, but saline intrusion, which will transform the freshwater ecosystem of the Broads, continues to slowly increase. A more flexible catchment-wide approach will have to be adopted following the European Union Water Framework Directive (European Commission 2000) and this may prove favorable to nature conservation in some respects, for example, the increased use of natural washland areas to contain peak flooding. On the other hand, an enhanced level of flood protection may stimulate renewed intensive agricultural activity with negative impacts in terms of water flow and eutrophication resulting in further pressure on the fenland and shallow lake ecosystems. Among the lessons learnt in the management of the Broads is the need for a flexible interpretation of European Union directives (such as the Birds and Habitats Directives) and for a more meaningful stakeholder dialogue process. The latter will need to encompass both consultation and stakeholder participatory and deliberative mechanisms.

When policy is reoriented to include shoreline managed retreat or realignment, as is the case, for example, in the North Norfolk area, where Shoreline Management Plans have suggested no active intervention, that is, no maintenance of existing defense works and managed realignment along parts of that coastline (Nicholson-Cole and O'Riordan 2009), future coastal protection may not be guaranteed. Many existing defense structures along the British coastline are coming to the end of their physical life and will soon require replacement or, at least, major maintenance works.

Managed coastal retreat or managed realignment provides a soft and, in many cases, more sustainable technique for flood defense. As existing seawalls are deliberately breached and the sea left free to flood the land behind, recreating salt marsh habitats, the construction of a new inland secondary defense could follow depending on the topography of the area. In other circumstances, the land elevation may provide the same level of flood protection as the hard defenses.

Previous research has concluded that managed realignment policy needs to be appraised across a more extensive spatial and temporal scale than has been the case in the traditional scheme-by-scheme coastal management system (Turner et al. 2007). In other words, whole estuaries or multiple coastal cells need to be treated as a single project encompassing a number of realignment sites. Recent case study analysis has focused on two estuaries, the Humber and the Blackwater (Luisetti et al. 2011b), which share physical similarities and risk profiles. Both estuaries are located on the east coast of England and are subject to the same legislative regime. A cost–benefit analysis was deployed in order to investigate the economic efficiency of managed realignment schemes in the context of more sustainable future coastal management scenarios than the traditional *hold the line* scenario. A complete *do nothing* strategy was not considered given the statutory duties imposed on coastal protection and sea defense agencies.

The methodologies applied in the Humber case study (Turner et al. 2007) and, subsequently, in the Blackwater study (Luisetti et al. 2011b) were similar. Both studies made use of GIS techniques in combination with the coastal management scenarios to identify possible realignment sites in which the opportunity costs of realignment involving significant social justice/ethical concerns were minimized (e.g., urban centers were assumed to be protected and, therefore, excluded from the analysis). For that reason, an efficiency-based cost–benefit analysis, considered to be capable of providing decisive information in the policy choice, was applied (Randall 2002; Turner 2007). In the cases where people, property, culture/historical assets, and designated freshwater conservation sites are part of the opportunity cost calculation, cost–benefit analysis cannot be as decisive and may have to be subsumed within a multicriteria decision support system and deliberative process in order to tackle the ethical and other value judgments that will be in dispute (Rhodes 1997; Sabatier 1998). The benefits provided by salt marshes were identified using an ecosystem services approach (Fisher et al. 2009; see Figure 18.5).

The two case studies differ in terms of the valuation method used to value the *composite environmental benefit* that includes the ecosystem services of amenity and recreation as well as biodiversity maintenance. In the Humber, benefit transfer values were used, while aggregated willingness to pay values elicited via stated preference techniques (choice experiment) were used in the Blackwater estuary study (Luisetti et al. 2011a, 2011b). Although the implementation of managed realignment schemes implies financial savings on

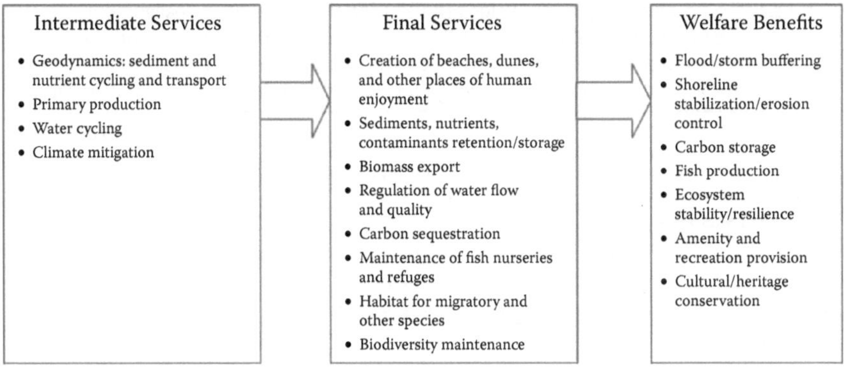

Figure 18.5 Classification of coastal and marine ecosystem services.

seawall maintenance, some costs are also involved, such as capital costs for realigning defenses whenever a secondary line of defense may be required further inland and opportunity costs associated with any agricultural land that is sacrificed as the old defenses are breached.

The results for both case studies provide a convincing case for managed realignment, for the Blackwater estuary analysis even more so (see Table 18.1). An anomaly is, however, found in the Blackwater case study: the managed realignment deep green scenario has the highest positive value over the 25 years time horizon compared to the other scenarios. A possible explanation lies in the length of defenses to be realigned. The extended deep green scenario, in which environmental protection takes priority over economic growth and which would be expected to have the highest net present values, has the longest length of defenses to be realigned. That implies higher costs of realignment for the extended deep green scenario, although most of the areas of realignment were chosen where the elevation of the land would not then require a secondary line of defense (Luisetti et al. 2011b).

The general public's perception of climate change risk and the need for adaptation measures may lag behind scientific and policy thinking. Therefore, although realignment at some points along a coastline may bring benefits in terms of intertidal habitat creation and associated recreation and other gains, the opportunity costs of such strategy are equally relevant. Politically and economically, the possibility of loss of or damage to urban areas and other assets is likely to weigh heavily against managed realignment schemes. If national strategic objectives require the withdrawal of existing defenses or no new defensive measures, the question of stakeholder/asset loss compensation will loom large. In the United Kingdom, there is no such compensation scheme (Nicholson-Cole and O'Riordan 2009), and a legacy of mistrust has built up with coastal communities increasingly calling for more accountability in the policy process. The need for adaptation

Table 18.1 Comparing net present values (NPVs) for three managed realignment scenarios and the *hold the line* (HTL) scenario for the Humber and the Blackwater estuary studies using conservative values (only use values) and a declining discount rate following Her Majesty's Treasury guidance (£ million)

	25 years		*50 years*		*100 years*	
Scenario	*Humber*	*Blackwater*	*Humber*	*Blackwater*	*Humber*	*Blackwater*
Policy Targets (PT)						
NPV PT	−73.23	68.41	−82.22	152.25	−92.27	307.19
NPV HTL	−70.40	−1.88	−86.01	−3.96	−100.93	−7.81
NPV(PT) – NPV(HTL)	**−2.83**	**70.29**	**3.79**	**156.21**	**8.66**	**315**
Deep Green (DG)						
NPV DG	−97.32	74.83	−101.42	185.35	−107.92	389.58
NPV HTL	−70.40	−1.88	−86.01	−3.96	−100.93	−7.81
NPV(DG) – NPV(HTL)	**−26.92**	**76.71**	**−15.41**	**189.31**	**−6.99**	**397.39**
Extended Deep Green (EDG)						
NPV EDG	−94.30	62.83	−74.48	186.22	−63.83	414.24
NPV HTL	−70.40	−1.88	−86.01	−3.96	−100.93	−7.81
NPV(EDG) – NPV(HTL)	**−23.90**	**64.71**	**11.53**	**190.18**	**37.10**	**422.05**

Source: Luisetti, T. et al., *Land Economics*, 87(2), 284–96, 2011a.

measures led the British Government to undertake an 18-month (2009–2011) scheme called Pathfinder that will support (with a GBP 11 million budget) 15 pathfinder authorities (County and District Councils) to explore new ways of adapting to coastal change (Department for Environment, Food and Rural Affairs 2010). The North Norfolk District Council received the largest amount, GBP 3 million, to implement the scheme in its local area, and it is the first to have put forward, in 2010, a compensation plan using some of the funding of the Pathfinder scheme. It is not yet clear precisely which kind of support mechanism will be taken up more generally by Pathfinder projects (North Norfolk District Council 2010).

CRITICAL BARRIERS AND OPPORTUNITIES FOR MAINSTREAMING CLIMATE CHANGE ADAPTATION

In principle, it seems clear that coastal policy in England must become much more long-term and strategic. Implementation will require new approaches to land-use planning that take into account enhanced coastal flood and erosion risk (O'Riordan et al. 2008). Currently, planning guidance by the British Government is set out in documents that are called Planning Policy Statements (previously, Planning Policy Guidance Notes). Planning Policy Statements are prepared after public consultation to provide guidance to local authorities and others on planning policy and the operation of the planning system (Nicholson-Cole and O'Riordan 2009). In terms of greater social justice, much more clarity is required over discretionary public defense of any person or property on the coast, together with the legal possibility for private owners to provide their own coastal protection. Moreover, a more formal legal position remains to be established over the provision and form of compensation to at-risk asset holders; the Department for Environment, Food and Rural Affairs seems to have accepted that there had to be some form of mechanism, like the Pathfinder scheme, to assist those affected by coastal changes. These practical challenges are formidable and have yet to be satisfactorily met.

RECOMMENDATIONS

In the United Kingdom, the process required to achieve a more sustainable coastline that can adapt to future changes has been initiated, but it is still in its formative phase. It will be a long process that will include more reliance on natural defenses, but deciding on the precise spatial location of each type of defense/protection remains a heavily contested issue.

Newly created intertidal habitats and their management are still experimental and their overall contribution to coastal management has yet to be determined. More and better climate and coastal processes science will help to reduce the inevitable uncertainties, but equally necessary will be agreed strategic decision-making framework and practice. Targeted financial support and coordination between stakeholders, recognizing the legitimate social justice interests and rights of individuals/communities living on the coast, are also essential. For any given coastal context, the process of the stakeholder mapping is crucial as it is the foundation for the assessment of those environmental change impacts and consequences that impinge most significantly on stakeholders. Finally, a much more participatory process needs to be embedded within the governance regime in order to tackle the range of contested issues that typically arise in coastal management contexts.

ACKNOWLEDGMENT

We are grateful to Sophie Nicholson-Cole for comments on an earlier draft of the chapter.

REFERENCES

Andrews, J., Burgess, D., Cave, R., Coombes, E., Jickells, T., Park, D., and Turner, R.K. (2006). "Biogeochemical value of managed realignment, Humber Estuary, UK," *Science of the Total Environment*, 371: 19–30.

Broads Authority. (2004). *Broads Plan: A Strategic Plan to Manage the Norfolk and Suffolk Broads*, Norwich: Broads Authority. http://www.broads-authority.gov.uk/index.html (accessed December 2011).

Brouwer, R., Turner, R.K., and Voisey, H. (2002). "Public perception of overcrowding and management alternatives in a multi-purpose open access resource," *Journal of Sustainable Tourism*, 9: 471–90.

Cave, R., Ledoux, L., Turner, R.K., Jickells, T., Andrews, J., and Davies, H. (2003). "The Humber catchment and coastal area," *Science of the Total Environment*, 314–316: 31–52.

Department for Environment, Food and Rural Affairs. (2005). *Making Space for Water: Taking Forward a New Government Strategy for Flood and Coastal Erosion Risk Management in England*, London: Department for Environment, Food and Rural Affairs. http://archive.defra.gov.uk/ (accessed December 2011).

Department for Environment, Food and Rural Affairs. (2009). *Data from the UK Climate Projection*, London: Department for Environment, Food and Rural Affairs. http://ukclimateprojections.defra.gov.uk/ (accessed December 2011).

Department for Environment, Food and Rural Affairs. (2010). *Pathfinder*, London: Department for Environment, Food and Rural Affairs. http://www.defra.gov.uk/ (accessed December 2011).

Environment Agency. (2009). *TE2100 Plan: Consultation Document*, London: Environment Agency. http://environment-agency.gov.uk (accessed December 2011).

Environment Agency. (2010). *Shoreline Management Plans*, London: Environment Agency. http://environment-agency.gov.uk/research/planning/104939.aspx (accessed December 2011).

European Commission. (1992a). *Directive 79/409/EEC (Birds Directive)*, Official Journal, April 25, 1979, Brussels, Belgium: European Commission.

European Commission. (1992b). *Directive 92/43/EEC (Habitats Directive)*, Official Journal, July 22, 1992, Brussels, Belgium: European Commission.

European Commission. (2000). *Directive 2000/60/EC (Water Framework Directive)*, Official Journal, December 22, 2000, Brussels, Belgium: European Commission.

Fisher, B., Turner, R.K., and Morling, P. (2009). "A systems approach to definitions and principles for ecosystem services," *Ecological Economics*, 68: 643–53.

Luisetti, T., Bateman, I.J., and Turner, R.K. (2011a). "Testing the fundamental assumptions of choice experiments: Are values absolute or relative?," *Land Economics*, 87(2), 284–96.

Luisetti, T., Turner, R.K., Bateman, I.J., Morse-Jones, S., Adams, C., and Fonseca, L. (2011b). "Coastal and marine ecosystem services valuation for policy and management: Managed realignment case studies in England," *Ocean & Coastal Management*, 54(3): 212–24.

Millennium Ecosystem Assessment. (2005). "Coastal Systems," in *Global Assessment Reports, Vol. 1: Current State and Trends*, World Resources Institute, Washington, DC: Island Press. http://www.maweb.org/en/Global.aspx (accessed December 2011).

Nicholson-Cole, S. and O'Riordan, T. (2009). "Adaptive governance for a changing coastline: Science, policy and the public in search of a sustainable future," in W.N. Adger, I. Lorenzoni, and K. O'Brien (eds.). *Adapting to Climate Change: Thresholds, Values and Governance*, Cambridge: Cambridge University Press.

Nissanke, M. and Thorbecke, E. (2010). "Globalization, poverty and inequality in Latin America: Findings from case studies," *World Development*, 38(6): 797–802.

North Norfolk District Council. (2010). *Programme Initiation Document: North Norfolk Pathfinder Programme*, North Norfolk: North Norfolk District Council. http://www.northnorfolk.org/ (accessed December 2011).

OECD (Organization for Economic Cooperation and Development). (1993). *Environmental Monographs No. 83: OECD Core set of indicators for environmental performance*, Paris, France: OECD.

O'Riordan, T., Nicholson-Cole, S., and Milligan, J. (2008). "Designing sustainable coastal futures, twenty-first century society," *Journal of the Academy of Social Sciences*, 3: 145–57

Parliamentary Office of Science and Technology. (2009). *Postnote: Coastal Management* (no. 342), London: POST. http://www.parliament.uk/ (accessed December 2011).

Parliament of the United Kingdom of Great Britain and Northern Ireland. (2010). *Flood and Water Management Act*, London: Parliament of the United Kingdom of Great Britain and Northern Ireland. http://www.legislation.gov.uk/ (accessed December 2011).

Randall, A. (2002). "Benefit-cost considerations should be decisive when there is nothing more important at stake," in D.W. Bromley and J. Paavola (eds.). *Economies, Ethics and Environmental Policy*, Oxford: Blackwell.

Rhodes, R. (1997). *Understanding Governance: Policy Networks, Governance, Reflexivity and Accountability*, Buckingham: Open University Press.

Royal Commission on Environmental Pollution. (2010). *Adapting Institutions to Climate Change*, London: Royal Commission on Environmental Pollution.

Sabatier, P.A. (1998). "The advocacy coalition framework: Revisions and relevance for Europe," *Journal of European Public Policy*, 5(1): 98–130.

Turner, R.K. (2007). "Limits to CBA in UK and European policy: Retrospect and future prospects," *Environmental and Resource Economics*, 37(1): 253–69.

Turner, R.K., Bateman, I.J., Georgiou, S., Jones, A., Langford, I., Matias, N., and Subramanian, L. (2004). "An ecological economics approach to the management of a multi-purpose coastal wetland," *Regional Environmental Change*, 4(2/3): 86–99.

Turner, R.K., Burgess, D., Hadley, D., Coombes, E., and Jackson, N. (2007). "A cost-benefit appraisal of coastal management realignment policy," *Global Environmental Change*, 17(2): 397–407.

Turner, R.K. and Fisher, B. (2008). "To the rich man the spoils," *Nature*, 451(7182): 1067–8.

Turner, R.K., Georgiou, S., Brouwer, R., Bateman, I.J., and Langford, I. (2003). "Towards an integrated environmental assessment for wetland and catchment management," *The Geographical Journal*, 169(2): 99–116.

Turner, R.K., Lorenzoni, I., Beaumont, N., Bateman, I.J., Langford, I.H., and McDonald, A.L. (1998). "Coastal management for sustainable development: Analysing environmental and socio-economic changes in the UK coast," *The Geographical Journal*, 164(3): 269–81.

Turner, R.K., van der Bergh, J., Soderquist, T., Barendregt, A., van der Straaten, J., Maltby, E., and van Ierland, E. (2000). "Ecological-economic analysis of wetlands: Scientific integration for management and policy," *Ecological Economics*, 35(1): 7–23.

United Nations Environment Programme. (2004). *Fresh Water in Europe*, a publication of the Division of Early Warning and Assessment, Office for Europe (DEWA Europe), Geneva, Switzerland: UNEP.

Chapter 19

Adaptation to change in the North Sea area

Maritime spatial planning as a new planning challenge in times of climate change

Andreas Kannen and Beate M.W. Ratter

Abstract: Impacts of climate change in the North Sea region include system changes such as rising water temperatures, increasing sea level, ocean acidification, an increase in extreme weather events, increased storm surges, and shifts in species composition. A further challenge for the North Sea region stems from new economic sectors such as offshore wind energy. In this context, Maritime Spatial Planning (MSP) gains increasing importance as a tool for integration of sectors and interests and for keeping the balance between sea use and environmental protection. The current experience with MSP in North Sea countries can offer important contextual insights for future climate adaptation governance.

INTRODUCTION

Human activities in marine areas are increasing in number and intensity, and patterns of sea use are changing as a result of political, economic, and societal developments. For the German North Sea, intensification of sea uses was assessed in several projects (e.g., Gee et al. 2006; Kannen et al. 2008, 2010). However, similar developments can be observed in all other countries around the North Sea: "The North Sea has some of the busiest shipping lanes in the world and maritime transport continues to increase. Construction activities have also been increasing …, with more coastal structures and wind farms being built and operated, and more tourist traffic" (OSPAR 2010: 154). Beside use intensification in general, the OSPAR Quality Status Report 2010 identifies the impacts of fisheries, hazardous substances (especially persistent organic pollutants), and nutrient inputs from land as highly important pressures for the North Sea region. Accompanying these trends, a range of environmental and spatial marine policies is currently evolving in Europe and around the North Sea. In addition, proposed mitigation and adaptation measures to climate change might alter the sea-use pattern and consequently the distribution and intensity of human impacts on marine ecosystems. A few examples are as follows:

- Large-scale development of coastal defense as a response to sea-level rise and erosion
- Use of old North Sea oil and gas fields as sites for sub-seabed storage of carbon dioxide
- Construction of large-scale sites for renewable energies (wind, wave, tidal, etc.)

In this chapter, we use the North Sea as a regional example to discuss current changes in coastal and marine use and in related responses from planning and management. We will focus on offshore wind farm development as a particular case of use change because of the following reasons: (1) this case is the most obvious challenge around the North Sea today; (2) it is directly linked to climate change; and (3) it has already led to the establishment of new planning mechanisms, in particular Maritime Spatial Planning (MSP), in North Sea countries. The account will first identify the challenges that might arise due to climate change and then characterize the current state and expected trends of offshore wind farm development in the North Sea. After briefly introducing the currently evolving policies concerning marine areas in Europe with a particular focus on MSP, we will discuss what can be learnt from currently evolving maritime policies for dealing with climate change.

LONG-TERM CHALLENGES: POTENTIAL IMPACTS FROM CLIMATE CHANGE

Impacts of climate change on marine and coastal areas are manifold. But the prediction of the precise rate and magnitude of change, and for some effects even the direction of change and local level impacts, are difficult to predict. Global climate change models are too general to predict specific regional effects. Therefore, the role of downscaling the models and regional-specific foci is crucial. Among general expected system changes, such as an increase in water temperature, rising sea level, and ocean acidification, there are effects in regional seas like the North Sea, which include an increase in extreme weather events, increased storm surges, and shifts in species composition. All these changes will have a direct impact not only on the marine areas but also on the coastal population and their activities.

For the North Sea, an increase in temperature over the past decades was observed. Future scenarios for northern Germany predict an increase in summer temperature of +0.7°C to +0.9°C by 2040, and a change in precipitation patterns of +1% to +20% during winter. By 2100, predictions result in an increase in temperature by +1.8°C to +5.1°C, increased wind speed of 14%, and a precipitation decrease in summer by –10% to –42% (von Storch et al. 2009).

Weisse et al. (2005) analyzed long-term changes in storm activity for the region and the northeastern North Atlantic. An increase was detected in storm activity from about 1960. Storm activity peaked around 1990–1995, after which a decrease was inferred. In particular, there was evidence found that the changes correspond to that of storm activity with increases in storm surges and wave heights between about 1960 and 1990, decreasing thereafter (Weisse et al. 2009). The changes in extreme storm surge levels are expected toward the end of the century. The calculations employ a mean sea-level rise of 9 cm for 2030 and of 29 and 33 cm (accounting for different scenarios) for 2085, respectively (Weisse et al. 2009). Although regional details differ among the different models and scenarios, all point toward a moderate increase in severe storm surge levels along most of the Netherlands, German, and Danish coastlines. When compared to the natural variability estimated from the hindcast based on the coastDat database (Weisse and Pluess 2006), climate change-related increases in storm surge heights are found to be smaller for most of the Netherlands and Danish coast, while they are significantly larger along most of the German coastline (Woth et al. 2006).

The expected changes demand adequate policies and measures in order to deal with climate change impacts along the North Sea coast, especially in terms of coastal protection but as well in terms of dealing with the use of coastal and marine areas. Whereas adaptation measures are related to dealing with future hazards under uncertain circumstances, mitigation measures include proactive activities in order to reduce the expected climate change effects. Societal response to climate change involves mitigation measures such as moving toward use of renewable energies or carbon dioxide storage as well as adaptation measures, for example, concerning coastal defence strategies (OSPAR 2009).

Adaptation measures become vital on different levels including governmental policies, private sector, and the population itself. In this context, the perception of what is considered to be a hazard (either slowly evolving like sea-level rise or sudden extreme events like storm surges) is crucial. The acceptance of spending money for increased coastal protection and the preparedness for personal adaptation measures depend on the awareness about hazards. A population survey along the German North Sea coast detected that 33% of the people questioned considered storm surges together with climate change as a major threat to their livelihood. The analysis also revealed that the majority felt that the responsibility for coastal protection measures should be put into the hands of the government. Forty-seven percent of the respondents saw themselves as *personally affected*, whereas 45% saw no personal threat from storm surges and climate change and 15% expressed the opinion that there is *no way of protecting against climate change impact* (Ratter et al. 2009). Lack of awareness that safety by protection measures is possible makes adaptation measures difficult.

Based on these findings, the mechanisms to adapt to climate change, including the associated uncertainties in the implementation of each of the different policies across management levels and sectors, might form a major challenge. In addition, due to the different timescales of natural processes affected by climate change and those of political and societal processes, the targets of planning and management are determined by and at the same time relevant for the perception of problems and issues. Alongside climate change there are other challenges arising for planning and management not only along the coastal areas but also in the adjacent offshore regions. Some of these will be discussed in the subsequent sections.

CURRENT CHALLENGES: DYNAMICS AND CONTEXT OF OFFSHORE WIND FARMING

With the exception of the Norwegian trench, the North Sea is a rather shallow sea, located on the European continental shelf with a mean depth of 90 m. Its generally shallow exposition makes the North Sea particularly suitable as a location for offshore wind farms, many of them located in water depths between 10 and 40 m, but at distances of more than 20–30 km and more away from the coast.

The southern coastline is characterized by extensive mudflats and estuaries. In particular, the Wadden Sea in the southeast of the North Sea comprises the largest area of intertidal mudflats in the world, hosting 10–12 million migrating birds every year and protected under several international and national regulations. Overall, the North Sea and its coastal areas are biologically highly productive areas and of high value for fisheries as well as for global ecological processes. Human population density around much of the North Sea is high, in particular in its eastern and southern parts. More than 500 people per km^2 live in some coastal areas and intensive farming covers up to 70% of the land that drains into this part of the ocean (OSPAR 2010), resulting in considerable input of nutrients (from agriculture) and hazardous substances (from industry).

Although overall fishing effort is decreasing, down 25% from 2000 to 2006 according to OSPAR (2010), a wide range of other human uses, most of them with dynamic growth rates, are relevant not only within the North Sea, but also for the national economies of its riparian states. These include, in particular, shipping, tourism, and recreational activities, but also wind farms (on- and offshore), cables and pipelines, coastal defense, dredging and dumping, and mineral extraction. The most challenging trend in North Sea use today is the planning and construction of offshore wind farms, which is an issue in all adjacent states, especially in the United Kingdom, the Netherlands, Belgium, Germany, Denmark, Norway, and Sweden.

Offshore wind farms form, on the one hand, for many people a symbol of an *industrialization of the sea*; on the other hand, they are a mitigation measure to reduce carbon dioxide emissions and therefore constitute a structural adaptation measure toward a carbon dioxide-free economy. Furthermore, they are part of an emerging and very dynamic economic sector, providing jobs and income in coastal locations as well as industrial areas inland. Even though offshore wind farming has become operational on some sites, its further development and long-term maintenance are still a developing and pioneering technology issue.

According to the Global Wind Energy Council (2009: 4), "wind power is on track to supply 10%–12% of global electricity demand by 2020, reducing carbon dioxide emissions by 1.5 billion tonnes per year." Global capacity (up to now mainly onshore) amounted to 120,798 megawatts in 2008, of which 65,946 megawatts were installed in Europe and 23,908 in Germany (Global Wind Energy Council 2009). Dedicated policy support has played a strong role in facilitating the steady growth of the sector, such as the recent European Union Renewable Energy Sources Directive from 2009 and national targets for renewable energies. Moving toward offshore areas is of increasing interest to developers because of higher and more predictable wind speeds, the possibility to construct higher turbines in larger areas and less likelihood of public resistance, at least if wind farms are sited far enough in the sea so that they are rarely visible from land. In 2009, 52 offshore wind farms had been fully consented in Europe, totalling more than 16,000 megawatts (European Wind Energy Association 2010a). As of June 2010, 16 offshore wind farms with 3,972 megawatts were under construction, while the installed capacity had already reached 2,396 megawatts (European Wind Energy Association 2010b).

Mid- and long-term policy targets behind the offshore wind farm development are quite ambitious. The European Wind Energy Association, an industrial lobby group, aims for 150,000 megawatts of installed offshore capacity in Europe by 2030, thereby meeting between 12.8% and 16.7% of the total electricity demand in the European Union (European Wind Energy Association 2009). In Germany, the federal government expects offshore wind energy to play a major role in reaching its national renewable energy targets. If current plans go ahead, offshore wind farms could provide between 20,000 and 25,000 megawatts by 2030, meeting about 15% of the German electricity demand (BMU 2002). Similarly ambitious national targets exist in the United Kingdom and other North Sea countries.

Promoting offshore wind farms is, however, more than looking for areas in which to place wind farms. In December 2009, nine European Union Member States (Belgium, the Netherlands, Luxembourg, Germany, France, Denmark, Sweden, the United Kingdom, and Ireland) and Norway signed a political declaration on the North Seas Countries, Offshore Grid Initiative with the objective to coordinate in particular the connection of offshore wind farms to the terrestrial electricity grid infrastructure in the North Sea (European Commission 2010). These countries will concentrate about 90% of all European Union offshore wind development. The offshore grid adds to the strong trans-boundary linkages within the North Sea, where European policies and other transnational (in particular environmental) regulations form additional levels to national policies.

The expansion of offshore wind interest can be described as a point of convergence of various push and pull factors emanating from several policy arenas and geographical scales (Kannen and Burkhard 2009). Major drivers are climate change policies and energy policy at an international and national level as well as regional economic development and employment for subnational regions. In addition to issues of local acceptance, often related to aesthetic (loss of free horizon) and ethical (e.g., the impact on birds and marine mammals) arguments (cf. Gee 2010; Licht-Eggert et al. 2008), offshore wind farming also enters an area, the sea, where other spatial uses exist such as shipping, fishing, tourism, and nature protection sites. All of them are related to today's regional identity and traditional economy in many coastal areas. Therefore, the foremost challenge for coastal and marine management in the North Sea is to govern a maritime system that is linked to several policy arenas and covers several interconnected scales. Translated in terms of system theory, offshore wind farming is part of a social-ecological space, which is characterized by emergent nonlinear system behavior and high uncertainty as far as social and ecological impacts are concerned (Kannen and Burkhard 2009). Dealing with this social-ecological system requires instruments that allow for the development of widely agreed visions and widely accepted priorities despite uncertain outcomes.

POLICY CHALLENGES: EVOLVING SPATIAL POLICIES FOR MARINE AREAS

Until now, offshore wind farming, despite its European and international drivers, is pushed and regulated under national jurisdictions and according to national planning schemes. Different planning rules and different mechanisms to provide incentives for investors are used in the different riparian states. For example, the financial support schemes of the United Kingdom and Germany vary considerably. In Germany, based on the Renewable Energy Sources Act (2009), direct feed-in tariffs are provided that grid operators must pay for renewable energy fed into the power grid. This model has been copied in many other countries of the world. But in the United Kingdom, based on the *Renewables Obligation*,* electricity

* For detail see https://www.ofgem.gov.uk/environmental-programmes/renewables-obligation-ro.

suppliers are required by law to provide a proportion of their sales from renewable sources (such as wind power) or pay a penalty fee. The supplier receives a Renewables Obligation Certificate for each megawatt hour of electricity purchased.

Similarly, the planning procedure can be very different according to the country in whose territorial waters or exclusive economic zone (EEZ) a wind farm is planned. In the United Kingdom, investors can bid for a license in predefined search areas, in which they need to identify a suitable location based on a consent procedure and impact assessments. In Germany, an investor can claim virtually any area in the EEZ which is not designated as a priority area for other uses within the spatial plan for the North Sea EEZ from 2009. (see BSH [2013] for text and map related to German spatial plans in the EEZ.) The investor has to follow an approval procedure including public hearings and an environmental impact assessment.

Given the size of offshore wind farming in all countries around the North Sea and the existing plans for a North Sea grid mentioned above, purely national planning and approval approaches will soon become insufficient. Different schemes of economic incentives prohibit to connect, for example, a wind farm in the German EEZ with the United Kingdom grid and vice versa. Up to now trans-boundary cooperation in the North Sea, for example, through OSPAR or the North Sea Conferences, has largely been limited to environmental issues and fishing. A particular driver for more cooperation is the European Union, which has strong political mandates in environmental policies (expressed in many binding regulations such as the Birds Directive or the Habitats Directive), but much less in spatial planning and land- or sea-use development.

At the European Union level, the Integrated Maritime Policy (IMP) and the Marine Strategy Framework Directive, the latter forming the environmental pillar of the IMP, are currently evolving as key policies for marine areas (Kannen et al. 2010). Integrated Coastal Zone Management and MSP are specified by the European Union as essential tools to implement these policies and to deal with the increasing diversity of marine uses. In particular, MSP is currently pushed at the European Union level as well as by several national governments. These policies and instruments include a request for stronger integration and more interactive ways of stakeholder involvement in the respective implementation processes.

MSP can be understood as a public process of analyzing and allocating the spatial and temporal distribution of human activities in marine areas to achieve ecological, economic, and social objectives that usually have been specified through a political process (see, e.g., www.unesco-ioc-marinesp.be). It is a normative approach for the development, ordering, and securing of space (Douvere and Ehler 2009). As MSP goes beyond internal waters and territorial seas, international legal frameworks such as the United Nations Convention on the Law of the Sea (UNCLOS), international conventions such as the Convention on Biological Diversity (CBD), and international policies, for example, on fisheries, also need to be considered (Maes 2008).

Rather recently, several North Sea countries, in particular, Germany, the Netherlands, and Belgium, have developed spatial plans or spatial policies covering their jurisdictional areas of the North Sea. For the United Kingdom, the Marine and Coastal Access Act (2009) defines eight areas for which marine plans should be set up. What is still missing in all of these national spatial policies and plans is the trans-boundary perspective or even more a coherent North Sea perspective. However, first experiences with transnational MSP have been gained in the European Union-funded BaltSeaPlan project for the Baltic Sea (www.baltseaplan.eu). BaltSeaPlan has developed a vision for MSP in 2030 for the Baltic Sea space (Gee et al. 2011), mentioning, in particular, three key principles:

1. Pan-Baltic thinking, which requires that MSP take an holistic approach, putting long-term objectives first, be guided by formulated objectives and targets, recognize spatial differences between different regions, fair distribution of advantages and disadvantages of human sea use, and harmonization of sea space planning and adjoining terrestrial areas
2. Spatial efficiency, which implies co-use of multiple activities within sea areas, avoidance of use of the sea as a repository for problematic land uses, and prioritization of immovable sea uses and functions
3. Connectivity thinking, which focuses on the connections that exist between areas and linear elements, for example, shipping lanes and ports and connections between habitats, breeding grounds, and feeding grounds (e.g., blue corridors and migration routes)

Although these principles aim to guide how MSP is implemented, the following key topics, which require transnational cooperation, have been identified:

- A healthy marine environment
- A coherent pan-Baltic energy policy
- Safe, clean, and efficient maritime transport
- Sustainable fisheries and aquaculture

Although a similar vision for the North Sea is still missing, the BaltSeaPlan vision can serve as a model, in particular as the key topics are virtually the same and the principles of spatial efficiency and connectivity thinking fit with the North Sea situation as much as they fit for the Baltic Sea. In this context, the role of actor networks and interaction mechanisms in integrative processes becomes obvious. This includes, in particular, bringing together actors of different sectors and countries, learning from the experiences of others, the incorporation of new information and knowledge and, ideally, the development of a common vision and development objectives. Perhaps the most developed example of such a network in the North Sea is the trilateral Wadden Sea Forum (www.waddensea-forum.org), in which local and regional representatives of authorities, associations, non-governmental organizations, and communities of the Wadden Sea regions of Germany, the Netherlands, and Denmark come together. Based on the analysis of current state and development scenarios, the forum has developed a joint cross-sectoral vision for the Wadden Sea region, action goals, and strategic action paths to development (Wadden Sea Forum, no date). However, a number of typical problems in communication and networking processes had to be overcome, in particular different sectoral interests, cultural differences, lack of or conflicting data, and mutual mistrust (De Jong and Vollmer 2005).

In summary, one challenge for the North Sea region is clearly spatial (Kannen et al. 2010). How much can be fitted into the available space? How can spatial efficiency be increased, and how can spatial conflicts be addressed? These questions are about scale: Do we want to look at it from national perspectives or a North Sea perspective? A related challenge is how to deal with the diverse environmental pressures that arise from changing patterns of use, in particular cumulative impacts (e.g., from looking at all wind farms in the North Sea versus looking at those within a particular administrative area or at a single wind farm scale as it is done in typical environmental impact assessments). Looking at the drivers behind these developments, it is clear that pressures on marine resources and space ultimately arise from particular constellations of interests and power within countries, but also transnationally. Furthermore, developments in the North Sea are not independent of global institutions and trends. As long as national policies prevail and common visions and trade-off systems are missing, solutions at the scale of a regional sea like the North Sea will be difficult to obtain and remain fragmented. There is an obvious need for transnational structures beyond OSPAR and other environmental regimes.

CONCLUSION

Neither long-term climate change impacts nor future socioeconomic and cultural developments can be precisely predicted. Based on climate change simulations, however, we can expect changes that will have an influence on the coastal zone as well as adjacent offshore areas. This might include increased risks, for example, from more intense storm surges and the need for either strengthened and/or adapted coastal defense measures. But it might as well include ecological changes, for example, the shift of commercial and noncommercial species, which could have impacts on the use of natural resources, but also on zoning for nature conservation areas such as marine protected areas. Thus, current patterns of human use and interest might change.

As a result, spatially integrated policies as well as planning and management processes gain increasing importance, but need to be informed by long-term monitoring of ecosystem processes, of changes in attitudes and problem perceptions of local communities and society in general, and of socioeconomic conditions in coastal areas. Planning under these circumstances can be characterized by the term *planning under uncertainty*. Adaptation implies then to (re)act upon changing contexts and to recognize that opportunities and risks associated with change are subject to societal interpretation and acceptance. The recommendation from the current observation of changes is that a move toward continuous, flexible, and transparent planning processes across several scales and administrative levels is necessary in order to improve adaptability. These processes should be inclusive and based on communication, social learning, and monitoring of the social as well as ecological systems in order to enhance the social capacity of coastal communities.

The development of the offshore wind industry in the North Sea Shelf areas is, on the one hand, an important contribution to the national and international mitigation policy of reducing carbon dioxide emissions. On the other hand, it requires new planning and management measures in marine areas. It is not only a key driver of the current change in sea-use patterns, but also raises several conflicts with other human demands for marine uses and resources such as shipping, fishing, and nature conservation. Furthermore, it brings new actors from large electricity companies into the current setting of stakeholders in the marine environment. A key recommendation from the dynamics of offshore wind farm development and from the current (and related) experiences with MSP throughout Europe is that spatially coherent policies, including strong trans-boundary modes of governance beyond environmental issues (where cooperation mechanisms and international conventions exist), need to be developed for the North Sea region in order to support sustainable regional development and use of marine areas. Experience from activities, such as the BaltSeaPlan Vision 2030 for the sustainable planning of Baltic Sea space, could be used to support efforts to distinguish between those issues requiring transnational cooperation and those which can be dealt with nationally or locally.

A wide range of economic actors as well as the affected local population need to be involved across management levels. Adaptation processes require awareness and acceptance for necessary changes. Experiences from existing networks such as the trilateral Wadden Sea Forum can be used to identify mechanisms for successful integration of different actors from diverse sectors and representatives of the local population. There is already an adaptation of governance structures and processes ongoing within Europe and in the North Sea. This adaptation is driven by evolving marine European Union policies and legislation. However, a major challenge remains, namely to harmonize different styles of planning and policy development without overriding nationally and regionally accepted provisions in process design and planning culture. It might take some time and need some further learning before the ongoing adaptation in marine policies becomes efficient and fully coherent. Adapting to climate change will add another layer of complexity and might stimulate additional learning leading to further modifications of policies and governance systems. However, the current experience with approaches such as MSP in the North Sea region can offer important contextual insights for a system of *climate adaptation governance* and could inform others based on this experience.

ACKNOWLEDGMENTS

This chapter is based on research in the project cluster, Zukunft Kueste—Coastal Futures, funded by the Federal Ministry for Education and Research, BMBF (FKZ 03F0404A, FKZ 03F0467A, FKZ 03F0476A) from 2004–2010 and in the KnowSeas project funded by the European Community's Seventh Framework Programme (FP7/2007–2013) under grant agreement number 226675 from 2009–2013.

REFERENCES

BMU (Bundesministerium für Umwelt, Naturschutz und Reaktorsicherheit). (2002). *Strategy of the German Government on the Use of Off-shore Wind Energy*, Berlin, Germany: BMU. http://www.bmu.de/english/renewable_energy/ (accessed December 2011).

BSH (Bundesamt für Seeschifffahrt und Hydrographie). (2013). *Spatial Planning in the German EEZ*. http://www.bsh.de/en/Marine_uses/Spatial_Planning_in_the_German_EEZ/ (accessed August 2013).

De Jong, F. and Vollmer, M. (2005). "The first steps: Stakeholder participation and ICZM in the international Wadden Sea Region," in *Bundesministerium für Verkehr, Bau- und Wohnungswesen und Bundesamt für Bauwesen und Raumordnung (Hrsg.) (2005): Nationale IKZM-Strategien—Europäische Perspektiven und Entwicklungstrends*, Konferenzbericht zur Nationalen Konferenz 28.02–01.03.2005 in Berlin, Bonn, Germany, pp. 27–32.

Douvere, F. and Ehler, C. (2009). "New perspectives on sea use management: Initial findings from European experience with marine spatial planning," *Journal of Environmental Management*, 90: 77–88.

European Commission. (2010). *Energy Infrastructure Priorities for 2020 and Beyond—A Blueprint for an Integrated European Energy Network*, [COM(2010)0677], Brussels, Belgium: European Commission.

European Wind Energy Association. (2009). *Oceans of Opportunity: Harnessing Europe's Largest Domestic Energy Resource*, Brussels, Belgium: EWEA.

European Wind Energy Association. (2010a). *The European Offshore Wind Industry—Key Trends and Statistics 2009*, Brussels, Belgium: EWEA.

European Wind Energy Association. (2010b). *The European Offshore Wind Industry—Key Trends and Statistics: 1st Half 2010*, Brussels, Belgium: EWEA.

Gee, K. (2010). "Offshore wind power development as affected by seascape values on the German North Sea coast," *Land Use Policy*, 27: 185–94.

Gee, K., Kannen, A., and Heinrichs, B. (2011). *BaltSeaPlan Vision 2030 for Baltic Sea Space*. http://www.baltseaplan.eu/index.php/BaltSeaPlan-Vision-2030;494/1 (accessed April 2012).

Gee, K., Kannen, A., Licht-Eggert, K., Glaeser, B., and Sterr, H. (2006). *The Role of Spatial Planning and ICZM in the Sustainable Development of Coasts and Seas*, Report prepared for the Research project of the Federal Ministry of Transport, Building and Urban Affairs (BMVBS) and Federal Office for Building and Spatial Planning (BBR), Integrated Coastal Zone Management (ICZM): Strategies for coastal and marine spatial planning, Berlin/Bonn, Germany: BMVBS, BBR.

Global Wind Energy Council. (2009). *Global Wind 2008 Report*. http://www.gwec.net/ (accessed July 2009).

Kannen, A. and Burkhard, B. (2009). "Integrated assessment of coastal and marine changes using the example of offshore wind farms: The coastal futures approach," *GAIA*, 18(3): 229–238.

Kannen, A., Gee, K., and Bruns, A. (2010). "Governance aspects of offshore wind energy and maritime development," in M. Lange, B. Burkhard, S. Garthe, K. Gee, H. Lenhart, A. Kannen, and W. Windhorst (eds.), *Analysing Coastal and Marine Changes—Offshore Wind Farming as a Case Study. Zukunft Kueste—Coastal Futures Synthesis Report*, Geesthacht, Germany: LOICZ Research & Studies No. 36.

Kannen, A., Gee, K., and Licht-Eggert, K. (2008). "Managing changes in sea use across scales: North Sea and North Sea coast of Schleswig-Holstein," in R.R. Krishnamurthy, B. Glavocic, A. Kannen, D.R. Green, A.L. Ramanathan, Z. Han, S. Tinti, and T.S. Agardy (eds.), *Integrated Coastal Zone Management: The Global Challenge*, Singapore; Chennai, India: Research Publishing.

Licht-Eggert, K., Gee, K., Kannen, A., Grimm, B., and Fuchs, S. (2008). "The human dimension in ICZM: Addressing people's perceptions and values in integrative assessments," in R.R. Krishnamurthy, B. Glavocic, A. Kannen, D.R. Green, A.L. Ramanathan, Z. Han, S. Tinti, and T.S. Agardy (eds.), *Integrated Coastal Zone Management: The Global Challenge*, Singapore; Chennai, India: Research Publishing.

Maes, F. (2008). "The international legal framework for marine spatial planning," *Marine Policy*, 32: 797–810.

Marine and Coastal Access Act. (2009). London: Parliament of the United Kingdom of Great Britain and Northern Ireland. http://www.legislation.gov.uk/ (accessed December 2011).

OSPAR. (2009). "Assessment of climate change mitigation and adaptation," *Monitoring and Assessment Series*, London: OSPAR Commission. http://www.ospar.org/ (accessed December 2011).

OSPAR. (2010). *Quality Status Report (QSR) 2010*, London: OSPAR Commission. http://qsr2010.ospar.org/en/index.html (accessed December 2011).

Ratter, B., Lange, M., and Sobiech, C. (2009). *Heimat, Umwelt und Risiko an der deutschen Nordseeküste: Die Küstenregion aus Sicht der Bevölkerung*, GKSS-Bericht 2009/10, Geesthacht, Germany: Institut für Küstenforschung, GKSS.

Renewable Energy Sources Act. (2009). Berlin: Federal Republic of Germany. http://www.bmu.de/english/renewable_energy/downloads/doc/42934.php (accessed December 2011).

Von Storch, H., Doerffer J., and Meinke, I. (2009). "Die deutsche Nordseeküste und der Klimawandel," in B.M.W. Ratter (ed.), *Küste und Klima* (Hamburger Symposium Geographie, Band 1), Hamburg, Germany: Institute for Geography, pp. 9–22.

Wadden Sea Forum. (no date). *Final Report: Breaking the Ice*, Wilhelmshaven, Germany: WSF Secretariat. http://www.waddensea-forum.org/ (accessed December 2011).

Weisse, R. and Pluess, A. (2006). "Storm-related sea level variations along the North Sea coast as simulated by a high-resolution model 1958–2002," *Ocean Dynamics*, 56(1): 16–25.

Weisse, R., von Storch, H., Callies, U., Chrastansky, A., Feser, F., Grabemann, I., Günther, H. et al. (2009). "Regional meteorological-marine reanalyses and climate change projections: Results for Northern Europe and potential for coastal and offshore applications," *Bulletin of the American Meteorological Society*, 90(6): 849–60.

Weisse, R., von Storch, H., and Feser, F. (2005). "Northeast Atlantic and North Sea storminess as simulated by a regional climate model 1958–2001 and comparison with observations," *Journal of Climate*, 18(3): 465–79.

Woth, K., Weisse, R., and von Storch, H. (2006). "Climate change and North Sea storm surge extremes: An ensemble study of storm surge extremes expected in a changed climate projected by four different regional climate models," *Ocean Dynamics*, 56(1): 3–15.

Chapter 20

Mainstreaming climate change adaptation with existing coastal management for the Mediterranean coastal region

Ailbhe Travers and Carmen Elrick-Barr

Abstract: This chapter highlights the imperative for regional collaboration to facilitate effective climate change adaptation in the Mediterranean coastal zone. The chapter summarizes work completed in 2010 under the auspices of the Mediterranean Action Plan (MAP), the first of the Regional Seas Programmes instigated by the United Nations Environment Programme, to inform planning efforts to support climate change adaptation within the region. These planning efforts acted as a backbone for subsequent initiatives within the region to mainstream climate change within the process of sustainable coastal development. Within the Mediterranean, the nature of the shared access to a common sea necessitates a trans-boundary approach to adaptation by the nations that share the coastal zone. This regional focus is particularly relevant in the face of increasing globalization which necessitates a *systems* approach to environmental management as a whole. The status of, and experience in, climate change adaptation within the Mediterranean was explored to identify lessons learned and formulate a set of recommendations for future initiatives. Particular attention was focused on the ongoing role of the Priority Actions Programme/Regional Activity Centre (PAP/RAC) of the United Nations Environmental Programme's Mediterranean Action Plan (UNEP/MAP). The information reported on in this chapter provides useful support to other regional organizations on approaches/barriers for climate change adaptation at this complex, trans-boundary scale.

INTRODUCTION

This chapter summarizes the work undertaken under the auspices of the Priority Actions Programme/Regional Activity Centre (PAP/RAC) of the United Nations Environmental Programme's Mediterranean Action Plan (UNEP/MAP) (Box 20.1) to support climate change adaptation planning within the Mediterranean coastal zone. The PAP/RAC undertook the work discussed here in 2010 and made aware that effective adaptation planning

BOX 20.1 THE MEDITERRANEAN ACTION PLAN

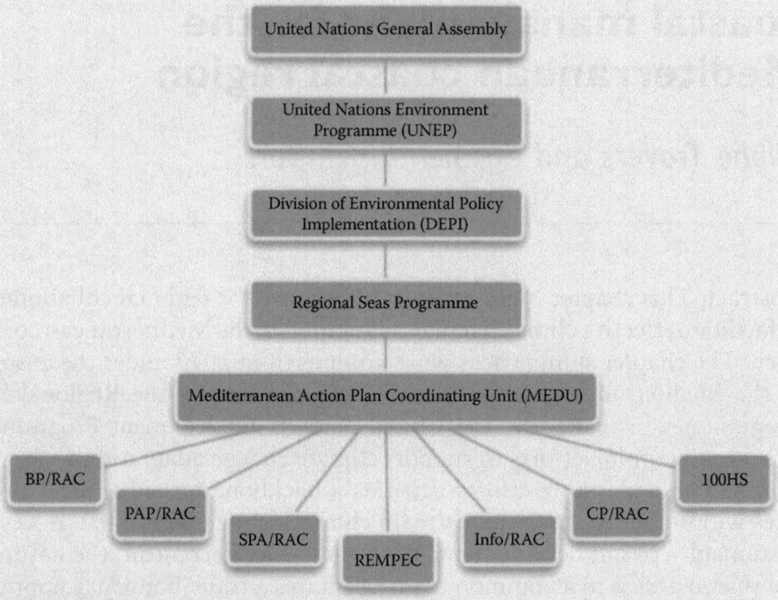

RAC—Regional Activity Centre
BP—Blue Plan
PAP—Priority Actions Programme
SPA—Specially Protected Areas
REMPEC—Regional Marine Pollution Emergency Response Centre for Mediterranean Sea
CP—Clean Production
100HS—Secretariat of the Programme for the Protection of Coastal Historic Sites

The Mediterranean Action Plan (MAP) was the first Regional Seas Programme established by UNEP in 1975. It has 22 Consulting Parties (CPs) with vested interest in the Mediterranean coastal zone and is governed by the Barcelona Convention. On its creation the main objectives of the MAP were to assist the Mediterranean countries to assess and control marine pollution, to formulate their national environment policies, to improve the ability of governments to identify better options for alternative patterns of development, and to optimize the choices for allocation of resources. Within the MAP umbrella, the specific objective of the Priority Actions Programme/Regional Activity Centre is to contribute to sustainable development of coastal zones and sustainable use of their natural resources. In this respect, PAP/RAC's mission is to carry out the tasks assigned to it in Article 32 of the 2008 ICZM Protocol.

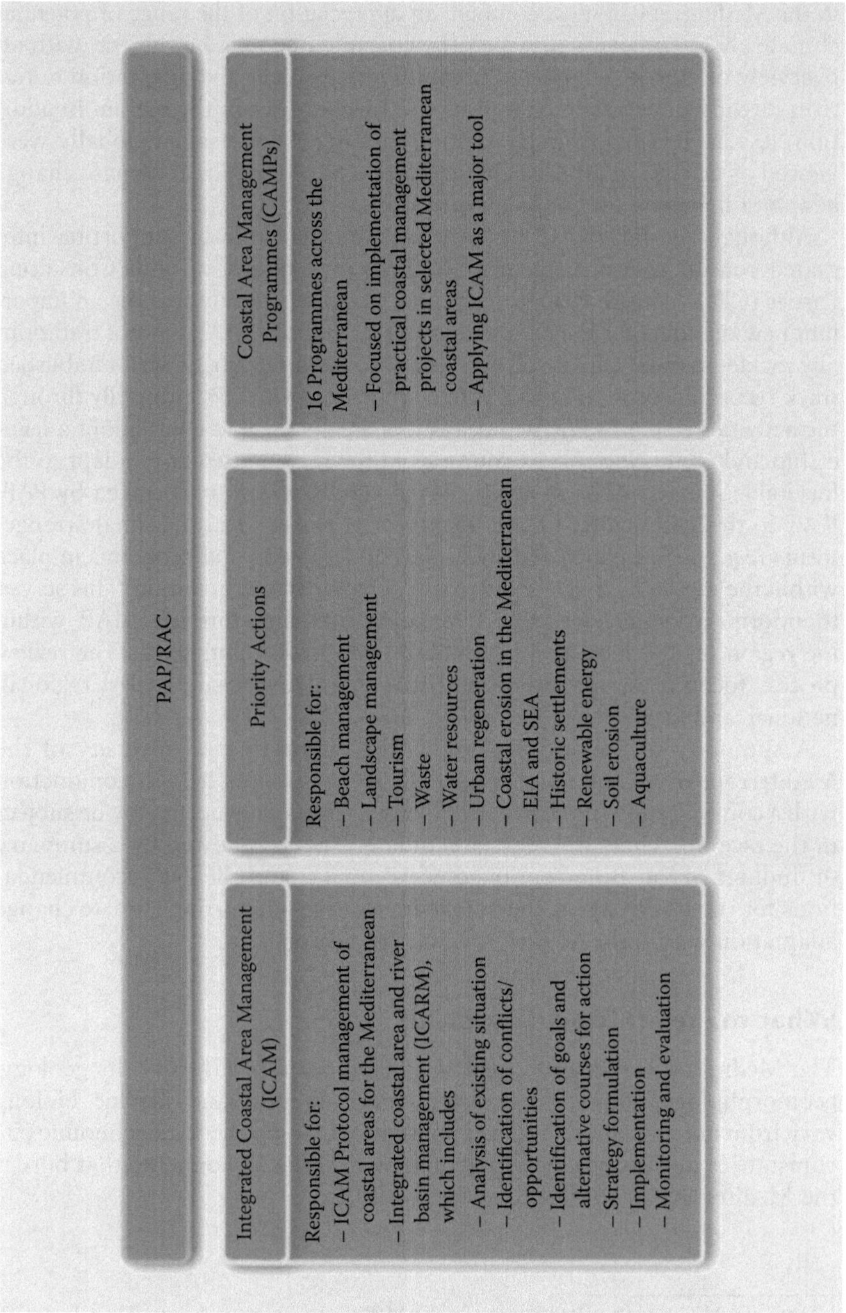
PAP/RAC
Integrated Coastal Area Management (ICAM)
Responsible for:
– ICAM Protocol management of coastal areas for the Mediterranean
– Integrated coastal area and river basin management (ICARM), which includes
– Analysis of existing situation
– Identification of conflicts/ opportunities
– Identification of goals and alternative courses for action
– Strategy formulation
– Implementation
– Monitoring and evaluation
Priority Actions
Responsible for:
– Beach management
– Landscape management
– Tourism
– Waste
– Water resources
– Urban regeneration
– Coastal erosion in the Mediterranean
– EIA and SEA
– Historic settlements
– Renewable energy
– Soil erosion
– Aquaculture
Coastal Area Management Programmes (CAMPs)
16 Programmes across the Mediterranean
– Focused on implementation of practical coastal management projects in selected Mediterranean coastal areas
– Applying ICAM as a major tool

in the Mediterranean must combine an appreciation of the range of potential climate change impacts to which the coastal zone may be subject, with an overview of climate change vulnerability assessment and adaptation initiatives already undertaken or in progress by countries in the region. In addition, lessons learned from adaptation initiatives undertaken globally were viewed as critical in guiding the development of a regional climate change adaptation agenda for the Mediterranean.

Although the PAP/RAC has a well-established role in supporting integrated coastal zone management (ICZM) in conjunction with Consulting Parties (CPs),* climate change adaptation has only recently become an important new mandate for the organization. Although the MAP has not traditionally tackled coastal climate change as a discrete program, its well-established track record of supporting ICZM in conjunction with CPs, primarily through the activities of the PAP/RAC, means it is ideally positioned to adopt a leadership and mentoring role as countries in the region attempt to adapt to the inevitable impacts of a changing coastal climate. Work undertaken by PAP/RAC to this end involved a consideration of pertinent institutional arrangements (e.g., policy, plans, and strategies) and projects and programs in place within the region to derive lessons and gaps for future attention. This served to inform recommendations for ongoing activities through MAP within the region (as the focal point for coastal adaptation initiatives). The review process focused on adaptation activities at three primary scales: regional, national, and local.

A summary of the socioeconomic and environmental diversity of the Mediterranean coastal zone (Figure 20.1) is presented here in conjunction with a consideration of the range of likely impacts to which it may be subject in the face of a changing regional climate. This is followed by a summary of findings of the aforementioned MAP review and the key recommendations for the PAP/RAC as they move forward in supporting climate change adaptation within the Mediterranean coastal zone.

What makes this place special?

The Mediterranean coastal zone is a macrocosm of diversity. Its geology, geomorphology, terrestrial ecology, oceanography, and marine biology vary from north to south and east to west. So too do the socioeconomic circumstances and governance structures within the 21 countries that border the Mediterranean coastal zone (Figure 20.1).

* There are 22 CPs with vested interest in the Mediterranean coastal zone. The CPs are, in alphabetical order, as follows: Albania, Algeria, Bosnia and Herzegovina, Croatia, Cyprus, Egypt, the European Community, France, Greece, Israel, Italy, Lebanon, Libya, Malta, Monaco, Montenegro, Morocco, Slovenia, Spain, Syria, Tunisia, and Turkey.

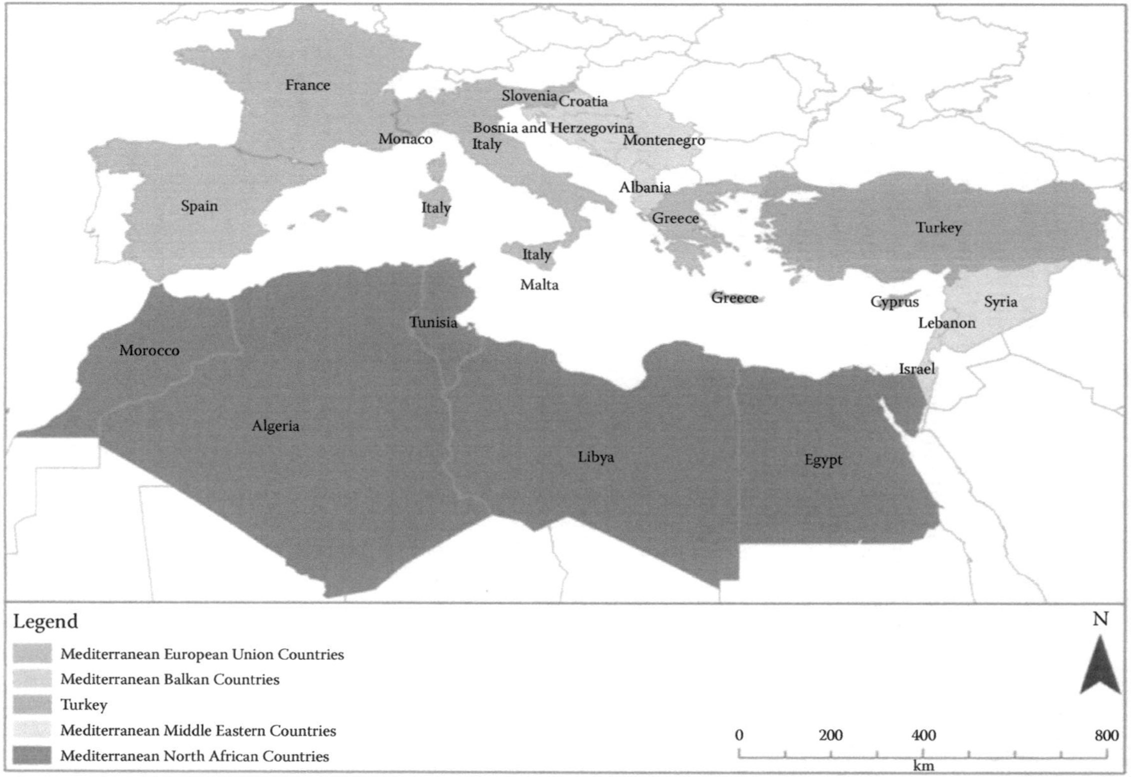

Figure 20.1 Mediterranean coastal countries. (Data from Travers, A. et al., *Position Paper: Climate Change in Coastal Zones of the Mediterranean*, Split, Priority Actions Programme, p. 73, 2010b.)

The Mediterranean Sea is one of the world's largest semi-enclosed seas covering an area of 2,542,000 km^2 with a coastline of 46,000 km (Grenon and Batiste 1989). The coastline displays a wide range of habitats including sandy beaches, sekhbas, salt marshes, and coastal plains (UNEP/MAP/PAP 2001). Rocky coastline makes up 54% while 46% is sedimentary coast, which includes important and fragile ecosystems such as reefs, lagoons, swamps, estuaries, and deltas. The physical, ecological, and socioeconomic conditions of the Mediterranean are summarized in Table 20.1.

In addition to the diverse regional variability in coastal geography, country-level coastal variability is also high. The long history of human use of the Mediterranean coastal zone has inevitably led to a wide range of pressures on the natural environment, exacerbated by the growing population and associated intense use of available resources. It is estimated that 50%–70% of the population in Mediterranean countries currently live within 60 km of the coast with this proportion continuing to rise (Caffyn et al. 2002). Activities such as fishing, industry, agriculture, and tourism have rapidly developed along the coastline with intensified urbanization leading to significant pollution threats (Di Castri et al. 1990; Grenon and Batisse 1989).

Within the Mediterranean, the very nature of the shared access to a common sea necessitates a trans-boundary approach to resource management by the nations that share the coastal zone. Coastal zone management throughout the region is currently addressed via a well-established and coordinated practice of ICZM, largely the initiative of MAP.

Table 20.1 Summary of geographic, biodiversity, and socioeconomic elements of the Mediterranean coastal zone

Attribute	*Details*
Geography	• Semi-enclosed seas covering an area of 2,542,000 km^2 • 46,000 km^2 long, with nearly 19,000 km^2 of island coastline. • Coastal geography is diverse incorporating sandy shores (46%) and rocky cliffs (54%) with vast area of low-lying coastal land.
Biodiversity	• 60% flora endemic. • 30% fauna endemic. • 7% of all the marine species known worldwide. • Nearly 19% of assessed species to date are considered threatened with extinction.
Socioeconomic	• 7% of the world's population with 460 million inhabitants. • 31% of international tourism, with 275 million visitors and every year. • 30% of international maritime freight traffic and some 20% to 25% of maritime oil transport transits the Mediterranean Sea. • 60% of the population of the world's "water-poor" countries.

Source: UNEP/MAP-Plan Bleu, *State of the Environment and Development in the Mediterranean*, UNEP/MAP, Athens, Greece, 2009.

Although it is clear that the effects of climate change (discussed in subsection "Looking forward, what makes this place at particular risk from future climate change and other potential threats?"), in keeping with existing coastal impacts, will not stop at national borders, the physical, social, and ecological complexity of the Mediterranean region makes implementation of a regional trans-boundary approach a challenging process. Countries within the region have differential priorities and it is unrealistic to assume that a *one-size-fits-all* approach to addressing ICZM, of which climate change adaptation is an integral component, will be successful. Fortunately, there is an opportunity to find regional synergies and to tap into the wealth of information that is available across the region to enhance the management of the potential impacts of climate change.

Looking forward, what makes this place at particular risk from future climate change and other potential threats?

Projected changes in temperature, precipitation, and sea level relevant to the Mediterranean are illustrated in Figures 20.2 through 20.4, respectively. The key impacts of climate change within the Mediterranean region are summarized in Table 20.2. Overall, the priority impacts in the region are freshwater shortage and sea-level rise. Sea-level rise is likely to affect parts of the coastline situated below 5 m elevation, resulting in a risk of coastal flooding. Under an IPCC high scenario (A1FI), up to an additional 1.6 million people each year in the Mediterranean, northern and western Europe, might experience coastal flooding by 2080 (Nicholls 2004). In addition, it is likely that large areas of coastal land will be affected by saltwater intrusion. This will have further implications for the availability of freshwater given that *dry* periods in the region are projected to increase in length and frequency.

General trends in impacts across the region can be inferred. For example, in areas of coastal subsidence or high tectonic activity, as in the Black Sea region, many coastal areas are increasingly being drawn within the range of influence of sea-level rise (Smith et al. 2000). Such areas are, in turn, more likely to suffer potential damage from storm surges and tsunamis (Gregory et al. 2001). Further, tectonic phenomena may work to raise certain coasts (e.g., in Algeria, Italy, Greece, and Turkey), which in turn will tend to locally mitigate the consequences of sea-level rise.

Erosion as a result of increases in mean sea level will impact soft sedimentary coasts in particular given they are more susceptible to change than *harder* rocky coastlines (Hall et al. 2007; Sánchez-Arcilla et al. 2000; Stone and Orford 2004). Deltas are particularly vulnerable to the impacts of erosion and inundation especially where dams upstream prevent the normal circulation of sediment leading to problems with delta consolidation. This is occurring in the major Mediterranean deltas such as the Nile and Rhône.

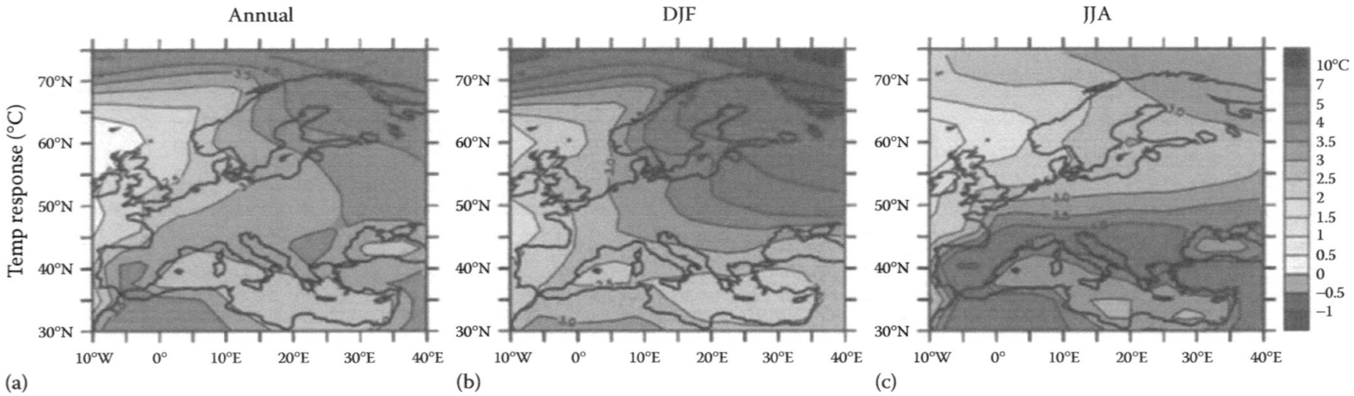

Figure 20.2 Temperature changes over Europe (A1B scenario) between 1980–1999 and 2080–2099. (Data from Christensen, J.H. et al. "Regional Climate Projections," in S. Solomon, D. Qin, M. Manning, Z. Chen, M. Marquis, K.B. Averyt, M. Tignor, and H.L. Miller, eds., *Climate Change 2007: The Physical Science Basis. Contribution of Working Group I to the Fourth Assessment Report of the Intergovernmental Panel on Climate Change*, Cambridge University Press, Cambridge; New York, 2007, p. 875. http://www.ipcc.ch/, Figure 11.5.)

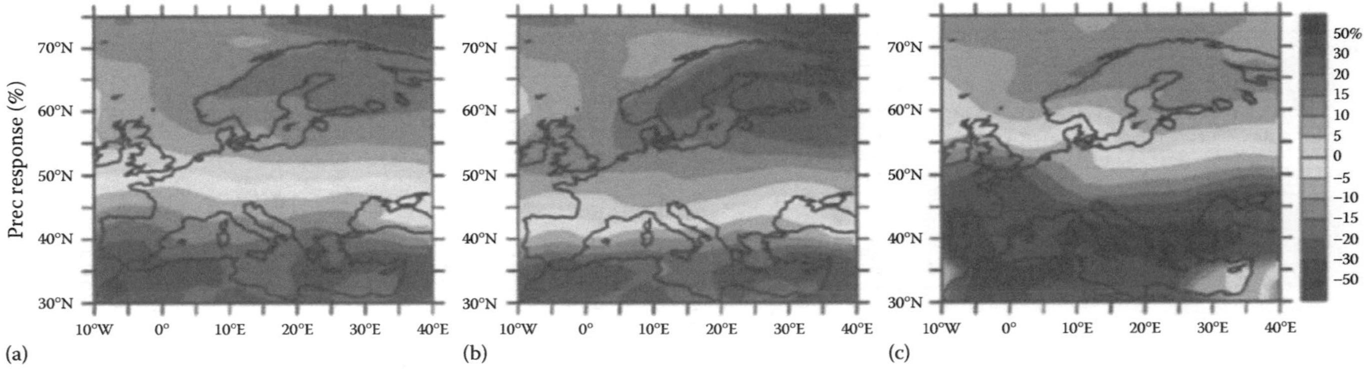

Figure 20.3 Precipitation changes over Europe (A1B scenario) between 1980–1999 and 2080–2099. Percent change in precipitation (a) annual, (b) DJF, and (c) JJA. (Data from Christensen, J.H. et al. "Regional Climate Projections," in S. Solomon, D. Qin, M. Manning, Z. Chen, M. Marquis, K.B. Averyt, M. Tignor, and H.L. Miller, eds., *Climate Change 2007: The Physical Science Basis. Contribution of Working Group I to the Fourth Assessment Report of the Intergovernmental Panel on Climate Change*, Cambridge University Press, Cambridge; New York, 2007, p. 875. http://www.ipcc.ch/, Figure 11.5.)

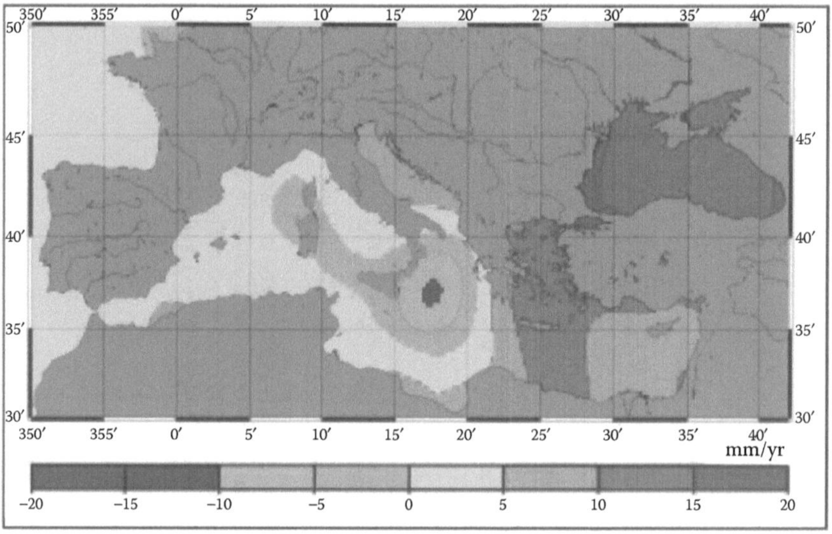

Figure 20.4 Sea level variations observed between 1992 and 1998 by the TOPEX/Poseidon programme. (Data from LEGOS-GRGS-CNES.)

Although this type of general information is a useful first step toward vulnerability assessment, details on the relative scale and severity of impacts at a country-specific scale are currently unavailable due in part to the lack of a regionally consistent approach to evaluating such impacts.

Adaptation architecture in the Mediterranean region: Existing measures to build community resilience, adaptive capacity, and sustainability

On a global stage, all CPs to MAP are signatories of the UNFCCC and have ratified the Kyoto Protocol. As such, they have indicated their commitment to combating climate change and have accepted an associated suite of obligations. Of the 22 CPs under consideration here, 13 are non-Annex I parties and 8 are Annex I parties. The European Union (EU) also has Annex I status. Five MAP CPs with non-Annex I status have yet to submit their initial national communication (Cyprus, Bosnia and Herzegovia, Montenegro, Syria, and Libya).

At a subregional level, seven Mediterranean countries belong to the EU (France, Spain, Italy, Greece, Cyprus, Malta, and Slovenia) with a further two (Turkey and Croatia) currently under consideration for admission. As members of the EU, countries are also impacted by a range of policies relating to the environment and, more recently, climate change.

Table 20.2 Summary of climate change projections for Mediterranean region

Attribute	*Details*	*Source*
Air temperature	• Average temperature increase between 2°C–6.3°C by 2100 (overall SRES scenarios) • Seasonal variations remain significant • Intraregional variations also important: W basin likely to have greater rise in autumnal temperature than elsewhere, while rise in temperature will be greatest in S and E of Basin in summer	Hallegatte et al. 2007; Van Grunderbeeck and Tourre 2008; Christensen et al. 2007
Precipitation	• 4% reduction in rain on northern coasts by 2030 (A1B Scenario) • 27% reduction in rain on southern coasts by 2030 (A1B Scenario) • Summer: –5% in SE and –30% W and NW • Winter: –20% in SW and slight increase in NW • Potential for one year out of two to be *dry* by 2100 • Fewer rainy days • Longest periods without rain extending • More water in winter • Less in spring and summer • Increased flooding associated with storm events	Christensen et al. 2007
Sea-level rise	• IPCC figures of 0.18–0.59 m by 2100 excluding small ice sheet melt (+0.2 m) • Mediterranean-specific projections are complicated by local subsidence effects, especially in deltas and coastal cities • Infraregional differences are apparent—more significant SLR in the East than to the West	Hansen 2007; Rahmstorf et al. 2007; Hallegatte et al. 2007

Regional/subregional projects and programs

A range of projects and programs relating to coastal climate change in the Mediterranean have been undertaken, each varying in scale and focus (see Travers et al. 2010b for summary of past initiatives). That is, initiatives undertaken may be regional, versus local and sector specific versus theme specific. Consideration of the range of projects in the region provides not only an overview of the diversity of issues that require attention, including future climate change, current climate variability, but also other environmental stresses (deterioration of biodiversity) and human stresses (unsustainable paths of development) (Hallegatte et al. 2009). Although a wide range of projects and programs have been launched in the region, interestingly, these initiatives are not all conducted from a common basis with divergent sectoral, intersectoral, and territorial actions carried out.

The review undertaken by MAP (Travers et al. 2010b) provides an overview of the status of vulnerability and adaptation assessment in the region:

- The range of projects currently underway demonstrates the lack of a coordinated approach to climate change action in the region.
- Few regional projects or programs are coastal or Mediterranean-specific.
- Implementation of adaptation action appears to have occurred based on regional directives (e.g., through the EU) or discrete countries proactively sourcing funding.
- The actions are not necessarily targeted based on priority needs of the region.
- There are a number of useful existing research activities that PAP/RAC can build upon and potentially leverage as they move toward coordinated adaptation planning on a region-wide basis.
- Key gaps include a regional program of works that can pull together the disparate activities and highlight key areas of focus for the PAP/RAC programme.

The information gained in the review of adaptation activity in the region has been used to infer a level of adaptation maturity associated with each of the MAP CPs (Travers et al. 2010a). That is, the level of adaptation action currently underway for each CP in terms of plans, strategies, projects, and programs was considered as a surrogate for *current capacity*. A further level of information used to inform this exercise was an appraisal of the coastal-specific legislative maturity within each CP. This was carried out because, in the absence of discrete climate change adaptation strategies and protocols, the key instruments to enable adaptation will be these legal frameworks.

It is clear that the Annex I EU countries have the most established adaptation *architecture* of all CPs under consideration. Conversely, the non-Annex I Mediterranean Balkan countries and Middle Eastern Mediterranean countries have a much lower number of projects, programs, policies, and strategies pertaining to climate change adaptation in the coastal zone. It is important to remember that the presence/absence of a project in a region cannot be directly translated into a measure of capacity. For example, Albania may have only one active project, but this project is extensive and is serving to build capacity both at regional and national levels. Rather, an appreciation of the existence of specific policy and program initiatives in conjunction with their overriding aims and objectives has been used here to create the following groupings of countries in terms of their capacities to adapt to coastal climate change (Table 20.3):

1. EU Annex I countries (France, Italy, Spain, Greece, Slovenia, and Monaco*)
2. Annex I non-EU countries (Croatia and Turkey)

* Although Monaco is not currently a member of the EU, its status in terms of vulnerability and capacity necessitate its inclusion in this group.

3. Non-Annex I EU countries (Malta and Cyprus)
4. Non-EU Balkan countries (Bosnia and Herzegovina, Montenegro, and Albania)
5. Middle East countries (Syria, Lebanon, and Israel)
6. North African countries (Egypt, Libya, Tunisia, Algeria, and Morocco)

Table 20.3 Categorization of countries by level of vulnerability (incorporates a consideration of capacity to adapt and expose to projected climate changes)

Prioritization ID[a]	*Grouping*	*Description*
1	EU Annex I countries	These countries have developed national adaptation strategies and/or completed a number of National Communications. They possess a heightened understanding of their vulnerability to climate change.
2	Annex I non-EU countries	Croatia and Turkey are Annex I countries, but are assigned special considerations. They are also candidate countries of the EU, and are therefore aligning themselves to EU policies and strategies. Consequently, EU CCA developments, such as the EU policy of CCA, have relevance to these countries, in more of a way than other non-EU countries.
3	Non-Annex I EU countries	These countries are highly sensitive to the projected impacts of climate change. While they are EU countries, their geographical nature disposes them to high impacts of climate change (they are small island nations). In addition, the countries have existing coastal management issues (such as erosion and water shortages) that are likely to be exacerbated by climate change. Finally, they are also very dependent upon tourism, which is likely to be highly affected by climate change.
4	Non-EU Balkan countries	These countries have not completed an initial national communication. Therefore, the understanding of their vulnerability to projected impacts of climate change is not as well understood as other Mediterranean countries.
5	Middle East countries	Countries are grouped, but on the recognition that Syria has not completed a national communication.
6	North African countries	While these countries are grouped together, Morocco is slightly less vulnerable to climate change because it has an active funded coastal adaptation project in the coastal zone and is also less physically susceptible to climate change due to geographic nature of the coast. Tunisia is very sensitive. All countries have issues of drought and desertification on top of all the other issues they have to deal with. The countries differ from European countries in that adaptation efforts are more development focused.

[a] Prioritization is presented from highest adaptive capacity (1) to lowest adaptive capacity (6).

It is important to note that within these groupings, there are subtleties in the categorization due to (a) the countries UN status and reporting position and (b) current projects underway in a country. For example, Albania and Egypt have large-scale GEF projects underway, which enhance their adaptive capacity, despite their grouping with other countries that do not have such projects. Further, Syria, for example, is more vulnerable than Lebanon and Israel due to its narrow coastal zone backed by mountain ranges, which inhibit landward ecosystem migration. Despite these subtleties, a rudimentary understanding of the range of adaptive capacities in the region will help support the development of recommendations for action. Following on from these observations, largely related to *gaps* in current knowledge, a series of specific recommendations for future action may be made:

- The five non-Annex I countries yet to complete their Initial National Communications should do so as a priority.
- The seven non-Annex I countries yet to submit their Second National Communications should focus attention to this task and provide an indicative timeframe for completion.
- The two Annex I countries yet to submit their Fifth National Communication should provide a status update.
- Turkey should be supported in completion of its Second National Communication (as the only Annex I country yet to do so).

All countries yet to produce specific climate change adaptation plans or strategies should do so as a priority or, at a minimum, timetable these activities into their programs of work.

WHAT WILL HELP AND WHAT WILL HINDER?

The preceding sections have highlighted a number of key issues and information gaps that inhibit a proactive and coordinated approach to climate change adaptation in the Mediterranean coastal zone. Recommendations for PAP/RAC to work to reduce the identified gaps and alleviate barriers to adaptation are summarized in this section after Travers et al. (2010a) under the headings of (1) mainstreaming climate change; (2) partnerships and cooperation; and (3) financing.

The recommendations align to the conceptual framework illustrated in Figure 20.5. Central to recommendations for effective climate change adaptation within the region is the establishment of a Regional Coastal Adaptation Framework (ReCAF) (Figure 20.5 and discussed further below). Key tools that will guide the development of this regional framework will be the ICZM Protocol and the Mediterranean Strategy for Sustainable Development (MSSD). In addition, a regional impact assessment and

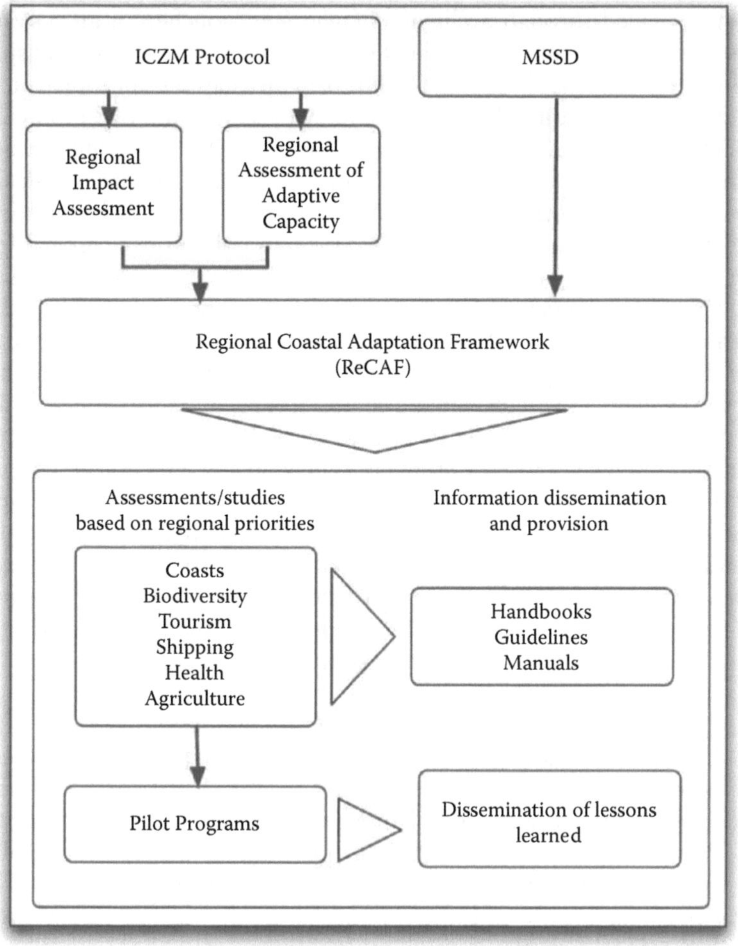

Figure 20.5 Conceptual framework guiding the development of PAP/RAC recommendations. (Data from Travers, A. et al., *Position Paper: Climate Change in Coastal Zones of the Mediterranean*, Split, Priority Actions Programme, p. 73, 2010b.)

assessment of adaptive capacity will guide the implementation of sector- or issue-specific programs in key vulnerable areas. Therefore, the completion of such assessments will be critical in the development of a ReCAF.

Mainstreaming climate change

ICZM should seek to integrate all key issues of critical importance to the management of coastal resources and resource use—including climate change adaptation. It is important that climate change is not seen as an issue outside of an ICZM framework. Rather effective ICZM can only be achieved by ensuring that it is viewed through a climate change *lens*, couched in an operational framework to facilitate implementation and ensure a coherent, transparent, and long-term approach. The ICZM Protocol provides the key tool to facilitate coastal climate change adaptation in the Mediterranean. The Protocol provides the legal framework for coastal management in the Mediterranean and was recognized by PAP/RAC as the key tool to enabling countries to deal with emerging coastal issues, including climate change. Further, to be effective, climate change adaptation must be embedded into policies at a regional level in order to reduce the long-term vulnerability of sensitive sectors such as agriculture, forests, biodiversity, energy, transport, water, and health. This is already occurring within the EU where, for example, climate change will be one of the main drivers that shape European agriculture and the Common Agricultural Policy. The degree to which climate change has been mainstreamed into policies, plans, and strategies at national and local scales has not been examined in detail. Although PAP/RAC can provide leadership and direction at the regional level through mainstreaming climate change into the ICZM Protocol and the MSSD, it will also be important for PAP/RAC to increase awareness of opportunities for mainstreaming at national and local scales through training and capacity building activities directed at CPs. Overall, PAP/RACs role in supporting the integration of climate change into coastal management policies and practices in the Mediterranean can be summarized as follows:

- Promote ratification of the ICZM protocol.
- Increase the awareness of CPs and other key stakeholders about the links between ICZM and climate change, as outlined in the ICZM protocol.
- Support mainstreaming of climate change adaptation into the MSSD.
- Provide support to CPs on how mainstreaming adaptation is applicable in the coastal context. It will be important to demonstrate climate change mainstreaming across a variety of tiers, from the regional (ICZM protocol and MSSD) to national and local (guidelines, etc.)
- Support capacity building in key areas, that is, climate change mainstreaming.
- When reviewing the ICZM protocol, recognize the important and intertwined nature of climate change in ensuring an integrated approach to coastal zone management.

Partnerships and cooperation: PAP/RACs role

Continued cooperation between the MAP, EU, national, subnational, and municipal authorities will be a prerequisite for successful coastal adaptation. The number and diversity of regional stakeholders necessarily results in a range of ICZM policy and practice. Consequently, it is important to ensure active and ongoing dialogue to develop a shared adaptation agenda.

The objective is not to decrease diversity in practice, but rather to ensure coordination of efforts, to reduce duplication, and to ensure efficient use of limited human and financial resources.

To promote cooperation on adaptation, the European Commission has proposed setting up a consultative forum known as the Impact and Adaptation Steering Group where EU Member States will play a role in developing the four key areas of the proposed EC adaptation framework. It is suggested that this type of forum be adopted for MAP countries so as to align with the overriding objectives of the ECs framework and expand their application to MAP CPs as a whole. Developing a ReCAF will be an important step in ensuring a coordinated approach to adaptation in the coastal zone.

The role of *PAP/RACs* is given as follows:

- Demonstrate leadership in climate change adaptation by promoting a coordinated approach to climate change adaptation in the Mediterranean through the development of a ReCAF.
- Strengthen existing regional ICZM cooperation mechanisms to enhance the flow/support of information through focal points to national and local decision-makers.

Financing

The financial costs of adaptation, and thus financial needs, are high. It is unlikely that many CPs will be able to meet these additional costs without substantial external support. UNFCCC Annex I countries have pledged to provide financial support for adaptation (Article 4 of the Convention, Article 11 of the Kyoto Protocol, Marrakesh Agreements) and should be encouraged to do so through well-coordinated, cost-effective regional initiatives. The establishment of such financial mechanisms requires in the Mediterranean region a continuous and informed dialogue, in line with regional and local specificities and related needs. Bilateral and multilateral development agencies will be critical in this regard. Their role is all the more important as fundamental synergies exist between development (through the Millennium Development Goals) and effective adaptation. These synergies are such that it is often difficult, and with limited practical meaning, to concretely distinguish between development-led and adaptation-led initiatives. The key issue is that a focus remains on coastal climate change adaptation initiatives with clear and concrete development co-benefits (and vice versa) leading to *win–win* outcomes. Consequently, it will be critical to align on-ground projects in targeted locations to the regional priorities of the Mediterranean ensuring that local and individual projects are contributing to the key objectives of a regional strategy. Development of a ReCAF would provide leadership and direction to national and local governments on the key issues that not only address development objectives at the local scale but also contribute to regional objectives. It will be important to engage national and local stakeholders in the development of a ReCAF to promote ownership and uptake at the local scale.

Finally, it is widely recognized that global financing needs will outstrip the available funds provided through such mechanisms as the UNFCCC. Consequently, there is a need to further consider a combination of policy measures and examine the potential use of funding measures, market-based instruments, guidelines, and public–private partnerships to ensure effective delivery of adaptation. There is scope for improving the uptake of adaptation action by CPs and for targeting better the use of available financial resources and instruments to encourage this. PAP/RACs role in supporting financing may include the following:

- Ensure up-to-date knowledge of the funding landscape for climate change adaptation and coastal management. Identify potential funding opportunities and disseminate this information to CPs as appropriate.
- Develop and/or support the development of targeted funding requests. Funding requests should be based on established understanding about regional priorities and local development objectives.
- Ensure collaboration in the development of any ReCAF. Although this is not a direct financial action, it provides the operational framework for financial activities, as finances should be targeted based on an understanding of regional priorities.
- Investigate opportunities to improve the uptake of adaptation action by CPs through better use of available financial resources and instruments and/or a combination of policy measures (i.e., funding measures, market-based instruments, guidelines, and public–private partnerships.
- Encourage public–private partnerships with a view to sharing investment, risk, reward, and responsibilities between the public and private sector in the delivery of adaptation actions.
- Countries in the Mediterranean are at different stages in understanding and taking action on the potential impacts of climate change. Although attempts to infer capacity levels have been made by Travers et al. (2010b), this differential capacity to adapt is not well understood. PAP/RAC could fill this gap through the following:
    - Commission an assessment of the relative impacts of climate change, and associated vulnerability, in the Mediterranean
    - Commission an assessment of the adaptive capacity of Mediterranean countries, to highlight focus areas for capacity building programs
    - Apply the outputs from the proposed vulnerability and adaptive capacity assessments into the Regional Coastal Adaptation Plan

Access to information differs considerably across regions and insight into the costs and benefits of different adaptation options and information on good coastal adaptation practice is limited. This results in disparity between the understanding of the potential impacts of climate change and understanding of the options available to treat any identified risks. The role of PAP/RAC in alleviating such gaps could include establishing information sharing mechanisms to, *inter alia*, promote collection, storage, and dissemination of climate information and to share lessons learned from climate change adaptation across the region.

Effective adaptation decision-making for coastal zones in the Mediterranean will necessarily involve a range of spatial scales from Mediterranean-wide to local, site-specific decisions. Advocating a blended approach for bottom-up (local level) and top-down (Mediterranean-wide) adaptation (with spatial scales in between) will be the key to successful adaptation. It will be crucial for MAP to build on existing partnerships, frameworks, and institutional arrangements within the region to facilitate this process. PAP/RAC should coordinate their efforts with such existing or proposed initiatives.

Information provision

PAP/RAC's recommended role in facilitating information provision across the region may include the following:

- Establish a well-coordinated coastal monitoring program to ensure that the signal of coastal change driven by climate change can be detected and evaluated with respect to broader anthropogenic changes to the coast. This requires leadership from PAP/RAC to establish a Mediterranean-wide agreement on the definition of key climate change indicators and the promotion of improved monitoring and reporting mechanisms.
- Support the development of climate information—downscaling climate projections and developing standardized approaches to allow subregional comparison of impacts. Such information is available for the EU but is not available for the entire Mediterranean region.

Information sharing

PAP/RAC's recommended role in facilitating information sharing across the region may include the following:

- Lead innovation and development of good practices through implementation of subnational adaptation projects. It would be important to ensure that the local adaptation projects are selected based on regional prioritization.
- Provide a facility for enhanced cooperation for climate change adaptation across the Mediterranean (north–south or from Annex I countries to others). As discussed earlier, some countries are more advanced in addressing the climate change issue than others. This differential

capacity not only leads to different climate change adaptation priorities but can also be utilized to share lessons learned and practical experience between countries. PAP/RAC could provide a vital information network through the already established CP Focal Points. In addition, a clearinghouse mechanism to share climate-relevant information could be developed. This may be an extension of the MAP online resources to include climate change or a new information sharing platform.
- Apply or develop new and innovative adaptation methodologies appropriate to the specificities of the Mediterranean coast including the adaptation approach, measures, options, and actions.

CONCLUSION

The foregoing recommendations focus on the information needs to support the delivery of technical assistance to countries as well as actions that can be taken to build capacity of relevant stakeholders at regional, national, and local levels. The activities of PAP/RAC will, by necessity, be proactive, and the overriding vehicle for MAP and PAP/RAC to achieve effective climate change adaptation mainstreaming is their ICZM Protocol. It will be important to recognize the very unique and context-specific challenges of climate change adaptation and make these a component part of this protocol in their own right. That said, adaptation in the Mediterranean needs to become integral to the existing, coastal management landscape.

Developing a Mediterranean Coastal Adaptation Framework is seen as a key priority. This will encompass the need for sharing experiences and tools for adaptation, which is increasingly important at the regional level as work develops in the field. Although vulnerability and adaptive capacity to climate change varies widely depending on the context, as do the initiatives to be undertaken, the need to share experiences and build capacity encourages the issue to be put on the regional agenda.

The approach to development of recommendations and the recommendations themselves presented in this chapter provide a sound basis for the PAP/RAC to provide this leadership role. The regional climate change adaptation journey for PAP/RAC is commencing and the effectiveness of the interventions outlined herein is yet to be tested. MAP and specifically PAP/RAC are ideally placed to play an active mentoring and leadership role as the countries of the Mediterranean move toward adapting to the distinctive climate change issues that manifest themselves at different scales, from the regional to local level. Ultimately, the effectiveness of this role will be shaped by political will within the region that should be cognizant of the need for foresight. This regional leadership will be essential to mitigate the potential for even greater expenditure of resources in the future should effective and *adaptive* adaptation not be instigated as a priority.

POSTSCRIPT

Since this chapter was written in 2010 the ICZM Protocol has officially entered into force (on March 24th 2011). During the CoP 17, held in Paris from February 8–10, 2012, the Action Plan for the implementation of the Protocol 2012–2019 was adopted by the Contracting Parties to the Barcelona Convention. The Protocol represents a crucial step in the history of MAP allowing the countries to better manage their coastal zones, as well as to deal with the emerging coastal environmental challenges, including in particular climate change.

ACKNOWLEDGMENTS

The information in this chapter draws from a review exercise commissioned by the PAP/RAC in 2009 to provide background information on the impacts of climate change on the Mediterranean coastal zone and an overview of the range of ongoing adaptation policies, projects, and programs within the region. The work was reported on in two papers published by the PAP/RAC (Travers et al. 2010a, 2010b). Key contributors to this review project were Dr. Robert Kay and Mr. Marko Prem from the PAP/RAC. The ideas, input, and feedback from these contributors and other supporting staff within CZM (Luke Dalton and Ania Neidzjerzak) are gratefully acknowledged.

REFERENCES

Caffyn, A., Prosser, B., and Jobbins, G. (2002). "Socio-economic framework: A framework for the analysis of socio-economic impacts on beach environments," in F. Scapini (ed.), *Baseline Research for the Integrated Sustainable Management of Mediterranean Sensitive Coastal Ecosystems*, Florence, Italy: IAO.

Christensen, J.H., Hewitson, B., Busuioc, A., Chen, A., Gao, X., Held, I., Jones, R. et al. (2007). "Regional Climate Projections," in S. Solomon, D. Qin, M. Manning, Z. Chen, M. Marquis, K.B. Averyt, M. Tignor, and H.L. Miller (eds.), *Climate Change 2007: The Physical Science Basis. Contribution of Working Group I to the Fourth Assessment Report of the Intergovernmental Panel on Climate Change*, Cambridge; New York: Cambridge University Press. http://www.ipcc.ch/ (accessed August 2012).

Di Castri, F., Hansen, A.J., and Debussche, M. (1990). *Biological Invasions in Europe and the Mediterranean Basin*, Monographiae Biologicae, vol. 65, Dordrecht, the Netherlands: Kluwer Academic Publishers.

Gregory, J.M., Church, J.E., Boer, G.J., Dixon, K.W., Flato, G.M., Jackett, D.R., Lowe, J.A. et al. (2001). "Comparison of results from several AOGCMs for global and regional sea-level change 1900–2100," *Climate Dynamics*, 18: 225–40.

Grenon, M. and Batisse, M. (1989). *Futures for the Mediterranean: The Blue Plan*, Oxford: Oxford University Press.

Hall, J., Dawson, R.J., Walkden, M.J.A., Nicholls, R.J., Brown, I., and Watkinson, A. (2005). A broad-scale analysis of morphological and climate impacts on coastal flood risk. *Proceedings of the International Conference on Coastal Dynamics*. Barcelona, Spain: American Society of Civil Engineers, April 4–8.

Hallegatte, S., Bille, R., Magnan, A., Garnaud, B., and Gemenne, F. (2009). *The Future of the Mediterranean: From Impacts of Climate Change to Adaptation Issues*, Paris, France: CIRED—IDDRI.

Hansen, J. (2007). "Scientific reticence and sea level rise," *Environmental Research Letter*, 2. doi:10.1088/1748-9326/2/2/024002.

Nicholls, R.J. (2004). "Coastal flooding and wetland loss in the 21st century: changes under the SRES climate and socio-economic scenarios," *Global Environmental Change*, 14: 69–86.

Nicholls, R.J. and de la Vega-Leinert, A.C. (2008). "Implications of sea-level rise for Europe's coasts: An introduction," *Journal of Coastal Research*, 24(2): 285–287.

Rahmstorf, S., Cazenave, A., Church, J.A., Hansen, J.E., Keeling, R.F., Parker, D.E., and Somerville, R.C.J. (2007). "Recent climate observations compared to projections," *Science*, 316(5825): 709.

Sánchez-Arcilla, A., Hoekstra, P., Jiménez, J.E., Kaas, E., and Maldonado, A. (2000). "Climate implications for coastal processes," in D.E. Smith, S.B. Raper, S. Zerbini, and A. Sánchez-Arcilla (eds.), *Sea Level Change and Coastal Processes: Implications for Europe*, Luxembourg: Office for Official Publications of the European Communities.

Smith, D.E., Raper, S.B., Zerbini, S., and Sánchez-Arcilla, A. (eds.) (2000). *Sea Level Change and Coastal Processes: Implications for Europe*, Luxembourg: Office for Official Publications of the European Communities.

Stone, G.W. and Orford, J.D. (eds) (2004). "Storms and their significance in coastal morpho-sedimentary dynamics," *Marine Geology*, 210: 1–5.

Travers, A., Elrick, C., and Kay, R. (2010a). *Background Paper: Climate Change in Coastal Zones of the Mediterranean*. Split, Priority Actions Programme. p. 121. http://www.pap-thecoastcentre.org/pdfs/CC%20BACKGROUND.pdf; http://www.pap-thecoastcentre.org/pdfs/CC%20POSITION.pdf.

Travers, A., Elrick, C., and Kay, R. (2010b). *Position Paper: Climate Change in Coastal Zones of the Mediterranean*. Split, Priority Actions Programme. p. 73.

UNEP/MAP/PAP. (2001). *White Paper: Coastal Zone Management in the Mediterranean*, Split, Priority Actions Programme.

UNEP/MAP-Plan Bleu. (2009). *State of the Environment and Development in the Mediterranean*, Athens, Greece: UNEP/MAP.

Van Grunderbeeck, P. and Tourre, Y.M. (2008). "Bassin Méditerranéen: Changement climatique et impacts au cours du XXIème siècle," in H.L. Thibault and S. Quéfélec (eds.), *Changement climatique et énergie en Méditerranée*, 1:1, 1.3–1.69.

Part VIII

Climate change and the coastal zone

Africa

Chapter 21

Climate change and the coastal zone of Mozambique

Supporting decision-making for community-based adaptation

Ailbhe Travers, Timoteo C. Ferreira, Jessica Troni, and Arame Tall

Abstract: Mozambique has one of the most exposed coastlines in the world to the potential impacts of climate change, including sea-level rise. The coast's biophysical vulnerability, especially the low-lying deltaic areas and fragile soft coasts, exposes the coastal population, among the world's least developed, to a long-term challenge in addition to their daily struggle to improve their quality of life. Mozambique, as a least developed country, with a recent history of conflict, has present-day chronic capacity constraints to manage its coastal resources especially at a local level. Against this backdrop, the Government of Mozambique is working with the United Nations Development Programme to seek the assistance of the Least Developed Country Fund for funds to pilot community-scale, ecosystem-based adaptation in parallel with national institutional capacity-building activities. This chapter outlines the process undertaken during 2011 to develop the project submission to the Least Developed Country Fund that uses an interlinked process of rapid biophysically based adaptations options analysis with stakeholder-driven consultations. The project proposal was endorsed in late 2011 and as of late 2013 is in the process of beginning initiation.

INTRODUCTION

This chapter provides an overview of the climate-related issues facing the coastal zone of Mozambique and provides an analysis of the barriers and opportunities for effective climate change adaptation at a community level. The key focus of the chapter is a discussion of the decision-making process adopted to inform robust and defensible adaptation planning at a coastal community level in one of the poorest countries in the world, with a human development index of 0.327 and ranked 185 out of 187 countries with comparable data (UNDP 2012). The basis for discussion stems from a project preparation process for the development of the first project to implement the Mozambique National Adaptation Programme of Action (NAPA) for submission to the Least Developed Countries Fund (LDCF), a United Nations

Framework Convention on Climate Change (UNFCCC) fund* administered by the Global Environment Facility. The United Nations Development Programme (UNDP) undertook the project preparation process during the early part of 2011 with funding from the LDCF that subsequently resulted in the successful application to the LDCF of the project proposal, Adaptation in the Coastal Zones of Mozambique. The project is currently (as of late 2014) in the early stages of implementation.†

The project addresses primarily NAPA Priority Three, to develop strategies to arrest coastal erosion and its impacts on livelihoods and the economy, expected to worsen under climate change. The LDCF project also addresses some of the priorities contained in NAPA Priority Two, to strengthen capacities of agricultural producers to cope with climate change, specifically in relation to activities around rainwater harvesting and irrigation, use of drought-tolerant crops, community management of forests, erosion management, conservation agriculture, and building sustainable livelihoods. Consequently, the project recognizes the complexity of poverty–environment relationships and seeks to build approaches to mitigate the potential for climate impacts to create destructive positive feedback loops that lead to a vicious cycle of impoverishment and environmental degradation.‡ Further information on the NAPA process is provided in Box 21.1.

The ultimate goal of the overall project will be to contribute to Mozambique's climate-resilience by integrating coastal adaptation in development policies, plans, projects, and actions. The project objective is to develop the capacity of communities living in the coastal zone to manage climate change risks. It is anticipated that the approach adopted in Mozambique could provide transferable principles and approaches for good practice for other least developed countries (LDCs), in particular, those within the African continent.

WHAT MAKES THIS PLACE SPECIAL?

Mozambique borders the Republic of Tanzania (N), Malawi, Zambia, Zimbabwe, South Africa and Swaziland (W), and Republic of South Africa (S) (Figure 21.1). Mozambique has the third longest coastline on the African continent extending about 2,700 km along the Indian Ocean. The coast is generally considered as comprising three discrete sections: northern, central, and southern coasts (Table 21.1).

* See the UNFCCC Website for further information on the LDCF: http://unfccc.int/cooperation_support/least_developed_countries_portal/ldc_fund/items/4723.php.

† The completed project proposal can be accessed from http://www.mz.undp.org/content/mozambique/en/home/operations/projects/environment_and_energy/adaptation-in-the-coastal-zones-of-mozambique/.

‡ For example, see the Poverty–Environment Initiative website for Mozambique (http://www.unpei.org/what-we-do/pei-countries/mozambique).

BOX 21.1 NATIONAL ADAPTATION PROGRAMMES OF ACTION

The NAPA process was mandated by the UNFCCC in the Marrakech Accords of 2001, the aim being to mainstream climate adaptation planning within national development planning in LDCs. All LDCs would prepare a NAPA, which would identify immediate and urgent adaptation priorities for funding, which, if not met without delay, would increase vulnerability or increase adaptation costs at a later stage. The need for action-oriented activities, which should be country-driven, flexible, and based on national circumstances, was recognized. By December 2011, 46 countries have officially submitted their NAPAs to the UNFCCC, using a range of stakeholder engagement processes during their preparation.

There is a range of NAPA guidelines and support materials, prepared by the LDCs Expert Group, that discuss the objectives and structure of NAPAs and provide a framework for their preparation. The flowchart below presents the set of steps undertaken through the NAPA process. Development of NAPAs is designed to be a time-effective process driven by expert elicitation and stakeholder assessments, in contrast to undertaking extensive new vulnerability assessments. The NAPA process strongly encourages analysis of existing vulnerability and adaptation (V&A) studies. Importantly, the open, consultative style of the NAPA framework has provided valuable lessons within LDCs for the development of the V&A components of their National Communications to the UNFCCC and these lessons have also been shared with several other developing countries for application within their National Communications. As an outcome of the 2011 UNFCCC Conference (COP17), held in Durban, Republic of South Africa, National Adaptation Plans (NAPs) will replace the NAPA process. The COP17 decision includes a "guideline" for the contents of NAPs. Though the NAP process relates to LDCs, other developing countries may also use the NAP guidelines, as appropriate, to guide their adaptation planning activities.

Assessing a selection of NAPAs, Ayers (2008) concludes that it is vital that NAPA projects receive the financial and institutional support they require from donors, governments, and climate change institutions. Not only do the projects identified respond to urgent needs but it is important for the LDCs that the long and much fought-for process of NAPA development is not wasted. Projects are country-owned and already have national support, having undergone participatory identification and design processes involving

(*Continued*)

many stakeholders. The projects are consistent with national development plans, which encourages mainstreaming and the scaling up of activities and project outcomes can feed into much needed evidence-based learning on how to actually do adaptation. Finally, although more strategic and programmatic NAPs are needed in developing countries, NAPAs can provide a basis on which to develop these.

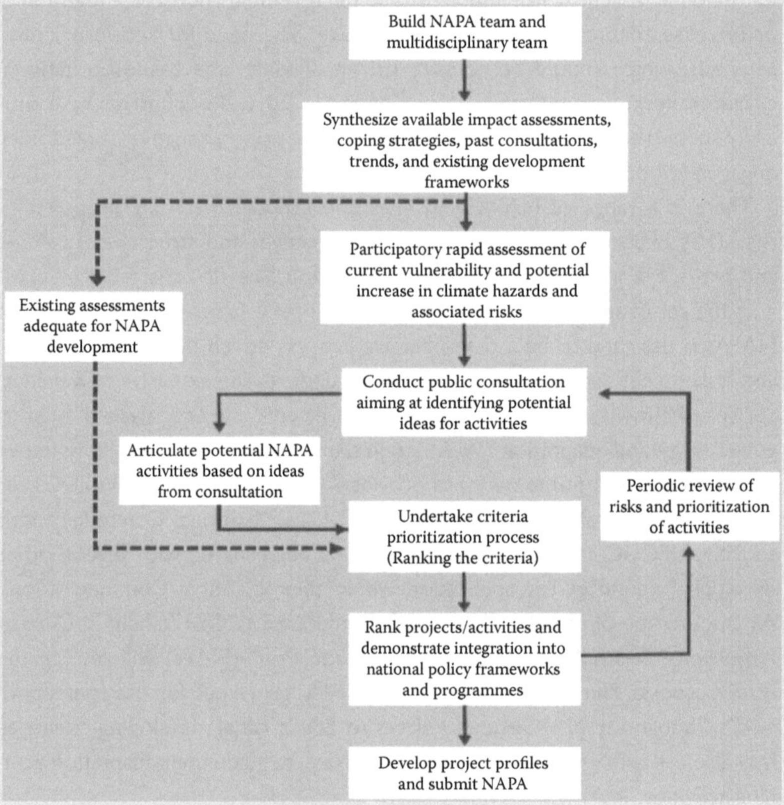

The range of conditions to which the coastal zone is exposed varies from north to south depending on regional-scale physical processes and local levels of coastal exposure. Of particular relevance to the current discussion is the exposure of the coastline to the impacts of natural disasters, particularly through cyclones and extreme rainfall events.

Over 90% of the coastline consists of low coastal plain with extensive mangrove forests. In addition to mangroves, the coast is characterized by a great variety of ecosystems such as estuaries, dunes, coastal lakes, banks

Figure 21.1 Location of Mozambique on the African continent. (Courtesy of Marcos Elias de Oliveira Júnior.)

and coral reefs, and swamps. These ecosystems represent critical habitats for various species of ecological and economic importance.

Mozambique has an estimated population of 22.9 million (World Bank 2010). Overall population density is low, although pockets of overpopulation exist within urban areas. Over 60% of the Mozambique population, estimated at over 15 million inhabitants, resides in the coastal zone.

Economic activities within urbanized coastal zones include fisheries and port exploitation. In addition, initiatives to mine oil and gas are currently under development and represent a significant economic value. Within rural coastal areas, community fisheries and agriculture remain the principal livelihoods, with aquaculture becoming important within several localities. Along parts of the coast, tourism and its supporting activities are becoming increasingly important to local economies. There is considerable scope for growth in the tourism sector throughout the region, the principal constraints being the absence of infrastructure for transport and other services, such as water and waste disposal.

Although Mozambique's economic potential is recognized, particularly in the context of its emerging democracy in the wake of decades of civil war, extreme poverty, and subsistence livelihoods remain the daily reality for the majority of Mozambicans. Over 55% of the population lives below the poverty line, 14.2% of babies born will die before reaching the age of five, and the average life expectancy is just 48 years. Adding to this,

Table 21.1 Geological and geomorphological attributes of the Mozambique coast

Attribute	*Northern coast*	*Central coast*	*Southern coast*
Area	600 km southwards from the Tanzania border	Maputo to Nacala covering 900 km	South of Maputo
Geology	Pleistocene sequence of cemented sands and mixed calcium carbonate rocks	Quaternary sediments, unconsolidated or poorly cemented deposits, deltaic alluvium, and recent poorly consolidated or unconsolidated layered sand and clay sequences	Quaternary sediments, unconsolidated or poorly cemented deposits, deltaic alluvium, and recent poorly consolidated or unconsolidated layered sand and clay sequences
Regional Geomorph-ology	Embayed coast with extensive cliff shorelines and fringing reef and reef platform development separated by low coral rock headlands Northern area from Ruvuma River to Mozambique Island is cut from cemented rock (limestone and sandstones) and reef rock	Dominated by deltas and large rivers; Zambezi, Maputo, and Save. Numerous deltas have developed extensive low lying plains with widths of over 100 km. Continental shelf is wide in the Bight of Sofala but becomes narrow near Nacala. Recent low lying deltaic and associated beach plains	Dominated by linear beaches or barrier swamp coasts and large fields of parabolic dunes in the backshore. Longshore drift large to the south near Save and Incomati rivers. Low-lying coastal terrain; cemented Pleistocene sand dunes fronted by a lagoonal marsh zone enclosed by a narrow, sea fronting belt of Holocene to recent dunes
Stability	Relatively stable shorelines except where recent deposits form lowlands overlying the lower coastal terraces	Coastal change is significant where recent unconsolidated sediments form spits, barriers islands, and coastal lowlands (e.g., 1 m/year accretion in Zambezi delta but rapid erosion around Beira)	Very vulnerable to shoreline change

Source: Travers, A., *Climate Change Adaptation in the Coastal Zone of Mozambique: Coastal Zone Management Expert Report*, Mozambique, LDCF Project0079862, April 2011.

malnutrition, HIV/AIDS, and endemic diseases restrict the adaptive capacity and capability of coastal communities.

The country is currently being rebuilt from virtual decimation with significant implications for future coastal activities. The livelihood realities of this situation include overexploitation of fisheries, illegal offshore fishing by foreign fleets, and conflicting uses between subsistence and commercial fishers near urban centers. Overexploitation of local resources in general is a serious concern, especially in the vicinity of urban areas with pollution, tourism-related development, and port infrastructure impacts also generating significant impacts.

Ongoing issues with respect to governance arrangements act to exacerbate existing threats. The Government of Mozambique has recognized these difficulties and in 2001 embarked on a Public Sector Reform (PSR) programme, covering a range of sectors including legislative, financial, and judicial sectors, and introduced decentralization and devolution of funds and decision-making to provinces and districts. PSR has included the progressive specialization of state employees and establishment of anticorruption measures.

Although the PSR programme, supported by UNDP and the World Bank, produced significant technical results, tangible differences in remote provinces have not been reported on. The state is still overstretched and faces difficulties in decision-making, service delivery, and enforcing legislation at the district level. Besides ongoing issues of accountability and transparency, it is apparent that a key stumbling block to achieve effective PSR and eventually governance in Mozambique, as in any other LDCs, is a lack of effective stakeholder participation on the part of the local population. This is largely attributable to high levels of poverty and illiteracy. Despite the fact that strenuous work has been carried out by the Government of Mozambique since independence in 1975, only half of the young population (>15 years old) has basic literacy skills and a large portion of the population (55%) still lives below the poverty line.

Decentralization of government has been another hurdle for better governance in Mozambique. Following independence in 1975, local government in Mozambique suffered both because decentralization was not operationalized due to either the state of civil conflict or a lack of sufficient cadres to support such initiatives. As a result, governance followed a centralized colonial model wholly inappropriate for the setting. Although a Local Government Reform Programme was sponsored by the World Bank in 1992 (Cuereneia 2001), the creation of municipalities, local governance, and deconcentration of resources is still in its embryonic phase in Mozambique.

A further facet in the poor governance arrangements in Mozambique is the growing gap in government expenditure between the Maputo and other regions. Government expenditure allocation in 2007 was 74% for central government in Maputo, 22% for the provinces, 3% for the districts, and 0.7% for the municipalities (Wiemar 2009). The disparity of these figures is highlighted by considering that most of the districts and municipalities account for 63.2% of the current rural population compared to an urban population of 36.8%,* less than 10% of which is in Maputo.

* http://www.tradingeconomics.com/mozambique/indicators.

It is unlikely that current poverty trends of the rural population will be overturned with current limited levels of investment. In addition, although the microfinance industry has grown, its outreach remains small, with a high concentration in Maputo and south of the country. All these factors lead to continuing natural resource challenges that will likely be aggravated by the impact of climate change.

FUTURE CLIMATE CHANGE RISKS AND OTHER POTENTIAL THREATS

Current and future climate-related risks to Mozambique and key areas of vulnerability have been analyzed in the country's First National Communication to the UNFCCC (MICOA 2003) and the NAPA (MICOA 2007). Climate risks are also considered to some extent in recent assessments of disaster risks, poverty, and vulnerability (INGC 2009; UNDP 2009).

The ability to predict the future behavior of the coastlines of Mozambique due to both climate-change and nonclimate-change drivers is strongly dependent on an understanding of contemporary process/response relationships within the coastal system. The different *types* of coast identified around Mozambique have associated levels of inherent susceptibility to change as a result of their geology and geomorphology. This susceptibility considered in the context of exposure to forcing factors dictates the level of contemporary *physical* vulnerability of a given coastal zone (Table 21.1).

Overall, the coastline of Mozambique is reported to be suffering from a general trend toward erosion. In the coastal region of southern Mozambique, the average erosion rate of the coastline has been 0.11 and 1.10 m/year between 1971–1975 and 1999–2004 in sheltered and exposed beaches, respectively. However, in certain areas, anthropogenic causes of these processes are dominant and include urban and port expansions, and more recently the expansion of tourism.

Climate change is expected to increase sea surface temperatures and increase the frequency and intensity of existing climate hazards particularly cyclones and long-term sea-level rise. A higher *launch point* for the surge increases both the areal extent of surge, all else being equal, and the depth of surge in areas already vulnerable to coastal storms (World Bank 2010). The risk of coastal impacts in low-lying and subsiding areas will significantly increase due to sea-level rise caused by climate change. Long-term effects of rising sea levels include increased shoreline erosion, saltwater intrusion into aquifers, and loss of coastal crop lands (Arndt et al. 2011). These climate change effects will undoubtedly challenge the existing coping mechanisms of communities living in coastal zones of Mozambique.

The impacts of sea-level rise on the coastline will likely be twofold: land lost directly through flooding, but also indirectly through coastal erosion. Furthermore, earlier results to compare the impacts of sea-level rise with storm surges found that a 10% future change in cyclone intensity could mean that the southeast African region (including Mozambique) could experience a loss of up to 40% of land within the low-lying coastal zone (Dasgupta et al. 2009). This assessment did not consider future changes in storm intensity or frequency, which could add to loss of land. Although these estimates of potential impacts are indicative only, it is clear that any of these future prospects for the Mozambican low-lying coastline will put an enormous pressure on communities and their livelihoods.

Tropical cyclones will remain a key threat, and their potential impact will possibly grow though an increase in their intensity and their interaction with the expected rates of sea-level rise. These climate change effects will aggravate underlying coastal erosion problems and increase the vulnerability of populations and settlements to strong winds, high waves, and flooding which are already detrimental to the livelihoods of more than 60% of the coastal population.

The Ministry for Coordination of Environmental Affairs (MICOA) (2011) recently reported that climate-related impacts have become more intense in the last two to four years, partially due to anthropogenic factors of which mangrove logging is the primary concern. Coral reefs are already stressed from increasing populations and marine pollution, coastal development, and marine-transported litter. Mining of coral and sand for use in construction is also damaging habitats. Moreover, intensive tourism will potentially damage reef habitats by pollution from boats, hotels, and other facilities, and by anchor damage, trampling and removal of coral as souvenirs.

An important finding of community consultations conducted during the project under discussion in this chapter was that most of the coastal communities in rural and remote villages reported unusual frequency flood events, strong winds (locally known as *Kussi*), sea water invasion, strong coastal erosion, and cyclones. These impacts mean that communities have to deal with situations that challenge traditional approaches to daily livelihoods, particularly farming and fishing (Tall 2011).

The likely impacts associated with a changing climate in the coastal zone of Mozambique will be exacerbated by inadequate development planning and land use (Table 21.2) (often associated with inherent high poverty levels). Development planning currently does not explicitly incorporate protection of coastal ecosystems for the essential services they provide. In the past, abundant native coastal vegetation, mangrove forests, coastal coconut plantations, extensive coastal sand ridges, and coral reef ecosystems provided natural protection to coastal areas against climate hazards. Today, these natural defenses are subject to ongoing degradation.

Table 21.2 Summary of the baseline problems and expected interaction with climate change selected pilot sites in Mozambique

Current issues	*Estimated directions of change (2011–2070)*	*Summary of likely impacts*
• Low-lying sandy dune area is subject to progressive erosion and undergoes inundation during high energy events. • Headland is eroding severely through a combination of terrestrial and marine pressure. • Livelihoods dependent on subsistence agriculture and fishing. • Pressures on livelihoods are due to (a) overfishing, (b) degradation of foreshore and dune environments, and (c) unstable coastline because of deposition of materials by rivers and erosion of river edges by strong currents. Shoreline change can be as much as 1 m/yr. • Communities live in transient dune system. Attempts at relocation in 2003 were unsuccessful. • Communities live in mangrove area. • High coconut tree mortality. • Degraded harbor infrastructure.	• SLR leads to increased bank erosion and instability of channel. • Marine erosion as a result of scouring and undercutting under elevated water levels combines with pressure from unregulated boat access on the channel banks and terrestrial pressure from run-off during the wet season to exacerbate alluvium wash out and create large-scale gullies. • Inundation of the relatively low lying areas adjacent to the shoreline (currently inhabited by fishers). • Continued damage and destruction of coastal infrastructure (e.g., remedial measures along the bank are currently ineffective and will be destroyed under projected SLR. Pier and adjacent make-shift walling will continue to be undermined and eventually undergo complete collapse.	• Erosion of infrastructure (private residences, tourist lodgings and facilities, and boat access/ pedestrian access points). • Degradation of mangrove ecosystem and associated services. • Damage and destruction of dune ecosystem and encroachment into backing wetland habitats. • Degradation of marine ecosystem (coral reef and associated protective function/ diving amenity; manta ray, whale, turtle, and fish populations). • Decreased beach recreational value. • Decrease in viability of subsistence fisheries.

Source: Travers, A., *Climate Change Adaptation in the Coastal Zone of Mozambique: Coastal Zone Management Expert Report*, Mozambique, LDCF Project0079862, April 2011.

Widespread concessions of coastal sand mining or coastal land reclamation results in the leveling of coastal sand ridges. Natural coastal habitats such as wetlands and mangroves are often converted for urban or agricultural uses, reducing the ability of such ecosystems to provide a natural barrier or buffer against wave action and storm surges, which results in further and increased erosion and other impacts such as flooding.

Mangroves are exploited for fuel wood (cooking and smoking fish) and for building purposes. In recent times, coastal ecosystems of Mozambique have been degraded because of widespread mangrove logging, lethal illness of coastal coconut trees, and beach sand mining, considerably lowering the natural coastal resilience (MICOA 2010, 2011). Population densities around urban and rural coastal areas often mean that critical physical infrastructure is sited in vulnerable locations. Current methods of controlling erosion and flooding rely on coastal engineering and hard physical structures such as seawalls and groynes, which are expensive and therefore difficult to maintain or replicate widely.

KEY SUCCESSES AND FAILURES EVIDENT IN EXISTING MEASURES TO BUILD COMMUNITY RESILIENCE, ADAPTIVE CAPACITY, AND SUSTAINABILITY

The Government of Mozambique and the general public have become increasingly aware of the extreme coastal erosion being experienced in low-lying areas of central Mozambique (Beira) and the south (Maputo), and the inherent vulnerability of these areas in the face of climate change. In response, the government has adopted several measures to protect the country's coral reefs, including a ban on coral mining, environmental safeguards on tourism development, and, more recently, the establishment of marine protected areas (INGC 2009). Despite these efforts to increase coastal resilience to the impending impacts of a changing climate, anthropogenic pressure on reefs and mangroves in particular continues.

Overall, the country faces a dichotomy between a drive to industrialization and associated impacts in urban areas and the impacts experienced by impoverished, subsistence coastal communities. Additionally, as outlined above, the government's capacity (at all levels) is extremely limited and the ability to govern and ensure compliance with adaptive and mitigative endeavors is inadequate. This challenge of dealing effectively with *scaling* governance efforts effectively to address adaptation from local through regional to national (and then regional) is well recognized (Osbahr et al. 2008).

For climate change adaptation within the coastal zone of Mozambique to be effective, it is essential to facilitate sustainable commercial and industrial development and sustainable livelihoods for ordinary Mozambicans. To achieve and retain sustainable livelihoods, the services provided by coastal ecosystems must be identified, acknowledged, and valued accordingly. Additionally, a participatory, collaborative, and community-driven approach is required to overcome the previously identified issues relating to governance arrangements, specifically, difficulties in public participation, decision-making, service delivery, and enforcing legislation at the district level.

In recognition of these issues, the project under discussion in this chapter adopted a two-pronged approach to scope out adaptation options to address current challenges posed by coastal erosion and address potential future climate change impacts. First, community-level Vulnerability and Capacity Assessments (VCAs) were undertaken in seven pilot communities, focusing on protection of livelihoods from the potential effects of climate change. Second, an assessment of adaptation options related to ecosystem protection and enhancement was undertaken, covering built and natural solutions, to address the expected effects of climate change on the coastline. This approach was adopted to ensure that both technical assessments and stakeholder consultations were undertaken (together with a detailed governance and institutional capacity assessment, outlined below) to build a thorough and well-justified project proposal for the LDCF. The aim was to develop a proposal that constitutes sound development practice—following UNDP processes—and recognizes the unique challenges in developing ambitious adaptation projects that seek to build community capacity and ecosystem resilience at a range of spatial scales (Bunce et al. 2010).

The analysis was undertaken for adaptation options at three pilot site locations; Pemba (Cabo Delgado Province, northern Mozambique), Pebane (Zambézia Province, central Mozambique), and Závora (Inhambane Province, southern Mozambique). The three sites were initially chosen from a shortlist of 27 potential locations identified through ongoing work carried out by the MICOA. The sites were considered representative of the discrete coastal ecosystems that characterize the coastal zone of Mozambique (Figure 21.2). Additionally, each site has a set of characteristic development and climate-related issues that allow them to be considered suitable proxies for the range of climate change issues that will face the coastal zone of Mozambique. The attributes that informed pilot site selection were as follows:

- Highly vulnerable to climate change
- Strong community leadership and social networks
- Willingness of communities and/or demand by communities to try new adaptation approaches
- Existing capacity development or coastal zone investments which the LDCF project could build on
- Return on investment likely to be greatest
- Accessibility in light of the need for ongoing monitoring and evaluation

The overall four-steps framework for the adaptations options analysis is summarized in Figures 21.3 and 21.4.

Information collected through a series of field visits and community consultations was used to profile the range of existing issues to which specific areas of the coastal zone within each of the pilot locations are susceptible. A consideration of these issues in the context of projections for climate change and an

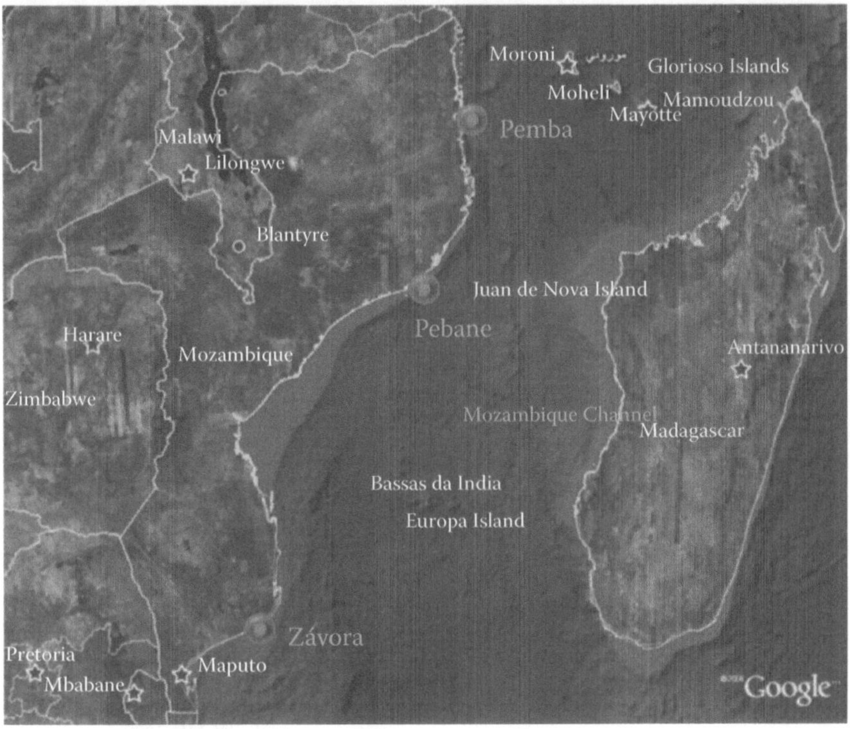

Figure 21.2 Geographic location of the selected vulnerability and capacity assessment (VCA) sites along Mozambique's coastline. (Courtesy of Daniel Zakarias; © 2009 Europa Technologies; © 2009 Tele Atlas; © 2009 Google; © 2009 LeadDog Consulting.)

appreciation of the sensitivity of discrete coastal landforms to change allowed identification of a list of likely future impacts at each site. The information derived at the end of step 2 (Figure 21.3) was subsequently used to inform development of a typology of adaptation options for identified sites at each pilot location. A consideration of the range of climate change impacts likely to affect each site was used as the basis to determine the perceived overarching management goal following categories used by USAID (2009), namely,

- Diversified livelihoods
- Functioning ecosystems
- Human safety
- Reduced exposure of built environment

Because each overarching goal can be associated with a series of complementary actions (USAID 2009), the output from this preliminary analysis

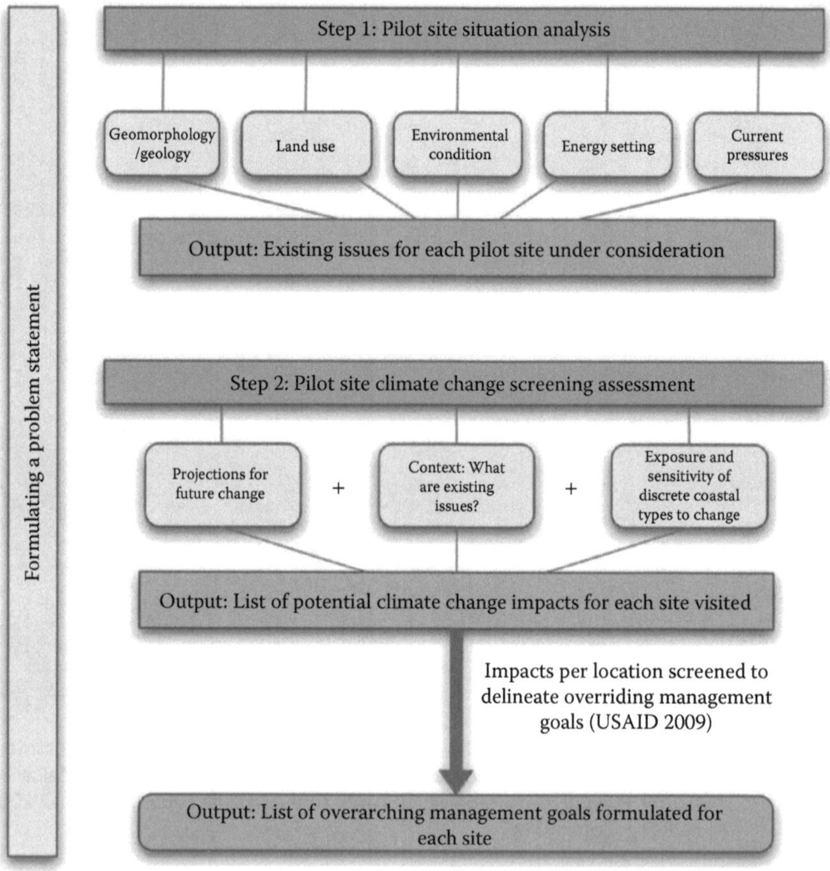

Figure 21.3 Approach to adaptation options analysis—Part I. (Data from Travers, A., *Climate Change Adaptation in the Coastal Zone of Mozambique: Coastal Zone Management Expert Report*, Mozambique, LDCF Project0079862, April 2011.)

was a list of adaptations options and associated detailed *actions* for each key area of interest at each of the pilot locations visited.

An analysis of the range of adaptation options for each site visited was carried out based on how appropriate options were to treat potential climate change impacts, their affordability, and the capacity of the local community to implement the identified options. The analysis was undertaken through expert judgment of the Project Team through the use of specific attributes that were considered for each option. Each attribute was ranked low, medium, and high by the expert team using qualitative thresholds as a guide:

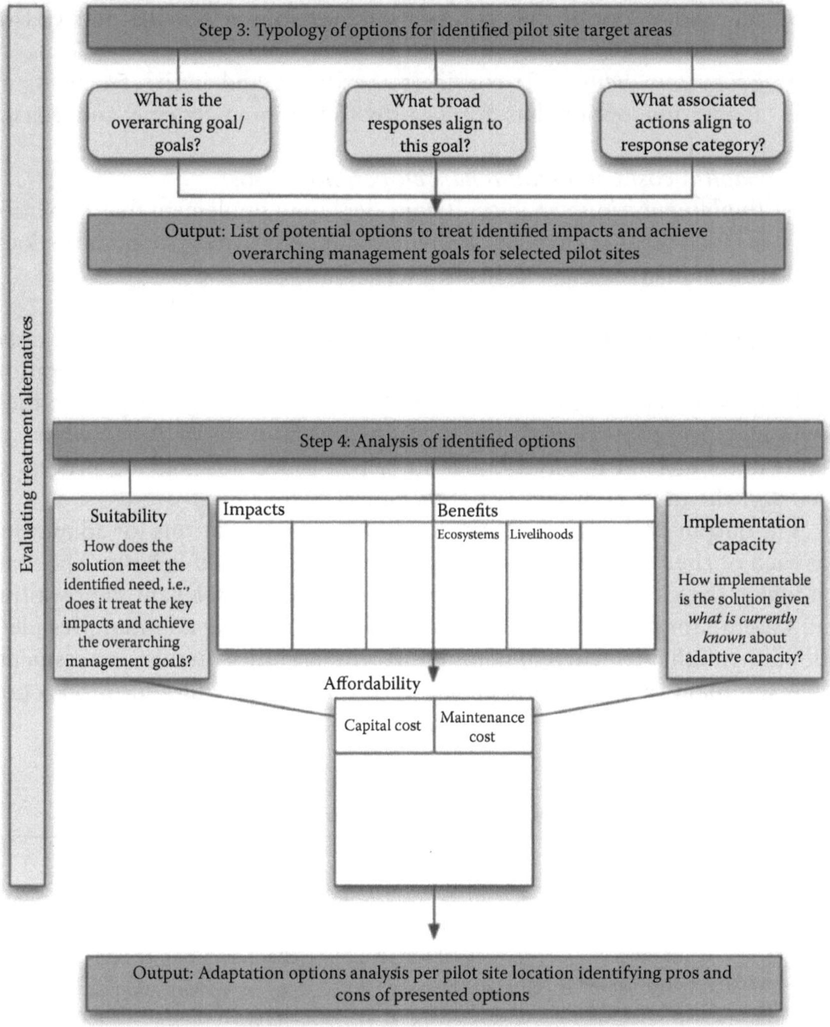

Figure 21.4 Approach to adaptations options analysis—Part II. (Data from Travers, A., *Climate Change Adaptation in the Coastal Zone of Mozambique: Coastal Zone Management Expert Report*, Mozambique, LDCF Project0079862, April 2011.)

- *Suitability:* It considers how does the solution meet the identified need? How suitable is the solution for dealing with the impacts identified for the site and does it assist in achieving the overall management goals?
- *Impact:* It considers the *negative* impacts on habitat, landscape, and physical processes.

- *Livelihood benefit:* It considers the direct and indirect benefits an adaptation option has for healthy/functioning marine and coastal livelihoods.
- *Ecosystem benefit:* It considers the direct and indirect benefits an adaptation option has for healthy/functioning marine and coastal ecosystems.
- *Capital costs and maintenance/operational cost.*
- *Implementation capacity:* It considers how implementable a solution is given what is currently known about adaptive capacity of the local community.

The options that were considered most suitable with lowest negative impact to the environment were initially selected for further consideration. Subsequently, a relative consideration of all of the rankings assigned to these options in the context of the problems observed in the field and what was known about land use/livelihoods informed a decision-making process to *shortlist* the most appropriate options for each location.

The output of this analysis was a set of recommendations for adaptation at each of the three pilot locations: Pemba, Zavora, and Pebane. The scale at which these recommendations were presented was relevant to prevailing and dominant physical processes and aligned to target communities identified through a concurrent community vulnerability analysis assessment.

A summary of key recommendations for the Pebane pilot location is presented in Table 21.3.

Table 21.3 Summary of coastal protection adaptation options for Pebane

Community	*Management concern*	*Problem summary*	*Options*
Malaua/ Porto Community, Pemba Harbour	Exposure of built environment/ ecosystems and livelihoods	Access issues: terrestrial runoff	Surface and install drainage channels along access
		Channel bank erosion/ undercutting	Logging/walling and reinforcement along eroding channel banks
		Mangrove degradation (access/cutting)	Mangrove replanting and custodianship program
		Fisher settlements located in coastal buffer adjacent to mangrove	Managed retreat and resettlement of coastal fisherman Fisheries management and diversification
		Damage to harbor infrastructure	Logging/walling and reinforcement along eroding channel banks Install climate resistant infrastructure

Source: Travers, A., *Climate Change Adaptation in the Coastal Zone of Mozambique: Coastal Zone Management Expert Report*, Mozambique, LDCF Project0079862, April 2011.

The pilot coastal communities interviewed in Pemba, Pebane, and Inharrime during the project design phase clearly expressed the need for a transition to alternative *climate-delinked* activities as the necessary condition for a successful adaptation to climate change impact on coastal livelihoods (Tall 2011). Priorities include the diversification of crops, the introduction of drought and flood-resilient crop options, and strengthening fishing capacity to adapt fishing practices to the changing patterns of climate variability.

BARRIERS AND OPPORTUNITIES

There are a number of constraints to modifying existing approaches to land-use planning, coastal protection, and development in the Mozambique. These are due to weak intersectoral policy coordination and development in the management of sea-level rise and coastal erosion, limited institutional and individual capacity for planning; and financial constraints. These are considered briefly in the subsequent paragraphs.

Weak intersectoral policy coordination and development

Mozambique has undergone a relatively politically stable period, but there has been, over the years, major restructuring of government ministries. Many policies are under development or review with insufficient intersectoral coordination to ensure overall policy coherence. Laws, regulations, and mandates can be inadequate, overlapping, or even in conflict with each other. This can result in a lack of understanding regarding the limits and responsibilities of individual agencies. Furthermore, laws that are enacted to protect and manage the coastal zone suffer from ad hoc enforcement regimes. In general, limited action has been taken to implement a sustainable and integrated coastal zone planning framework, such that the national integrated coastal zone management plan has not been endorsed with clear budget allocations, responsible institutions, or accountabilities. The National Environmental Management Programme is the master plan for environmental management in Mozambique and provides a framework for coordination, including the National Coastal Zone Management Committee led by MICOA. However, this committee has largely been inactive, with members lacking clear roles and responsibilities.

Coordination of climate change adaptation strategies is currently weak and/or ad hoc. MICOA, INGC (Instituto Nacional de Gestã das Calamidades), and the Eduardo Mondlane University have all carried out projects in the coastal zone in a stand-alone way. In addition, some ministries (such as Transport and Public Works) not directly related to the environment are still of the view that climate change *belongs* to the environmental ministry only. Centrally placed Ministries of Planning and Development, Ministry of State Administration, and Ministry of Finance realize that climate change (including adaptation) is central to their interests but their capacity is too limited to take a leadership role. Finally, this lack of cross-sectoral coordination in government and among other development actors is reflected at the local level, often resulting in weak/inadequate organizational structure to implement and enforce adaptation-relevant legislation.

Ministries have traditionally worked in silos and there is limited culture of knowledge sharing. For example, a recent comprehensive risk analysis that has integrated information from across sectors (INGC 2009) has not been readily absorbed across government. There are insufficient mechanisms in place for data and information exchange, which have resulted in a potential mischaracterization of climate change-related threats. Two key missing elements for overcoming the existing barriers are (1) a focal point within Mozambique for climate change adaptation information where the public, development partners, and other interested parties can access and share information; and (2) the emergence of an effective institutional *champion* to promote adaptation.

Limitations in institutional and individual capacities to plan for climate change

There is a severe shortage of skilled and professional staff within the environment sector in Mozambique overall, and this shortage is particularly severe for addressing climate risks on the coast. As in any LDC, specialized training programs are limited, particularly on climate change, although Mozambique has recently introduced several higher education degrees in environmental sciences. If at national-level institutional and individual capacities are lacking, the capacity for climate change adaptation planning at the subnational and municipal level is even lower.

The Government of Mozambique is aware that urgent action is needed to address the threats posed by climate change to the country's population and to continued sustainable development. In the Poverty Reduction Programme (PARP), the Government acknowledges that "by preventing disasters, we can make communities and territories less vulnerable to the various threats" (Republic of Mozambique 2007: 70). The Government Action Plan has also acknowledged the role of good environmental stewardship in poverty reduction, but they have not yet been explicit in articulating a climate change adaptation imperative. This can be partly attributed to the fact that the many agencies responsible for coastal zone management lack the climate risk assessment abilities needed to identify and integrate climate risks and appropriate adaptation response measures into relevant policy and regulatory frameworks. Decision-makers in the Ministries of

Planning and Development, and Finance are currently not equipped with skills that can effectively negotiate and coordinate adaptation investments through a common framework. This has led to development partners funding different climate change interventions with different sectoral ministries in an uncoordinated way, resulting in a diminished impact on the target communities. Priorities for funding have also been biased toward short-term goals (e.g., focusing on relief efforts or service delivery in sectors such as education and health) as opposed to preparedness, mitigation measures, and adaptation strategies that are longer term in nature. Thus, awareness of the short- and long-term consequences of climate change for key ministries, such as transport, agriculture, fisheries, health and public works, and impacts on gender relations in relation to climate change are still weak and a matter for concern as a potential barrier to effective adaptation.

Financial constraints

Like other LDCs, Mozambique has high adaptation costs relative to gross domestic product (GDP). Adaptation costs are especially high, because of the long coastline (>2,700 km) and the scattered distribution of more than 60% of total population across many small coastal towns and villages. Currently, the country is facing a range of economic problems including the impacts of the global recession and the country's dependence on imports of food, oil, and manufactured products. Therefore, budgetary resources for the country's development plan for the next five years are already severely constrained and there are limited resources to meet the additional costs of adaptation. The country has shown impressive GDP growth over the past decade (6.4% in 2009 to 7.7% in 2011) with the poverty rate declining from 69.4% of the population in 1997 to 55% in 2010. Even so, poverty remains widespread and is now worsening, highlighting the weak linkages between macroeconomic performance and the bulk of the population in Mozambique.* Widespread rural poverty limits the adaptive capacity and capability of individuals, farmers, and villagers to respond to natural disasters, such as flooding and droughts. Poor farmers/fishermen have limited opportunities to improve yields, increase income, and/or to develop alternative, appropriate farming systems with in-built resilience to climate hazards. Consequently, the potential role of external donors in both providing funding to support the implementation of adaptation options and in supporting the institutional capacity required to mainstream climate change in government decision-making at all levels (including development assistance) is vitally important (Sietz et al. 2011).

* African Economic Outlook, 2011: http://www.africaneconomicoutlook.org/en/countries/southern-africa/mozambique/.

PRACTICAL RECOMMENDATIONS FOR BUILDING COMMUNITY RESILIENCE, ADAPTIVE CAPACITY, AND SUSTAINABILITY

If effective climate change adaptation is to be achieved within the coastal zone of Mozambique, it is essential that the important role of ecosystems services be realized and the benefits of it shared with local communities. The LDCF project preparation process recognized the interrelationship between ecosystems degradation and poverty. The adaptation prioritization process followed two interlinked methodologies: one people-centered (the VCA) and the other ecosystem-centered (the coastal management assessment report). The project approach promotes adaptation options at the household and community levels, recognizing that the success of each relies on the implementation of both.

Key recommendations stemming from this work are as follows:

- Capacity development for adaptation planning is critical to the sustainability and replication of adaptation measures. An extensive program of capacity building should accompany the implementation of climate change adaptation measures and site demonstrations of adaptation techniques and practices in a learning-by-doing approach. This will assist in building a cadre of skills and experience at subnational level that will be able to support ongoing adaptation beyond the project lifetime. The capacity building activities through stakeholder consultations, mobilization, networking, and field-level presence aim to help achieve social sustainability of the project.
- A participatory planning process by communities is critical to promoting ownership and sustainability of adaptation measures. Adaptation priorities were solicited during the VCAs. These will be further prioritized to fit into an adaptation budget allocated to each pilot community. Communities will be trained to set up and maintain their adaptation investments. Climate risk mapping and assessments will be coproduced between local communities and scientists to improve the accuracy and utility of the climate risk information. Communities will be involved in the monitoring and evaluation schemes to gauge the actual effectiveness of the *soft* coastal stabilization measures.
- Women will likely lead the farming/fishing interventions, which will be critical for sustainability. Microfinancing success in Mozambique is dependent on women's participation. The project indicators are to be tracked with data that are disaggregated by gender.
- Financial sustainability and affordability of adaptation measures underpins the sustained delivery of adaptation benefits and makes it more likely that the adaptation measures can be replicated.

- Developing a strategy to replicate and scale-up the project's activities will be critical for getting adaptation implemented both at scale and at speed if vulnerable communities are to be enabled to cope with the impacts of climate change.

Another important strategy for replication would be the adjustment of regulatory and fiscal frameworks to guide private investments, including household-level investments, into climate-resilient practices. Evidence about what works best for adaptation, costs, and the ways in which private sector finance could be drawn into climate-resilient practices and technologies would be necessary to make the case for these adjustments. In the context of the LDCF project, development of regulation for land-use planning will include the following:

- Risk zoning for the design and construction of infrastructure
- Definition of shoreline setbacks or buffer zones around vulnerable coastlines to avoid loss of human life as well as damage to infrastructure in case of natural hazards
- Minimum height level restrictions for development of coastal infrastructure/services to reduce impacts of sea-level rise and storm surges

In summary, the key for Mozambique is development of alternative livelihoods that are resilient to change and surprise in a world of climate change. In this context, there is a need to distinguish challenges of next 50 years that are about the immediacy of livelihoods from longer-term considerations (next 50–100 years) when climate change impacts are likely to become even more pronounced. Choices about infrastructure and at-risk communities need to be made now given the anticipated impacts of climate change. Finally, the governance capacity developed to sustain livelihoods in coming decades and specific adaptation project activities, such as the LDCF project outlined in this chapter, must be designed to support both the short-term development imperatives in Mozambique and the long-term adaptive journey faced by this vulnerable country.

REFERENCES

Arndt, C., Strzepeck, K., Tarp, F., Thurlow, J., Fant, C., and Wright, L. (2011). "Adapting to climate change: An integrated biophysical and economic assessment for Mozambique," *Sustainability Science*, 6: 7–20.

Ayers, J. (2008). "Progress implementing NAPAs," *Tiempo*, 69: 15–19.

Bunce, M., Brown, K., and Rosendo, S. (2010). "Policy misfits, climate change and cross-scale vulnerability in coastal Africa: How development projects undermine resilience," *Environmental Science & Policy*, 13: 485–97.

Cuereneia, A. (2001). "The process of decentralization and local governance in Mozambique: Experiences and lessons learnt," *Proceedings from the participatory symposium on Decentralization & Local Governance in Africa*, Cape Town, Republic of South Africa, March 26–30, UNCDF Publications and Reports.

Dasgupta, S., Laplante, B., Murray, S., and Wheeler, D. (2009). *Climate Change and the Future Impacts of Storm-Surge Disasters in Developing Countries*, Washington, DC: Center for Global Development.

Instituto Nacional de Gestão das Calamidades (INGC). (2009). *Study on the Impact of Climate Change on Disaster Risk in Mozambique: Main Report*, Maputo, Mozambique: INGC.

Ministry for the Coordination of Environmental Affairs (MICOA). (2003). *Initial National Communications to the UNFCCC*, Mozambique: MICOA.

Ministry for the Coordination of Environmental Affairs (MICOA). (2007). National Adaptation Programme of Action (NAPA). Mozambique: MICOA.

Ministry for the Coordination of Environmental Affairs (MICOA). (2010). *Plano de Gestão Ambiental do Município de Pemba*, Moçambique: Conselho Municipal da Cidade de Pemba Ministério para a Coordenação da Acção Ambiental, DNGA & DPCA-CD.

Ministry for the Coordination of Environmental Affairs (MICOA). (2011). *ANEXO: Perfil Ambiental do Município de Pemba*, Moçambique: Ministério para a Coordenação da Acção Ambiental, DNGA & DPCA-CD.

Osbahr, H., Twyman, C., Adger, W.N., and Thomas, D.S.G. (2008). "Effective livelihood adaptation to climate change disturbance: Scale dimensions of practice in Mozambique," *Geoforum*, 39(6): 1951–64.

Republic of Mozambique. (2007). *Republic of Mozambique: Action Plan for the Reduction of Absolute Poverty, 2006–2009*, (PARPA II) Poverty Reduction Action Plan (PARP) January 2007, International Monetary Fund Country Report No. 07/37, IMF.

Sietz, D., Boschütz, M., and Klein, R.J.T. (2011). "Mainstreaming climate adaptation into development assistance: Rationale, institutional barriers and opportunities in Mozambique," *Environmental Science & Policy*, 14: 493–502.

Tall, A. (2011). *Summary of Results from the Vulnerability and Capacity Assessments along the Coast of Mozambique*, UNDP Report, UNDP.

Travers, A. (2011). *Climate Change Adaptation in the Coastal Zone of Mozambique: Coastal Zone Management Expert Report*, Mozambique, LDCF Project0079862, April 2011.

United Nations Development Programme (UNDP). (2009). *Mozambique Prodoc Final: Climate Change Adaptation Action and Mainstreaming in Mozambique*, UNDP, UNDP Africa Adaptation Programme. http://www.undp.org/content/dam/undp/documents/projects/MOZ/00058248/Africa_Adaptation_Programme_-_Mozambique_-_Prodoc_Final_signed_1.pdf.

United Nations Development Programme (UNDP). (2012). *Mozambique Country Profile: Human Development Indicators*, UNDP. http://hdrstats.undp.org/en/countries/profiles/MOZ.html (accessed November 2013).

USAID. (2009). *Adapting to Coastal Climate Change: A Guidebook for Development Planners*. http://www.usaid.gov/ (accessed March 2012).

Wiemar, B. (2009). "Decentralization of the African state—or state building through local governance—a paradox? Challenges to governance and decentralization in Mozambique," Conference report, FAU—Foreningen af Udviklingsforskere i Danmark, The Association of Development Researchers, Denmark.

World Bank. (2010). *Economics of Adaptation to Climate Change: Country Report, Mozambique.* http://climatechange.worldbank.org/ (accessed March 2012).

Chapter 22

Climate change and the coasts of Africa

Durban case study

Andrew A. Mather and Debra C. Roberts

Abstract: Durban surrounds a natural estuarine bay and over the last 150 years has developed into one of the top three port cities in the southern hemisphere. With a population of approximately 3.6 million, one of the most pressing sustainability challenges facing the city is the need to meet developmental aspirations, while still ensuring the protection and management of the natural resource base. The city has a diverse society which faces a complex mix of social, economic, environmental, and governance challenges. Among these challenges are high levels of unemployment, poverty, and the HIV/AIDS pandemic that make communities particularly vulnerable. This vulnerability will be exacerbated under projected climate change conditions, as will the risk facing other key sectors such as water and food security, infrastructure provision and maintenance, health services, and disaster management. In order to respond to the multiple challenges of climate change, a Municipal Climate Protection Programme (MCPP) was initiated in 2004. Work done within the program has looked to identify the developmental risk posed to the city by existing climate variability and incremental climate change and to address the possible impacts of climate change through a number of targeted and programmatic interventions aimed at improving the overall resilience of the city. This chapter describes the approaches, challenges, and lessons learnt in Durban during the development and implementation of the MCPP and reflects on the possible way forward in terms of the further development of the program. The lessons learnt in Durban are applicable to similar third world cities struggling with high levels of unemployment, poverty, and the HIV/AIDS pandemic as well as having to deal with an exposed high energy coastline.

INTRODUCTION

Durban is the largest city on the east coast of Africa and one of the top three ports in the southern hemisphere (Figure 22.1). As such, it acts as a gateway for imports and exports for the southern African region. It has a population of

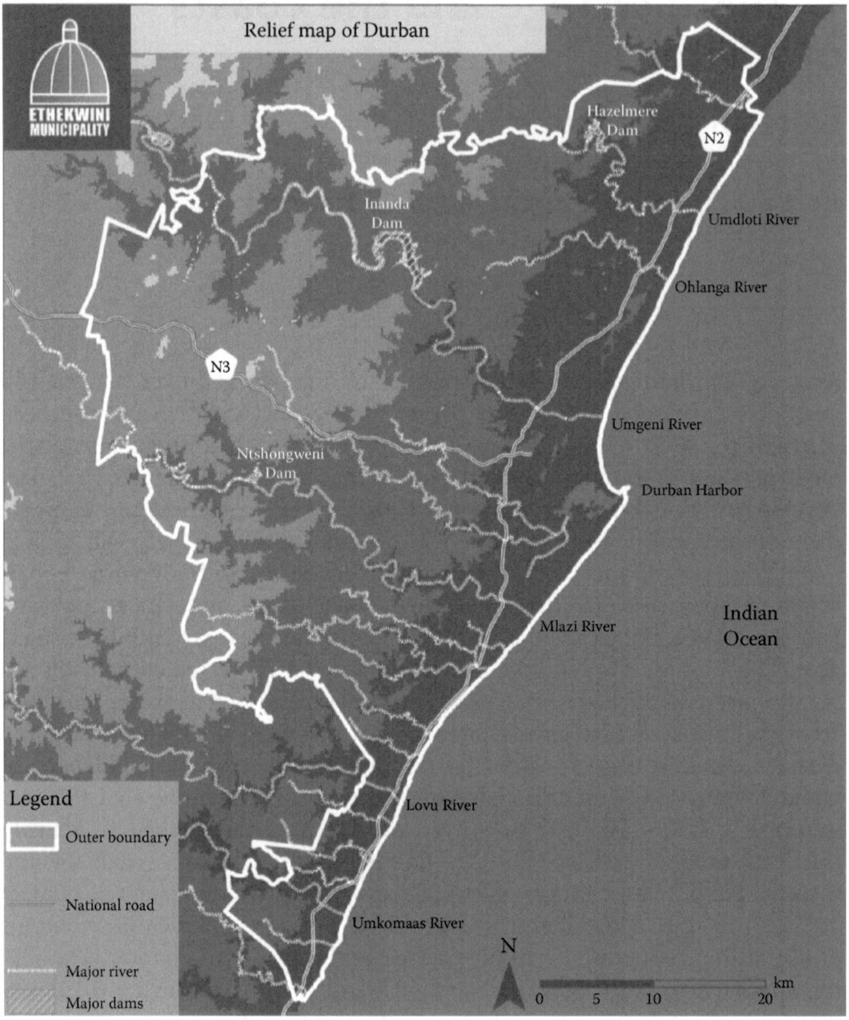

Figure 22.1 Locality map of Durban.

approximately 3.6 million people and a municipal area of approximately 2,300 km^2. Durban is a relative old city in the South African context, emerging as a rudimentary settlement in the 1820s and flourishing to become one of the three largest metropolitan cities in the Republic of South Africa. It is a critical part of the regional economy of the province of KwaZulu-Natal, being responsible for two-thirds of the provincial gross value added (a measure of economic value used in the calculation of GDP) and approximately one-tenth of the national economy (eThekwini Municipality 2010a). Durban is renowned for its sandy beaches and is South Africa's beach playground, attracting thousands of local, regional, and international visitors to the city's shores every year. Tourism thus plays a vital role in the regional economy with 73% of visitors to the city visiting Durban's beaches (KwaZulu-Natal Tourism Authority 2009).

Durban's coastline is sandy with a median grain size of 300 microns with very few rocky outcrops. The coastline is relatively straight with a net northerly longshore sediment transport rate of 460,000 m^3 per year (Mather et al. 2003). As the coast is open, storm surges are relatively small (~0.5 m during significant events) but large waves (wave heights of ~14 m) are generated in summer during the cyclone season and in winter by the Southern Ocean swells. As Durban is located on a sandy coastline with a relatively narrow continental shelf facing the ocean, this makes it vulnerable to wave attack and erosion (Mather 2008). The Port of Durban interrupts the longshore sediment transport and a sand bypass has been in place in some form since 1935 (Barnett 1999). Generally, the coastline has a relatively steep profile and there are two significant areas of inundation located in a residential area and in the port. The sandy coast has been relatively stable in recent years but erosional trends have been observed along portions of the coast that have been attributed to sand winning and dam construction in the catchments (Council for Scientific and Industrial Research 2008).

Durban's local government body, the eThekwini Municipality, has adopted a vision that states that Durban will be Africa's most *caring and liveable city* by 2020. Although there has been significant progress since the move to democracy in 1994, there remain a number of significant challenges that need to be addressed (eThekwini Municipality 2010b), including the following:

- Low economic growth and high rates of unemployment
- Access to basic household and community services are less than optimal
- Relatively high levels of poverty
- Low levels of literacy and skills development
- Sick and dying population affected by HIV/AIDS
- Exposure to unacceptable high levels of crime and risk
- Many development practices are still unsustainable
- Ineffectiveness and inefficiency of inward-looking local government still prevalent in the municipality

BIOPHYSICAL, SOCIOECONOMIC, AND GOVERNANCE SETTING

Biophysical setting

Durban has a particularly rich natural resource base as the city is located within a global biodiversity hot spot. It also has an extensive coastline (97 km) and is an important catchment area with 17 rivers and 16 estuaries.

The coastal zone is under particular pressure for development for both residential and commercial purposes but has been negatively affected by poor water quality in both river and estuarine ecosystems. A 2007 river health survey indicated that out of 61 sample sites in 33 different river systems, only two were considered to have a *natural* eco-status (GroundTruth Consulting 2007). Similarly, a 2008 estuarine survey showed that only three out of the 16 estuaries were regarded as being in *good* condition (Forbes and Demetriades 2008), while none fell into the *excellent* category. A more recent report on the state of national wastewater treatment works indicated that only 7.4% of municipal wastewater treatment works in South Africa met the required standard. Within eThekwini Municipality, only 11 of the 29 wastewater treatment works met the national criteria (Department of Water Affairs 2009).

Historically, the city has prioritized the protection of its natural assets through the design and implementation of the Durban Metropolitan Open Space System. This system was designed to protect the city's key biodiversity assets and related ecosystem services. The system is designed around river catchments and the city's coastline and currently accounts for approximately 28% of the municipal area. Despite the fact that the open space system provides ZAR 3.1 billion worth of free ecosystem services every year (e.g., flood prevention and water supply), it is under continuous pressure and threat from development (both formal and informal) and is extensively utilized by local inhabitants who harvest raw materials for a range of uses from basic building materials to medicinal plants (eThekwini Municipality 2005). A more recent valuation of ecosystem services provided by the estuaries and beaches in the municipality put a value on just this part of the open space system at ZAR 5.13 billion per year (Council for Scientific and Industrial Research 2008). Similar pressures affect the coastline and access for the collection of food is still an important issue for the subsistence communities living along the coast.

There is also demand for increased port capacity through the establishment of a new *dig out* port (a port constructed on land with future water areas dug out to form basins for shipping) in conjunction with increased recreation opportunity and access for previously excluded and landlocked communities. As a result, the current era is one of severe environmental degradation and transformation of the coastal zone. It is also marked by increasing infrastructure failure as more infrastructure is needed closer to the coastal hazard zone in order to meet human recreational and service requirements. As a result, these interventions are often placed in the areas in which the natural variations to sea storms and coastal erosion need to be accommodated. If this natural buffer is removed, the inherent capacity of the environment to deal with the negative impacts of human utilization is reduced or impaired.

It has now become clear that climate change will in all likelihood significantly affect the city's biodiversity and the resulting ecosystem services. This change in the natural resource base will have both social and developmental repercussions and work is currently being undertaken to better understand the impact of this change.

Socioeconomic setting

As with all South African cities, Durban faces the challenge of transforming its previously segregated, racially divisive urban form into a more integrated, equitable and mixed social and economic landscape. This has not been an easy task as the structural changes required to effect social and economic reform are costly. Furthermore, the process of social integration is still viewed with suspicion by some groups and the process of transformation is, therefore, likely to take several generations to complete. The serious economic challenges facing the city also remain persistently high. These include high levels of unemployment (30%–40%), poverty, large wealth disparities (Gini coefficient = 0.64), and a high incidence of HIV/AIDS (eThekwini Municipality 2010b).

Governance setting

eThekwini Municipality is one of the three largest and most powerful local government bodies in the country and is well managed financially, maintaining a credit rating of AA^- for long-term loans and A1 for short-term loans (eThekwini Municipality 2010b). Since the advent of democracy in 1994, the municipality has been developmentally focused and people oriented, with a strong concern for sustainability. This is reflected in the key strategic planning document of the city, the Integrated Development Plan, which includes an eight-point action plan to address key challenges. The eight plans are as follows:

- *Plan 1*: Sustaining our natural and built environment
- *Plan 2*: Economic development and job creation
- *Plan 3*: Quality living environments
- *Plan 4*: Safe, healthy, and secure environments
- *Plan 5*: Empowering citizens
- *Plan 6*: Celebrating our cultural diversity
- *Plan 7*: Good governance
- *Plan 8*: Financial viability

Plan 1 includes the requirement for the development of a Municipal Climate Protection Programme (MCPP). The development of this program was initiated in 2004 and the approach adopted has been very similar to the climate change programs of other major cities (e.g., London and New York) which incorporate both an assessment of local-level impacts and the development of locally focused response strategies. The adaptation work stream of the MCPP was initiated in 2006 (following the completion of an initial climate change impact analysis) with the development of a Headline Climate Change Adaptation Strategy (HCCAS). A key reason for adaptation achieving such prominence in Durban is that adaptation or resilience-focused interventions offer the potential for developmentally linked co-benefits that are responsive to a context of poverty and underdevelopment.

The key objectives of the HCCAS were to identify which key municipal sectors would be impacted by incremental climate change and to highlight appropriate and practicable adaptation options. The sectors reviewed included human health, water and sanitation, solid waste, coastal zone, biodiversity, infrastructure (i.e., electricity and transportation), food security/agriculture, strategic planning, economic development, and disaster risk reduction. This case study will focus particularly on the coastal zone.

CLIMATE CHANGE RISK ASSESSMENT

Local and national-level research has indicated that climate change in the Durban region will result in higher (mean, maximum, and minimum) temperatures (Golder Associates 2010), more intense and erratic rainfall (Schulze et al. 2010), and increasing sea levels (Mather 2007; Mather et al. 2009). All of these will impact to a lesser or greater extent on the coast; however, sea-level rise is probably the most significant among them. Climate change impacts, such as sea-level rise and increased storminess, are thus likely to result in accelerated erosion of the shoreline. Current sea-level rise observations indicate that levels are increasing at a rate of 2.7 mm per year (Mather 2007). Of particular concern is the rate of future sea-level rise, particularly if this accelerates. Increased temperatures and in turn increasing storminess have the potential to accelerate changes in the coastal zone. Currently, the wave climate is driven by two major weather patterns, cyclones during the summer season and storm swell from the southern oceans during the winter season. As a result of the narrow continental shelf and the high wave energy experienced, the sandy shoreline is vulnerable to erosion, which can penetrate inland by 100 m in a single event. Observations appear to support changes in storminess as the winter swells have been increasing slowly, but they also reveal a change in direction that has the potential to realign the current position of the sandy shoreline (Corbella 2010). Any increase in extreme events is likely to result in significant damage to coastal infrastructure, particularly the tourism-related assets.

Climate change risks at the coast are compounded by inland and coastal land-use practices. The construction of dams to supply potable water in the region has had a significant negative effect on the supply of sediment to the coast. The sandy coastline relies on the continual input of river sediment to maintain the sandy beaches. The siltation of dams and the extraction of sand for building purposes have, however, reduced the original sand supply to one-third of the natural volume, thereby reducing the natural buffering capacity of the coast against storm events (Council for Scientific and Industrial Research 2008). Inappropriate land-use practices are evident along the coastline and are the result of historically poor and inappropriate planning. For example, the city's main wastewater treatment works, which deals with 60 million liters of effluent per day, is located on a sandy manmade terrace on the beach (as shown in Figure 22.2). Many buildings and beach facilities are also located too close to the high water mark and in time they will need to be either relocated inland or defended at a considerable cost. To assist in planning for sea-level rise, the municipality has undertaken a study based on three scenarios: sea-level rise of 300, 600, and 1000 mm. The first scenario of 300 mm was selected as the extrapolation of the current observed sea-level rise of 270 mm per year (Mather 2007; Mather et al. 2009) over this century. The second scenario of 600 mm was based on a doubling of the current rate of sea-level rise and corresponds closely with the IPCC (Pachauri et al. 2007) report maximum of 590 mm. The third scenario was based on accelerated ice melt, which was excluded from the IPCC 2007 report, and 1000 mm was selected. Higher scenarios for sea-level rise have been published, however, as this was a first assessment the higher levels and changes in wind direction and wave heights were excluded for now but will be undertaken in a future reassessment. All scenarios were rounded to the nearest 100 mm as this was the accuracy of the digital elevation maps available for mapping. These three scenarios were then mapped along the coastline (Figure 22.2).

This study has identified a number of vulnerable areas in terms of both built infrastructure and the provision of ecosystem services. For example, at a projected 1000 mm of sea-level rise between 30% and 50% of the coastal forest along the southern half of the coastline will be lost (Figure 22.3).

Given that the coastline is already vulnerable due to poor planning and existing climate variability, incremental climate change can only serve to amplify this vulnerability. In Durban, these physical vulnerabilities are exacerbated by social vulnerabilities such as high levels of unemployment and poverty. This dramatically raises the risk profile of the city and limits the opportunities to adapt to human-induced climate change. All the indications are that the municipality will need to strengthen its adaptation capacity at all levels if it is to limit the negative impacts of future climate change.

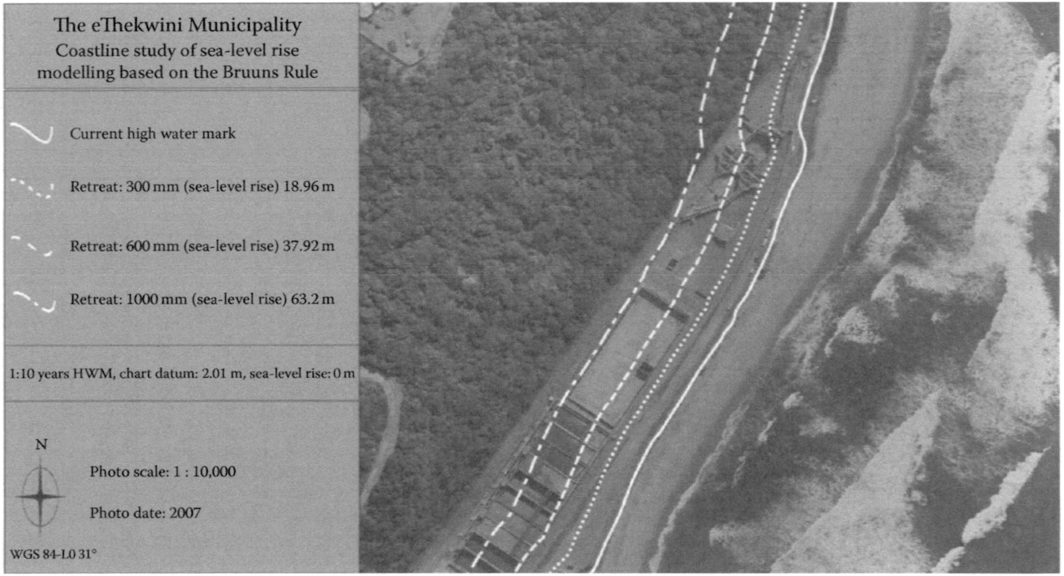

Figure 22.2 Sea-level rise scenarios showing the potential impacts on the Southern Waste Water Treatment Works using the Bruun's rule. (Data from and compiled by the eThekwini Municipality 2009.)

Figure 22.3 Sea-level rise scenario (1000 mm) showing the potential impacts on the natural coastal forest. (Data from and compiled by the eThekwini Municipality 2009.)

BARRIERS AND OPPORTUNITIES

Durban's climate protection program (MCPP) is distinct from many other similar initiatives around the world because of its strong and early focus on adaptation planning. It recognizes that there are limited mitigation interventions that Durban, and in fact most third world cities, can undertake and so the main focus has been to understand the future impacts and to proactively plan a range of adaptation interventions. In this process, there have been a number of important lessons learned in the development of the municipality's adaptation approach. As indicated above, the process was initiated with the development of the climate change adaptation strategy (HCCAS). Although this provided a useful way of engaging municipal sectors in a discussion around climate change impacts and possible responses, it ultimately stimulated no new adaptation actions (other than in the biodiversity sector). To address this problem, it was decided that a more successful approach would be to embed the adaptation planning process through the development of sector-specific adaptation plans that were more fully aligned with existing business plans, development objectives, and available funding and skills. Two pilot sectors, health and water, were initially selected to test this more focused approach. The adaptation needs of the component line functions were identified within each sector. In the case of the water sector, these included water and sanitation, coastal, stormwater and catchment management, and coastal policy. The key actions of the resulting municipal adaptation plan for the coastal sector are shown in Table 22.1.

Several challenges and opportunities emerged through this planning process and these are discussed subsequently.

Political will

Experience has shown that many politicians have a short-term focus, largely determined by the length of the electoral cycle (in Durban this is five years). This poses significant challenges to climate change adaptation interventions which require action now to address changes that may or may not happen at some unknown point in the future. This is further complicated in a city such as Durban where urgent and immediate development challenges exist and where resources are limited. As a result, the hard decisions regarding adaptation are often put off for fear of a loss of votes and not being reelected. In this regard, the leadership by the city's Mayor, drawing political attention to the significance of climate change and stressing the need for leadership to begin addressing this issue, has been a key factor in providing a positive environment in which to pursue municipal adaptation planning.

Technical integration

Given the multisectoral impact of climate change in Durban, effective adaptation planning could not be undertaken by a single department alone and also posed the risk of *turf wars* if key stakeholders were excluded. In order to engage sectors effectively, the lead department responsible for the climate change program faced the choice of either learning the *language* of each discipline or *converting* existing key staff in those departments, so that they became *adaptation champions*. The latter has proved to be the most successful as the process of institutional change is much easier to effect when it is internally motivated. Creation of sectoral adaptation champions has prevented the emergence of turf wars and has enabled the adaptation work to roll out quickly where these *converts* exist.

Working within existing constraints

One of the most significant challenges faced at the local level is the lack of adequate human and financial resources. In acknowledgment of this fact,

Table 22.1 Coastal municipal adaptation plan

Subcategory	*Impact*	*Intervention*	*Implementation plan (including policy framework for addressing issue)*	*Outcome*	*Priority*	*Responsible parties (1st listed = lead)*	*Alignment with IDP*
Infrastructure protection (new)	Sea-level rise	Revise coastal set back lines	Determine and demarcate the High Water Mark based on sea-level rise modeling and revise coastal set back lines accordingly	Coastal set back lines modified for climate change influence on sea level and storm surges	High	*Coastal Policy* Coastal stormwater and catchment management	Plan 1: Sustaining our Natural and Built Environment
Infrastructure protection (new)	Sea-level rise	Develop council policy and by-laws or scheme controls covering development within coastal set back lines	• Coastal policy officials to draft policy • Discuss potential bylaws or scheme controls • Report to be submitted to infrastructure committee and Exco and published for public comment	• Developments potentially at risk through a climate induced increase in sea levels and storm damage required to adhere to the requirements of the council Coastal Policy and become resilient to climate change • Will ensure the implications of climate change are placed at the forefront of spatial land-use planning considerations and will help ensure that planning proposal or development approvals complement the adaptation interventions listed in this plan	Medium	*Coastal Policy* Legal Dept	Plan 1: Sustaining our Natural and Built Environment

(*Continued*)

Table 22.1 (Continued) Coastal municipal adaptation plan

Subcategory	*Impact*	*Intervention*	*Implementation plan (including policy framework for addressing issue)*	*Outcome*	*Priority*	*Responsible parties (1st listed = lead)*	*Alignment with IDP*
Infrastructure protection (new and existing)	Sea-level rise	Prepare Coastal Management Plans for the entire Durban coastline	Identify and prioritize coastal areas at highest risk from storm damage and flooding using sea-level rise model. Prepare Shoreline and Estuary Management Plans for whole Durban coastline (focusing initially on Central Beachfront, Amanzimtoti, Bluff dunes, and Umdloti)	• Detailed understanding of the impacts of coastal storms and flooding on coastline • Mitigation of risk through prevention, reduction, and adaptation • Rehabilitation plans following storm events • Better management of estuaries	High	*Coastal Policy*	Plan 1: Sustaining our Natural and Built Environment
Infrastructure protection (existing)	Sea-level rise	Ensure assets management plans consider revised sea-level rise scenarios in assessment of the conditions of coastal assets	Develop asset management plans and program for replacement in order of priority	• Drainage infrastructure capable of managing increased run up • Prioritization across projects with regards to urgency and impact	Medium	*Coastal Policy*	Plan 3: Quality living environments

Infrastructure protection (existing)	Flooding and sea-level rise	Relocate informal settlements which are highly vulnerable to flooding and sea-level rise	• Housing dept to be provided with revised flood lines and coastal set back data • Housing dept to review priority informal settlements and low cost housing for relocation in light of revised flood and coastal set back lines • Review to be carried out every five years based on population growth and subsequent revisions to flood lines and coastal set back lines	• Priority relocation list more accurately reflects populations at risk • Reduced number of people living in the flood risk area	High	*Housing*, Coastal Stormwater and Catchment Management, Coastal Policy	Plan 1: Sustaining our Natural and Built Environment Plan 3: Quality living environments Plan 4: Safe, Health and Secure Environment
Infrastructure protection (existing)	Flooding and sea-level rise	Protection of municipal infrastructure (e.g., transport, storm water, sewerage, electric, etc.)	Identify key assets at risk following the development of: • Master Drainage Plans (W5) • Shoreline Management Plans (W8) • Asset Management Plans	Better understanding of highly vulnerable assets/ infrastructure • Reduced risk to infrastructure • Opportunities to implement new, efficient, low emissions technology • Simultaneous protection of private properties	High	*Coastal Stormwater and Catchment Management*, Coastal Policy	Plan 1: Sustaining our Natural and Built Environment Plan 3: Quality living environments

Source: Environmental Resources South Africa (ERM), 2009, http://www.erm.com/.

existing work streams and business plans in the pilot sectors were examined and staff encouraged to identify work streams that already had adaptation co-benefits. They were also encouraged to look for opportunities to change or expand these (without excessive cost or effort) to improve the adaptative outcome without necessarily having to label it as climate change adaptation. At this stage, some line functions realized that this type of approach was not that alien to them as all municipal practitioners take into account a spectrum of possible project risks and impacts, and by modifying this approach slightly to include climate change considerations, the twin goals of adaptation and good design could be achieved. This encouraged the affected line functions to look at the rest of their day-to-day activities in order to assess what additional adaptation measures could be mainstreamed into the design processes.

Funding

A key consideration in all changes to delivery systems is cost and determining who will pay. The real question, however, is *What is the cost of inaction?* In other words, can the city afford to lose the affected infrastructure and what are the broader social, environmental, and economic costs associated with such a failure? This question is often difficult to quantify given the diversity of viewpoints by officials, politicians, and the public and the differing methods of evaluation. However, if one adopts an asset management approach that assesses the costs and risks of maintenance and failure of the infrastructure over its lifespan; this methodology can provide an objective and balanced view in evaluating *costs*. Nevertheless, the question of who should bear the burden of payment for adaptation remains a larger question linked to the issue of who caused the climate change problem in the first place. The biggest difficulty appears to be trying to determine future adaptation costs of changes that are not well understood at a local level. The work undertaken so far by Durban has been to understand the likely extent of local impacts on the ground at a microcommunity level. The evaluation of adaptation measures is more meaningful and realistic based on an understanding of local impacts. A process is thus underway to commission a study that will examine the economic and financial costs of the sectoral adaptation plans in order that this information can be used to lobby for external funding to support local-level implementation.

Integrating climate change concerns into the development process

Incorporating climate change considerations into the planning and development process has been a difficult process as every architect or developer requires an explanation for the new requirements. This places additional burden on line function staff. Continued erosion and loss of the shoreline and river banks, for example, raises complex questions about how the local authority can provide services (particularly those which rely on gravity and which are most often placed on the lower side of the property). The obligation to provide services, for example, in a location where such services could become unviable due to loss of land along the coast or river, raises complex and uncertain legal issues that need resolution.

Human resources

A shortage of technical staff with capacity to implement climate change work has been a challenge even in a relatively well-capacitated municipality such as Durban. It also requires time and effort to understand climate change science and the possible impacts of climate change on a city. More often than not, staff are already overworked and involved in crisis management and, therefore, do not have the time available to develop this knowledge base. The shortage of technical staff means that most of the work is reactive and there is therefore little time for proactive and strategic planning.

The science of climate change

The largest challenge encountered revolved around the quantum of climate change likely to be experienced locally, that is, *What is the size of the expected change?* and *When is this going to happen?* One of the difficulties facing the city's engineers was determining what scenarios should be used, which of these need to be planned for, and what factor of change to include in their planning. It required time to develop an understanding that it was not possible to exactly predict the future and that achieving resilience would require the ability to deal with a range of possible futures. There was intense debate about the variations and uncertainly in sea-level rise predictions. Mindful that these two issues have the potential to derail any worthwhile progress on the project, it was decided to base the adaptation planning work on a reasonable evaluation of potential sea-level rise scenarios given the current state of the science. A choice of a figure that was significantly higher than IPCC projections for sea-level rise (e.g., 5 m) was likely to be dismissed as unbelievable. It is clear that this type of work will be iterative as more information about sea-level rise becomes available and, therefore, it was more important to have the stakeholders on board than to alienate them at the start of the process.

Mainstreaming

Early integration of the need for climate protection planning into the city's key strategic planning document, the Integrated Development Plan, enabled officials to introduce the issue of climate change into mainstream municipal planning. This has resulted in corresponding institutional change through the establishment of a dedicated climate protection branch within the municipality. This section will be able to drive and coordinate climate protection efforts.

LESSONS LEARNT

The most notable failure in the adaptation planning undertaken under the auspices of the MCPP was the lack of success of the HCCAS in generating sectoral action. Although no formal analysis of the failure of the strategy process was undertaken, past experience suggests a number of factors were probably critical, *inter alia*: the high-level and generic nature of the strategy; excessive existing workloads; urgent development challenges/pressures that result in issues perceived as less urgent being ignored; the perception of climate change as a distant and unlikely threat; and a shortage of skills and funds. This situation is exacerbated by the implicit (and often explicit) assumption that environmentally related issues such as climate change will be dealt with by the environmental department, so there is no need to engage with them in any depth.

A key success has been adopting a more sectoral (rather than a cross-sectoral or integrated) approach to the development of the municipal adaptation plans. This has allowed *champions* to emerge within line functions, facilitated the development of more realistic work plans, and (somewhat ironically) encouraged more cross-sectoral dialogue and engagement than emerged in the more inclusive and integrative HCCAS process. Nevertheless, converting plans into action remains an ongoing challenge.

A further shortfall of the MCPP process to date has been the limited involvement of nonmunicipal stakeholders in the adaptation planning process. In order to encourage a broader base of engagement, the climate protection branch is in the process of establishing a *Durban Climate Change Partnership* that will bring a broader range of stakeholders together on a regular basis to encourage communication and action in the area of climate protection.

An ongoing challenge in a municipality that has been effectively restructuring for over 16 years is the fact that changes and turnover in leadership and key personnel means that processes can be substantially delayed. This has resulted in slow progress being made in the adaptation planning process in two of the pilot sectors.

The use of a multicriteria analysis approach was found to be particularly useful in the prioritization of the broad range of climate change adaptation options that emerged for each of the pilot sectors. This also allowed a broad range of evaluating factors to be brought into consideration when identifying and short listing adaptation interventions.

THE WAY FORWARD

The work outlined here is still in progress and will be subject to review in terms of the Integrated Development Plan on yearly and five-yearly bases. Even at this stage in the project, however, three key lessons have emerged from the Durban experience:

Leadership

Good, directed leadership at a number of levels was a signature of this process. At a political level, this was achieved through the championing of the climate change debate by the Mayor and supported by strong technical input by officials in the fields of environment, engineering, and disaster management.

Technical excellence

The technical staff, consisting of a mix of academics, consultants, and in-house specialists, all of whom were recognized leaders in their respective fields, formed a multidisciplinary team that encouraged debate and action. Arising out of this partnership, new research into the local downscaled impacts of climate change led to the development of innovative new tools such as the freeware GIS viewer showing future changes in sea level as well as temperature and rainfall changes, biodiversity and food crop changes resulting from temperature and rainfall changes, and disease changes such as the increase in the distribution of malaria (Golder Associates 2010). This GIS viewer has been distributed to the public enabling them to visualize what future changes could bring, capacitating them and allowing them to provide input into adaptation options.

Communication

Early in the process, communication of these issues was limited by the capacity of the internal staff resources. Recognizing this, Durban has instituted a Durban Climate Change Partnership process with internal departments, nongovernmental organizations, the public, ratepayer bodies, and the Chamber of Business. This process of continuous engagement with stakeholders provided the platform to communicate issues of concern and to gain public trust in the process. Improving and providing the right type of information in a useable form is critical in helping communities to understand the risks posed by climate change. Cross-sector dialogue is also critical in ensuring that action in one sector does not have a negative effect on another sector—resulting in maladaptation.

REFERENCES

Barnett, K.A. (1999). "The management of Durban's beaches," Paper presented to the *Fifth International Conference on Coastal and Port Engineering in Developing Countries*, Cape Town, Republic of South Africa, April 19–23.

Corbella, S. (2010). *A Review of Durban's Wave Climate and Storm Induced Shoreline Changes*, Unpublished MSc thesis, University of KwaZulu-Natal, Durban, Republic of South Africa.

Council for Scientific and Industrial Research. (2008). *Sand Supply from Rivers within the eThekwini Jurisdiction, Implications for Coastal Sand Budgets and Resource Economics*, Stellenbosch, Republic of South Africa: CSIR.

Department of Water Affairs. (2009). *Green Drop Report (version 1): South African Waste Water Quality Management Performance*, Pretoria, Republic of South Africa: Department of Water Affairs, Republic of South Africa.

Environmental Resources South Africa (ERM). (2009). Climate Change Municipal Adaptation Plan: Health and Water, Unpublished report for eThekwini Municipality, 42pp. http://www.erm.com/ (accessed August 2013).

eThekwini Municipality. (2005). *State of the Environment Report 2004*, Durban, Republic of South Africa: eThekwini Municipality, Environmental Management Department.

eThekwini Municipality. (2010a). *Integrated Development Plan*, Durban, Republic of South Africa: eThekwini Municipality. http://www.durban.gov.za/ (accessed November 2011).

eThekwini Municipality. (2010a). *State of Biodiversity Report 2008/2009*, Durban, Republic of South Africa: eThekwini Municipality, Environmental Planning and Climate Protection Department.

Forbes, A.T. and Demetriades, N.T. (2008). *Estuaries of Durban (second edition)*, Durban, Republic of South Africa: eThekwini Municipality.

Golder Associates. (2010). *eThekwini Integrated Assessment Tool for Climate Change*, Prepared for eThekwini Municipality, Durban, Republic of South Africa: Golder and Associates.

GroundTruth Consulting. (2007). *eThekwini Municipality State of Rivers Report*, Unpublished reported prepared for eThekwini Municipality, Durban, Republic of South Africa: GroundTruth Consulting.

KwaZulu-Natal Tourism Authority. (2009). *Statistics of our Tourism Industry*, Durban, Republic of South Africa: KwaZulu-Natal Tourism Authority.

Mather, A.A. (2007). "Linear and nonlinear sea level changes at Durban, Republic of South Africa," *South African Journal of Science*, 103(11/12): 509–12.

Mather, A.A. (2008). "Coastal erosion and sea-level rise: Are municipalities prepared?," *Journal of the Institute of Municipal Engineering of Southern Africa*, March 2008: 49–70.

Mather, A.A., Kasserchun, R.K., and Wenlock, H.G. (2003). "City of Durban sand bypass scheme: 20 year performance evaluation," Paper presented to the *Sixth International Conference on Coastal and Port Engineering in Developing Countries*, Colombo, Sri Lanka, September 15–19.

Mather, A.A., Stretch, D.D., and Garland, G.G. (2009). "Southern African sea level: Corrections, influences and trends," *African Journal of Marine Science*, 31(2), 145–56.

Pachauri, R.K. and Reisinger, A. (eds.); IPCC. (2007). *Climate Change 2007: Synthesis Report. Contribution of Working Groups I, II and III to the Fourth Assessment Report of the Intergovernmental Panel on Climate Change*, Geneva, Switzerland: IPCC. http://www.ipcc.ch/ (accessed August 2013).

Schulze, R.E., Knoesen, D.M., Kunz, R.P., and van Niekerk, L.M. (2010). *Impacts of Projected Climate Change on Design Rainfall and Streamflows in the eThekwini Metro Area*, Report prepared for the eThekwini Municipality, Pietermaritzburg, Republic of South Africa: University of KwaZulu-Natal.

Part IX

Conclusion and practical steps for adapting to climate change

Chapter 23

Toward reflexive adaptation and resilient coastal communities

Bruce C. Glavovic, P. Mick Kelly, Robert Kay, and Ailbhe Travers

Abstract: In this chapter, we summarize the main lessons to emerge from the case studies presented in this book. A number of common themes have been highlighted as important characteristics of an effective response to climate change that proactively builds adaptive capacity and resilience and fosters sustainable coastal development. On the evidence of these case studies, we conclude that adaptation at the coast would benefit from being responsive, deliberative, transformative, holistic, integrative, inclusive, equitable, and empowering. We develop these characteristics into a conceptual framework that we term *reflexive adaptation* that, we believe, will help communities better understand and address distinctive features of the challenge presented by climate change. A set of priority actions, also drawn from the case studies, is outlined to help guide coastal communities translate this concept of reflexive adaptation into practical reality. A final, and particularly encouraging, conclusion that can be drawn from the case studies is that the process of social learning required to adapt to a changing climate is well underway. The challenge is to convert adaptation barriers into enablers of change. This is a monumental task, but, as the case studies show, momentum is growing to chart new adaptation pathways.

INTRODUCTION

As clearly demonstrated in the preceding case studies, climate change presents a challenge to society that is without precedent. It is wide ranging in geographical scale and impacts current and future generations. It is insidious, developing progressively but unpredictably on decadal and longer timescales that create difficulties for conventional planning and political processes. It is complex, because of its multidimensional and evolving character and nonlinear interactions and feedback, with impacts that are differentially distributed but manifest in specific localities. Climate change compounds preexisting coastal risks and exacerbates prevailing patterns of inequitable and unsustainable coastal development. Managing these multifaceted risks is vexed due to incomplete understanding of the physical and social processes that are involved and various legitimate but contending views about appropriate courses of action. The development of an effective adaptive response that fosters community resilience and sustainability must take full account of these characteristics. Failure to do so can lead, at best, to imperfect solutions and, in the worst case, maladaptation, aggravating rather than ameliorating the already dire situation in many coastal localities. Coastal communities are on the frontline of the global sustainability struggle and will chart the course for humanity as we face the challenges and opportunities of the Anthropocene.

In this chapter, we outline the lessons that we have drawn from the case studies in this book. A number of common themes have been highlighted in the case studies that could prove key components or characteristics of an effective response to climate change that would proactively build adaptive capacity and resilience. In reviewing the case studies, a conceptual framework emerged founded on lessons from the case studies and drawing on previous academic work as well as our practical experience. We believe that this framework is well suited to addressing the distinct features of the climate change *problématique*. Other conceptualizations of adaptation do, of course, exist and some will, no doubt, be more valuable in specific circumstances. The *reflexive adaptation* framework presented here does, however, provide an effective summary of our learning during the course of producing this book and locates the adaptation endeavor in the context of resilience and sustainability thinking and current theories of societal development more broadly.

COASTAL ADAPTATION IN THE TWENTY-FIRST CENTURY

Over the past two decades, adaptation thinking and practice has evolved considerably. It has moved from a linear, science-driven process intended to predict future climate change, resultant impacts, and subsequent adaptation solutions, toward more nuanced approaches (see Chapter 3). There has been a growing recognition of the benefits of *soft protection* and of *retreat* and *accommodate strategies*, an increasing reliance on technologies to develop and manage information and an enhanced awareness of the need for coastal adaptation to reflect local natural and socioeconomic conditions (Klein et al. 2001; see also EPA 2009; USAID 2009; UNEP 2010). Traditional, linear models of decision-making, whereby the goal is clearly, rationally, and unambiguously defined, assume that the effects of any action can be assessed based on firm predictive knowledge and the power to control implementation. Such models have, however, proven ineffective in the face of the complexity and uncertainty that characterize many sustainability problems (Voss et al. 2007). In their place, more adaptive and collaborative approaches are being adopted. To take just one example, rather than focusing solely on trying to determine exactly what climate change might take place, it has proved constructive to develop a range of scenarios of climate change so that attention can be focused on understanding the implications of different possible futures. In this way, decision-makers and stakeholders from the community to national government level can begin to think through their vulnerability and identify a range of adaptation pathways or adaptation options that are available under different circumstances.

A key conclusion of the 2007 IPCC assessment of adaptation in the coastal zone was that efforts to address climate-related risks are not effective if they are reactive and standalone; they need to be a mainstreamed into integrated coastal management (ICM) efforts. ICM has the potential to overcome a number of barriers to effective adaptation in the coastal zone because

- It is a cross-sectoral approach
- Enhancing adaptive capacity is an important aspect
- It locates the climate response in the broader context of and wider objectives of coastal planning and management
- It focuses on integrating and balancing multiple objectives in the planning process
- Generation of equitably distributed social and environmental benefits is a key factor
- Due attention is paid to legal and institutional frameworks that support integrative planning on local and national scales (Nicholls et al. 2007)

Finally, ICM is, in theory if not always in practice, an inclusive process and this is critical in ensuring sustainable adaptation (see Chapters 1 and 3). The need to integrate adaptation endeavors into coastal governance thinking and practice is now widely recognized (see Chapter 3; Box 23.1).

Notwithstanding the potential of ICM as a basis for coastal adaptation, climate change introduces important dimensions that deserve more focused attention: notably, greater recognition of and attention to uncertainty in coastal planning and decision-making; a longer planning horizon; and the need to address adaptation and mitigation opportunities in an integrated manner (Tobey et al. 2010; USAID 2009). A number of impediments need to be addressed. Nicholls et al. (2007) cite a lack of data, information and understanding,* divergent information management systems, insufficient or inappropriate shoreline protection measures, fragmented and ineffective

* For example, data, information, and understanding is limited with regard to dynamic predictions of landform migration, the response of coastal systems to climate drivers and adaptation initiatives, key indicators and thresholds relevant to coastal managers, and knowledge of coastal conditions and appropriate management measures (Nicholls et al. 2007).

BOX 23.1 MAINSTREAMING ADAPTATION INTO COASTAL MANAGEMENT

USAID (2009) presents a five-stage coastal adaptation roadmap based on the ICM management cycle: (1) assess vulnerability; (2) select course of action; (3) mainstream coastal adaptation; (4) implement adaptation; and (5) evaluate for adaptive management. Tobey et al. (2010), based on USAID (2009), identify best practices and lessons transferable to the challenge of adapting to climate change in the coastal zone that have emerged from experience in developing ICM.

First, the best available knowledge should be used to address the complexity and uncertainty of climate change. Expert and local knowledge needs to be harnessed to understand the range of possible climate change scenarios so that stakeholders can evaluate the implications of the different scenarios and choose appropriate courses of action within broader economic and social development pathways rather than being preoccupied in trying to discern precisely how climate will change. A precautionary approach is appropriate in the face of uncertainty.

Second, it is important to strategically define priorities, goals, and objectives for adaptation. Given the complexity and scope of climate change, adaptation efforts are more likely to be successful if they focus on a limited set of climate risks and adaptation issues that resonate with key interested and affected parties, including key government agencies at relevant levels, community-based and private sector organizations, user groups, university and other research organizations, and the general public. Clearly defined, broadly agreed goals and unambiguous, time-bound objectives for coastal adaptation need to be specified.

Third, coastal adaptation should be tailored to address local conditions. Coastal adaptation efforts, like ICM more generally, need to address the circumstances that shape local reality (environmental, cultural, social, economic, technical, and political) and be framed by prevailing institutional capacity. Key aspects of such capacity include leadership, government resources and capacity, levels of awareness and public support, and commitment to shared goals and objectives. Responses to climate change will necessarily vary from place to place because of different circumstances. Coastal adaptation measures seldom involve one-off or *silver bullet* actions. Rather, a combination of measures needs to be developed to facilitate sequenced and complementary actions.

Fourth, it is necessary to develop reinforcing linkages among the many possible adaptation entry points that exist in policy arenas, sectoral institutions,

(*Continued*)

and place-based actions. To overcome fragmented decision-making, ICM seeks to coordinate actions within and between different levels of government and sectors across interconnected spatial settings. An enabling national policy mandate is needed to secure the necessary mandate and funding to undertake locality-specific place-based ICM. In the same way, climate adaptation needs to be mainstreamed into relevant policies, plans, and programs from the national to local level.

Fifth, an inclusive, participatory and ongoing process is essential. The ICM approach involves an inclusive, strategic, and adaptive process that can be used to secure commitment and funding for coastal adaptation, assess climate change risks and vulnerability, plan and select a preferred course of adaptation that can be implemented, evaluated, and adjusted over time. Crucially important, coastal stakeholders should be meaningfully engaged in the planning and decision-making process to ensure salience and ownership of adaptation pathways. Constituency support is vital for successful coastal adaptation.

Finally, ongoing monitoring, evaluation, and adaptation are necessary. A strategic, learning-based management approach is key to ICM. Adaptation measures will need to be monitored and evaluated to discern efficacy and where appropriate adapted to meet expectations.

institutional arrangements, weak governance, and societal resistance to change; and, as Kay (2012) points out, political decision-making biases must also be taken into account.

Broader challenges must be met if sustainable adaptation is to be secured (Adger et al. 2007; Chapter 3). There may be physical and ecological limits to adaptation. The rate and scale of climate change may surpass critical thresholds beyond which ecological or human systems may no longer be viable. There may be technological limits. Although there is considerable potential for the development of adaptation technology, uncertainties may inhibit the process of development and adoption. Technologies may not be economically feasible or may be socially undesirable. They may prove to be location-specific and not widely transferable. Financial barriers are inevitable. According to the World Bank (2006), the overall cost of *climate proofing* development could be as high as USD 40 billion a year. There are informational and cognitive barriers. Perceptions of the risks associated with climate change, and thus the need for timely adaptation, vary and this will affect decisions regarding adaptation priorities. Finally, social and cultural barriers exist, resulting from the different ways in which people and groups experience, interpret, and respond to the climate change.

Tolerance to risk varies across and within communities, as will preferences for particular adaptive options.

Chapter 3 underscores how formidable these barriers are. Adaptation plans and actions have mushroomed in recent years, yet effective implementation is nascent at best. Five decades of ICM experience shows surprisingly disappointing progress in establishing the enabling conditions for sustained implementation, let alone transitioning beyond this point of departure. Climate risk at the coast is escalating, compounding pervasive unsustainable coastal practices and jeopardizing coastal livelihoods and lifestyles. Yet, despite long-standing efforts to understand and address risk, traditional approaches paradoxically tend to exacerbate risk problems. Coastal communities face *waves of adversity* and need to build *layers of resilience* but face a governance impasse. Coastal issues are locked into a vortex of conflict with problems displaced from one social choice arena to another without securing outcomes that foster community resilience and sustainability.

In Chapter 3, the review of scholarship on adaptation, ICM and coastal governance, and risk governance, buttressed by governance scholarship more generally, reveals a clear and compelling answer to the question: *How might the rationality that "locks" in path-dependent unsustainable and maladaptive practices be transmogrified and institutions reformed and, if necessary, transformed to overcome adaptation barriers and foster adaptive capacity and resilience?* A key ingredient to breaking this impasse, and addressing the *super-wicked* problem of climate change at the coast, is to create opportunities for reflexive deliberation by stakeholders and communities. Such a process needs to be inclusive, authentic, and consequential. It needs to be founded on respect for and reconciliation of the divergent ethical, emotional, spiritual, and relational perspectives on and connections to the coast, and the different kinds of knowledge(s), understanding(s), and underlying norms that frame the moral and practical justification of alternative adaptation pathways. Adaptation governance processes need to be constructed that enable coastal communities to understand and manage climate risk at the coast, map out adaptation pathways, and make strategic choices that build layers of resilience in the context of intractable conflict, complexity, structural uncertainty, ambiguity, and surprise. The case studies in this book provide valuable insights into the key features of a reflexive approach that enables translation of adaptation rhetoric into practical reality.

A REFLEXIVE APPROACH TO ADAPTATION BY COASTAL COMMUNITIES

These case studies have revealed a diversity of approaches to climate change adaptation in the coastal zone, reflecting a wide variety of geographical, institutional, and socioeconomic contexts. All represent work in progress

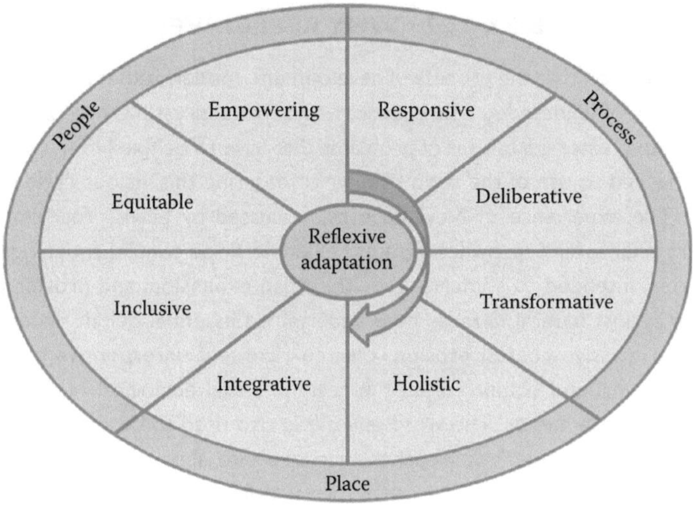

Figure 23.1 Characteristics of reflexive adaptation.

and it would be unwise to judge degrees of success at this stage. Nevertheless, important lessons have been highlighted, many of which are of broad applicability. A number of common themes have emerged that indicate the opportunities, barriers, and on-ground realities in progressing adaptation at the coast. These themes are presented in Figure 23.1 and concern three key dimensions: *process*, the means by which adaptive measures are designed and delivered; *place*, as adaptation is context-specific; and *people* whose potential must be released for adaptation to be successful. Coastal communities are a complex combination of all three dimensions—social-ecological systems in which people make social and livelihood choices that shape the places they live.

As convenient shorthand, we term the conceptual framework that this assembly of themes or characteristics suggests reflexive adaptation. The word *reflexive* is used to capture the notion of critical self-reflection and self-correction in the face of adversity, uncertainty, surprise, tension, and conflict. Whereas a physical reflex is an involuntary response to an external stimulus, the process of reflexive adaptation is the capacity to consciously reflect on prevailing circumstances, contemplate future prospects, including the prospect of shocks and surprise, and deliberately choose pathways that are attuned to changing circumstances.

Box 23.2 outlines the wider use of the term *reflexive* in the context of theories of societal development. In the following sections, we discuss the main themes or characteristics that have emerged from the preceding case studies, citing examples from particular chapters.

BOX 23.2 WHY REFLEXIVE?

The recognition that the process of development, modernization, has resulted in a variety of unintended consequences, and that addressing these problems often then creates a fresh set of problems that have to be fixed in an on-going cycle, has led to use of the term *reflexive* to describe this vicious cycle (Beck 1994). The experience of New Orleans, discussed by Lewis, Yoachim, and Meffert in Chapter 6, provides a graphic example: flood control and navigation problems, intended to stimulate growth, urban expansion, and provide protection against natural hazards have undermined essential deltaic processes, triggered massive coastal erosion, created rigid and unresponsive management systems, and compromised the region's social-ecological resilience to hurricane storm surges. The word *reflexive* is also used to describe attempts to develop management approaches that take into account the inevitability of unforeseen side effects and other forms of failure. An important aspect of what is termed *reflexive modernization* (Beck 1994; Beck et al. 2003) is a reassessment of conventional means of problem-solving and the development of management approaches that are, for example, iterative, interdisciplinary, cross-sectoral, and learning-oriented, taking into account the fact that, particularly when dealing with complex systems in the absence of complete knowledge, unintended consequences cannot be avoided.

Voss and Kemp (2005), in discussing reflexive governance, identify three key characteristics of sustainability problems that have to be taken into account when considering how society might approach the challenge of sustainable development. The first characteristic is complexity. Complexity suggests that the traditional, reductionist approach to problem-solving, based on specialized, disciplinary expertise, should be integrated with a broader kind of knowledge derived, for example, from the practical experience of actors outside the science system. To develop this form of integrated knowledge production will mean transcending boundaries both within science and between science and society in "interactive settings in which knowledge is coproduced by scientists and actors from respective fields of societal practice" (Voss and Kemp 2005: 10). The second characteristic of sustainability problems in general, of which climate change is a prime example, is uncertainty. Uncertainty leads to a strategic requirement of, what Voss and Kemp term *adaptivity*. The capacity to respond to the unexpected must be developed, as must the ability to learn from experience and correct errors through adaptive management (an iterative process of decision-making based on learning through system monitoring

(Continued)

and evaluation). Experimentation, monitoring, and evaluation should be key elements of the response to climate change. The third characteristic is path-dependence: future developments will be influenced, enabled, and constrained by what has gone before. Decisions taken early on may limit, or extend, what is possible at a later date; different pathways may lead to different destinations. Path-dependence means that foresight is required in anticipating and assessing the long-term implications of the range of alternative pathways available in any situation.

Voss and Kemp (2005) also consider how sustainability problems might be addressed in a world of reflexive modernization where the traditional linear, hierarchical approach to decision-making no longer holds. That adaptation pathways may lead to different destinations raises the critical issue of how adaptation goals are selected. Voss and Kemp argue that it is inevitable that goals will be contested and that social conflict is an unavoidable aspect of addressing any sustainability problem. They emphasize the importance of participatory processes that involve all affected actors, the fourth strategic element of reflexive governance. Relatedly, given the broad distribution of control in modern society, they consider that effective implementation will depend, not on institutionalized hierarchies as is the tradition, but on interactive strategic development, the fifth strategic element, networks in which the perceptions, interests, and knowledge of stakeholders are linked together. Finally, the need for congruence between the decision-making space and the problem space is cited by Voss and Kemp as the sixth strategic element of reflexive governance. The importance attached to community-based adaptation (Reid and Huq 2007), and as shown by many case studies in this book, including the Arctic (Chapter 4), India (Chapter 7), Bangladesh (Chapter 8), Vietnam (Chapter 9), and Mozambique (Chapter 21), is an example of the recognition of congruence as an important aspect of society's response to climate change.

As a point of departure, as explained in Chapter 1, it is important to recognize that adaptation takes many forms, including autonomous and planned adaptation. The latter is the primary focus here but does not imply a *blueprint* that is implemented in a top-down, *one-size-fits-all* fashion. Rather, the case studies in this book have underscored the importance of *progressive* planned adaptation, that is, advancing with foresight and a strategic view. Planned adaptation needs to be staged, nested, or step-by-step (see Box 23.1). It needs to cater for changing circumstances, including the potential for transformation where a change of state is inevitable or necessary, acknowledging that climate change represents a fundamental challenge to societal systems, as well as significant opportunities for advancement. For example, Nagy, Gómez-Erache, and Kay, describing the Uruguayan case study in Chapter 16, conclude that a step-by-step planning and implementation process of affordable, easy, cost-effective, and accepted measures can pave the way for greater change. Reisinger and his coauthors, discussing prospects for New Zealand in Chapter 13, consider the strategy of retreat as a long-term option in the context of present-day planning based on a static perspective of *hold the line*. They compare the potential for retreat in two contrasting areas, highlighting the importance of strategic and consistent local authority decisions, community support, and a commitment to implementation decisions over periods of decades, echoing conclusions reached by Turner and Luisetti, in Chapter 18, in their study of adaptation options along the eastern coastline of England. The merits of a progressive planned adaptation approach to address the perennial challenge of reconciling immediate and longer-term development benefits in the face of climate change is illustrated in the comparative island nation study of Elrick-Barr, Glavovic, and Kay, in Chapter 14.

The first set of reflexive adaptation characteristics (Figure 23.1) centers on the *process* dimension.

Responsive

A responsive process is tailored to the distinctive needs and circumstances of particular coastal communities and to the evolution of these needs and circumstances over time. It is first and foremost responsive to *people* and *places*. With respect to the former, in contrast to being prescriptive, as in a blueprint, responsive processes are elicitive, drawing out the views, preferences, knowledge, experiences, feelings, and hopes of different coastal stakeholders and communities. With respect to the latter, responsive processes are context specific, attuned to the distinctive characteristics of particular coastal localities. Responsive thus means taking full consideration of people and places as social-ecological systems in adaptation planning and implementation. Not only are the impacts of climate change variable with location, but the most effective forms of adaptation are dependent on local circumstances, not only geography but also social and economic considerations (including the availability of human, technical, and financial resources). Although much can be learnt from experience in other areas, adaptation must, therefore, take account of the local context. The importance of context has been drawn out in a number of case studies in this book. Armitage, for example, in Chapter 4, describes the *triple squeeze* faced by the inhabitants of the Canadian Arctic as global environmental change, economic globalization, and cultural transition combine. In Chapter 14, Elrick-Barr, Glavovic, and Kay, in a meta-analysis of potential adaptation pathways in Kiribati and the Maldives, demonstrate how geography, history, local culture, and external funding priorities are shaping adaptive options. Similarly, in Chapter 10, the comparative analysis of pathway possibilities in southeast Queensland, Australia, and the Yangtze delta region, People's Republic of China, by Choy and colleagues, reveals the necessity to chart pathways that are responsive to cultures and institutions, even when the same fundamental challenge (i.e., urban agglomeration) is being faced and the primary adaptation planning tool is the same (i.e., regional strategic planning).

Responsive processes are also able to react effectively and constructively to changing circumstances in social-ecological systems, lessons learnt from experience (both success and failure), and evolving understanding. The adaptive process must, therefore, be based on intentional monitoring and evaluation and incorporate flexibility, for example, through staged or nested implementation, and redundancy (e.g., back-up systems for critical infrastructure) without sacrificing robustness. It should be learning-oriented, innovative, at times consciously exploratory and experimental in nature but always reflective. It is not responsive in a simple *knee-jerk* fashion. Responsiveness enables tailor-made decisions and actions that can be modified over time in the face of evolving understanding and changing circumstances. It is contingent and iterative. In Chapter 5, Solecki and his coauthors, discussing prospects for New York City, emphasize the importance of regular monitoring and reassessment of climate indicators and measures of climate-related coastal impacts in sustaining adaptation strategies. They also recommend monitoring of advances in scientific understanding, technology, and adaptation strategies. They highlight the distinctive challenges and tailor-made responses required in coastal megacities where there is high dependency on a massive network of at-risk critical infrastructure. Turner and Luisetti, in Chapter 18, illustrate the complex evolution of stakeholder perceptions in response to a policy of managed realignment, providing a cogent example of reflexivity in action and demonstrating the importance of ongoing consideration of the social consequences of adaptation measures. Many case studies highlight the importance of seizing *windows of opportunity* for political action and institutional change that may arise (e.g., Chapter 4) especially in the aftermath of extreme events, as indicated in the New York City (Chapter 5) post-Katrina coastal Louisiana (Chapter 6), Vietnam (Chapter 9), and Caribbean (Chapter 16) cases. These experiences underscore the need to keep options open to be able to respond flexibly in the face of uncertainty, rapid change, and surprise. Such changes might be driven by climate change impacts or by the emergence of new coastal activities, as discussed by Kannen and Ratter (Chapter 19) in their study of the role that maritime spatial planning can play in enabling reflexive adaptation to emerging coastal–marine activities such as offshore wind farms in the already intensively used North Sea. In Chapter 20, Travers and her coauthors similarly point out that adaptation responses should be attuned to the distinctive needs and conditions of different Mediterranean countries but point out the importance of coordinating regional responses.

Deliberative

Deliberation is a noncoercive process of information sharing, reflection, dialogue, and negotiation over societal values, preferences, and opinions; and it is foundational for coastal governance and adapting to climate change (see Chapter 3). Deliberation takes place in a range of formal and informal public settings—consider the diversity of such settings in the case studies described in this book, ranging from the Arctic to New York City, Tamil Nadu, India, the Red River delta, Vietnam, and Durban, Republic of South Africa. Always contested, and invariably requiring significant investments of time and resources, such processes can, nonetheless, build the human, social, and political capital necessary to better understand and address vexed adaptation challenges. Deliberation can help to overcome the limits of traditional science to understand wicked problems as well as the shortcomings of adversarial uses of knowledge to win *adaptation battles*. It can help to strengthen community spirit, improve problem-solving and public decision-making, resolve conflict, and foster collaboration. It requires inclusive, meaningful, and consequential participation, not only so that stakeholder views can be taken into account but also to ensure that the broadest range of knowledge and understanding is accessed and reflected upon so that adaptive measures can be negotiated to secure community approval and support. Deliberation bolsters community awareness raising and social learning and is thus vital for reducing climate risk and charting adaptive pathways.

Armitage, in Chapter 4, describes how sharing knowledge, social learning, and conflict resolution in the Canadian Arctic has been facilitated by the development of new governance arrangements, such as multilevel comanagement. Bridging organizations have played an important role in linking indigenous knowledge and conventional science in shared processes of learning, though emphasis needs to switch from *integration* of these forms of understanding to *coproduction*. Community-based research has revealed a significant historic adaptive capacity on the part of coastal communities that should frame efforts to develop future adaptive strategies. In Chapter 15, Chambers and Roberts-Hodge discuss the role of climate education in ensuring citizen engagement in the adaptation process in the Caribbean. Turner and Luisetti, in Chapter 18, consider the mesh of issues emerging from the transition from *holding the line* to an approach based on risk management on the east coast of England. Managed realignment raises deep-seated questions about governance reform, compensation, social justice, and equity that require public deliberation. Public participation and opportunities for deliberation were key to the recovery planning process in post-Katrina Louisiana and are central to building resilience in this climate change *hot spot* (Chapter 6). Such processes are not merely *talk fests.* They can range from grassroots activities, such as the participatory irrigation management process in the India case study (Chapter 7), to multi-jurisdictional stakeholder–scientist processes, as in the New York City case (Chapter 5), that employed state-of-the art scientific projections and risk mapping as an important part of the deliberative process. The case studies presented in this book reveal that participatory processes are time consuming and require persistence, patience, and prudence, with local champions playing crucial roles.

In Chapter 9, Kelly's study of adaptation challenges and prospects in the Red River delta of Vietnam reveals the inadequacy of top-down technocentric planning and decision-making, and the need to draw upon and reinvigorate local customs and decision-making processes that have dealt effectively with historic adversity. Smith and colleagues, in their account of adaptation challenges and opportunities in southeast Queensland, Australia (Chapter 12), argue that participatory and collaborative processes are key to social learning in the face of uncertainty, and diverse needs, interests and aspirations and the imperative to identify new pathways to adapt to a changing climate. Constructive public deliberation is especially important for developing a shared understanding of the implications of alternative pathways for different coastal stakeholders, resolving conflicting interests and choosing pathways that are adaptive to changing circumstances (Chapter 13). Such deliberative processes need to be embedded in coastal governance regimes to enable communities to resolve contested issues on the frontline of the sustainability challenge in the Anthropocene.

Transformative

Transformative means supporting the potential for fundamental change, acknowledging that climate change represents a substantive challenge to coastal social-ecological systems as well as significant opportunities for advancement. Chapter 3 and the case studies in this book show that the dominant coastal development trajectory is unsustainable and that innovative development strategies and fundamental institutional reforms are needed to secure social equity and ecological and livelihood sustainability. From a resilience perspective, adaptation pathways range from persistence to transformation, and these pathways need to be adjusted to changing circumstances.

In practice, climate change is already leading to transformative change in social-ecological systems in climate-sensitive localities. The Arctic; low-lying islands, such as the Maldives and Kiribati; deltas such as the Mississippi and Red River; and communities reliant on coral reef ecosystems, among others, provide graphic examples. Biophysical change in the Arctic is unprecedented in modern human history and transformative changes to the institutional structures and processes that govern social choices in this region are needed (Chapter 4). Extreme events, such as Hurricane Sandy, can transform attitudes to climate change and stimulate unprecedented political action and policy changes, a compelling necessity given that the critical infrastructure that supports New York City needs to be *future proofed* (Chapter 5). The potential for such a sea change in thinking and action is writ large in the Louisiana narrative—a region devastated by disasters and undergoing transformation in part because of climate change (Chapter 6). Lewis and his coauthors poignantly ask "will transformation lead to change that is positive—driven by foresighted leadership or business as usual based on myopic crisis management?" Climate change will transform local reality, and, as explained in the Indian case study (Chapter 7), necessitates a shift in mind-sets from dependence on government to greater local autonomy, a theme reiterated in the Bangladesh case wherein the need to transform vulnerability into resilience was emphasized. However, as highlighted by Kelly in the Vietnam example (Chapter 9), rapid transformation due to a combination of socioeconomic, institutional, and environmental change can ratchet up stress levels not readily resolvable by reliance on top-down, technocratic modalities of governance. Ultimately, transformational adaptive governance processes are needed to overcome poverty and inequity, as shown in the Mozambique (Chapter 21) and Durban (Chapter 22) case studies. The reality of transformation is perhaps most obvious and compelling in the case of communities and even whole nations, such as the Maldives and Kiribati (Chapter 14), that are extremely low-lying and likely to be significantly impacted by projected sea-level rise and other climate change impacts.

The next two sections explore the *place* dimension, with a focus on holism and integration.

Holistic

Holism recognizes that the whole is greater than the sum of the parts. Coastal social-ecological systems cannot be fully understood solely on the basis of analysis of their constituent parts. Although reductionism is a valuable research tool for understanding component parts, a system-wide perspective is necessary to understand the coastal zone. Moreover, given the complexity of climate change and its insidious nature, both in terms of biophysical and societal effects, a holistic approach is essential for charting adaptation pathways (see Chapter 3). This requirement can present a challenge to conventional approaches to problem-solving that are often focused on sectoral and localized issues and reliant on traditional science that is reductionist in nature. Many case studies in this book demonstrate the need to foster holism, recognizing and taking into account synergistic interconnections across time, space, and institutions. The need for holism is clearly shown, for example, in efforts to build adaptive capacity and resilience in the Canadian Arctic (Chapter 4) and the Mississippi (Chapter 6) and Red River (Chapter 9) deltas. Coastal livelihoods are woven into the histories, cultures, and coastal environments of these settings. Navigating waves of adversity compounded by climate change necessitates holistic thinking and actions that take into account the synergistic interactions that together create the distinctive character of these places. Failure to do so can lead to unintended perverse consequences, as clearly illustrated by the Mississippi River Gulf Outlet in coastal Louisiana (Chapter 6). This example reiterates the need to adopt holistic approaches to adaptation planning that take into account a wide range of interconnected social, political, economic, technical, and environmental considerations. In Chapter 10, Low Choy, Chen, and Serrao-Newmann discuss the role that regional strategic planning can play in providing a suitable context for adaptive governance in the context of two urbanizing regions in Australia and the People's Republic of China. They highlight the importance of understanding regional adaptive capacity, and recognizing the distinctive challenges and opportunities that pervade different cultural and institutional settings.

Integrative

Integrative means capable of coordination across space and within society. Sectoral barriers and compartmentalization of institutional structures and processes mitigate against effective planning and implementation. Both horizontal and vertical integration are, therefore, necessary to enable coordinated action. An integrative approach recognizes the interrelationships between component parts of coastal social-ecological systems and the need to bring together disparate elements into a coherent whole through institutional coordination, inter- or trans-disciplinary analysis and praxis, and assimilation of science and local and traditional knowledge. The challenge of enabling coordination between sectors and strata of society is explored in many case studies.

Kannen and Ratter, in Chapter 19, propose that maritime spatial planning, deployed in the context of offshore wind farming in the North Sea region, can provide an integrative model for climate adaptation governance. In Chapter 16, Nagy, Gómez-Erache, and Kay consider building climate considerations into existing coastal governance arrangements in Uruguay, specifically, ICZM frameworks. Participatory approaches are being used to embed climate change into key institutions. Along with other case studies, they underline the importance of local *champions*. In the Uruguay example, local champions proved significant in creating bridges between local communities, academia, and national agencies. Mather and

Roberts, in Chapter 22, discussing the experience of adaptation planning in Durban, find that local champions have raised awareness and facilitated internal motivation within institutions, facilitating the process of change. In this case, a sectoral approach has been beneficial in the early stages of planning; it minimized *turf wars* between departments and, ultimately, promoted cross-sectoral dialogue and engagement. Ramesh and his coauthors, in Chapter 7, describing an approach to irrigation management in a coastal area of India, also stress the importance of local champions. As mentioned above, Armitage, in Chapter 4, highlights the importance of bridging organizations as a forum for integrating different domains of practice across scales, sectors, and institutions by enabling trust-building, sense-making, social learning, and strengthening conflict resolution capacity. In a markedly different setting, New York City, Solecki and his colleagues (Chapter 5) demonstrate the necessity to develop mixed integrative strategies that include the use of structural protection and soft engineering measures as well as policy and planning provisions, and building regulations that recognize linkages across scales and that are sequenced and able to be adjusted over time.

More generally, climate change considerations need to be mainstreamed into policies, plans, and practice, with many case studies reiterating the increasingly recognized need to integrate climate change adaptation into coastal governance (see Chapter 3; Box 23.1). In Chapter 9, for example, Kelly describes efforts underway in Vietnam to develop and implement climate change and coastal zone management policy, noting that horizontal and vertical integration is key to progressing adaptive governance. This finding is reiterated in many other cases from so-called developed (e.g., Canadian Arctic, the United States, the United Kingdom, New Zealand, Australia) and developing countries (e.g., Caribbean, India, Bangladesh, Brazil, Mozambique, South Africa, and Uruguay). In southeast Queensland, for example, Smith and his coauthors (Chapter 12) point out that climate change considerations need to be integrated into policies, plans, and day-to-day operations among other things such as critical infrastructure, emergency management, and coastal zone management, taking into account cross-scale interconnections, a finding reiterated in many other cases (e.g., Durban, Chapter 22). The integrative imperative is powerfully illustrated in the comparative and regional case studies, for example, regional planning in Australia and People's Republic of China (Chapter 10), the comparative study of the Maldives and Kiribati (Chapter 14), maritime spatial planning in the North Sea (Chapter 19), and the Mediterranean (Chapter 20).

The human dimension of adaptation—*people*—is critical as far as the success or failure of the adaptation process is concerned and is central to the sustainability of the process. The sheer scale of the challenge posed by the process of adaptation requires the committed efforts of society as a whole, hence the importance of inclusivity, equity, and empowerment.

Inclusive

Inclusive means that adaptation planning and decision-making processes involve coastal stakeholders and interested and affected parties in matters of importance to them in ways that are meaningful to them. Inclusive does not imply that every person has to be involved in every decision. Rather, it recognizes that public decisions need to be informed by those affected, without excluding any social groups. Mention has already been made of the importance of authentic participation. It is critical that adaptation is an inclusive and authentic process and this requires a concerted effort and, more likely than not, mediation as conflicting interests typically emerge. In some cases, where, for example, relocation is the only option, it will be impossible to meet all interests, and inclusive participation in charting adaptation pathways becomes all the more important. Lewis, Yoachim, and Meffert in Chapter 6, in the context of New Orleans, and Turner and Luisetti, in Chapter 18 in the case of the United Kingdom, consider the issues of relocation support and compensation, respectively. Relocation is also being confronted in settings like the Maldives and Kiribati, whose communities face daunting prospects (Chapter 14). In fact, as Smith and his coauthors note in Chapter 12, the range of stakeholder perceptions in relation to climate change means that the development of adaptation strategies must reflect a diversity of needs, aspirations, and capacities. One of the challenges is to incorporate and reconcile different types of knowledge and understanding, which range from scientific and technical knowledge to political, policy, professional, local and traditional discourses, and cultures of knowledge. In Chapter 7, Ramesh and colleagues explain the need to dove-tail bottom-up (i.e., local community) and top-down processes (i.e., government) for building shared knowledge and enabling adaptive action. The many different vantage points on and ways for achieving inclusivity is illustrated in the diversity of cases here; cogently, in the comparative analysis by Choy and colleagues of regional strategic planning in Australia and People's Republic of China (Chapter 10). In addition to differences between different coastal settings, prospects for inclusive adaptation planning change over time as shown in the Red River delta, Vietnam, case study (Chapter 9). Importantly, inclusive adaptation planning should take place early on—a point highlighted by Reisinger and his coauthors in their analysis of barriers and opportunities to explore managed retreat in different areas of Auckland, New Zealand (Chapter 14).

Equitable

Equitable means impartial, fair, and just. That the adaptation process be equitable is essential not only from the viewpoint of social justice, but also because inequity will inevitably lead to social tension and vulnerability and erode sustainability prospects. The sheer scale of the challenge posed by the process of adaptation requires the committed efforts of society as a whole.

The Bangladesh National Adaptation Programme of Action, discussed by Huq and Rabbani, in Chapter 8, provides a valuable example of a pro-poor framework for adaptation initiatives, which also includes gender equality as a cross-cutting criterion. The merit of a pro-poor approach is underscored in the Mozambique case study by Travers and her coauthors (Chapter 21). Equity is a central concern in this case because Mozambique is one of the poorest countries in the world, with pervasive coastal poverty; communities are still struggling after decades of war, many coastal localities are exposed to coastal erosion and climate change impacts, and rapid development is taking places in some localities. A two-pronged approach was developed to scope out adaptation options—community-level Vulnerability and Capacity Assessments coupled with an assessment of options to protect coastal ecosystems. Prospects for more equitable adaptation were enhanced by integrating technical assessments and stakeholder participation, informed by a detailed governance and institutional capacity assessment, and securing resources from the Least Developed Country Fund to pilot community-scale, ecosystem-based adaptation in parallel with national institutional capacity-building activities. Equity is not just a matter of concern for so-called developing countries. Turner and Luisetti's case study of managed coastal realignment in eastern England (Chapter 18) clearly reveals the need for coastal policies to be more strategic and long-term in their orientation. In addition, transformative governance reforms are needed to take into account the resolution of social conflict, compensation, social justice, and equity considerations because fundamental legal and property right considerations need to be clarified, including discretionary public defense of coastal properties and compensation of at-risk asset holders.

The Arctic case study (Chapter 4) brings to the fore the imperative to consider equity in seeking to protect indigenous cultures and practices that are at risk of being subsumed by conflicting emerging and future uses that imperil traditional ways of life but may also open up new livelihood opportunities. Inter-generational equity considerations are also raised in these case studies. For example, the costs, benefits, and risks associated with climate proofing critical infrastructure in New York City will be borne by both current and future generations and hence the necessity for adaptation planning to have very long time frames (Chapter 5). Furthermore, equity considerations underpin questions about historical attribution of and responsibility for climate change, which is implicit in the comparative study of the Maldives and Kiribati. Finally, this ensemble of case studies clearly shows that adapting to a changing climate will inevitably result in winners and losers: equity is, therefore, central to charting adaptive pathways that minimize the burden on those already exposed and vulnerable to unsustainable and maladaptive practices.

Empowering

Empowering means capable of reducing vulnerabilities and developing strengths in individuals and communities. It is both the process and outcome of acquiring influence or power over the affairs of one's life in a community and is, therefore, central to enhancing adaptive capacity.

In Chapter 7, Ramesh and his coauthors, discuss participatory irrigation management as a contribution to community adaptation in a coastal area of India, aimed at building up the capacity of local communities in operating and maintaining their own resources. In Chapter 12, Smith and his coauthors consider key challenges and opportunities facing coastal adaptation initiatives in South East Queensland. They underline the importance of assessing adaptive capacity within exposed communities and describe an innovative set of participatory approaches used to determine adaptive capacity and potential adaptive pathways. Political and governance drivers were perceived as particularly important in enabling the successful design and implementation of adaptation measures. Power and politics are central to shaping empowerment prospects as well as the securing outcomes that are inclusive and equitable. These normative imperatives must be institutionalized if they are to be realized in practice, hence the recommendation to adopt both bottom-up and top-down approaches in the Indian case study (Chapter 7). In the Canadian Arctic (Chapter 4), institutionalized recognition of indigenous rights, including settlement of land claims, lays a foundation for empowering Arctic communities in charting adaptive pathways in the face of abrupt climate-driven change.

REFLEXIVE ADAPTATION IN PRACTICE

Translating these intentions into practical reality lies at the heart of reflexive adaptation. It is beyond the scope of this concluding chapter to provide a reflexive adaptation cookbook, but we can highlight key ingredients evident in the case studies. As discussed earlier, in the context of the conceptual framework portrayed in Figure 23.1, adapting to climate change comprises three interlinked dimensions.

First, a reflexive adaptation process needs to be designed and implemented in a manner that is responsive, deliberative, and transformative. Initiating and sustaining such a process requires that coastal communities move beyond path-dependent practices that can lead to outcomes that are inequitable and unsustainable or maladaptive. In practice, extreme weather-related events prompt reflection on prevailing practices and create a window of opportunity to move beyond business as usual, as clearly shown in the New York City, Mississippi, India, Bangladesh, Vietnam, Caribbean, and Durban case studies. Making this transition requires sustained civic leadership that is translated into innovative institutional reform. In Chapter 22, Mather and Roberts explain how the Mayor of Durban, with good technical support, played a pivotal role in elevating the climate change issue in the city and fostering institutional changes in the municipality, supported by the emergence of line function champions, notwithstanding the significant day-to-day demands of life in a post-apartheid coastal city. The crucial leadership role that elected representatives and other governance actors have to play is highlighted case after case in this book—from Durban to coastal communities in the Mississippi delta and the shores of India and Uruguay. Leadership is crucial in building support for reflexive adaptation processes that can be legitimately institutionalized, upscaled and replicated, and enabled with sustained resourcing, including financial, human, and technical resources.

A responsive, deliberative, and transformative process of adaptation helps to foster *good governance* under dynamic conditions of complexity, uncertainty, surprise, and contestation. It helps to strengthen institutional capabilities, creating an enabling environment for government decision-making, private sector investment as well as household livelihood choices. As Travers and her coauthors illustrate, the process of household and institutional capacity building is vital in an impoverished country like Mozambique (Chapter 21), where the immediate focus is on meeting pressing basic human needs but, at the same time, vital coastal ecosystem functions, goods, and services need to be secured because they sustain livelihoods and help to build resilience in the face of escalating climate risk. A reflexive adaptive process is attuned to the distinctive characteristics of climate risk at the coast. This imperative permeated many of the case studies mentioned above, but requires breaking out of reliance on traditional handling of risk that treats all risk as a probability–consequence calculus (see also Chapter 3).

The case studies show that reflexive, adaptive, and resilience-building processes needed to be institutionalized, within enabling frameworks and, if necessary, with legally binding mandates, to make the difficult decisions provoked by chronic climate change, with the threat of unexpected and rapid change ever present. Given these circumstances, adaptive processes need to be exploratory and experimental in nature, iterative, and able to chart new pathways at critical thresholds and inflection points. The Uruguay case (Chapter 16), among others, shows that such a process is more likely to be supported if it is founded on affordable, cost-effective, and broadly accepted measures that progress multiple goals. Securing the requisite buy-in is more achievable when the tangible and intangible impacts, costs, benefits, and risks of climate change are clearly articulated. Chambers and Roberts-Hodge showed that exposing the cost of *poor planning* decisions, and portraying the real consequences of climate change impacts in visceral and visual terms, can help to garner the necessary attention to take adaptive actions that might otherwise be put off (Chapter 15). It is important to recognize the limits of static *hold the line* responses when making the *tough choices* that adaptation may require. As the United Kingdom (Chapter 18) and New Zealand (Chapter 14) case studies show, inflexible institutional provisions can prove maladaptive in the long run. Even along developed coastlines, such as Auckland, New Zealand, consideration needs to be given to managed retreat and this requires far-sighted leadership to catalyze *conversations* at multiple levels to develop a shared understanding of options, choose pathways, and align institutional mandates for adaptation in the face of inexorable change. Reflexive adaptation processes take time—on the order of decades and more, as Kay and colleagues demonstrate in more than two decades of vulnerability assessments and adaptation planning in Australia (Chapter 11). Reflecting on efforts to build local-level adaptive capacity in India, in Chapter 7, Ramesh and colleagues wryly observe: "it requires patience, persistence and prudence." It is a process that is fraught and conflict ridden, requiring the mediation of conflicting interests. Such a process must be tailor-made for particular localities taking into account their distinctive social-ecological characteristics.

Second, the process of reflexive adaptation needs to be tailored for particular places, founded on holism and integrative thinking and action. Building a holistic understanding of climate risk at the coast is founded on social learning and is enabled by integrative institutional structures and processes. The need to mainstream climate considerations in policy and day-to-day public decision-making is reiterated case after case from the Arctic to New Zealand, with bridging organizations, civic leaders, and local champions playing vital roles. Spatial planning is an important practical mechanism for tailoring reflexive adaptation processes to particular places, enabling social learning and horizontal and vertical coordination. Spatial planning takes place at multiple scales and can enable integration of formal and informal institutions: for example, post-Katrina recovery planning in the Mississippi delta (Chapter 6); maritime spatial planning in the North Sea (Chapter 19); integrated development planning in Durban, the Republic of South Africa (Chapter 22); and regional strategic planning in South East Queensland and the Yangtze delta region in the People's Republic of China (Chapter 10). Spatial planning enables consideration to be given to the configuration of coastal settlements and the locality-specific impacts and risks of intersecting climate and nonclimate risks through, among other things, the use of innovative visual- and scenario-generating tools, such as computer-mapping software, as highlighted in settings as diverse as New York City (Chapter 5), South East Queensland, Australia (Chapter 12), Antigua in the Caribbean (Chapter 15), and Durban, Republic of South Africa (Chapter 22). Spatial planning for climate change adaptation can enable information sharing, social learning, envisaging of alternative futures, mediation of conflicting interests, and institutional integration. Planning processes that stimulate reflexive adaptation facilitate sequenced actions that are adjusted in light of experience and anticipated change. Such planning processes need to be sustained over time by investment of financial, human, and technical resources—on decadal timescales, notwithstanding waxing and waning political salience—with enabling higher level directives to ensure consistent and context-relevant choices at the local level (see, e.g., Chapter 11). Kannen and Ratter (Chapter 19), writing about the potential of maritime spatial planning to chart adaptive pathways in the North Sea, note that, "Planning under uncertainty requires continuous, flexible and transparent processes that cut across scales and engage governance actors and networks in a manner that is inclusive and based on communication, social learning and monitoring of social-ecological systems." In other words, reflexive adaptation planning is a process that is tailored for particular places based on meaningful public participation.

Finally, reflexive adaptation is people-centered; it is inclusive, equitable and empowering so that coastal communities can chart adaptation pathways that foster resilience and sustainable coastal development. Translating this intention into reality is far from simple. A starting point is to identify coastal stakeholders and to focus particular attention on engaging vulnerable groups and communities in the adaptation planning process. Post-disaster windows of opportunity need to be seized upon and are more likely to result in societal outcomes if pre-event risk-reduction and resilience-building plans have been prepared. Mechanisms for resolving conflicting interests within and between coastal communities and current and future generations can be made visible, resolved, and institutionalized, as demonstrated in the Canadian Arctic (Chapter 4). Inclusive, equitable, and empowering processes are founded on authentic and consequential participatory planning and decision-making that conjoin bottom-up and top-down processes, as shown in the participatory irrigation management initiative in Tamil Nadu (Chapter 7) and many other case studies in this book.

Critical reflection on past and present adaptation barriers and opportunities, as well as future prospects, tipping points, and different climate change scenarios, is foundational for building adaptation pathways that are empowering and enduring. On the one hand, the institutional reforms necessary to empower coastal communities may have to overcome troubled legacies of past conflict, racism, and an array of marginalizing practices that render some groups especially vulnerable, consider the legacy of apartheid in the Republic of South Africa (Chapter 22), post-conflict Vietnam (Chapter 9) and Mozambique (Chapter 21), poverty and discrimination in Louisiana (Chapter 6), and the very different realities facing countries in the Mediterranean (Chapter 21) and North Sea (Chapter 19) regions. On the other hand, climate change presents coastal communities with unprecedented and unpredictable future challenges and opportunities.

IN CONCLUSION

Uncertainty, ambiguity, contestation, and surprise now characterize the development process and will shape how we make social choices, with profound consequences for both current and future generations living, working, and recreating in coastal zones. Even in the unlikely event that substantial progress is made to mitigate climate change, impacts are unavoidable. Adaptation is imperative, and coastal communities are on the frontline of the global sustainability challenge in the Anthropocene. The first step in responding to climate change must be to address those unsustainable practices that are rendering coastal communities and associated social-ecological systems increasingly vulnerable to climate stress. Beyond that, the case studies in this book demonstrate that innovative approaches to managing long-term climate change are being taken. The obstacles to be overcome are formidable, but the potential for coastal communities to respond to climate change by averting negative impacts and taking advantage of beneficial effects are clear. We hope that, by sharing experience, this book can make an effective contribution to that process.

REFERENCES

Adger, W.N., Agrawala, S., Mirza, M.M.Q., Conde, C., O'Brien, K., Pulhin, J., Pulwarty, R., Smit, B., and Takahashi, K. (2007). "Assessment of adaptation practices, options, constraints and capacity," in M.L. Parry, O.F. Canziani, J.P. Palutikof, P.J. van der Linden, and C.E. Hanson (eds.), *Climate Change 2007: Impacts, Adaptation and Vulnerability. Contribution of Working Group II to the Fourth Assessment Report of the Intergovernmental Panel on Climate Change*, Cambridge: Cambridge University Press. http://www.ipcc.ch/ (accessed August 2013).

Beck, U. (1994). "The reinvention of politics: Towards a theory of reflexive modernization," in U. Beck, A. Giddens, and S. Lash (eds.), *Reflexive Modernization*, Cambridge: Polity Press.

Beck, U., Bonss, W., and Lau, C. (2003). "The theory of reflexive modernization: Problematic, hypotheses and research programme," *Theory, Culture & Society*, 20(2): 1–33.

EPA (US Environmental Protection Agency). (2009). *Synthesis of Adaptation Options for Coastal Areas*, Washington, DC: Environmental Protection Agency. http://water.epa.gov/type/oceb/cre/upload/CRE_Synthesis_1-09.pdf (accessed August 2013).

Kay, R.C. (2012). "Adaptation by ribbon cutting: time to understand where the scissors are kept," *Climate and Development*, 4(2): 75–7.

Klein, R.J.T., Nicholls, R.J., Ragoonaden, S., Capobianco, M., Aston, J., and Buckley, E.N. (2001). "Technological options for adaptation to climate change in coastal zones," *Journal of Coastal Research*, 17(3): 531–43.

Nicholls, R.J., Wong, P.P., Burkett, V.R., Codignotto, J.O., Hay, J.E., McLean, R.F., Ragoonaden, S., and Woodroffe, C.D. (2007). "Coastal systems and low-lying areas," in M.L. Parry, O.F. Canziani, J.P. Palutikof, P.J. van der Linden, and C.E. Hanson (eds.), *Climate Change 2007: Impacts, Adaptation and Vulnerability. Contribution of Working Group II to the Fourth Assessment Report of the Intergovernmental Panel on Climate Change*, Cambridge; New York: Cambridge University Press. http://www.ipcc.ch/ (accessed August 2013).

Reid, H. and Huq, S. (2007). *Community-Based Adaptation: A Vital Approach to the Threat Climate Change Poses to the Poor*, International Institute for Environment and Development Briefing Paper, London: IIED. http://pubs.iied.org/17005IIED.html (accessed August 2013).

Tobey, J., Rubinoff, P., Robadue, Jr., D., Ricci, G., Volk, R., Furlow, J., and Anderson, G. (2010). "Practicing coastal adaptation to climate change: Lessons from integrated coastal management," *Coastal Management*, 38(3): 317–35.

UNEP (United Nations Environment Programme). (2010). *Technologies for Climate Change Adaptation—Coastal Erosion and Flooding*, Roskilde, Denmark: UNEP. http://www.unep.org/pdf/TNAhandbook_CoastalErosionFlooding.pdf (accessed August 2013).

USAID (US Agency for International Development). (2009). *Adapting to Coastal Climate Change: A Guidebook for Development Planners*, Washington, DC: USAID. http://www.crc.uri.edu/download/CoastalAdaptationGuide.pdf (accessed August 2013).

Voss, J.-P. and Kemp, R. (2005). *Reflexive Governance for Sustainable Development—Incorporating Feedback in Social Problem Solving*, Paper presented at *ESEE Conference*, June 14–17, Lisbon, Portugal. http://kemp.unu-merit.nl/ (accessed August 2013).

Voss, J.-P., Newig, J., Kastens, B., Monstadt, J., and Nölting, B. (2007). "Steering for sustainable development: A typology of problems and strategies with respect to ambivalence, uncertainty and distributed power," *Journal of Environmental Policy & Planning*, 9 (3/4): 193–212.

World Bank. (2006). Clean Energy and Development: towards an investment framework, Washington, DC: World Bank. http://siteresources.worldbank.org/DEVCOMMINT/Documentation/20890696/DC2006-0002(E)-CleanEnergy.pdf (accessed August 2013).

Index

Note: Locators followed by "*f*" and "*t*" denote figures and tables in the text